GENERAL CONTINUUM MECHANICS

In this book, a new approach is pioneered in providing a unified theory in continuum mechanics. *General Continuum Mechanics* is intended for the beginner, but it develops advanced material covering interdisciplinary subjects. With applications of convective, Lagrangian, and Eulerian coordinates and the first and second laws of thermodynamics, the first-year graduate student will learn solid mechanics and fluid mechanics as an integrated subject. Electromagnetic continuum and relativistic continuum are included. The conservational properties of mass, momentum, and energy on earth and in the universe constitute the ingredients of this book. They are the monumental contributions of Newton, Maxwell, and Einstein, a beautiful panorama of universal laws that evolved over the past four centuries. No boundaries are needed to separate them, but rather they are integrated in harmony and placed in perspective. This is the book for interdisciplinary studies to carry out the modern scientific projects in which engineering, physics, and applied mathematics must be combined.

T. J. Chung is Distinguished Professor of Mechanical and Aerospace Engineering, the University of Alabama in Huntsville. He has also authored *Computational Fluid Dynamics* and *Applied Continuum Mechanics*, both published by Cambridge University Press.

General Continuum Mechanics

T. J. CHUNG

The University of Alabama in Huntsville

CAMBRIDGE UNIVERSITY PRESS
Cambridge, New York, Melbourne, Madrid, Cape Town, Singapore, São Paulo

Cambridge University Press
32 Avenue of the Americas, New York, NY 10013-2473, USA

www.cambridge.org
Information on this title: www.cambridge.org/9780521874069

First published 2007

Printed in the United States of America

A catalog record for this publication is available from the British Library.

Library of Congress Cataloging in Publication Data

Chung, T. J., 1929–
General continuum mechanics / T. J. Chung. – 2nd ed.
p. cm.
Originally published as : Continuum mechanics : Englewood Cliffs, N.J. : Prentice-Hall, 1988.
Includes bibliographical references and index.
ISBN-10: 0-521-87406-8 (pbk.)
ISBN-13: 978-0-521-87406-9 (hardback)
1. Continuum mechanics. I. Chung, T.J., 1929– Continuum mechanics. II. Title.
QA808.2.C55 2007
531–dc22 2006025606

ISBN 978-0-521-87406-9 hardback

To my family

Contents

Preface *page* xiii

PART I. BASIC TOPICS

1 Introduction 3

1.1 General 3
1.2 Vectors and tensors in Cartesian coordinates 5
1.3 Domain and surface integrals and boundary conditions 17
Problems 23

2 Kinematics 25

2.1 General 25
2.2 Coordinate systems 26
2.2.1 Lagrangian coordinates 26
2.2.2 Eulerian coordinates 30
2.2.3 Relationships between Lagrangian and Eulerian coordinates 34
2.3 Strain tensor 37
2.4 Rate-of-deformation tensor 40
2.5 Coordinate transformations for strains 41
2.6 Dilatational and deviatoric properties 51
Problems 53

3 Kinetics 55

3.1 General 55
3.2 Forces and stresses 55
3.3 Basic conservation laws 57
3.3.1 Conservation of linear momentum 57
3.3.2 Conservation of angular momentum 59
3.4 Coordinate transformations for stresses 60
3.5 The deviatoric stress tensor 66
Problems 68

4 Linear elasticity 70

4.1 Constitutive equations for linear elastic solids 70
4.1.1 Three-dimensional solids 70
4.1.2 Plane stress and plane strain 80

4.2 Navier equations 82
4.3 Energy principles 83
4.3.1 Variational principle, virtual work, and total potential energy 83
4.3.2 Boundary conditions 86
4.4 Thermodynamics of solids 90
4.4.1 General 90
4.4.2 The first law of thermodynamics 91
4.4.3 The second law of thermodynamics 93
4.4.4 Constitutive theory for thermodynamics of solids 94
4.4.5 Thermomechanically coupled equations of motion and heat conduction 96
4.5 Fiber composites 101
4.5.1 General 101
4.5.2 Coordinate transformations for transversely isotropic materials 101
4.5.3 Characterization of material constants for orthotropic and transversely isotropic composites 103
Problems 106

5 Newtonian fluid mechanics 108

5.1 Introduction 108
5.2 Constitutive equations 110
5.2.1 Historical review 111
5.2.2 The stress tensor 112
5.3 The Navier–Stokes system of equations 117
5.3.1 The first law of thermodynamics 117
5.3.2 The conservation form of the Navier–Stokes system of equations and control-volume–control-surface equations 121
5.3.3 Initial and boundary conditions 126
5.3.4 The second law of thermodynamics 126
5.4 Compressible viscous flow 130
5.5 Ideal flow 132
5.5.1 General 132
5.5.2 Existence of the stream function in two dimensions 134
5.5.3 Existence of the stream function in three dimensions 136
5.5.4 The Bernoulli equation 138
5.6 Rotational flow 139
5.6.1 General 139
5.6.2 The vorticity transport equation 140
5.7 Turbulence 142
5.7.1 Time and ensemble averages 142
5.7.2 Correlations 144

5.7.3 Isotropic and homogeneous turbulence 147
5.7.4 Transport equations of Reynolds stresses, turbulent kinetic energy, and dissipation rates 148
5.7.5 Phenomenological theories 149
5.7.6 Other approaches to turbulence 150
5.8 The boundary layer 151
5.8.1 Laminar boundary layer 151
5.8.2 Turbulent boundary layer 154
5.9 Convective heat transfer 155
5.9.1 General 155
5.9.2 Forced and natural convection 156
5.10 High-speed aerodynamics 158
5.10.1 General 158
5.10.2 Full potential equation 158
5.10.3 Shock waves 162
5.11 Acoustics 165
5.11.1 General 165
5.11.2 Monochromatic waves 167
5.11.3 Pressure-mode acoustics 171
5.11.4 Vorticity-mode acoustics 172
5.11.5 Entropy-mode acoustics 174
5.12 Reacting flows 176
5.12.1 General 176
5.12.2 Conservation of mass for mixture and species 179
5.12.3 Conservation of momentum 180
5.12.4 Conservation of energy 180
5.12.5 Physical derivation of multicomponent diffusion 182
5.12.6 Shvab–Zel'dovich approximations 184
Problems 187

PART II. SPECIAL TOPICS

6 Curvilinear continuum 191

6.1 Tensor properties of curvilinear continuum 191
6.1.1 General 191
6.1.2 Covariant derivatives 198
6.1.3 Physical components of tensors 200
6.1.4 Cylindrical coordinates 201
6.1.5 Spherical coordinates 203
6.1.6 Divergence and curl of vectors 204
6.1.7 Strain tensors 206
6.2 Linear elasticity 209
6.2.1 Conservation of linear and angular momenta 209
6.2.2 Navier equations 212

6.2.3 Compatibility equations 214
6.2.4 Torsion 216
6.3 Newtonian fluid mechanics 221
6.3.1 Navier–Stokes system of equations 221
6.3.2 Stream functions for cylindrical and spherical coordinates 226
Problems 228

7 Nonlinear continuum 231

7.1 Nonlinear elasticity 231
7.1.1 Stresses with large strains 231
7.1.2 Equations of motion for large strains 236
7.1.3 Finite elasticity 237
7.2 Non-Newtonian fluid mechanics 240
7.2.1 General 240
7.2.2 Kinematics 242
7.2.2.1 Stretch and rotation 242
7.2.2.2 Rates of strain: Stretching and spin 246
7.2.2.3 Convective coordinate approach 248
7.2.2.4 Constant-stretch history 250
7.2.3 Simple fluids 255
7.2.3.1 General 255
7.2.3.2 Fading memory 256
7.2.4 Constitutive equations 257
7.2.4.1 Integral equations 257
7.2.4.2 Differential equations 258
7.2.4.3 Convective coordinate approach 259
7.2.4.4 Illustrations 263
7.3 Viscoelasticity 266
7.3.1 Spring–dashpot models 267
7.3.2 Multiple-parameter spring–dashpot models 271
7.3.3 General form of the differential equations 277
7.3.4 Constitutive equations for generalized models 278
7.3.5 Superposition integral 280
7.3.6 Correspondence principle 288
7.3.7 Viscoelasticity based on irreversible thermodynamics 290
7.4 Plasticity 296
7.4.1 General 296
7.4.2 Plasticity theories 297
7.4.3 Stress-space representation 304
7.4.4 Plastic potential function 307
7.4.5 Isotropic hardening 312
7.4.6 Kinematic hardening 314

7.4.7 Multisurface models of flow plasticity 316
7.4.8 The strain trajectory models 318
7.5 Thermoviscoelastoplasticity 320
Problems 322

8 Electromagnetic continuum 323

8.1 Introduction 323
8.2 The electrodynamics 323
8.2.1 The field variables in electrodynamics 323
8.2.2 Maxwell's equations 328
8.2.3 Electromagnetic stress and energy 330
8.3 Magnetohydrodynamics 331
8.3.1 General 331
8.3.2 Magnetohydrodynamic equations 332
8.4 Electromagnetic waves 335
8.4.1 Alfvén velocity 335
8.4.2 Electromagnetic waves in a compressible conducting fluid 336
Problems 338

9 Differential geometry continuum 339

9.1 Fundamental tensors 339
9.1.1 Curvilinear coordinates and base vectors 339
9.1.2 Derivatives of base vectors 342
9.2 Curvature tensors 343
9.2.1 Riemann–Christoffel tensor 343
9.2.2 Gaussian and Codazzi curvatures 345
9.3 Applications to flexural continuum (shell geometries) 346
9.3.1 Kinematics of shells 346
9.3.2 Membrane strain tensor 349
9.3.3 Bending strain tensor 350
9.3.4 Plates derived from shell theory 353
9.4 Applications to relativistic continuum (spacetime geometries) 355
9.4.1 Special relativity 356
9.4.1.1 Lorentz transformation 358
9.4.1.2 The energy-momentum tensor 361
9.4.1.3 Ideal fluids in special relativity 362
9.4.1.4 Nonideal fluids in special relativity 365
9.4.1.5 Relativistic electrodynamics 369
9.4.1.6 Relativistic charged particles in external magnetic fields 372
9.4.2 General relativity 377
9.4.2.1 Tangent vectors 377
9.4.2.2 Metric tensors 378

9.4.2.3 The Riemann curvature tensors 382
9.4.2.4 Geodesic deviation and tidal forces 385
9.4.2.5 The nonlinear theory of gravitation 386
9.4.2.6 Kinematics and dynamics of general relativity 390
9.4.3 Conservation hydrodynamic equations in numerical relativity 395
9.4.3.1 Ideal fluids 396
9.4.3.2 Nonideal fluids 397
9.4.3.3 Magnetohydrodynamic general relativity 400
Problems 402

Epilogue 403
References 405
Index 417

Preface

Continuum mechanics is one of the most important interdisciplinary studies in our pursuit toward physical phenomena and mathematical aspects of universal laws. In the past, however, continuum mechanics was regarded as a discipline mainly associated with various branches of engineering dictated by Newton's theory of mechanics. As such, one focuses on solid mechanics and fluid mechanics governed by the speed of sound. However, the aim of this book is to include in continuum mechanics the physical phenomena governed by the speed of light as well. Doing so involves Maxwell's electromagnetic continuum and Einstein's relativistic spacetime continuum.

My previous books *Continuum Mechanics* (Prentice-Hall, 1988) and *Applied Continuum Mechanics* (Cambridge University Press, 1996) were written primarily for engineers. The present book is intended for a broader audience, including engineers, physicists, and applied mathematicians. Although the material presented is diversified, my desire is to address the importance of unifying and placing in perspective the concept of mass, momentum, and energy associated with all physical phenomena on earth and beyond, thus assisting the reader to be interdisciplinary.

This book is divided into two parts: Part I, Basic Topics, and Part II, Special Topics. Part I covers mathematical preliminaries, kinematics, kinetics, linear elasticity, and Newtonian fluid mechanics. Only Cartesian tensors are used in Part I. Part II includes curvilinear continuum, nonlinear continuum, electromagnetic continuum, and differential geometry continuum. Curvilinear tensors are used extensively in Part II.

Part I consists of five chapters. The emphasis in Part I is on showing how mass conservation and constitutive equations in solid mechanics and fluid mechanics are intimately related, a notion seldom stressed in other textbooks. Chapter 1 provides the basic mathematical tools for vectors and Cartesian tensors. Concepts of boundary conditions as related to the integrals of domain and boundary surfaces are elaborated on. We discuss in Chapter 2 the kinematics including general Cartesian coordinate systems, relationships between convective Lagrangian and Eulerian coordinates, strain tensor, the rate-of-deformation tensor, coordinate transformations for strains, dilatational strains, and deviatoric strains. Although the concept of metric tensor is formally discussed in curvilinear coordinates in Chapter 6, Part II, it is shown that the Lagrangian coordinates including the curvilinear tangent vectors, known as the convective coordinates, do provide a rigorous derivation of volume changes and the strain tensor for large deformations from which the small strain tensor can simply be extracted. More important, by use of these convective coordinates and substantial Jacobian derivatives, the conservation of mass for fluids is shown to arise from the conservation of mass of solids by a combination of the Lagrangian and Eulerian coordinates. Similarly, the rate-of-deformation tensor for fluids is obtained by use of the Eulerian coordinates from the definition of the strain tensor in Lagrangian coordinates. This process also provides the basis for the rigorous derivation of the Navier–Stokes system of equations in fluid mechanics. Thus the close relationship between solid mechanics and fluid mechanics is greatly emphasized. Chapter 3 presents kinetics. We study, in this chapter, forces and stresses, conservation of linear momentum, angular momentum, coordinate transformations for stresses, and dilatational and deviatoric stresses. Linear elasticity is discussed in Chapter 4. We examine constitutive equations, demonstrating how stresses of any anisotropic material, whether solids or fluids, can be related to the strain tensor or rate-of-deformation tensor through a series of coordinate transformations from anisotropy to isotropy. This is followed by Navier equations, energy principles, and thermodynamics of solids. A combination of the first and second laws of thermodynamics is shown to resolve relationships among Helmholtz free energy, internal energy, and entropy, leading to the thermomechanically coupled equations of motion and heat conduction. In this chapter we also study fiber composites. Coordinate transformations for transversely isotropic materials and characterization of material constants for orthotropic and transversely isotropic composites are discussed. Chapter 5 presents Newtonian fluid mechanics. Constitutive equations, derivations of the Navier–Stokes system of equations through the first and second laws of thermodynamics, the conservation form of the Navier–Stokes system of equations, integrals of control volume and control surfaces, compressible flows, ideal flows, rotational flows, turbulence, boundary layers, convective heat transfer, high-speed aerodynamics, acoustics, and reacting flows are included in this chapter.

Thus Part I is marked by the uniqueness of this text, that the convective, Lagrangian, and Eulerian coordinates with the first and second laws of thermodynamics are shown to connect solid mechanics and fluid mechanics under one roof, truly a manifestation of the spirit of continuum mechanics.

Part II contains four chapters. A curvilinear continuum is discussed in Chapter 6. The basic definitions of covariant and contravariant components of a vector, tangent vectors, derivatives of tangent vectors, and Christoffel symbols are elaborated on. Curvilinear coordinates for small deformations for solids and Newtonian fluid mechanics are described for cylinders and spheres. Curvilinear characteristics of large deformations involved in compatibility equations and their applications to torsion are also included. Chapter 7 presents nonlinear continua. The curvilinear large deformation tangent vectors and metric tensors of Chapter 6 are utilized in nonlinear elasticity and non-Newtonian fluid mechanics, followed by viscoelasticity, plasticity, and thermoviscoelastoplasticity. In Chapter 8 we study electromagnetic continua, including electrodynamics, magnetohydrodynamics, and electromagnetic waves. One of the purposes of this chapter is to apply electromagnetics to gravitohydromagnetics forthcoming in Chapter 9. Finally, in Chapter 9 we discuss differential geometry continuum. The differential geometries have many applications in engineering, physics, and applied mathematics. The concept of differential geometries through fundamental tensors and curvature tensors is shown to be a basic tool to derive governing equations not only for shell structures in engineering, but also for motions of particles and the universe as related to Einstein's theory of relativity. This is a good example illustrating that the curvatures of shells (flexural continuum) and the curvatures of spacetime (relativistic continuum) can be treated under a common subject, the differential geometry continuum. Thus our intention in Chapter 9 is not to delve into the full shell theory or the full theory of general relativity, but to demonstrate that the differential geometry can constitute a regime of continuum mechanics. The flexural continuum and relativistic continuum are used as applications. It is demonstrated that the fundamental tensors arising from metric tensors and Riemann–Christoffel tensors form a basis for the deformations of shell surfaces as well as the spacetime surface geometry of Einstein's theory of relativity. Moreover, a conservation form of hydrodynamic equations in numerical relativity is given, demonstrating that the continuum mechanics approach leads to important research areas such as relativistic gravitohydromagnetics in line with the conservation form of the Navier–Stokes system of equations for shock-wave turbulent boundary-layer interactions in fluid mechanics.

The subject areas covered in this book are the outgrowth of teaching in the classroom and my funded and unfunded research projects in the past at the University of Alabama in Huntsville. Some of the details and emphasis, not usually found in other textbooks, are motivated by the need in the classroom for additional elaboration. Thus it is hoped that this book will be useful also for self-study and to practitioners as a reference book. Because of my work involved in numerical simulations of differential equations, I was often guided by the desire to arrange the equations conducive to numerical solutions. They are reflected in my emphasis on the conservation form of differential equations both in fluid mechanics (Section 5.3) and relativistic mechanics (Section 9.4).

I have endeavored to make this book suitable for engineers, physicists, and applied mathematicians, and convenient for interdisciplinary work. The basic principles applied to deformations of solids, flows of fluids and heat, field lines of electromagnetism, and changes in metric surfaces of the universe are similar. Thus it is intended that this book serve as a bridge among engineers, physicists, and applied mathematicians alike. The material covered in this book will be adequate for two-semester courses. For the curriculum allowing only one semester, the instructor may choose chapters as desirable for the audience.

Coincidently, this text may be considered as a companion book for my previous publication, *Computational Fluid Dynamics* (Cambridge University

Press, 2002), which provides numerical simulations of most of the governing equations introduced in *General Continuum Mechanics*.

In my work in continuum mechanics, undoubtedly, I have been deeply influenced by pioneers of the past four centuries. Equally, I am indebted to students in my continuum mechanics classes for the past thirty years. A great many of them contributed to the improvement of this book. Each year I accumulated valuable information when students asked questions. I could not have done so without them, and I am grateful to all of them.

I would like to express my sincere appreciation to Peter Gordon, Senior Editor of Cambridge University Press; Peter Katsirubas of Techbooks; and Vicky Danahy for their excellent contributions to the editorial process of this book.

Anna Mulvany typed the entire manuscript with great care and skill. To her I am truly thankful.

T. J. Chung

PART I

Basic Topics

1 Introduction

1.1 General

Trends have been seen in recent years in which engineers, physicists, and applied mathematicians perform research with interdisciplinary interactions. Vectors and tensors are the common language used in these interactions. Continuum mechanics combines mathematical operations and physical processes in continuous media for various subjects of engineering, physics, and applied mathematics. This is contrary to the notion that continuum mechanics is predominantly solid mechanics, as conceived in the past.

Thus our objective is to explore how the theory of continuum mechanics can combine deformations of solids, flows of inviscid and viscous fluids, electromagnetic waves, and motions of astrophysical objects in a single book. Common to all of these physical phenomena are the concepts of mass, velocity, acceleration, stress, momentum, and energy. We shall examine them as conceived in engineering disciplines and in spacetime of our universe, thus placing them all in proper perspective. Indeed, it shall be shown that the conservation forms of the electromagnetic continuum and Einstein's relativistic equations are similar to the conservation form of the Navier–Stokes system of equations such that the numerical solution schemes are similar for both cases, referring to a process of continuum mechanics dealing with speed of sound, speed of light, or both, in a similar fashion.

Newton's theory prevails today and will continue to govern our daily lives on earth. As the velocity of a particle increases, reaching nearly the speed of light, however, the effect of gravitation becomes significant and Newton's law must be replaced with Einstein's relativity theory. The word "continuum" implies "continuous media." Therefore, as long as mass, velocity, acceleration, stress, momentum, and energy are continuous functions in continuous media, our task of describing the behavior of these functions will be the same on earth and in the universe except as affected by relativity in spacetime and gravitation. Thus in this book our task is to become aware of both analogies and differences between them.

In continuum mechanics we are concerned with a macroscopic view, thus excluding the microscopic treatments such as in quantum mechanics and rarefied gases. A concept of fundamental importance here is that of the *mean free path*, which can be defined as the average distance a molecule travels between successive collisions with other molecules. The ratio of the mean free path λ to the characteristic length S of the physical boundaries of interest, called the *Knudsen number* Kn, may be used to determine the dividing line between the macroscopic and microscopic models:

$$\text{Kn} = \frac{\lambda}{S} < 1, \quad \text{macroscopic,} \tag{1.1.1}$$

$$\text{Kn} = \frac{\lambda}{S} \geq 1, \quad \text{microscopic,} \tag{1.1.2}$$

where $\lambda \cong 10^{-7}$ cm for solids and liquids and $\lambda \cong 10^{-6}$ cm for gases. Thus the Knudsen number is smaller than 1 for continuum mechanics as the characteristic length is larger than the mean free path. This is the case for solid mechanics, fluid dynamics, electromagnetic continuum, and relativistic continuum. It is known as the macroscopic problem as opposed to the microscopic problem such as in quantum mechanics (small characteristic length, $S \cong 10^{-33}$cm, large Knudsen number) or in rarefied gases (large mean free path, $\lambda > 10^{-6}$cm, large Knudsen number). Thus, in our study in continuum mechanics, rarefied gases and quantum mechanics are excluded.

For the macroscopic model, mass m is defined as a continuous function of volume Ω, such that density ρ is determined by the relation

$$\rho = \frac{dm}{d\Omega}, \tag{1.1.3}$$

whereas in the microscopic model, we define

$$\rho = \sum_{i=1}^{N} \rho_i = \sum_{i=1}^{N} m_i n_i, \tag{1.1.4}$$

where n_i denotes the number density of molecules per unit volume of a gas composed of a chemical species i.

The unified approach to the study of the global behavior of materials consists of, first, a thorough study of the basic principles common to all media and, second, a clear demonstration of the properties specific to the medium under consideration. The basic principles include the conservation of mass, the conservation of linear and angular momentum, the conservation of energy, and the principle of entropy. The underlying assumption of the unified theory is that these principles are valid for all materials irrespective of their constitution. Thus, to account for the nature of different materials – the various types of solids, liquids, or gases – we require additional equations, known as the equations of state, to describe the basic characteristics of the body and its response to the external agent under consideration.

This book is divided into two parts: Part I, Basic Topics, and Part II, Special Topics. We begin with the basic operations of vectors, matrices, Cartesian tensors, and domain and boundary surface integrals in Chap. 1, followed by kinematics, kinetics, linear elasticity, and Newtonian fluid mechanics in subsequent chapters of Part I. Curvilinear continuum, nonlinear continuum, electromagnetic continuum, and differential geometry continuum are included in Part II.

1.2 Vectors and Tensors in Cartesian Coordinates

A vector is determined in a given reference frame by a set of components. If a new coordinate system is introduced, the same vector is determined by another set of components, and these new components are related, in a definite way, to the old ones. The law of transformation of components of a vector is the essence of the vector representation.

Tensors are founded on a notion similar to that of vectors, but are much broader in conception. Tensor analysis is concerned with the study of *abstract objects*, called *tensors*, whose properties are independent of or invariant with the reference frames used to describe an object. A tensor is represented in a particular reference frame by a set of functions, termed its *components*, just as a vector is determined by a set of components. Tensor analysis deals with entities and properties that are independent of the choice of reference frames. Thus it forms an ideal tool for the study of natural laws because tensor equations are invariant with respect to a given category of coordinate transformations. Tensors are capable of delineating a variety of objects, ranging from scalars to multiple components of matter encountered in various physical phenomena. To discuss this subject further, however, it is necessary to introduce some notation and rules that will be applied to tensors and also to other topics in continuum mechanics.

Index Notation

A vector is denoted by a boldfaced letter symbol. A vector may be written in terms of its components by use of indices. For example, consider a vector in a right-handed rectangular three-dimensional Cartesian coordinate system (Fig. 1.2.1):

$$\mathbf{A} = A_1\mathbf{i}_1 + A_2\mathbf{i}_2 + A_3\mathbf{i}_3, \tag{1.2.1a}$$

where each $\mathbf{i}_i$ denotes one of the unit vectors and the indices $i = 1, 2, 3$ have a range of 3. This expression may be written as

$$\mathbf{A} = \sum_{i=1}^{3} A_i\mathbf{i}_i, \tag{1.2.1b}$$

where A_i indicates the components of the vector $\mathbf{A}$ at point p. Henceforth we shall dispense with the summation sign and write Eq. (1.2.1) in the form

$$\mathbf{A} = A_i\mathbf{i}_i \quad (i = 1, 2, 3)\,. \tag{1.2.1c}$$

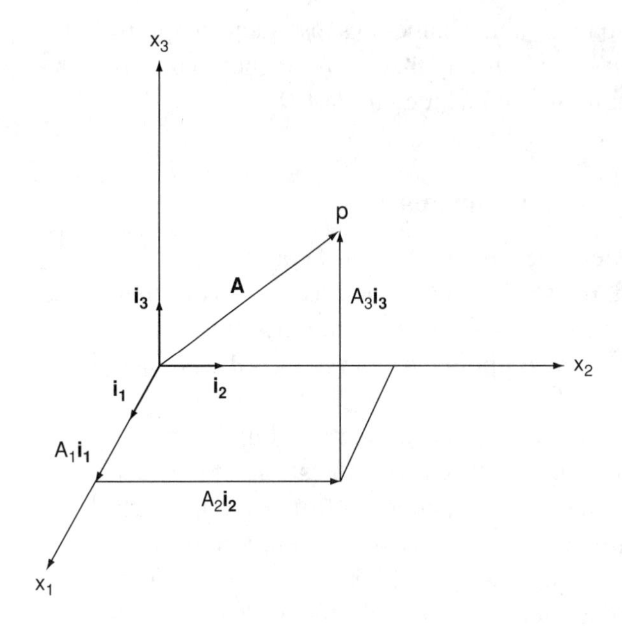

Figure 1.2.1. Three-dimensional representation of the vector **A** and its components.

Note that *repeated indices* (sometimes called *dummy indices*) imply summing over the range of the index. For example, let us consider

$$x'_i = a_{ij}x_j = a_{i1}x_1 + a_{i2}x_2 + a_{i3}x_3 \quad (i, j = 1, 2, 3)\,. \tag{1.2.2a}$$

Here, j is the repeated index and i must change independently of x_j to give x'_1, x'_2, x'_3, which indicates that Eq. (1.2.2a) represents three equations. The index i, which is not repeated here, is called a *free index*, thus allowing the free index i in (1.2.2a) to assume $i = 1,\ 2,\ 3$ one at a time and to obtain a total of three equations:

$$\begin{aligned} x'_1 &= a_{11}x_1 + a_{12}x_2 + a_{13}x_3, \\ x'_2 &= a_{21}x_1 + a_{22}x_2 + a_{23}x_3, \\ x'_3 &= a_{31}x_1 + a_{32}x_2 + a_{33}x_3. \end{aligned} \tag{1.2.2b}$$

Vector Multiplication

The Kronecker delta is defined as

$$\delta_{ij} = \begin{cases} 1 & \text{if} \quad i = j \\ 0 & \text{if} \quad i \neq j \end{cases}. \tag{1.2.3a}$$

This represents the 3×3 unit matrix,

$$\delta_{ij} = \begin{bmatrix} \delta_{11} & \delta_{12} & \delta_{13} \\ \delta_{21} & \delta_{22} & \delta_{23} \\ \delta_{31} & \delta_{32} & \delta_{33} \end{bmatrix} = \begin{bmatrix} 1 & 0 & 0 \\ 0 & 1 & 0 \\ 0 & 0 & 1 \end{bmatrix} = [\mathbf{I}], \tag{1.2.3b}$$

where [**I**] is the unity matrix.

The permutation symbol is given by

$$\epsilon_{ijk} = \begin{cases} 1 & \text{for even permutations of } ijk \text{ (123, 231, 312),} \quad \text{clockwise rotation} \\ -1 & \text{for odd permutations of } ijk \text{ (132, 213, 321),} \quad \text{counterclockwise rotation.} \\ 0 & \text{for two or more equal indices (112, 111, etc.)} \end{cases} \tag{1.2.4}$$

Note that the permutation symbol has an array (9 × 3) of 27 terms, with only 6 terms being nonzero and 21 terms zero.

A dot product of any two vectors in index notation reads

$$\mathbf{A} \cdot \mathbf{B} = A_i \mathbf{i}_i \cdot B_j \mathbf{i}_j = A_i B_j \delta_{ij} = A_i B_i = A_j B_j = \lambda. \tag{1.2.5}$$

Because of the orthogonal or orthonormal coordinate system, a dot product of the unit vectors produces a Kronecker delta:

$$\mathbf{i}_i \cdot \mathbf{i}_j = \delta_{ij}. \tag{1.2.6}$$

The role of the Kronecker delta is to interchange the index of a component of a vector, as demonstrated in Eq. (1.2.5). In this process, as a consequence of the Kronecker delta, only the nonzero terms are allowed to remain, the zero terms being removed. Furthermore, the dot product of two vectors results in a scalar λ, which does not have a free index.

On the other hand, a cross product of any two vectors in index notation reads

$$\begin{aligned} \mathbf{A} \times \mathbf{B} &= A_i \mathbf{i}_i \times B_j \mathbf{i}_j \\ &= A_i B_j \mathbf{i}_i \times \mathbf{i}_j = A_1 B_1 \mathbf{i}_1 \times \mathbf{i}_1 + A_1 B_2 \mathbf{i}_1 \times \mathbf{i}_2 + A_2 B_1 \mathbf{i}_2 \times \mathbf{i}_1 + \cdots + \\ &= 0 + A_1 B_2 \mathbf{i}_3 - A_2 B_1 \mathbf{i}_3 \\ &\quad + \cdots + (\text{27 terms, 21 of them zero, only 6 terms nonzero}) \\ &= A_i B_j \epsilon_{ijk} \mathbf{i}_k \\ &= (A_2 B_3 - A_3 B_2)\mathbf{i}_1 + (A_3 B_1 - A_1 B_3)\mathbf{i}_2 + (A_1 B_2 - A_2 B_1)\mathbf{i}_3, \end{aligned} \tag{1.2.7}$$

where the cross product of unit vectors produces a permutation symbol ϵ_{ijk} such that

$$\mathbf{i}_i \times \mathbf{i}_j = \epsilon_{ijk} \mathbf{i}_k. \tag{1.2.8}$$

It is interesting to note that the result obtained in Eq. (1.2.7) can be written as

$$\begin{aligned} \mathbf{A} \times \mathbf{B} &= \begin{vmatrix} \mathbf{i}_1 & \mathbf{i}_2 & \mathbf{i}_3 \\ A_1 & A_2 & A_3 \\ B_1 & B_2 & B_3 \end{vmatrix} \\ &= (A_2 B_3 - A_3 B_2)\mathbf{i}_1 - (A_1 B_3 - A_3 B_1)\mathbf{i}_2 + (A_1 B_2 - A_2 B_1)\mathbf{i}_3, \end{aligned} \tag{1.2.9}$$

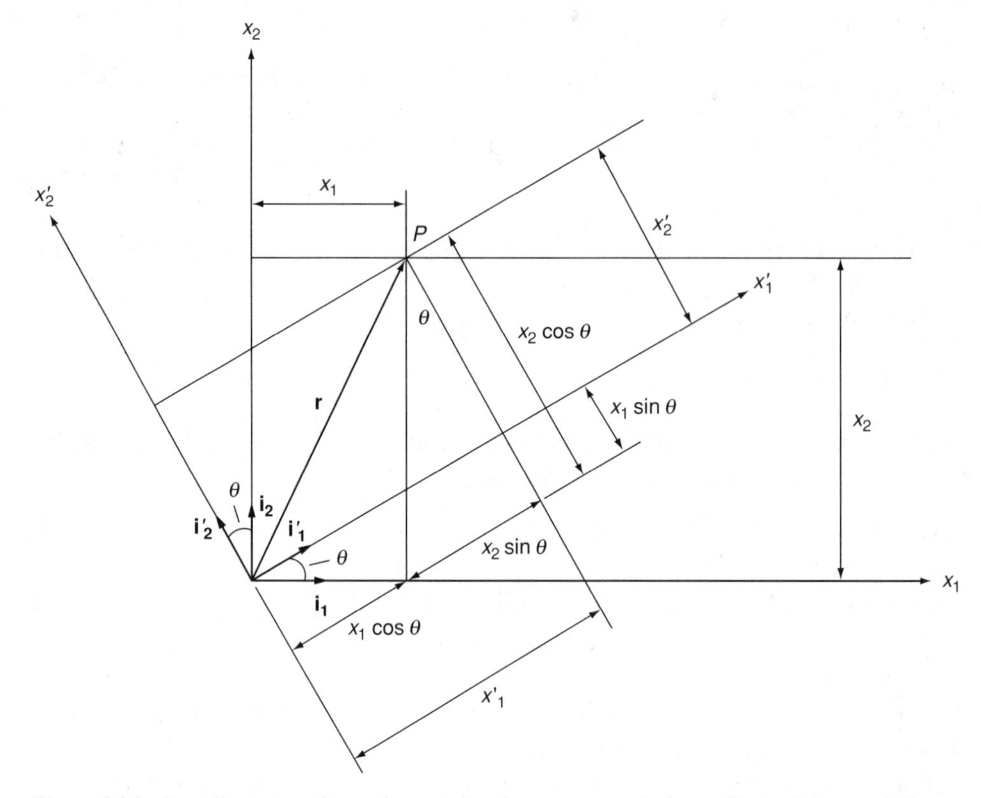

Figure 1.2.2. Coordinate transformation rotation between x_i and x_i' coordinates by an angle θ.

indicating that the application of a permutation symbol results in the determinant of a 3×3 matrix array. Notice that Eq. (1.2.7) indicates a logical sequence of derivation whereas Eq. (1.2.9) is merely a definition resulting from Eq. (1.2.7).

It can easily be shown that $\mathbf{A} \cdot \mathbf{B} = \mathbf{B} \cdot \mathbf{A}$ (commutative law) and $\mathbf{A} \times \mathbf{B} = -\mathbf{B} \times \mathbf{A}$. The reader is encouraged to prove these relations by using the index notation as an exercise.

Tensors

The quantities that appear in the foregoing paragraphs are identified as tensors because they satisfy the basic properties set forth at the beginning of this section. To provide a specific example, let us examine Eqs. (1.2.2b). Let $\mathbf{r}$ be a position vector in Fig. 1.2.2:

$$\mathbf{r} = x_i \mathbf{i}_i \quad (i = 1, 2, 3), \tag{1.2.10a}$$

where the standard right-hand rule is used. If the old coordinates x_i are rotated by an angle θ about the x_3 axis to a set of new coordinates x_i', then the position vector can be represented by

$$\mathbf{r} = x_i' \mathbf{i}'_i \quad (i = 1, 2, 3). \tag{1.2.10b}$$

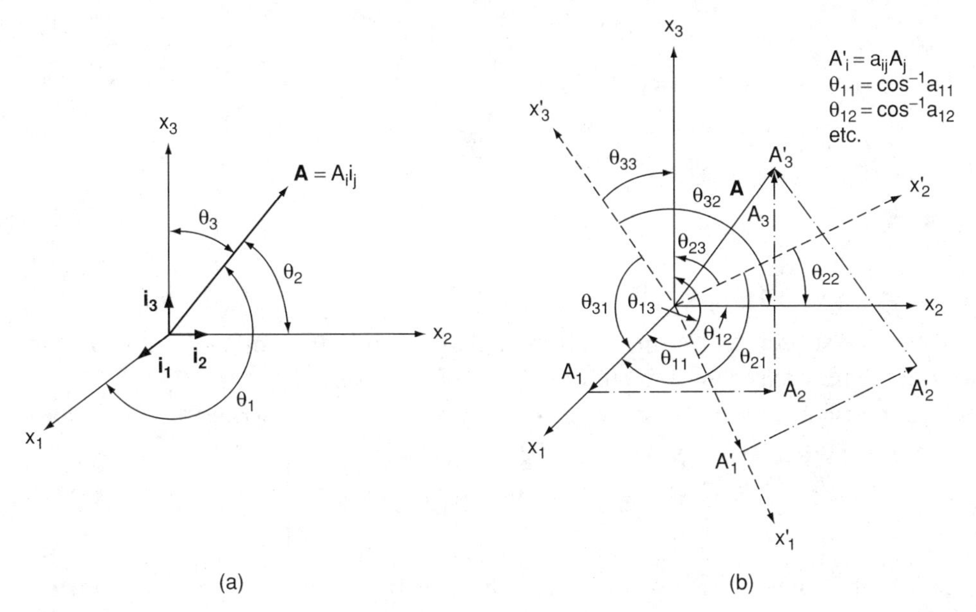

Figure 1.2.3. Coordinate transformations: (a) representation of a vector **A**, (b) representation of a vector **A** in two sets of right-handed Cartesian axes with different orientations.

Figure 1.2.2 shows that the old coordinates x_i are rotated about the x_3 axis counterclockwise by an angle θ to a set of new coordinates x_i'. We can calculate the new coordinates in terms of the old coordinates by adding and subtracting coordinate lengths,

$$
\begin{aligned}
x_1' &= (\cos\theta)x_1 + (\sin\theta)x_2, \\
x_2' &= -(\sin\theta)x_1 + (\cos\theta)x_2, \\
x_3' &= x_3.
\end{aligned}
\tag{1.2.11}
$$

These relationships may be obtained in a more general and systematic manner. To this end, we write Eqs. (1.2.11) in a form given by Eq. (1.2.2a) and as shown in Fig. 1.2.3:

$$
x_i' = a_{ij}x_j \quad (i, j = 1, 2, 3), \tag{1.2.12}
$$

with

$$
\begin{aligned}
a_{11} &= \cos\theta_{11} = \cos\theta, \\
a_{12} &= \cos\theta_{12} = \cos\left(\frac{\pi}{2} - \theta\right) = \sin\theta, \\
a_{13} &= \cos\frac{\pi}{2} = 0, \\
a_{21} &= \cos\theta_{21} = \cos\left(\frac{\pi}{2} + \theta\right) = -\sin\theta,
\end{aligned}
$$

$$\begin{aligned}
a_{22} &= \cos\theta_{22} = \cos\theta, \\
a_{23} &= \cos\frac{\pi}{2} = 0, \\
a_{31} &= \cos\theta_{31} = \cos\frac{\pi}{2} = 0, \\
a_{32} &= \cos\theta_{32} = \cos\frac{\pi}{2} = 0, \\
a_{33} &= \cos\theta_{33} = \cos 0 = 1,
\end{aligned} \tag{1.2.13}$$

where the subscripts on the rotation angle θ_{ab} imply that an angle is measured from the new axis a (x'_i axis) to the old axis b (x_j axis).

In matrix notation, using the results of Eqs. (1.2.13), we write Eq. (1.2.12) as

$$\begin{bmatrix} x'_1 \\ x'_2 \\ x'_3 \end{bmatrix} = \begin{bmatrix} \cos\theta & \sin\theta & 0 \\ -\sin\theta & \cos\theta & 0 \\ 0 & 0 & 1 \end{bmatrix} \begin{bmatrix} x_1 \\ x_2 \\ x_3 \end{bmatrix}. \tag{1.2.14}$$

This is the same as in Eqs. (1.2.11), which were obtained from a geometrical deduction. The approach given in Eqs. (1.2.13) is much more efficient, particularly when rotations about various axes are performed many times consecutively, as shown in Example 1.2.1. Note that the old coordinates x_j are transformed into the new coordinates x'_i by means of the quantities called the *transformation matrix* a_{ij}, whose components are the cosines of angles measured from the new coordinates x'_i to the old coordinates x_j. We have seen that the position vector $\mathbf{r}$ remains invariant through coordinate transformations. The quantities x'_i, x_j, and a_{ij} are the abstract objects whose properties remain invariant with the coordinate transformations. Therefore they are all tensors, as defined earlier.

Similarly, we may consider a vector $\mathbf{A} = A_i\mathbf{i}_i$ oriented at θ_1, θ_2, and θ_3 from the respective rectangular Cartesian coordinate axes [Fig. 1.2.3(a)] so that

$$A_i = An_i \quad (i = 1, 2, 3),$$

where

$$A = \sqrt{A_1^2 + A_2^2 + A_3^2},$$

$$n_1 = \frac{A_1}{A} \quad n_2 = \frac{A_2}{A} \quad n_3 = \frac{A_3}{A}.$$

Here $n_i(n_1, n_2, n_3)$ are the direction cosines whose properties must satisfy

$$\mathbf{n}\cdot\mathbf{n} = n_i n_i = n_1^2 + n_2^2 + n_3^2 = 1.$$

Note also in Eq. (1.2.12),

$$\begin{aligned}
\mathbf{n}^{(1)}\cdot\mathbf{n}^{(1)} &= a_{11}^2 + a_{12}^2 + a_{13}^2 = \cos^2\theta + \sin^2\theta = 1, \\
\mathbf{n}^{(2)}\cdot\mathbf{n}^{(2)} &= a_{21}^2 + a_{22}^2 + a_{23}^2 = (-\sin\theta)^2 + \cos^2\theta = 1, \\
\mathbf{n}^{(3)}\cdot\mathbf{n}^{(3)} &= a_{31}^2 + a_{32}^2 + a_{33}^2 = 1.
\end{aligned}$$

Proceeding in a similar manner, as shown in Fig. 1.2.3(b), we may write

$$A'_i = a_{ij} A_j,$$

where A'_i refers to a set of new coordinates so that the vector **A** is now measured in terms of the new coordinates and a_{ij} acts as a set of direction cosines, with i and j indicating the new and old coordinates, respectively.

Again, the vector components A'_i, A_j as well as a_{ij}, are tensors. They are abstract objects invariant with the frame of reference. Once a quantity is determined to be a tensor, then the number of free indices indicates the order of the tensor. Thus we define, in general, for any abstract object B,

$$\begin{aligned} B &= \text{zero-order tensor}, \\ B_i &= \text{first-order tensor}, \\ B_{ij} &= \text{second-order tensor}, \\ B_{ijk} &= \text{third-order tensor}, \\ B_{ijkl} &= \text{fourth order tensor}, \\ &\vdots \\ &\text{etc.} \end{aligned}$$

Tensors using the Cartesian coordinates are called "Cartesian tensors," whereas those based on the curvilinear coordinates are called "curvilinear tensors," which will be introduced in Chap. 6. Note that the second-order tensor refers to the 3×3 matrix as identified such as in Eqs. (1.2.3b) and (1.2.14).

EXAMPLE 1.2.1. Consider a set of new axes x'_i as obtained by rotating the old axes x_i through a 60° angle counterclockwise about the x_2 axis. What are the components of a vector **A** in the new coordinates if A_i in the old coordinates are (2, 1, 3)?

Solution. The direction cosines a_{ij} between the old and new axes are

$$a_{ij} = \begin{bmatrix} \cos 60^\circ & 0 & -\sin 60^\circ \\ 0 & 1 & 0 \\ \sin 60^\circ & 0 & \cos 60^\circ \end{bmatrix}.$$

From the transformation law,

$$A'_i = a_{ij} A_j,$$

we obtain

$$A'_1 = a_{11}A_1 + a_{12}A_2 + a_{13}A_3 = \frac{2 - 3\sqrt{3}}{2},$$

$$A'_2 = a_{21}A_1 + a_{22}A_2 + a_{23}A_3 = 1,$$

$$A'_3 = a_{31}A_1 + a_{32}A_2 + a_{33}A_3 = \frac{2\sqrt{3} + 3}{2},$$

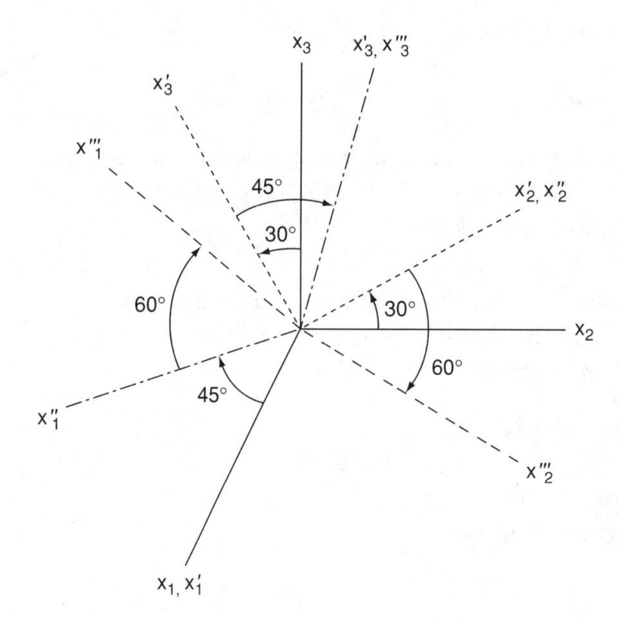

Figure 1.2.4. Successive rotations in the right-handed system, first about x_1 counterclockwise ($\theta_x = 30°$), then about x_2' clockwise ($\theta_y = 45°$), and finally about x_3'' clockwise ($\theta_x = 60°$) (Example 1.2.2).

which demonstrates that the vector **A** remains invariant, as subsequently shown:

$$\text{old coordinates } |\mathbf{A}| = \sqrt{A_1^2 + A_2^2 + A_3^2} = \sqrt{14},$$

$$\text{new coordinates } |\mathbf{A}| = \sqrt{A_1'^2 + A_2'^2 + A_3'^2} = \sqrt{14}.$$

EXAMPLE 1.2.2. Let the coordinate transformations be carried out successively by first rotating about the x_1 axis counterclockwise with $\theta_x = 30°$, then about the x_2' axis clockwise with $\theta_y = 45°$, and finally, about the x_3'' axis clockwise with $\theta_z = 60°$, as shown in Fig. 1.2.4. Determine the resulting direction cosines and calculate the components of **A** (2,1,3), ($|\mathbf{A}| = \sqrt{14}$) in terms of these rotations.

Solution. First, rotation about the x_1 axis gives

$$A_i' = a_{ij}' A_j.$$

Second, following the rotation about the x_2' axis,

$$A_i'' = a_{ij}'' A_j' = a_{ij}'' a_{jk}' A_k.$$

The third rotation about the x_3'' axis gives

$$A_i''' = a_{ij}''' A_j'' = a_{ij}''' a_{jk}'' a_{km}' A_m = a_{ij} A_j$$
$$= \begin{bmatrix} \dfrac{3}{4\sqrt{2}}(1 - \sqrt{2} + \sqrt{3} - \sqrt{6}) \\ \dfrac{1}{4\sqrt{2}}(3\sqrt{3} + \sqrt{6} + 9 + 3\sqrt{2}) \\ \dfrac{1}{2\sqrt{2}}(-5 + 3\sqrt{3}) \end{bmatrix}.$$

It should be noted that no more than one set of repeated indices is allowed to occur as we make substitutions from the previous step by changing indices repeatedly to avoid conflicts.

We observe that the summing of the repeated indices is equivalent to the matrix multiplications:

$$a_{ij} = \begin{bmatrix} c_z & -s_z & 0 \\ s_z & c_z & 0 \\ 0 & 0 & 1 \end{bmatrix} \begin{bmatrix} c_y & 0 & s_y \\ 0 & 1 & 0 \\ -s_y & 0 & c_y \end{bmatrix} \begin{bmatrix} 1 & 0 & 0 \\ 0 & c_x & s_x \\ 0 & -s_x & c_x \end{bmatrix}$$
$$= \begin{bmatrix} c_z c_y & -c_z s_y s_x - s_z c_x & c_z s_y c_x - s_z s_x \\ s_z c_y & -s_z s_y s_x + c_z c_x & s_z s_y c_x + c_z s_x \\ -s_y & -c_y s_x & c_y c_x \end{bmatrix},$$

with $s_x = \sin\theta_x$, $s_y = \sin\theta_y$, $s_z = \sin\theta_z$, $c_x = \cos\theta_x$, $c_y = \cos\theta_y$, $c_z = \cos\theta_z$, and

$$a_{ij} = \frac{1}{\sqrt{2}} \begin{bmatrix} \dfrac{1}{2} & \dfrac{-1 - 3\sqrt{2}}{4} & \dfrac{\sqrt{3} - \sqrt{6}}{4} \\ \dfrac{\sqrt{3}}{2} & \dfrac{-\sqrt{3} + \sqrt{6}}{4} & \dfrac{3 + \sqrt{2}}{4} \\ -1 & -\dfrac{1}{2} & \dfrac{\sqrt{3}}{2} \end{bmatrix}.$$

Thus the components of $\mathbf{A}$ based on the new coordinates x_i''' are

$$A_i''' = a_{ij} A_j,$$

with

$$A_1''' = \frac{3}{4\sqrt{2}}(1 - \sqrt{2} + \sqrt{3} - \sqrt{6}),$$

$$A_2''' = \frac{1}{4\sqrt{2}}(3\sqrt{3} + \sqrt{6} + 9 + 3\sqrt{2}),$$

$$A_3''' = \frac{1}{2\sqrt{2}}(-5 + 3\sqrt{3}).$$

Once again, the vector **A** remains invariant with the coordinate transformation ($|\mathbf{A}| = \sqrt{14}$).

It is interesting to note that the old coordinates x_j may be solved from Eq. (1.2.14) in terms of the new coordinates x'_i in the form

$$x_j = a_{ij} x'_i \tag{1.2.15a}$$

or

$$x_i = a_{ji} x'_j \tag{1.2.15b}$$

Substitute Eq. (1.2.15a) into Eq. (1.2.14) to give

$$x'_i = a_{ij} a_{kj} x'_k, \tag{1.2.16}$$

which requires that

$$a_{ij} a_{kj} = \delta_{ik}. \tag{1.2.17a}$$

Substitute this into Eq. (1.2.16) to obtain

$$x'_i = \delta_{ik} x'_k = x'_i.$$

In matrix notation Eq. (1.2.17a) can be written as

$$[a]\,[a]^T = [I], \tag{1.2.17b}$$

where I is the unity matrix. The transformation matrix of this kind is called the *proper orthogonal matrix* and has the properties

$$[a]^T = [a]^{-1}, \tag{1.2.18a}$$

$$\det[a] = |a| = 1. \tag{1.2.18b}$$

Note that the transpose arises in Eqs. (1.2.15a) and (1.2.15b) from the repeated index occurring with the first index in a_{ij} rather than the second index, as in Eq. (1.2.12). The matrix form of a_{ij} in Eq. (1.2.14) is the proper orthogonal matrix because it satisfies the property given by Eq. (1.2.18a).

A tensor that has the same components in all orthogonal coordinate systems is referred to as an *isotropic tensor*. The Kronecker delta is an isotropic tensor of order 2:

$$a_{ik} a_{jn} \delta_{ij} = a_{jk} a_{jn} = \delta_{kn}. \tag{1.2.19}$$

A scalar such as λ in Eq. (1.2.5) is an isotropic tensor of order 0. There is no first-order isotropic tensor. The permutation symbol in orthogonal coordinate systems is the isotropic tensor of order 3. Isotropic tensors of a higher order arise in the descriptions of material constants, which will be discussed in Chap. 4 [see Eq. (4.1.24)].

The permutation symbol is related to the Kronecker delta as

$$\begin{aligned}\epsilon_{ijk} &= \epsilon_{rst}\delta_{ir}\delta_{js}\delta_{kt} \\ &= \delta_{i1}\delta_{j2}\delta_{k3} - \delta_{i1}\delta_{j3}\delta_{k2} + \delta_{i2}\delta_{j3}\delta_{k1} \\ &\quad - \delta_{i2}\delta_{j1}\delta_{k3} + \delta_{i3}\delta_{j1}\delta_{k2} - \delta_{i3}\delta_{j2}\delta_{k1}.\end{aligned}$$

Thus ϵ_{ijk} becomes the determinant of the form

$$\epsilon_{ijk} = \begin{vmatrix} \delta_{i1} & \delta_{j1} & \delta_{k1} \\ \delta_{i2} & \delta_{j2} & \delta_{k2} \\ \delta_{i3} & \delta_{j3} & \delta_{k3} \end{vmatrix}. \tag{1.2.20}$$

We can then compute a product $\epsilon_{ijk}\epsilon_{mnp}$ in this manner, noting that the determinant of a matrix is equal to the determinant of its transposed matrix so that $|A||B| = |A^T||B| = |[A]^T[B]|$, with $[A]$ and $[B]$ being the 3×3 square matrices. It follows from Eq. (1.2.20) that the following identities can be derived:

$$\begin{aligned}\epsilon_{inm}\epsilon_{jpq} &= \delta_{ij}\delta_{np}\delta_{mq} + \delta_{ip}\delta_{nq}\delta_{mj} + \delta_{iq}\delta_{nj}\delta_{mp} \\ &\quad - \delta_{ij}\delta_{nq}\delta_{mp} - \delta_{ip}\delta_{nj}\delta_{mq} - \delta_{iq}\delta_{np}\delta_{mj}, \\ \epsilon_{inm}\epsilon_{ipq} &= \delta_{np}\delta_{mq} - \delta_{nq}\delta_{mp}, \\ \epsilon_{inm}\epsilon_{inq} &= 2\delta_{mq}, \\ \epsilon_{inm}\epsilon_{inm} &= 6.\end{aligned}$$

Derivatives of Vectors

A spatial rate of change of a tensor field is obtained by the *del operator* ∇, which is a vector quantity, defined as

$$\nabla = \mathbf{i}_i \frac{\partial}{\partial x_i} = \mathbf{i}_1 \frac{\partial}{\partial x_1} + \mathbf{i}_2 \frac{\partial}{\partial x_2} + \mathbf{i}_3 \frac{\partial}{\partial x_3}. \tag{1.2.21}$$

For a scalar field ψ, we write the del of ψ as

$$\nabla\psi = \mathbf{i}_i \frac{\partial\psi}{\partial x_i} = \mathbf{i}_i \psi_{,i}. \tag{1.2.22}$$

Here, the partial derivative of ψ with respect to x_i is written in a compact form,

$$\frac{\partial\psi}{\partial x_i} = \psi_{,i},$$

where the comma between ψ and i denotes a partial derivative with respect to an independent variable x_i.

The dot product of $\nabla\psi$ and a vector normal to a surface may be written in the form

$$\mathbf{n} \cdot \nabla\psi = n_i \mathbf{i}_i \cdot \mathbf{i}_j \frac{\partial\psi}{\partial x_j} = \psi_{,j} n_i \delta_{ij} = \psi_{,i} n_i = \frac{\partial\psi}{\partial x_1} n_1 + \frac{\partial\psi}{\partial x_2} n_2 + \frac{\partial\psi}{\partial x_3} n_3 = \frac{\partial\psi}{\partial n},$$

where the alternative notation $\partial/\partial n$, although used often in the literature, is regarded an undesirable practice, as the notation n is neither the vector $\mathbf{n}$ nor the components n_i of the vector $\mathbf{n}$.

The divergence of a vector $\mathbf{V}$ is given by

$$\nabla \cdot \mathbf{V} = \mathbf{i}_i \frac{\partial}{\partial x_i} \cdot V_j \mathbf{i}_j = \frac{\partial V_j}{\partial x_i} \delta_{ij} = \frac{\partial V_i}{\partial x_i} = V_{i,i}. \tag{1.2.23}$$

Likewise, the curl of a vector $\mathbf{V}$ takes the form

$$\nabla \times \mathbf{V} = \mathbf{i}_i \frac{\partial}{\partial x_i} \times V_j \mathbf{i}_j = \frac{\partial V_j}{\partial x_i} \epsilon_{ijk} \mathbf{i}_k = V_{j,i} \epsilon_{ijk} \mathbf{i}_k. \tag{1.2.24}$$

Similarly, the curl of the cross product of two vectors $\mathbf{V}$ and $\mathbf{W}$ reads

$$\begin{aligned} \nabla \times (\mathbf{V} \times \mathbf{W}) &= \mathbf{i}_j \frac{\partial}{\partial x_j} \times (V_k \mathbf{i}_k \times W_n \mathbf{i}_n) \\ &= \mathbf{i}_j \frac{\partial}{\partial x_j} \times (V_k W_n \epsilon_{knm} \mathbf{i}_m) = (V_k W_n)_{,j} \epsilon_{knm} \epsilon_{jmi} \mathbf{i}_i \\ &= (V_{k,j} W_n + V_k W_{n,j}) (\delta_{ki} \delta_{nj} - \delta_{kj} \delta_{ni}) \mathbf{i}_i \\ &= (V_{i,j} W_j + V_i W_{j,j} - V_{j,j} W_i - V_j W_{i,j}) \mathbf{i}_i \\ &= (\mathbf{W} \cdot \nabla)\mathbf{V} + \mathbf{V}(\nabla \cdot \mathbf{W}) - \mathbf{W}(\nabla \cdot \mathbf{V}) - (\mathbf{V} \cdot \nabla)\mathbf{W} \\ &= A_1 \mathbf{i}_1 + A_2 \mathbf{i}_2 + A_3 \mathbf{i}_3 = A_i \mathbf{i}_i, \end{aligned} \tag{1.2.25}$$

where

$$\begin{aligned} A_1 &= V_{1,2} W_2 + V_{1,3} W_3 + V_1 W_{2,2} + V_1 W_{3,3} - V_{2,2} W_1 - V_{3,3} W_1 \\ &\quad - V_2 W_{1,2} - V_3 W_{1,3}, \\ A_2 &= V_{2,1} W_1 + V_{2,3} W_3 + V_2 W_{1,1} + V_2 W_{3,3} - V_{1,1} W_2 - V_{3,3} W_2 \\ &\quad - V_1 W_{2,1} - V_3 W_{2,3}, \\ A_2 &= V_{3,1} W_1 + V_{3,2} W_2 + V_3 W_{1,1} + V_3 W_{2,2} - V_{1,1} W_3 - V_{2,2} W_3 \\ &\quad - V_1 W_{3,1} - V_2 W_{3,2}. \end{aligned}$$

It is important to realize that in the preceding example one of the powerful features of tensor analysis has been demonstrated; that is, we have shown the vector identity:

$$\nabla \times (\mathbf{V} \times \mathbf{W}) = (\mathbf{W} \cdot \nabla)\mathbf{V} + \mathbf{V}(\nabla \cdot \mathbf{W}) - \mathbf{W}(\nabla \cdot \mathbf{V}) - (\mathbf{V} \cdot \nabla)\mathbf{W}. \tag{1.2.26}$$

Note that this identity is a consequence of the routine tensorial manipulations in the summing of repeated indices. It is convenient to be able to shift back and forth between vector and tensor notation. The reader is encouraged to verify the components A_1, A_2, and A_3 in Eq. (1.2.25) by using the standard determinant calculations.

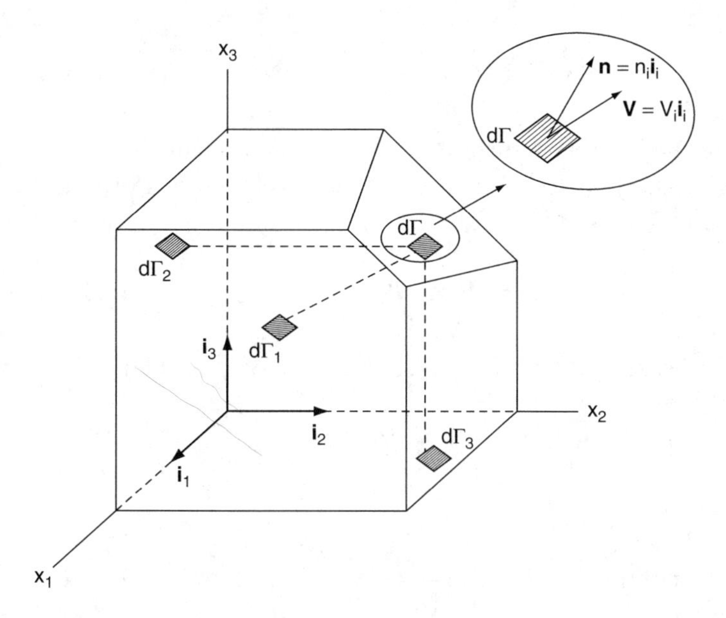

Figure 1.3.1. Projections of an inclined boundary surface area $d\Gamma$ on each of the three orthogonal planes of the three-dimensional body are $d\Gamma_1$, $d\Gamma_2$, and $d\Gamma_3$, which may be regarded as shadows or images of $d\Gamma$ on the x_1, x_2, and x_3 planes, respectively.

Using a similar procedure as in Eq. (1.2.25), we can show that

$$\nabla \times (\nabla \times \mathbf{V}) = \nabla(\nabla \cdot \mathbf{V}) - \nabla^2 \mathbf{V}. \tag{1.2.27}$$

Both Eqs. (1.2.26) and (1.2.27) are important relations used in fluid mechanics (Chap. 5). They are derived by manipulation of tensorial indices rather than by use of definitions of determinants of matrices. Additional discussions of vectors and tensors will appear in later chapters. It will be seen that they play an indispensable role in continuum mechanics.

1.3 Domain and Surface Integrals and Boundary Conditions

One of the most important roles of continuum mechanics is to derive the governing differential equations for any physical phenomena of interest. To solve such differential equations, then, we must have appropriate boundary conditions. In this section, given a certain type of differential equation, it is demonstrated how required boundary conditions can be derived. To this end, it is shown that tensor algebra plays a crucial role.

Let us consider a domain characterized by the cube with an inclined surface exposed to the external disturbances, as shown in Fig. 1.3.1. The domain and surface integrals are defined by the Green–Gauss theorem, relating the surface integral for a vector $\mathbf{V}$ in the direction normal to an inclined surface with the volume integral of the divergence of $\mathbf{V}$:

$$\int_{\Gamma} \mathbf{V} \cdot \mathbf{n} \, d\Gamma = \int_{\Omega} \nabla \cdot \mathbf{V} \, d\Omega \tag{1.3.1a}$$

where Γ and Ω denote the boundary surface (area) and the domain (volume), respectively, and $\mathbf{n}$ is the unit vector normal to the surface. In terms of index notation, the Green–Gauss theorem is written as

$$\int_{\Gamma} V_i n_i \, d\Gamma = \int_{\Omega} V_{i,i} \, d\Omega. \tag{1.3.1b}$$

To prove a more general case of the Green–Gauss theorem, we consider

$$\int_{\Gamma} \rho \mathbf{V} \cdot \mathbf{n} \, d\Gamma = \int_{\Omega} \nabla \cdot (\rho \mathbf{V}) \, d\Omega \tag{1.3.2a}$$

or

$$\int_{\Gamma} \rho \mathbf{V} \cdot \mathbf{n} \, d\Gamma = \int_{\Omega} \rho \nabla \cdot \mathbf{V} d\Omega + \int_{\Omega} (\mathbf{V} \cdot \nabla) \rho \, d\Omega, \tag{1.3.2b}$$

where both ρ and $\mathbf{V}$ are the dependent variables (ρ is the density and $\mathbf{V}$ is the velocity vector). Writing Eq. (1.3.2b) in index notation, we obtain

$$\int_{\Gamma} \rho V_i n_i d\Gamma = \int_{\Omega} \rho V_{i,i} d\Omega + \int_{\Omega} V_i \rho_{,i} d\Omega. \tag{1.3.2c}$$

To see that ρV_i acting on $d\Gamma$ on the inclined surface may be broken into three components on the plane surfaces of three orthogonal walls (Fig. 1.3.1), we integrate the first term of the right-hand side of Eq. (1.3.2c) by parts:

$$\begin{aligned} \int_{\Omega} \rho V_{i,i} d\Omega &= \int_{\Omega} \rho \left(\frac{\partial V_1}{\partial x_1} + \frac{\partial V_2}{\partial x_2} + \frac{\partial V_3}{\partial x_3} \right) dx_1 dx_2 dx_3 \\ &= \int_{\Gamma} \rho V_1 dx_2 dx_3 + \int_{\Gamma} \rho V_2 dx_3 dx_1 + \int_{\Gamma} \rho V_3 dx_1 dx_2 \\ &\quad - \int_{\Omega} \left(V_1 \frac{\partial \rho}{\partial x_1} + V_2 \frac{\partial \rho}{\partial x_2} + V_3 \frac{\partial \rho}{\partial x_3} \right) dx_1 dx_2 dx_3. \end{aligned} \tag{1.3.3}$$

Notice that $dx_2 dx_3$, $dx_1 dx_3$, and $dx_1 dx_2$ are the infinitesimal areas of images or "shadows" of $d\Gamma$ projected in the direction normal to the boundary surface:

$$dx_2 dx_3 = d\Gamma_1 = n_1 d\Gamma, \tag{1.3.4a}$$

$$dx_3 dx_1 = d\Gamma_2 = n_2 d\Gamma, \tag{1.3.4b}$$

$$dx_1 dx_2 = d\Gamma_3 = n_3 d\Gamma, \tag{1.3.4c}$$

where n_i (n_1, n_2, n_3) are the direction cosines denoting the components of the unit vector normal to the surface, as shown in Fig. 1.3.1. These unit vector components normal to the surface can be calculated from the coordinates given on the boundary surface (see Problems 1.9 and 1.10). Let us now rearrange Eq. (1.3.3):

$$\int_\Gamma \rho(V_1 n_1 + V_2 n_2 + V_3 n_3)d\Gamma = \int_\Omega \rho(V_{1,1} + V_{2,2} + V_{3,3})d\Omega + \int_\Omega (V_1\rho_{,1} + V_2\rho_{,2} + V_3\rho_{,3})d\Omega \quad (1.3.5a)$$

or

$$\int_\Omega \rho V_{i,i}d\Omega = \int_\Gamma \rho V_i n_i d\Gamma - \int_\Omega \rho_{,i} V_i d\Omega. \quad (1.3.5b)$$

These are identical to Eqs. (1.3.2). We conclude that integration by parts of an equation in the form given by Eq. (1.3.3) is equivalent to an application of the Green–Gauss theorem, as shown in Eq. (1.3.2a). Note that Eqs. (1.3.4) show how to construct the appropriate direction cosines for inclined surfaces for use in numerical models of boundary-value problems with irregular multidimensional boundary geometries.

The proof of (1.3.2a) may be obtained alternatively as follows:

$$\begin{aligned}\int_\Omega (\rho V_i)_{,i}\, d\Omega &= \int\int\int [(\rho V_1)_{,1} + (\rho V_2)_{,2} + (\rho V_3)_{,3}]dx_1 dx_2 dx_3 \\ &= \int\int \rho V_1 dx_2 dx_3 + \int\int \rho V_2 dx_1 dx_3 + \int\int \rho V_3 dx_1 dx_2 \\ &= \int\int \rho V_1 n_1 d\Gamma + \int\int \rho V_2 n_2 d\Gamma + \int\int \rho V_3 n_3 d\Gamma \\ &= \int_\Gamma \rho V_i n_i d\Gamma, \end{aligned} \quad (1.3.6)$$

where the integration "*by parts*" is avoided with the product (ρV_i) being integrated at once.

Physically, the preceding equation implies that the spatial rate of change of the mass flow within the domain is equal to the mass flow normal to the boundary surface, leading to the steady-state conservation of mass in fluids. This subject is discussed further in Subsection 2.2.2 for transient conservation of mass for fluids.

Note that the proof of the Green–Gauss theorem given by Eqs. (1.3.1) is now trivial, as we have just seen for the more general case. The approach given in Eq. (1.3.3) is to show the procedure of integration by parts for the product term whereas in Eq. (1.3.6) the "*partial*" integration is avoided. The most significant aspect of integration is to identify boundary surfaces in terms of the components of a vector normal to the surface or direction cosines, n_i, as given in Eqs. (1.3.4), so that, in all other applications henceforth, the integration from volume to surface can be written at once by inspection rather than going through all intermediate steps as in Eq. (1.3.5) or (1.3.6).

To demonstrate a further application of integration by parts and boundary surface integrals, we consider the Laplace equation

$$\nabla^2\psi = 0 \tag{1.3.7}$$

or

$$(\nabla \cdot \nabla)\psi = \left(\mathbf{i}_i \frac{\partial}{\partial x_i} \cdot \mathbf{i}_j \frac{\partial}{\partial x_j}\right)\psi$$

$$= \delta_{ij}\frac{\partial^2\psi}{\partial x_i \partial x_j} = \frac{\partial^2\psi}{\partial x_i \partial x_i} = \psi_{,ii} = 0. \tag{1.3.8}$$

Let us now take the inner product of Eq. (1.3.8) and a scalar ϕ and twice integrate by parts. The use of indices is seen to be convenient in performing integrations:

$$(\phi, \nabla^2\psi) \equiv \int_\Omega \phi\psi_{,ii} d\Omega = \int_\Gamma \phi\psi_{,i} n_i \, d\Gamma - \int_\Omega \phi_{,i}\psi_{,i} \, d\Omega$$

$$= \int_\Gamma \phi\psi_{,i} \, n_i d\Gamma - \int_\Gamma \phi_{,i}\psi \, n_i d\Gamma + \int_\Omega \phi_{,ii}\psi \, d\Omega. \tag{1.3.9}$$

Here, $\psi_{,i}$ and ψ, which appear in the integrand of the surface integral, represent the gradients of ψ and the value of ψ normal to the surface, respectively. They are the boundary data, called *boundary conditions*, to be specified on the boundary surfaces. This implies that the solution of governing equation (1.3.7) is subject to the boundary conditions consisting of derivatives of all orders lower than the highest derivatives in the governing differential equation. The first-order derivative $\psi_{,i}$ is called the *Neumann* or *natural boundary condition*, whereas the zero derivative of ψ, which is the ψ itself, is known as the *Dirichlet* or *essential boundary condition*. The solution process of a differential equation that depends on specifying the Neumann and Dirichlet boundary conditions is known as the *Cauchy problem*. The physical significance of the inner product, integration by parts, and boundary surface integrals are given in another subsequent example.

The following example shows an even more pronounced advantage of using the tensor algebra in performing integration by parts. Let us consider the biharmonic differential equation [see Eq. (5.6.10) and Eq. (9.3.1)],

$$\nabla^4\psi = f, \tag{1.3.10}$$

representing the equilibrium equation for a plate bending in solid mechanics (ψ is the transverse displacement, f is the force that is due to transverse loads) or the steady-state momentum equation for viscous incompressible flows in fluid mechanics (ψ is the stream function, f is the force gradient that is due to convection). These topics are described in Subsection 9.3.3 on the plate-bending theory and in Subsection 5.6.2 on the vorticity transport theory. We examine what kinds of boundary conditions are needed to solve the differential equation given by Eq. (1.3.10), but the main intention is to provide further exercises on manipulations of tensorial indices, rather than the physics of boundary conditions.

We expect that the successive integrations of the fourth-order partial differential equation produce boundary conditions with their third, second, first, and zero derivatives of ψ. To see this, we consider the inner product of a biharmonic equation (in two dimensions) and a scalar ϕ, physically representing the energy or work. This implies that the product of the displacement ($\phi = \psi$) and the force ($\nabla^4\psi$) for the plate bending, for example, becomes

$$
\begin{aligned}
(\phi, \nabla^4\psi) &\equiv \int_\Omega \phi\nabla^4\psi \, d\Omega = \int_\Omega \phi\psi_{,iijj} d\Omega \\
&= \int_\Omega \phi(\psi_{,1111} + \psi_{,1122} + \psi_{,2211} + \psi_{,2222}) d\Omega \\
&= \int_\Omega \phi\left(\frac{\partial^4\psi}{\partial x^4} + 2\frac{\partial^4\psi}{\partial x^2\partial y^2} + \frac{\partial^4\psi}{\partial y^4}\right) dx dy \quad (i, j = 1, 2). \qquad (1.3.11)
\end{aligned}
$$

This must be integrated by parts four consecutive times. Having seen the integration by parts performed by using index notation, we should now find it straightforward that successive integrations can be carried out simply by inspection with proper manipulations of indices as follows:

$$
\begin{aligned}
\int_\Omega \phi\psi_{,iijj} \, d\Omega &= \int_\Gamma \phi\psi_{,iij} n_j d\Gamma - \int_\Gamma \phi_{,j}\psi_{,ii} \, n_j d\Gamma + \int_\Gamma \phi_{,jj}\psi_{,i} n_i d\Gamma \\
&\quad - \int_\Gamma \phi_{,jji}\psi n_i d\Gamma + \int_\Omega \phi_{,jjii}\psi \, d\Omega \quad (i, j = 1, \ 2), \qquad (1.3.12)
\end{aligned}
$$

with one set of boundary conditions emerging at each integration step. If integrations by parts are performed without indices, then the mixed derivative term in Eq. (1.3.11) must be split into two terms:

$$
2\frac{\partial^4\psi}{\partial x^2\partial y^2} = \frac{\partial^4\psi}{\partial x^2\partial y^2} + \frac{\partial^4\psi}{\partial y^2\partial x^2}. \qquad (1.3.13)
$$

Integrations by parts are carried out for the first term on the right-hand side of Eq. (1.3.13), twice with respect to x and then twice with respect to y. This process is reversed for the second term on the right-hand side. It is seen that this approach is rather cumbersome. It is clearly evident that integration by parts with tensor algebra by use of indices as in Eq. (1.3.12) is much more efficient.

To distinguish between the Neumann and Dirichlet boundary conditions, in general, we may consider a $2m$th-order differential equation, where m is an integer (i.e., $m = 1$ and $m = 2$ correspond to the second- and fourth-order differential equations, respectively). The Neumann boundary conditions then consist of derivatives of the order $2m - 1, 2m - 2, ..., m$, whereas the Dirichlet boundary conditions consist of derivatives of the order $m - 1, m - 2, \ldots, 0$. Thus the Neumann boundary conditions in Eq. (1.3.12) are $\psi_{,iij}$ and $\psi_{,ii}$, which represent, respectively, the boundary shears and moments for the bending of plates and boundary forces (stress gradients) and velocity gradients (stresses) for incompressible viscous flows, respectively. The Dirichlet boundary conditions are given by $\psi_{,i}$ and ψ, which represent, respectively, the boundary slopes and boundary displacements for the plate bending and the boundary velocity and boundary stream function for incompressible viscous flows.

If $\phi = \psi$ in Eq. (1.3.11) represents the transverse displacement of a plate in bending, then the inner product, $(\psi, \nabla^4\psi)$, is the energy or work mobilized in undergoing the transverse displacement. Thus each boundary term integral in Eq. (1.3.12) implies that the energy exerted at the boundaries that are due to the corresponding boundary conditions is in balance with the work done in the domain of the plate. Obviously, for free boundaries, all boundary terms must vanish and the left-hand side is equal to the right-hand side for $\phi = \psi$, representing the physical energy $(\psi, \nabla^4\psi)$ that corresponds to the original differential equation defined within the domain only, with no boundary conditions involved. Similar arguments can be applied to the momentum equation with a stream-function description for the viscous incompressible flow.

In general, Dirichlet (essential) boundary conditions must be specified with at least one boundary point when Neumann (natural) boundary conditions are given everywhere else. Otherwise the system is unstable. Dirichlet boundary conditions can be assigned throughout the boundaries with no Neumann boundary conditions specified anywhere. In fact, this represents the most stable condition for the solution of a differential equation. The converse is not true. The system with Neumann boundary conditions everywhere without Dirichlet boundary conditions is most unstable, and in this case the solution of a differential equation cannot be obtained.

The main emphasis of the last example was not necessarily the physics of the boundary conditions per se but to demonstrate the rigor of tensor algebra that produces various boundary conditions corresponding to a given differential equation. We carry this out simply by inspection with manipulations of indices rather than integrating term by term as required in Eq. (1.3.3). In particular, the identification of boundary surfaces as given by Eqs. (1.3.4a)–(1.3.4c) in terms of components of a vector normal to the surface related to particular types of boundary conditions is a significant feature of tensor algebra. It is intended that the beginner attain a sufficient practice for tensor index operations before undertaking the forthcoming chapters. Throughout the rest of this book it will be seen that tensor analysis leads to the remarkable physical consequences that sometimes cannot easily be shown otherwise. Cartesian tensors are used in Part I, whereas curvilinear tensors dominate Part II.

The reader may wish to examine other textbooks on continuum mechanics. Among them are Prager (1961), Eringen (1962), Truesdell (1965), Frederick and Chang (1965), Eringen (1967), Leigh (1968), Malvern (1969), Fung (1977), Gurtin (1981), Hunter (1983), Lai, Rubin, and Krempl (1995), and McDonald (1996).

PROBLEMS

1.1 Let $\mathbf{A} = 3\mathbf{i}_1 + 2\mathbf{i}_2 + 4\mathbf{i}_3$. Compute the direction cosines with respect to the (x_1, x_2, x_3) axes of the line containing the vector $\mathbf{A}$. Calculate the corresponding angles between $\mathbf{A}$ and each component line (A_1, A_2, A_3).

1.2 Let x_i be rotated 30° counterclockwise about the x_1 axis to x'_i through transformations a_{ij} in Problem 1.1 such that

$$A'_i = a_{ij} A_j \quad (i, j = 1, 2, 3).$$

Determine the new components A'_i of this vector.

1.3 Let x_i be rotated 60° clockwise successively about the x_2, x'_1, and x''_3 axes. Calculate the components of a vector $\mathbf{A}$ in terms of the final set of coordinates x'''_i if the components of $\mathbf{A}$ based on the original coordinates x_i are $(1, 1, 1)$. Prove that your calculation is correct by verifying $|\mathbf{A}| = \sqrt{3}$.

1.4 Calculate the component of a vector $\mathbf{A}$ $(2, 3, 1)$ in a direction normal to the plane $x_1 = x_2$.

1.5 Prove the following equations:

(a) $\delta_{ii} = 3$,

(b) $\epsilon_{inm}\epsilon_{jpq} = \delta_{ij}\delta_{np}\delta_{mq} + \delta_{ip}\delta_{nq}\delta_{mj} + \delta_{iq}\delta_{nj}\delta_{mp} - \delta_{ij}\delta_{nq}\delta_{mp} - \delta_{ip}\delta_{nj}\delta_{mq} - \delta_{iq}\delta_{np}\delta_{mj}$,

(c) $\epsilon_{inm}\epsilon_{ipq} = \delta_{np}\delta_{mq} - \delta_{nq}\delta_{mp}$,

(d) $\epsilon_{inm}\epsilon_{inq} = 2\delta_{mq}$,

(e) $\epsilon_{inm}\epsilon_{inm} = 6$.

1.6 Show that

(a) $(\mathbf{V} \cdot \nabla)\,\phi = V_i \phi_{,i}$,

(b) $\nabla(\nabla \cdot \mathbf{V}) = V_{j,ji}\mathbf{i}_i$,

and expand completely for $i, j = 1, 2, 3$. Write the final results using the standard derivative notation with

$$x_1 = x,\ x_2 = y,\ x_3 = z,\ V_1 = u,\ V_2 = v,\ V_3 = w.$$

1.7 Using index notation, prove the relation

$$(\mathbf{V} \cdot \nabla)\mathbf{V} = \frac{1}{2}\nabla(\mathbf{V} \cdot \mathbf{V}) - \mathbf{V} \times (\nabla \times \mathbf{V}).$$

1.8 Prove Eq. (1.2.27), showing all components explicitly.

1.9 Consider an inclined boundary surface identified by the coordinates at A $(4, 4, 3)$, B $(0, 4, 2)$, and C $(4, 2, 4)$. Determine the unit vector normal to the surface ABC satisfying $\mathbf{n} \cdot \mathbf{n} = 1$.

1.10 Let an angle θ be measured from the x_1 axis counterclockwise to a line drawn normal to the boundary surface $d\Gamma$ in a two-dimensional domain (x_1x_2 plane). What are the direction cosines n_i (n_1, n_2), the components of a vector normal to the surface according to Eq. (1.3.9), reduced to two dimensions? From these direction cosines, write the boundary surface integral $\int_\Gamma \phi\psi_{,i} n_i d\Gamma$ in terms of the rotated angle θ.

1.11 Prove Eq. (1.3.12), showing integration by parts each step at a time.

1.12 Carry out the integration by parts of Eq. (1.3.11) without using indices, with the mixed derivatives given by the right-hand side of Eq. (1.3.13). Show that the results are the same as those obtained by expansion of Eq. (1.3.12). Identify the boundary data by using the completely expanded results.

2 Kinematics

2.1 General

One of the basic variables in continuum mechanics is displacement, the geometrical change of a point in the continuum. In a uniaxial tension or compression test of a solid bar, strain is defined as the change in length or displacement per unit of initial length. Thus, if a bar of 10 cm in length is elongated to 10.1 cm, then the strain is $+0.01$ or 1% in tension. If, instead, this bar is shortened to 9.9 cm, the strain is -0.01 or 1% in compression. However, if the bar is elongated to 10.2 cm and is subsequently compressed to 10.1 cm, a statement that the final strain is still 1% in tension does not adequately describe what has happened. If this deformed state is the same as in the case of the specimen that was merely strained to a length of 10.1 cm, such deformation is referred to as *elastic*. On the other hand, some materials exhibit *inelastic* rearrangements of particles under applied loads (viscoelasticity, plasticity, in Part II) and the final states of deformation in these two cases are different. Definitions of strain also become complicated when displacements are very large (nonlinear elasticity in Part II) and when they are caused by sustained loading, cyclic loading, torsional loading, or flexural (bending) loading, as discussed in Part II.

In fluids, on the other hand, we cannot visualize the strain as defined in solids. We prefer to measure the velocity of a fluid particle that passes through a given point in space rather than to keep track of the distance of a particle traveling downstream at high speed. Thus, in fluid mechanics, we are concerned with the rate of change of velocity (velocity gradients), known as *deformation rate*, which may be regarded as the *time rate of change of strain*. The stress in fluids may be proportional to velocity gradients (Newtonian fluids) or nonlinearly proportional to velocity gradients (non-Newtonian fluids, as discussed in Part II).

The mechanics of deformations, strains, deformation rates, and accelerations is referred to as *kinematics*. In this chapter, we discuss the concept of strain, the deformation rate, and related topics, including the various coordinate systems, derivatives of variables on these coordinates, strain-displacement relations, coordinate transformations, and dilatational and deviatoric strains. Studies of these subjects will prepare us for the kinetics, linear elasticity, and Newtonian fluid mechanics, which will be discussed in Chaps. 3–5.

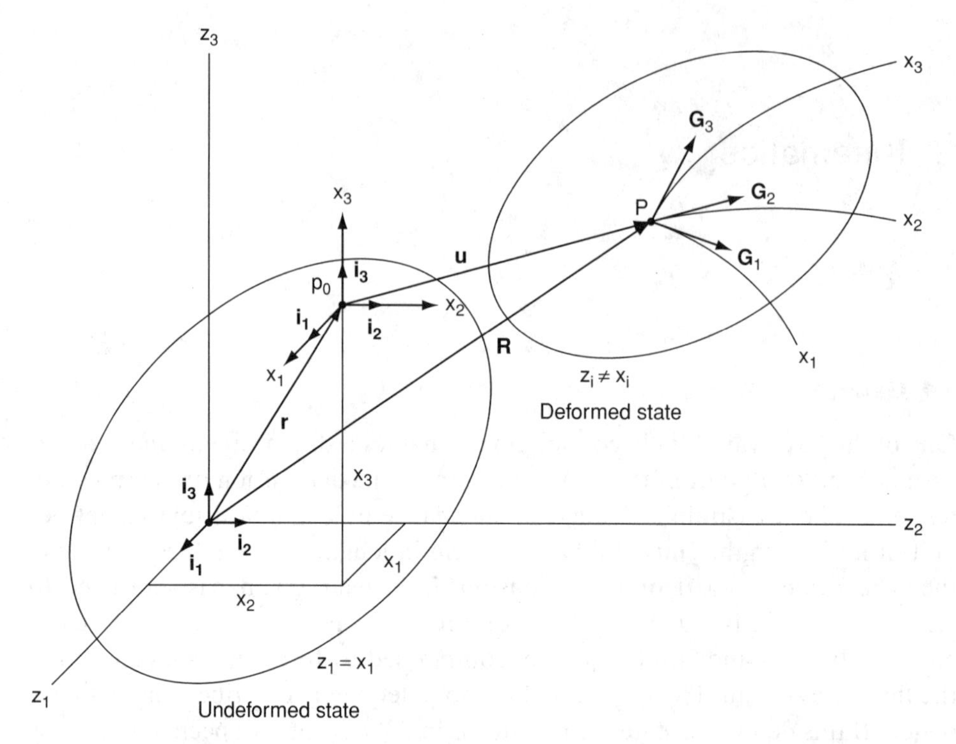

Figure 2.2.1. Lagrangian coordinates. The undeformed state is defined by rectangular Cartesian coordinates and the deformed state by arbitrary curvilinear (convected) coordinates.

2.2 Coordinate Systems

2.2.1 Lagrangian Coordinates

If a body undergoes geometric changes, we need, in addition to the reference Cartesian coordinates, a coordinate system that follows the deformed shape. Such a coordinate system is called the *Lagrangian* coordinates, more specifically known as the *convective Lagrangian* coordinates. On the other hand, in fluid mechanics, we are interested in measuring the velocity of fluid particles as they pass through a fixed point in space. The coordinate system required for this purpose is referred to as *Eulerian* coordinates, which is discussed in the next subsection.

The Lagrangian coordinate system (Fig. 2.2.1) refers to particle motion given by

$$\mathbf{R} = \mathbf{R}(\mathbf{r}, t), \tag{2.2.1}$$

where $\mathbf{r}$ denotes the position vector from the origin of the rectangular Cartesian coordinates to a point P_0 at time $t = t_0$ in the undeformed state, whereas $\mathbf{R}$ is the position vector to the point P in the deformed state at time $t = t$. Note that P_0 is located at x_i $(i = 1, 2, 3)$, where the x_i's are called the *material labels*. As deformation takes place, P_0 moves to P and the coordinate lines initially given in rectangular Cartesian coordinates at P_0 may be convected arbitrarily in nonrectangular form (similar to the standard curvilinear coordinates as introduced in Chap. 6). For this reason, such a coordinate system is referred to as the *convective coordinate* system. We shall show that it is the convective Lagrangian coordinates that will provide a link between solids and fluids.

The position vectors $\mathbf{r}$ and $\mathbf{R}$ are defined as

$$\mathbf{r} = x_i \mathbf{i}_i, \tag{2.2.2}$$

$$\mathbf{R} = z_i \mathbf{i}_i. \tag{2.2.3}$$

Let us consider the infinitesimal change in $\mathbf{R}$ such that

$$d\mathbf{R} = \frac{\partial \mathbf{R}}{\partial x_i} dx_i = \frac{\partial z_k \mathbf{i}_k}{\partial x_i} dx_i = \mathbf{G}_i dx_i, \tag{2.2.4}$$

where

$$\mathbf{G}_i = \frac{\partial z_k \mathbf{i}_k}{\partial x_i} = z_{k,i} \mathbf{i}_k \tag{2.2.5}$$

is known as the tangent vector or the base vector and $\partial z_k / \partial x_i$ is called the *deformation gradient*. The fact that the tangent vector is tangent to the curvilinear lines that are deformed from the original straight lines refers to "convective" Lagrangian coordinates. Note that commas are used to represent partial derivatives only when differentiation is performed with respect to the independent variables x_i. The role of $\partial z_k / \partial x_i$ in Eq. (2.2.5) is to transform the unit vectors into tangent vectors.

The displacement vector $\mathbf{u}$ is given by the relation (Fig. 2.2.1)

$$\mathbf{u} = \mathbf{R} - \mathbf{r}.$$

Thus, from Eq. (2.2.4),

$$d\mathbf{R} = \frac{\partial}{\partial x_i} (\mathbf{r} + \mathbf{u}) dx_i = \left(\frac{\partial x_k \mathbf{i}_k}{\partial x_i} + \frac{\partial u_k \mathbf{i}_k}{\partial x_i} \right) dx_i. \tag{2.2.6}$$

With $\partial x_k / \partial x_i = \delta_{ki}$, which is the Kronecker delta, it follows that

$$d\mathbf{R} = \left(\delta_{ki} + \frac{\partial u_k}{\partial x_i} \right) \mathbf{i}_k dx_i. \tag{2.2.7}$$

Equate Eqs. (2.2.4) and (2.2.7) to yield

$$\mathbf{G}_i = \left(\delta_{ki} + \frac{\partial u_k}{\partial x_i} \right) \mathbf{i}_k \tag{2.2.8}$$

or

$$\frac{\partial z_k}{\partial x_i} = \delta_{ki} + \frac{\partial u_k}{\partial x_i}. \tag{2.2.9}$$

In the Lagrangian coordinate system, we regard $\mathbf{r}$ and t as independent variables and $\mathbf{R}(\mathbf{r}, t)$ and $\mathbf{u}$ as dependent variables. The particle velocity $\mathbf{v}(\mathbf{r}, t)$ is the Lagrangian velocity, defined as

$$\mathbf{v}(\mathbf{r}, t) = \frac{\partial \mathbf{R}}{\partial t} = \frac{\partial}{\partial t} [\mathbf{r} + \mathbf{u}(\mathbf{r}, t)] = \frac{\partial \mathbf{u}(\mathbf{r}, t)}{\partial t} = \dot{\mathbf{u}}. \tag{2.2.10}$$

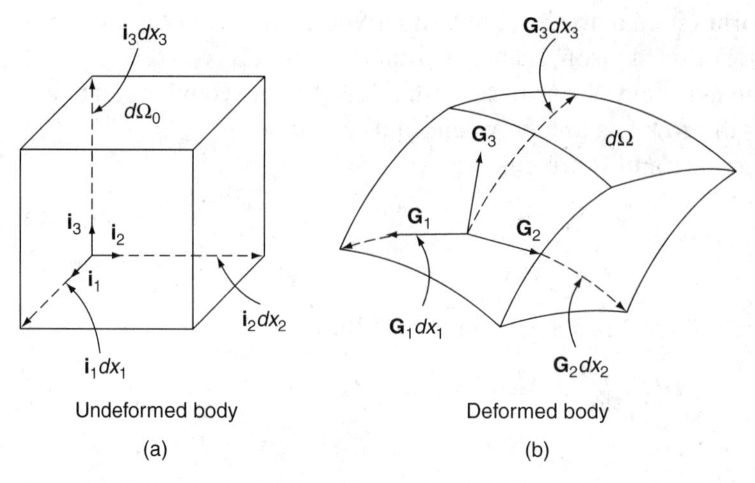

Figure 2.2.2. Infinitesimal (a) undeformed and (b) deformed configurations represented by unit vectors and tangent vectors, respectively.

Note here that $\mathbf{r}$ is independent of time and the overdot denotes the time derivative. Likewise, the particle acceleration or Lagrangian acceleration is

$$\mathbf{a}(\mathbf{r}, t) = \dot{\mathbf{v}}(\mathbf{r}, t) = \frac{\partial^2 \mathbf{R}}{\partial t^2} = \frac{\partial^2 \mathbf{u}}{\partial t^2} = \ddot{\mathbf{u}}. \tag{2.2.11}$$

A fundamental requirement in mechanics is the conservation of mass. In Lagrangian coordinates, the mass-conservation principle may be established by examination of the undeformed and the deformed infinitesimal bodies, as depicted in Fig. 2.2.2. First, consider the undeformed cube with its volume given by

$$d\Omega_0 = (dx_1 \mathbf{i}_1 \times dx_2 \mathbf{i}_2) \cdot dx_3 \mathbf{i}_3 = dx_1 dx_2 dx_3. \tag{2.2.12}$$

The deformed volume is then calculated as

$$\begin{aligned} d\Omega &= (\mathbf{G}_1 dx_1 \times \mathbf{G}_2 dx_2) \cdot \mathbf{G}_3 dx_3 \\ &= (\mathbf{G}_1 \times \mathbf{G}_2 \cdot \mathbf{G}_3) dx_1 dx_2 dx_3. \end{aligned} \tag{2.2.13}$$

Write the tangent vectors in the form

$$\begin{aligned} \mathbf{G}_1 &= \frac{\partial z_m \mathbf{i}_m}{\partial x_1} = \frac{\partial z_1 \mathbf{i}_1}{\partial x_1} + \frac{\partial z_2 \mathbf{i}_2}{\partial x_1} + \frac{\partial z_3 \mathbf{i}_3}{\partial x_1}, \\ \mathbf{G}_2 &= \frac{\partial z_m \mathbf{i}_m}{\partial x_2} = \frac{\partial z_1 \mathbf{i}_1}{\partial x_2} + \frac{\partial z_2 \mathbf{i}_2}{\partial x_2} + \frac{\partial z_3 \mathbf{i}_3}{\partial x_2}, \\ \mathbf{G}_3 &= \frac{\partial z_m \mathbf{i}_m}{\partial x_3} = \frac{\partial z_1 \mathbf{i}_1}{\partial x_3} + \frac{\partial z_2 \mathbf{i}_2}{\partial x_3} + \frac{\partial z_3 \mathbf{i}_3}{\partial x_3}, \end{aligned} \tag{2.2.14}$$

and substitute each of Eqs. (2.2.14) into Eq. (2.2.13) to obtain

$$d\Omega = J d\Omega_0, \tag{2.2.15}$$

where J is called the Jacobian, which is the determinant of the deformation gradient $\partial z_i/\partial x_j$:

$$J = \left|\frac{\partial z_i}{\partial x_j}\right| = \begin{vmatrix} \frac{\partial z_1}{\partial x_1} & \frac{\partial z_1}{\partial x_2} & \frac{\partial z_1}{\partial x_3} \\ \frac{\partial z_2}{\partial x_1} & \frac{\partial z_2}{\partial x_2} & \frac{\partial z_2}{\partial x_3} \\ \frac{\partial z_3}{\partial x_1} & \frac{\partial z_3}{\partial x_2} & \frac{\partial z_3}{\partial x_3} \end{vmatrix}. \tag{2.2.16}$$

To prove the relation given by Eq. (2.2.15), write Eq. (2.2.13) in the form

$$\begin{aligned} d\Omega &= \frac{\partial z_m}{\partial x_1}\frac{\partial z_n}{\partial x_2}\frac{\partial z_p}{\partial x_31}\mathbf{i}_m \times \mathbf{i}_n \cdot \mathbf{i}_p dx_1 dx_2 dx_3 \\ &= \epsilon_{mnp}\frac{\partial z_m}{\partial x_1}\frac{\partial z_n}{\partial x_2}\frac{\partial z_p}{\partial x_31} dx_1 dx_2 dx_3 \\ &= \left|\frac{\partial z_i}{\partial x_j}\right| dx_1 dx_2 dx_3 = \sqrt{\left|\frac{\partial z_k}{\partial x_i}\frac{\partial z_k}{\partial x_j}\right|} dx_1 dx_2 dx_3 \\ &= \sqrt{|G_{ij}|} dx_1 dx_2 dx_3 = \sqrt{G} dx_1 dx_2 dx_3 = J dx_1 dx_2 dx_3 \\ &= J d\Omega_0, \end{aligned} \tag{2.2.17}$$

where

$$J = \sqrt{G} = \sqrt{|G_{ij}|}, \tag{2.2.18}$$

$$G_{ij} = \mathbf{G}_i \cdot \mathbf{G}_j = \frac{\partial \mathbf{R}}{\partial x_i} \cdot \frac{\partial \mathbf{R}}{\partial x_j} = \frac{\partial z_m}{\partial x_i}\frac{\partial z_m}{\partial x_j}. \tag{2.2.19}$$

Here, G_{ij} is symmetric and called the *metric tensor*. It will be demonstrated that the metric tensor is responsible for distinguishing large strains from small strains (Section 2.3.), conservation of mass of fluids from that of solids (Eq. 2.2.29), nonlinear elasticity from linear elasticity (Chap. 7), and even the spacetime curvature of Einstein's theory of relativity (Section 9.4) from Newtonian mechanics. The metric tensor arises because of the convective Lagrangian coordinates.

The physical significance of the existence of the Jacobian is that the mass is conserved during the transition of the volume change from $d\Omega_0$ to $d\Omega$, as signified by the fact that the motion

$$z_i = z_i(x_i, t)$$

is continuously differentiable. We also note that

$$d\Omega_0 = J^{-1} d\Omega, \tag{2.2.20}$$

which implies that the matrix in Eq. (2.2.16) is nonsingular.

We now point out that a continuous density function ρ exists:

$$\rho = \lim_{\Delta\Omega \to 0} \frac{\Delta m}{\Delta \Omega} = \frac{dm}{d\Omega}, \tag{2.2.21}$$

where m is the mass. By the law of conservation of mass, we find

$$m = \int_{\Omega} \rho \, d\Omega = \int_{\Omega} \rho_0 \, d\Omega_0. \tag{2.2.22a}$$

Substitute Eq. (2.2.15) into Eq. (2.2.22a) to yield the global form of the conservation of mass:

$$\int_{\Omega} (\rho J - \rho_0) \, d\Omega_0 = 0. \tag{2.2.22b}$$

For this equation to be valid for all arbitrary volumes $d\Omega_0$, it is required that the integrand vanish. Thus,

$$\rho J = \rho_0 \tag{2.2.23a}$$

or

$$\rho = \frac{1}{\sqrt{G}} \rho_0. \tag{2.2.23b}$$

Equation (2.2.22b) is the local form of the conservation of mass as it applies to motion in solid mechanics.

2.2.2 Eulerian Coordinates

A fluid particle may travel with such high velocity that the human eye is unable to trace it downstream. The Lagrangian coordinates described in Subsection 2.2.1 are obviously inadequate because our aim is to determine the velocity of fluid particles at any point in space rather than to calculate the displacement of fluid particles. For this purpose, we introduce Eulerian coordinates (see Fig. 2.2.3), in which we consider the velocity vector $\mathbf{V}(\mathbf{R}, t)$ and density ρ to be dependent variables and $\mathbf{R}$ and t to be independent variables.

We use the uppercase $\mathbf{V}$ to indicate the Eulerian velocity in fluid mechanics and the lowercase $\mathbf{v}$ to indicate the Lagrangian velocity in solid mechanics. The velocity $\mathbf{V}$ is measured at the current position P located at $\mathbf{R}$. Thus the initial position vector $\mathbf{r}$, located at P_0 used in Lagrangian coordinates, is irrelevant in Eulerian coordinates. Because the dependent variables change with respect to both time t and spatial coordinates z_i in the Eulerian coordinates, we introduce a special form of a derivative given by

$$\frac{D}{Dt} = \frac{\partial}{\partial t} + \frac{\partial}{\partial z_i}\frac{\partial z_i}{\partial t} = \frac{\partial}{\partial t} + V_i \frac{\partial}{\partial z_i}, \tag{2.2.24a}$$

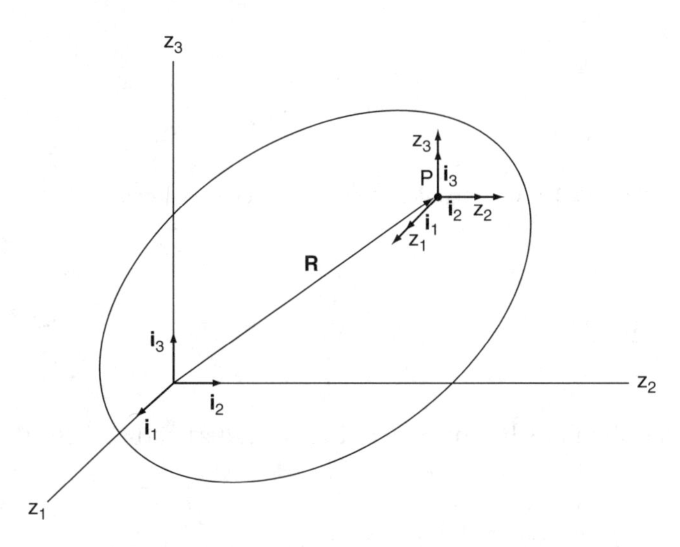

Figure 2.2.3. Eulerian coordinates – Cartesian representation. Velocity $V_1 = \partial z_i/\partial t$ is measured at P.

where $\partial z_i/\partial t$ is defined as the Eulerian velocity V_i. The preceding expression may be written in vector notation as

$$\frac{D}{Dt} = \frac{\partial}{\partial t} + \mathbf{V} \cdot \nabla. \tag{2.2.24b}$$

Here the symbol D/Dt is called the *substantial derivative* or the *material derivative*, with $\mathbf{V}(\mathbf{R}, t) = \partial \mathbf{R}/\partial t$. Therefore the Eulerian acceleration $\mathbf{A}(\mathbf{R}, t)$ is given by

$$\mathbf{A}(\mathbf{R}, t) = \frac{D\mathbf{V}(\mathbf{R}, t)}{Dt} = \frac{\partial \mathbf{V}}{\partial t} + V_j \frac{\partial \mathbf{V}}{\partial z_j} = \frac{\partial \mathbf{V}}{\partial t} + (\mathbf{V} \cdot \nabla)\mathbf{V}$$

$$= \left(\frac{\partial V_i}{\partial t} + V_{i,j} V_j \right) \mathbf{i}_i. \tag{2.2.25}$$

Note that z_i is the dependent variable with respect to time, but is an independent variable with respect to the velocity. Therefore we can use the comma to represent the derivative of the velocity with respect to z_j ($V_{i,j} = \partial V_i/\partial z_j$), contrary to the case of Lagrangian coordinates ($u_{i,j} = \partial u_i/\partial x_j$).

Let us now consider a domain Ω with an infinitesimal boundary surface $d\Gamma$ with a unit vector $\mathbf{n}$ normal to $d\Gamma$ (see Fig. 2.2.4). The mass rate of flow out of $d\Gamma$ is

$$\frac{dm}{dt} = \int_\Gamma \rho \mathbf{V} \cdot \mathbf{n}\, d\Gamma = \int_\Gamma \rho V_i n_i \, d\Gamma. \tag{2.2.26a}$$

On the other hand, the mass rate of flow in Ω is

$$\frac{dm}{dt} = -\int_\Omega \frac{\partial \rho}{\partial t} d\Omega. \tag{2.2.26b}$$

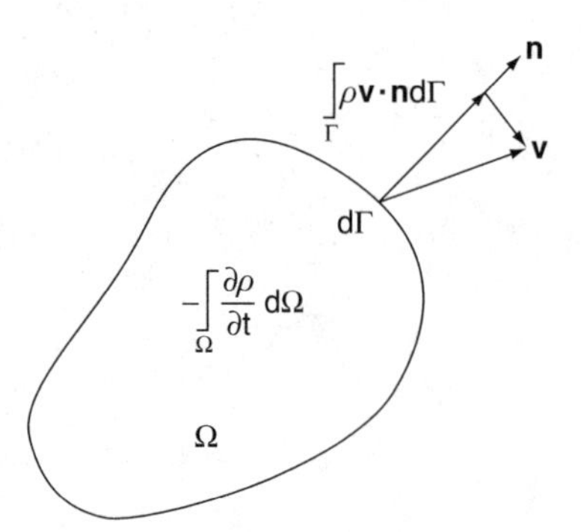

Figure 2.2.4. Conservation of mass in Eulerian coordinates.

Because it is necessary that the mass be conserved, Eqs. (2.2.26a) and (2.2.26b) must be equal:

$$-\int_\Omega \frac{\partial \rho}{\partial t} d\Omega = \int_\Gamma \rho \mathbf{V} \cdot \mathbf{n} \, d\Gamma.$$

Use the Green–Gauss theorem or Eq. (1.3.2a) to write

$$\int_\Gamma \rho \mathbf{V} \cdot \mathbf{n} \, d\Gamma = \int_\Omega \nabla \cdot (\rho \mathbf{V}) \, d\Omega. \tag{2.2.27}$$

It follows that

$$\int_\Omega \left[\frac{\partial \rho}{\partial t} + \nabla \cdot (\rho \mathbf{V}) \right] d\Omega = 0. \tag{2.2.28a}$$

For the preceding integral equation to remain valid for all arbitrary volumes $d\Omega$, it is necessary that the integrand vanish. Therefore,

$$\frac{\partial \rho}{\partial t} + \nabla \cdot (\rho \mathbf{V}) = 0 \tag{2.2.28b}$$

or

$$\frac{\partial \rho}{\partial t} + (\rho V_i)_{,i} = 0. \tag{2.2.28c}$$

Once again, we may write

$$(\rho V_i)_{,i} = \frac{\partial}{\partial z_i} (\rho V_i),$$

which implies that the comma represents partial derivatives with respect to z_i instead of x_i, contrary to the case of Lagrangian coordinates. Equations (2.2.28a) and (2.2.28b) represent, respectively, the global form and the local form of the conservation of mass in Eulerian coordinates or for fluid mechanics.

The conservation of mass as shown in Eq. (2.2.28a) is arrived at by means of intuition. However, the conservation of mass for fluids can be obtained rigorously by use of the convective coordinates for the undeformed and deformed volumes of Fig. 2.2.2. To this end, we take a substantial derivative of Eq. (2.2.22b) representing the conservation of mass for solids, which allows a transformation from Lagrangian coordinates to Eulerian coordinates as

$$
\begin{aligned}
\frac{D}{Dt}\int_{\Omega}(\rho d\Omega - \rho_0 d\Omega_0) &= \frac{D}{Dt}\int_{\Omega}(\rho d\Omega)\\
&= \int_{\Omega}\left(\frac{D\rho}{Dt}d\Omega + \rho\frac{Dd\Omega}{Dt}\right) = \int_{\Omega}\left(\frac{D\rho}{Dt}d\Omega + \rho\frac{DJ}{Dt}d\Omega_0\right)\\
&= \int_{\Omega}\left(\frac{D\rho}{Dt}d\Omega + \rho J\nabla\cdot\mathbf{V}d\Omega_0\right) = \int_{\Omega}\left(\frac{D\rho}{Dt} + \rho\nabla\cdot\mathbf{V}\right)d\Omega\\
&= \int_{\Omega}\left(\frac{\partial\rho}{\partial t} + (\mathbf{V}\cdot\nabla)\rho + \rho\nabla\cdot\mathbf{V}\right)d\Omega\\
&= \int_{\Omega}\left(\frac{\partial\rho}{\partial t} + \nabla\cdot(\rho\mathbf{V})\right)d\Omega = 0.
\end{aligned}
\tag{2.2.29}
$$

This is the conservation of mass for fluids in the global form, identical to the conservation of mass given by Eq. (2.2.28a), which arises merely from physical intuition. The tensor algebra involved in previously calculating DJ/Dt is given by

$$
\begin{aligned}
\frac{DJ}{Dt} &= \frac{\partial J}{\partial t} + \frac{\partial J}{\partial z_i}\frac{\partial z_i}{\partial t}\\
&= \frac{\partial}{\partial t}\left(\epsilon_{mnp}\frac{\partial z_m}{\partial x_1}\frac{\partial z_n}{\partial x_2}\frac{\partial z_p}{\partial x_3}\right) + V_j\frac{\partial}{\partial z_j}\left(\epsilon_{mnp}\frac{\partial z_m}{\partial x_1}\frac{\partial z_n}{\partial x_2}\frac{\partial z_p}{\partial x_3}\right)\\
&= \epsilon_{mnp}\left(\frac{\partial V_m}{\partial z_k}\frac{\partial z_k}{\partial x_1}\frac{\partial z_n}{\partial x_2}\frac{\partial z_p}{\partial x_3} + \frac{\partial z_m}{\partial x_1}\frac{\partial V_n}{\partial z_k}\frac{\partial z_k}{\partial x_2}\frac{\partial z_p}{\partial x_3} + \frac{\partial z_n}{\partial x_1}\frac{\partial z_m}{\partial x_2}\frac{\partial V_p}{\partial z_k}\frac{\partial z_k}{\partial x_3}\right)\\
&\quad + V_j\epsilon_{mnp}\left[\frac{\partial}{\partial x_1}(\delta_{mj})\frac{\partial z_n}{\partial x_2}\frac{\partial z_p}{\partial x_3} + \frac{\partial z_m}{\partial x_1}\frac{\partial}{\partial x_2}(\delta_{nj})\frac{\partial z_p}{\partial x_3} + \frac{\partial z_m}{\partial x_1}\frac{\partial z_n}{\partial x_2}\frac{\partial}{\partial x_3}(\delta_{pj})\right]\\
&= \epsilon_{123}\left(\frac{\partial V_1}{\partial z_1}\frac{\partial z_1}{\partial x_1} + \frac{\partial V_1}{\partial z_2}\frac{\partial z_2}{\partial x_1} + \frac{\partial V_1}{\partial z_3}\frac{\partial z_3}{\partial x_1}\right)\frac{\partial z_2}{\partial x_2}\frac{\partial z_3}{\partial x_3} + \cdots +\\
&\qquad \text{(a total of 54 terms)}\\
&= \left(\frac{\partial V_1}{\partial z_1} + \frac{\partial V_2}{\partial z_2} + \frac{\partial V_3}{\partial z_3}\right)\left(\frac{\partial z_1}{\partial x_1}\frac{\partial z_2}{\partial x_2}\frac{\partial z_3}{\partial x_3} - \cdots -\right)\\
&\qquad \text{(a total of 18 terms)}\\
&= \frac{\partial V_k}{\partial z_k}\left|\frac{\partial z_i}{\partial x_j}\right| = \frac{\partial V_k}{\partial z_k}J = J\nabla\cdot\mathbf{V}.
\end{aligned}
\tag{2.2.30}
$$

Note that the spatial variation of J vanishes because

$$
\frac{\partial}{\partial z_j}\left(\frac{\partial z_m}{\partial x_1}\right) = \frac{\partial}{\partial x_1}\left(\frac{\partial z_m}{\partial z_j}\right) = \frac{\partial}{\partial x_1}(\delta_{mj}) = 0, \text{ etc.}
$$

In view of these results, we have reconfirmed the conservation of mass shown in Eq. (2.2.28a) as a consequence of transforming the conservation of mass in solids from Lagrangian to Eulerian coordinates for the conservation of mass in fluids. This is a remarkable result obtained from the rigor of tensor algebra. In this process, it is important to recognize that $\partial z_i/\partial x_j$ is an expression that is valid only in Lagrangian coordinates, whereas $\partial V_i/\partial z_j$ can be defined only in Eulerian coordinates, thus requiring Eq. (2.2.30) to participate in both Lagrangian and Eulerian coordinates in this transformation process. The close relationships between solid mechanics and fluid mechanics through the Lagrangian and Eulerian coordinates are discussed again in Chap. 5.

Multiplying Eq. (2.2.30) by $d\Omega_0$ and integrating over the domain, we obtain

$$\frac{\partial}{\partial t}\int_\Omega J\, d\Omega_0 = \int_\Omega (\nabla\cdot\mathbf{V}) J\, d\Omega_0 = \int_\Omega \nabla\cdot\mathbf{V}\, d\Omega$$

or

$$\frac{\partial}{\partial t}\int_\Omega d\Omega = \int_\Gamma \mathbf{V}\cdot\mathbf{n}\, d\Gamma, \tag{2.2.31}$$

which represents an important process in temporal moving boundary problems, known as the *space-conservation law*. This equation is also the basis for the so-called *volume-of-fluid* (VOF) computations in multiphase flow.

Another consequence of Eq. (2.2.30) is the condition of incompressible flow,

$$\nabla\cdot\mathbf{V} = \frac{1}{J}\frac{\partial J}{\partial t} = \frac{1}{d\Omega}\frac{\partial d\Omega}{\partial t}. \tag{2.2.32}$$

For an incompressible flow, $\nabla\cdot\mathbf{V} = 0$, coinciding with the requirement $J =$ constant or $d\Omega =$ constant, implying no volume change, the basic criterion for the incompressible flow's being satisfied in due course of the derivation of the conservation of mass for fluids from solids.

Once again, the physical and mathematical significance of Eq. (2.2.29) showing the conservation of mass that transforms from solids to fluids highlights the beauty of continuum mechanics. This is achieved through the combined convective Lagrangian and Eulerian coordinates founded on delineating the undeformed and deformed bodies shown in Fig. 2.2.2 and Eq. (2.2.15). The basic ingredient is the Jacobian J that constitutes the square root of the determinant of the metric tensor G_{ij}.

2.2.3 Relationships between Lagrangian and Eulerian Coordinates

In continuum mechanics the use of material description (Lagrangian coordinates) or spatial description (Eulerian coordinates) is dictated by the nature of the continuum. Although these coordinate systems are distinctly different from each other, they are related in such a way that properties in one coordinate system can be expressed in an alternative way in the other. If $\mathbf{R}(\mathbf{r}, t)$ denotes particle paths, then $\mathbf{R}$ is the solution of the ordinary differential equation:

$$\frac{d\mathbf{R}}{dt} = \mathbf{v}(\mathbf{r}, t).$$

However, particle paths, in general, are unimportant in fluid mechanics. Rather, the streamlines, which are the lines drawn parallel to the velocity field, are defined by

$$d\mathbf{R} = \mathbf{V}(\mathbf{R}, t)\, d\lambda,$$

where $d\lambda$ is a time parameter. This can be written in the Pfaffian form (Sneddon, 1957):

$$\frac{dz_1}{V_1(\mathbf{R}, t)} = \frac{dz_2}{V_2(\mathbf{R}, t)} = \frac{dz_3}{V_3(\mathbf{R}, t)} = d\lambda, \tag{2.2.33}$$

which indicate streamlines. For a steady-state condition with the streamlines remaining constant, we may set $\lambda = t$. For steady flows, we have $\mathbf{V}(\mathbf{R}, t) = \mathbf{V}(\mathbf{R}) = \partial\mathbf{R}(\mathbf{R}, t)/\partial t = \partial z_i \mathbf{i}_i/\partial t = \mathbf{V}(\mathbf{r}, t)$ and the streamlines in fluids and the particle paths in solids coincide. As mentioned earlier, we designate $\mathbf{v}$ (lowercase) and $\mathbf{V}$ (uppercase) as the Lagrangian and Eulerian velocities, respectively.

To illustrate the relationship between Lagrangian and Eulerian coordinates, let us assume that the motion is given by

$$z_1 = \frac{x_1}{1 + t x_1}, \qquad z_2 = x_2, \qquad z_3 = x_3. \tag{2.2.34}$$

The Lagrangian velocities are determined as

$$\mathbf{v}(\mathbf{r}, t) = \frac{\partial \mathbf{R}(\mathbf{r}, t)}{\partial t}$$

or

$$v_1 = \frac{dz_1}{dt} = -\frac{x_1^2}{(1 + t x_1)^2},$$
$$v_2 = v_3 = 0. \tag{2.2.35}$$

We calculate the Eulerian velocities by first solving for x_1 from the Lagrangian motion equations (2.2.34),

$$x_1 = \frac{z_1}{1 - t z_1}, \tag{2.2.36}$$

and then by substituting Eq. (2.2.36) into Eqs. (2.2.35):

$$\mathbf{V}(\mathbf{R}, t) = \mathbf{v}\left[\mathbf{r}(\mathbf{R}, t)\right]$$

or

$$V_1 = \frac{-x_1^2}{(1 + t x_1)^2} = \frac{-[z_1/(1 - t z_1)]^2}{[1 + t z_1/(1 - t z_1)]^2} = -z_1^2,$$
$$V_2 = V_3 = 0. \tag{2.2.37}$$

The Lagrangian velocity v_1 and the Eulerian velocity V_1 are shown in Figs. 2.2.5(a) and 2.2.5(b), respectively. The Eulerian velocities, Eqs. (2.2.37), may be transformed back into the Lagrangian velocities if we solve for z_1 from the differential

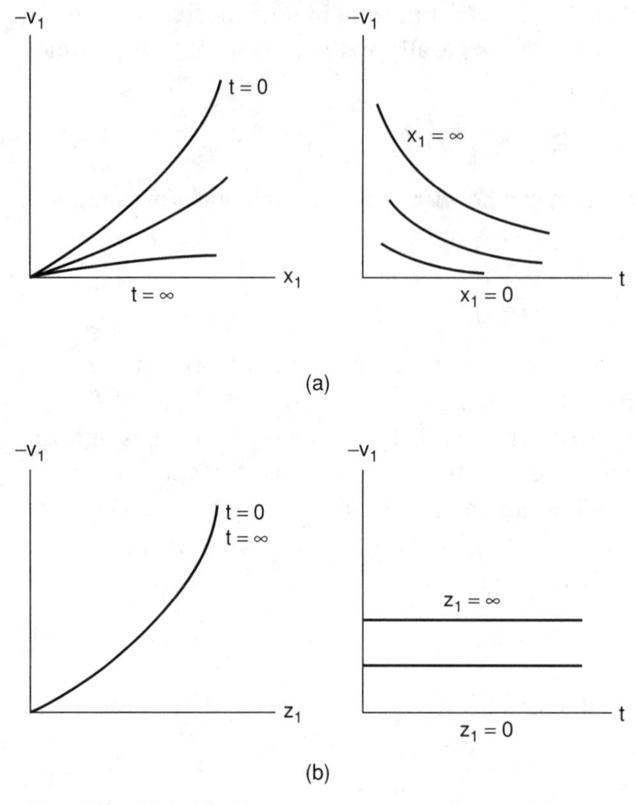

Figure 2.2.5. Relationship between Lagrangian and Eulerian velocities for the motion $z_1 = x_1/(1 + tx_1)$, $z_2 = x_2$, $z_3 = x_3$: (a) velocity distributions in material coordinates and time (Lagranian coordinates); (b) velocity distributions in space, independent of time (Eulerian coordinates).

equation resulting from Eqs. (2.2.37),

$$\frac{dz_1}{dt} + z_1^2 = 0, \tag{2.2.38}$$

with $V_1 = dz_1/dt$, subject to the initial condition $z_1 = x_1$ at $t = 0$. The solution of Eq. (2.2.38) gives the Lagrangian motion, Eqs. (2.2.34). Problem 2.3 at the end of this chapter illustrates a reverse process in which Eulerian velocities are given initially and Lagrangian velocities are to be calculated.

Remarks

Even though we draw a clear line by saying that Lagrangian descriptions are applicable to solids and Eulerian descriptions to fluids, engineering applications often call for a reversal of that rule or a combined use of both systems in a given analysis. For example, in hypervelocity impact problems in which phase changes are involved between solids and fluids, it is advantageous to invoke both Lagrangian and Eulerian descriptions. Another example is the particle tracking of spray combustion, in which, for convenience, both coordinate systems are used. Furthermore, convective terms in fluid mechanics (see Chap. 5) may be eliminated by coordinate transformation from Eulerian to Lagrangian descriptions, which results in numerical expediency.

2.3 Strain Tensor

It was shown in Subsection 2.2.1 that the Jacobian is considered a measure of deformation:

$$J = \left| \frac{\partial z_i}{\partial x_j} \right|,$$

which represents the deformation gradient in the convective coordinates, as defined in Eq. (2.2.5). However, it is more convenient to define the deformation specifically through strain-displacement relationships. Toward this end, we consider squares of infinitesimal line segments on the undeformed and deformed surfaces (see Fig. 2.2.1):

$$ds_0^2 = d\mathbf{r} \cdot d\mathbf{r} = dx_i dx_i, \tag{2.3.1a}$$

$$ds^2 = d\mathbf{R} \cdot d\mathbf{R} = \frac{\partial \mathbf{R}}{\partial x_i} \cdot \frac{\partial \mathbf{R}}{\partial x_j} dx_i dx_j = G_{ij} dx_i dx_j, \tag{2.3.2a}$$

where G_{ij} is the metric tensor defined earlier in Eq. (2.2.19) on the deformed geometry,

$$G_{ij} = \mathbf{G}_i \cdot \mathbf{G}_j = z_{m,i} z_{m,j} = (x_{m,i} + u_{m,i})(x_{m,j} + u_{m,j}) = (\delta_{mi} + u_{m,i})(\delta_{mj} + u_{m,j})$$

or

$$G_{ij} = \delta_{ij} + u_{i,j} + u_{j,i} + u_{m,i} u_{m,j}. \tag{2.3.2}$$

The difference between ds^2 and ds_0^2 is the measure of strain:

$$ds^2 - ds_0^2 = (G_{ij} - \delta_{ij})\, dx_i dx_j = 2\gamma_{ij} dx_i dx_j, \tag{2.3.3}$$

where γ_{ij} is called the strain tensor,

$$\gamma_{ij} = \frac{1}{2}(G_{ij} - \delta_{ij}). \tag{2.3.4}$$

Note that coefficient 2 in Eq. (2.3.3) is used to produce a standard definition for the strain components for the normal strains, but this definition provides one-half of the so-called *total shear strain* (or engineering shear strain). Substitution of Eq. (2.3.2) into Eq. (2.3.4) yields

$$\gamma_{ij} = \frac{1}{2}(u_{i,j} + u_{j,i} + u_{m,i} u_{m,j}). \tag{2.3.5}$$

It should be recognized that the strain tensor is symmetric, that is,

$$\gamma_{ij} = \gamma_{ji},$$

where the symmetry is due to the property of the commutative nature of the dot product associated with the metric tensor $G_{ij} = G_{ji}$ and the Kronecker delta

$\delta_{ij} = \delta_{ji}$. In the strain tensor γ_{ij} given by Eq. (2.3.5), the first and second indices imply, respectively, the plane and direction in which the strain components arise.

Let $u_1 = u$, $u_2 = v$, $u_3 = w$, $x_1 = x$, $x_2 = y$, and $x_3 = z$; thus we can write the relation in Eq. (2.3.5) as

$$\gamma_{11} = \frac{\partial u}{\partial x} + \frac{1}{2}\left[\left(\frac{\partial u}{\partial x}\right)^2 + \left(\frac{\partial v}{\partial x}\right)^2 + \left(\frac{\partial w}{\partial x}\right)^2\right], \tag{2.3.6a}$$

$$\gamma_{22} = \frac{\partial v}{\partial y} + \frac{1}{2}\left[\left(\frac{\partial u}{\partial y}\right)^2 + \left(\frac{\partial v}{\partial y}\right)^2 + \left(\frac{\partial w}{\partial y}\right)^2\right], \tag{2.3.6b}$$

$$\gamma_{33} = \frac{\partial w}{\partial z} + \frac{1}{2}\left[\left(\frac{\partial u}{\partial z}\right)^2 + \left(\frac{\partial v}{\partial z}\right)^2 + \left(\frac{\partial w}{\partial z}\right)^2\right], \tag{2.3.6c}$$

$$\gamma_{12} = \frac{1}{2}\left(\frac{\partial u}{\partial y} + \frac{\partial v}{\partial x} + \frac{\partial u}{\partial x}\frac{\partial u}{\partial y} + \frac{\partial v}{\partial x}\frac{\partial v}{\partial y} + \frac{\partial w}{\partial x}\frac{\partial w}{\partial y}\right), \tag{2.3.6d}$$

$$\gamma_{23} = \frac{1}{2}\left(\frac{\partial v}{\partial z} + \frac{\partial w}{\partial y} + \frac{\partial u}{\partial y}\frac{\partial u}{\partial z} + \frac{\partial v}{\partial y}\frac{\partial v}{\partial z} + \frac{\partial w}{\partial y}\frac{\partial w}{\partial z}\right), \tag{2.3.6e}$$

$$\gamma_{31} = \frac{1}{2}\left(\frac{\partial w}{\partial x} + \frac{\partial u}{\partial z} + \frac{\partial u}{\partial z}\frac{\partial u}{\partial x} + \frac{\partial v}{\partial z}\frac{\partial v}{\partial x} + \frac{\partial w}{\partial z}\frac{\partial w}{\partial x}\right). \tag{2.3.6f}$$

These equations represent the strain-displacement relationship, indicating the contributions from small strains (linear terms) and from large strains (nonlinear terms). Note that normal strains are $\gamma_{11} = \gamma_{xx}$, $\gamma_{22} = \gamma_{yy}$, and $\gamma_{33} = \gamma_{zz}$, and shear strains are $2\gamma_{12} = \gamma_{xy}$, $2\gamma_{23} = \gamma_{yz}$, and $2\gamma_{31} = \gamma_{zx}$. Here, γ_{12}, γ_{23}, and γ_{31} are the *tensor shear strains*, whereas γ_{xy}, γ_{yz}, and γ_{zx} are the total shear strains. Example 2.3.2 explains the significance of tensor shear strains and total shear strains.

It should be noted that the use of convective Lagrangian coordinates associated with the metric tensor G_{ij} is responsible for the nonlinear terms in the strain tensor. If only small strains are considered we need not invoke the convective Lagrangian coordinate system. They can be derived by use of simple geometries as shown in the following example problems.

EXAMPLE 2.3.1. SMALL EXTENSIONAL STRAIN. Consider $d\mathbf{R} = (dx_1, 0, 0)$ and determine the strain γ_{11}, assuming that its magnitude is small.

Solution

$$ds^2 - ds_0^2 = 2\gamma_{ij}dx_i dx_j = 2\gamma_{11}(dx_1)^2 = 2\gamma_{11}ds_0^2.$$

Solve for the axial strain γ_{11}; thus we have

$$\gamma_{11} = \frac{1}{2}\left(\frac{ds^2}{ds_0^2} - 1\right) = \frac{1}{2}(\beta^2 - 1),$$

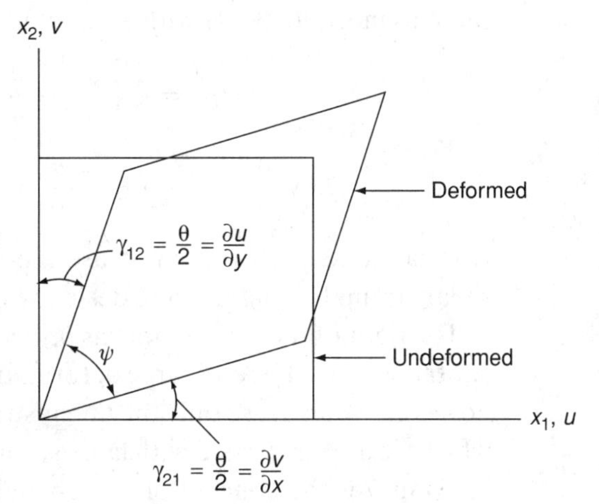

Figure 2.3.1. Shear deformation. Engineering or total shear strain $\gamma_{xy} = 2\gamma_{12} = 2\gamma_{21} = \theta$.

where β is defined as an *extension ratio*:

$$\beta = \frac{ds}{ds_0} = (1 + 2\gamma_{11})^{1/2} = 1 + \gamma_{11} - \frac{1}{2}\gamma_{11}^2 + \cdots + .$$

Neglect higher-order terms to obtain

$$\gamma_{11} \approx \beta - 1 = \frac{ds - ds_0}{ds_0} = \frac{du_1}{dx_1},$$

which is the same as Eq. (2.3.6a) with all the nonlinear terms neglected, and identified as the normal strain in undergraduate mechanics. It refers to a special case (small strain) that results from the general definition of the strain (large strain), as derived in the form shown in Eqs. (2.3.6a)–(2.3.6f).

EXAMPLE 2.3.2. SMALL SHEAR STRAIN. The undeformed square block in Fig. 2.3.1 undergoes a shear deformation. Calculate the shear strain, assuming that its magnitude is small. Here, $d\mathbf{R} = (dx_1, dx_2, 0)$. From Eq. (2.3.4.), the shear strain γ_{12} becomes

$$\gamma_{12} = \frac{1}{2}(G_{12} - 0),$$

$$\cos\psi = \frac{\mathbf{G}_1 dx_1 \cdot \mathbf{G}_2 dx_2}{|\mathbf{G}_1 dx_1 \cdot \mathbf{G}_2 dx_2|} = \frac{G_{12} dx_1 dx_2}{\sqrt{G_{11}}\sqrt{G_{22}} dx_1 dx_2}$$

$$= \frac{2\gamma_{12}}{\sqrt{1 + 2\gamma_{11}}\sqrt{1 + 2\gamma_{22}}}.$$

For small strains, $\gamma_{11} \ll 1$ and $\gamma_{22} \ll 1$, the shear strain takes the form

$$\gamma_{12} \approx \frac{1}{2}\cos\psi = \frac{1}{2}\cos\left(\frac{\pi}{2} - \theta\right) = \frac{1}{2}\sin\theta \approx \frac{\theta}{2}.$$

This implies that the shear strain γ_{12} is the tensor shear strain and is equal to one-half of θ, which is known as the total shear strain (total angle change that is due to shears, $\gamma_{xy} = \gamma_{yx} = 2\gamma_{12} = 2\gamma_{21}$). Once again, the small shear strain is the same as in Eq. (2.3.6d) with all nonlinear terms neglected:

$$\gamma_{21} = \gamma_{12} = \frac{1}{2}\left(\frac{\partial u}{\partial y} + \frac{\partial v}{\partial x}\right), \quad \text{tensor shear strain;}$$

$$\gamma_{xy} = \gamma_{21} + \gamma_{12} = \frac{\partial u}{\partial y} + \frac{\partial v}{\partial x}, \quad \text{total shear strain.}$$

It is cautioned that one must use the total shear strain, rather than the tensor shear strain, for engineering design work.

The rigor of tensor algebra has provided the most general form of strain tensor containing the effects of large deformations, as shown in Eqs. (2.3.6). Henceforth, however, we shall use only the linear strain components of (2.3.6) until Chap. 7, in which we are concerned with large strains or nonlinear mechanics. Discussed also in Chap. 7 is the concept of stretch and rotation needed for large deformations associated with non-Newtonian fluid mechanics.

2.4 Rate-of-Deformation Tensor

The strain tensor, which is a measure of deformation in Lagrangian coordinates in solid mechanics, is not applicable to fluid mechanics. In fluid mechanics, because the dependent variables are velocity $\mathbf{V}(\mathbf{R}, t)$ and density $\rho(\mathbf{R}, t)$ with the independent variables being $\mathbf{R}$ and t (see Fig. 2.2.3), we determine the time rate of change of $ds^2 - ds_0^2$ in the convective coordinates as follows.

Take the substantial derivative of $ds^2 - ds_0^2$, which gives

$$\begin{aligned} \frac{D}{Dt}\left(ds^2 - ds_0^2\right) &= \frac{Dds^2}{Dt} = \frac{D}{Dt}(d\mathbf{R}\cdot d\mathbf{R}) = \frac{D}{Dt}(dz_i dz_j \delta_{ij}), \\ &= \delta_{ij} dz_j \frac{Ddz_i}{Dt} + \delta_{ij} dz_i \frac{Ddz_j}{Dt}, \end{aligned} \tag{2.4.1}$$

where $Dds_0^2/Dt = 0$ as ds_0 is independent of time and space, implying that the velocity measurement is made at $t = t$, not at $t = t_0$. Here, we must perform the time rates of change of dz_i and dz_j separately, making use of the Kronecker delta so that the multidimensionality of velocity gradients can arise. The substantial derivative of dz_i becomes

$$\begin{aligned} \frac{Ddz_i}{Dt} &= \frac{\partial}{\partial t}\left(\frac{\partial z_i}{\partial x_j} dx_j\right) + \frac{\partial}{\partial z_k}\left(\frac{\partial z_i}{\partial x_j} dx_j\right)\frac{\partial z_k}{\partial t} \\ &= \frac{\partial}{\partial x_j}\left(\frac{\partial z_i}{\partial t}\right) dx_j + \frac{\partial}{\partial x_j}(\delta_{ik})\frac{\partial z_k}{\partial t} dx_j \\ &= \frac{\partial V_i}{\partial x_j} dx_j = \frac{\partial V_i}{\partial z_k}\frac{\partial z_k}{\partial x_j} dx_j = \frac{\partial V_i}{\partial z_k} dz_k = V_{i,k} dz_k = V_{i,j} dz_j, \end{aligned} \tag{2.4.2a}$$

where the transformation between the reference coordinates dx_i and the spatial coordinates dz_i leads to the derivative of the Eulerian velocity V_i with respect to the spatial coordinates z_i. This is the process of transformation from the strain tensor in solids to its time rate of change in fluids. It is possible to skip this process, however, by working with dz_i alone directly as follows:

$$\begin{aligned}\frac{Ddz_i}{Dt} &= \frac{\partial}{\partial t}(dz_i) + \frac{\partial}{\partial z_j}(dz_i)\frac{\partial z_j}{\partial t} \\ &= dV_i + d(\delta_{ij})V_j \\ &= \frac{\partial V_i}{\partial z_j}dz_j = V_{i,j}dz_j. \end{aligned} \tag{2.4.2b}$$

Proceeding similarly as in Eqs. (2.4.2), we obtain

$$\frac{Ddz_j}{Dt} = V_{j,i}dz_i. \tag{2.4.3}$$

Substituting Eqs. (2.4.2) and (2.4.3) into Eq. (2.4.1) and following the procedure similarly as in the case of the strain tensor in solids, we arrive at

$$\frac{D}{Dt}\left(ds^2 - ds_0^2\right) = (V_{i,j} + V_{j,i})dz_i dz_j = 2d_{ij}dz_i dz_j,$$

where d_{ij} is called the *rate-of-deformation tensor* for fluids, implying the time rate of change of the strain tensor in the Eulerian coordinates

$$d_{ij} = \frac{1}{2}(V_{i,j} + V_{j,i}) = \frac{1}{2}\left(\frac{\partial V_i}{\partial z_j} + \frac{\partial V_j}{\partial z_i}\right), \tag{2.4.4}$$

in which partial derivatives are with respect to the spatial coordinates z_i. This is in contrast to the case of the strain tensor [Eq. (2.3.5)], where the partial derivatives are with respect to the undeformed reference coordinates x_i. Note that in Eq. (2.4.4) the strain tensor for solids in Lagrangian coordinates has been transformed to the rate-of-deformation tensor for fluids in Eulerian coordinates. The rate-of-deformation tensor as derived in Eq. (2.4.4) will play a fundamental role in viscous flow, as will be seen in Chap. 5.

2.5 Coordinate Transformations for Strains

It has been and will repeatedly be demonstrated that we make use of the invariant properties of tensors in developing both constitutive and governing equations in general. In this section, we study additional properties of tensors related to coordinate transformation and strain invariants.

We begin with a simple example of coordinate transformation between the old coordinates x_i and new coordinates x_i', as shown in Fig. 2.5.1,

$$dx_i' = a_{ij}dx_j, \tag{2.5.1a}$$

$$dx_j = a_{ij}dx_i'. \tag{2.5.1b}$$

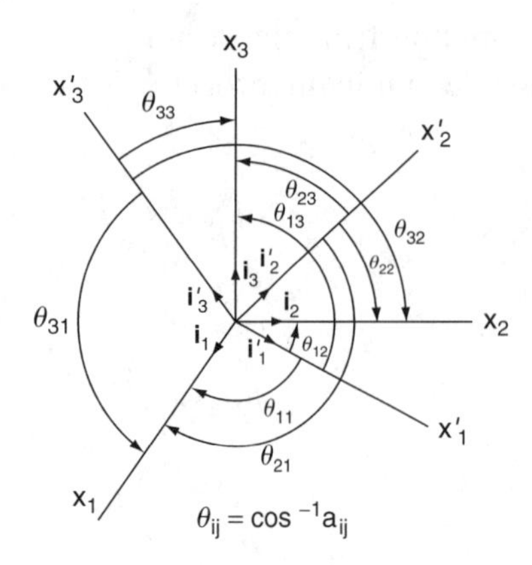

Figure 2.5.1. Coordinate transformation.

In index notation, the transpose of a matrix arises when, as in Eq. (2.5.1b), the first index of a_{ij} is repeated, not the second index, contrary to Eq. (2.5.1a), as pointed out in Chap. 1. Thus, from Eq. (2.3.3) written in the new coordinates, we have

$$ds^2 - ds_0^2 = 2\gamma'_{ij} dx'_i dx'_j, \tag{2.5.2a}$$

or, in the old coordinates,

$$ds^2 - ds_0^2 = 2\gamma_{rs} dx_r dx_s. \tag{2.5.2b}$$

Applying the transformation given by Eq. (2.5.1b) to Eq. (2.5.2b), we obtain

$$ds^2 - ds_0^2 = 2\gamma_{rs} a_{ir} dx'_i a_{js} dx'_j. \tag{2.5.2c}$$

Equating Eqs. (2.5.2a) and (2.5.2c) leads to

$$\gamma'_{ij} = a_{ir} a_{js} \gamma_{rs}. \tag{2.5.3}$$

This represents the transformation of the strain tensor from the old coordinates to the new coordinates, indicating that it requires the fourth-order tensor, $a_{ir}a_{js}$, for this transformation.

Principal Strains

At a given point in the body, it is always possible to choose a special set of axes through the point so that the shear strain components vanish. These special axes are called *principal axes* of the strain or *principal directions*. There are three planes through the point perpendicular to the three principal axes, called *principal planes*. The normal strain components (γ_{11}, γ_{22}, γ_{33}) on the three principal planes are called *principal strains* and are denoted by γ_1, γ_2, and γ_3 with a single index in which all shear strain components are zero and normal strains become maximum.

If a laboratory cube specimen is loaded perpendicular to one or more planes of this cube, the shear strains as well as the normal strains of different magnitudes develop on various planes rotated 0°–90° from the principal axes. Consider strain components γ_{ij} on a plane arbitrarily inclined from the principal planes. Let n_j be the direction cosines that identify the inclined plane. Then the relationship between the principal strain components γ_i and the γ_{ij} on the inclined plane is given by

$$\gamma_i = \gamma_{ij} n_j. \tag{2.5.4a}$$

If we assign a scalar $\gamma = \lambda$ for one of the principal strain components, then the principal strain γ_i may be written in an alternative form:

$$\gamma_i = \lambda n_i = \lambda \delta_{ij} n_j. \tag{2.5.4b}$$

Let Eqs. (2.5.4a) and (2.5.4b) be equated to yield

$$(\gamma_{ij} - \lambda \delta_{ij}) n_j = 0. \tag{2.5.5}$$

We can derive this equation from Eq. (2.5.3) by considering a scalar in the direction n_i normal to the surface of a principal plane or, equivalently, by multiplying Eq. (2.5.4a) by n_i,

$$\gamma_i n_i = \gamma(n) = n_i n_j \gamma_{ij}, \tag{2.5.6}$$

where the direction cosines n_i have the property

$$\mathbf{n} \cdot \mathbf{n} = n_i n_i = 1. \tag{2.5.7}$$

We now construct a scalar function f from the superposition of Eqs. (2.5.6) and (2.5.7):

$$f = \gamma_{ij} n_i n_j - \lambda(n_i n_i - 1), \tag{2.5.8}$$

where λ is known as the *Lagrange* multiplier, providing a compatibility with reference to Eq. (2.5.6). Physically, f represents the work required for satisfying Eq. (2.5.6) and that required for enforcing the constraint condition, Eq. (2.5.7). The sign for the constraint energy is arbitrary, plus or minus, depending on convenience, as required by the final physical consequence. The question here is this: In which direction or for what normal vector $\mathbf{n}$ will $\gamma(n)$ be a maximum? To obtain the condition of extrema, we proceed with differentiating the function f with respect to direction cosines and setting the result equal to zero, a standard procedure for finding an extremum. Differentiation of tensorial quantities with respect to another tensorial quantity requires careful manipulation:

$$\frac{\partial}{\partial n_k}(\gamma_{ij} n_i n_j) = \gamma_{ij}\left(\frac{\partial n_i}{\partial n_k} n_j + n_i \frac{\partial n_i}{\partial n_k}\right) = \gamma_{ij}(\delta_{ik} n_j + n_i \delta_{jk})$$

$$= \gamma_{kj} n_j + \gamma_{ik} n_i = 2\gamma_{kj} n_j.$$

Likewise,

$$\frac{\partial}{\partial n_k}(\lambda n_i n_i) = \lambda\left(\frac{\partial n_i}{\partial n_k} n_i + n_i \frac{\partial n_i}{\partial n_k}\right) = \lambda(\delta_{ik} n_i + n_i \delta_{ik}) = 2\lambda n_k.$$

Note that in both of these cases the free index is k. Thus, if k is changed to i, then

$$\frac{\partial f}{\partial n_i} = 2(\gamma_{ij} - \lambda\delta_{ij})n_j = 0$$

or

$$(\gamma_{ij} - \lambda\delta_{ij})\, n_j = 0,$$

which is the same as Eq. (2.5.5). This is known as the Lagrange multiplier method of determining the extremum condition for γ_{ij}.

Because n_j is arbitrary, for nontrivial solutions of Eq. (2.5.5) to exist, we must have

$$|\gamma_{ij} - \lambda\delta_{ij}| = 0. \tag{2.5.9}$$

An evaluation of the determinant of this matrix gives a cubic equation of the form, known as the Cayley–Hamilton equation,

$$-\lambda^3 + \lambda^2 I_1 - \lambda I_2 + I_3 = 0, \tag{2.5.10}$$

in which

$$I_1 = \gamma_{ii}, \tag{2.5.11a}$$

$$I_2 = \frac{1}{2}(\gamma_{ii}\gamma_{jj} - \gamma_{ij}\gamma_{ij}), \tag{2.5.11b}$$

$$I_3 = |\gamma_{ij}|, \tag{2.5.11c}$$

and the functions I_1, I_2, and I_3 have the same value in any rectangular Cartesian coordinate system. For this reason, they are called invariants or, more specifically, the *principal strain invariants*, with I_1, I_2, and I_3 being the first, second, and third principal strain invariants, respectively. Physically, the principal strain invariants play important roles in nonlinear solid mechanics. Specifically, for large deformations (geometric nonlinearity or finite elasticity), the principal strain invariants are the determining factors for calculating correct stress distributions. This will be discussed in Section 5.2 (constitutive theory for fluids) and Section 7.1 (nonlinear elasticity). For small deformation problems, however, the principal strain invariants have no role to play except that they appear in the process of determining the principal strains and principal planes, as shown in Example 2.5.1.

The expression given by Eq. (2.5.9) is a standard eigenvalue problem. Because γ_{ij} is a 3×3 matrix, we expect to obtain three eigenvalues. They represent the three principal strains (major, intermediate, and minor). For each eigenvalue there exist three eigenvector components. Thus there will be a total of nine eigenvector components corresponding to each of the three eigenvalues. These eigenvectors are the direction cosines of the 3×3 matrix that determine the angle of rotation required for identifying the location of principal axes or planes.

EXAMPLE 2.5.1. DETERMINATION OF PLANES OF PRINCIPAL STRAINS. Given the strain tensor measurements obtained in the laboratory,

$$\gamma_{ij} = \begin{bmatrix} 1 & \sqrt{3} & 0 \\ \sqrt{3} & 0 & 0 \\ 0 & 0 & 1 \end{bmatrix},$$

determine (1) the principal strain invariants, (2) the principal strains, and (3) the principal directions.

Solution

(1) Principal strain invariants:

$$I_1 = \gamma_{ii} = 2,$$

$$I_2 = \frac{1}{2}(\gamma_{ii}\gamma_{jj} - \gamma_{ij}\gamma_{ij}) = -2,$$

$$I_3 = |\gamma_{ij}| = -3.$$

(2) Principal strains:

$$\gamma_{ij} - \lambda\delta_{ij} = \begin{bmatrix} 1-\lambda & \sqrt{3} & 0 \\ \sqrt{3} & -\lambda & 0 \\ 0 & 0 & 1-\lambda \end{bmatrix},$$

$$|\gamma_{ij} - \lambda\delta_{ij}| = 0 = (1-\lambda)\left(\lambda - \frac{1+\sqrt{13}}{2}\right)\left(\lambda - \frac{1-\sqrt{13}}{2}\right),$$

$$\gamma_{(1)} = \lambda_{(1)} = \frac{1+\sqrt{13}}{2},$$

$$\gamma_{(2)} = \lambda_{(2)} = 1,$$

$$\gamma_{(3)} = \lambda_{(3)} = \frac{1-\sqrt{13}}{2}.$$

This gives the major principal strain $= (1+\sqrt{13})/2$, the intermediate principal strain $= 1$, and the minor principal strain $= (1-\sqrt{13})/2$.

(3) Principal directions:

Eigenvectors for $\lambda_{(1)} = (1+\sqrt{13})/2$ give the principal directions in $\mathbf{n}^{(1)}$:

$$\begin{bmatrix} 1-\frac{1+\sqrt{13}}{2} & \sqrt{3} & 0 \\ \sqrt{3} & -\frac{1+\sqrt{13}}{2} & 0 \\ 0 & 0 & 1-\frac{1+\sqrt{13}}{2} \end{bmatrix} \begin{bmatrix} n_1^{(1)} \\ n_2^{(1)} \\ n_3^{(1)} \end{bmatrix}$$

$$= \begin{bmatrix} (1-\frac{1+\sqrt{13}}{2})n_1^{(1)} + \sqrt{3}n_2^{(1)} \\ \sqrt{3}n_1^{(1)} - (\frac{1+\sqrt{13}}{2})n_2^{(1)} \\ (1-\frac{1+\sqrt{13}}{2})n_3^{(1)} \end{bmatrix} = \begin{bmatrix} 0 \\ 0 \\ 0 \end{bmatrix}.$$

Solving for the eigenvector components $n_i^{(1)}$, we obtain

$$n_1^{(1)} = \frac{1+\sqrt{13}}{2\sqrt{3}} n_2^{(1)},$$

$$n_3^{(1)} = 0.$$

From

$$\mathbf{n}^{(1)} \cdot \mathbf{n}^{(1)} = \left(\frac{1 + 2\sqrt{13} + 13}{12} + 1 \right) \left[n_2^{(1)} \right]^2 = 1,$$

we obtain

$$n_2^{(1)} = n_1^{(1)} \sqrt{\frac{12}{12 + (1+\sqrt{13})^2}} = \frac{1+\sqrt{13}}{2\sqrt{3}} \sqrt{\frac{12}{12 + (1+\sqrt{13})^2}}.$$

Thus,

$$\mathbf{n}^{(1)} = \left[n_1^{(1)} \quad n_2^{(1)} \quad n_3^{(1)} \right] = [0.8 \quad 0.6 \quad 0].$$

Similarly, we can calculate eigenvectors for $\lambda_{(2)} = 1$, noting that $\mathbf{n}^{(2)} \cdot \mathbf{n}^{(2)} = 1$. Thus,

$$\mathbf{n}^{(2)} = [0 \quad 0 \quad 1].$$

Eigenvectors for $\lambda_{(3)} = (1 - \sqrt{13})/2$ give the principal directions in $\mathbf{n}^{(3)}$, which we may obtain in the same manner, but perhaps more easily by taking a cross product of $\mathbf{n}^{(1)}$ and $\mathbf{n}^{(2)}$:

$$\mathbf{n}^{(3)} = \mathbf{n}^{(1)} \times \mathbf{n}^{(2)} = \begin{bmatrix} \mathbf{i}_1 & \mathbf{i}_2 & \mathbf{i}_3 \\ 0.8 & 0.6 & 0 \\ 0 & 0 & 1 \end{bmatrix} = 0.6\mathbf{i}_1 - 0.8\mathbf{i}_2.$$

Therefore,

$$a_{ij} = \begin{bmatrix} \mathbf{n}^{(1)} \\ \mathbf{n}^{(2)} \\ \mathbf{n}^{(3)} \end{bmatrix} = \begin{bmatrix} 0.8 & 0.6 & 0 \\ 0 & 0 & 1 \\ 0.6 & -0.8 & 0 \end{bmatrix}.$$

These results can be checked by the strain tensor coordinate transformation relation given by Eq. (2.5.3):

$$\gamma'_{ij} = a_{ir} a_{js} \gamma_{rs} = \begin{bmatrix} \lambda_{(1)} & 0 & 0 \\ 0 & \lambda_{(2)} & 0 \\ 0 & 0 & \lambda_{(3)} \end{bmatrix} = \begin{bmatrix} 2.3 & 0 & 0 \\ 0 & 1 & 0 \\ 0 & 0 & -1.3 \end{bmatrix},$$

which, as shown in Eq. (2.5.3), indicates that the direction cosines a_{ij} transform the strain tensor γ_{ij} (old coordinates) into the principal directions (new coordinates), leading to the principal strains (γ'_{ij}), or $\lambda_{(1)}$, $\lambda_{(2)}$, and $\lambda_{(3)}$. It is interesting to note that the summing implied in $a_{ir}a_{js}\gamma_{rs}$ is the same as the matrix multiplication:

$$[a][\gamma][a]^T = \begin{bmatrix} 0.8 & 0.6 & 0 \\ 0 & 0 & 1 \\ 0.6 & -0.8 & 0 \end{bmatrix} \begin{bmatrix} 1 & \sqrt{3} & 0 \\ \sqrt{3} & 0 & 0 \\ 0 & 0 & 1 \end{bmatrix} \begin{bmatrix} 0.8 & 0 & 0.6 \\ 0.6 & 0 & -0.8 \\ 0 & 1 & 0 \end{bmatrix}$$

$$= \begin{bmatrix} 2.3 & 0 & 0 \\ 0 & 1 & 0 \\ 0 & 0 & -1.3 \end{bmatrix}.$$

These exercises indicate that if laboratory strain measurements for specimens at any orientation are provided, then it is possible to locate the planes of the principal directions on which shear strains are zero and only principal normal strains prevail.

Maximum Shear Strains

The shear strains in the directions x_1' and x_2' can be expressed in terms of the principal strains $\lambda_{(i)}$ by

$$\gamma_{12}' = a_{1i}a_{2i}\lambda_{(i)}, \tag{2.5.12}$$

where the index (i) simply follows the repeated index i in the repeated index summing process with (i) itself being suppressed from summing. Because x_1' and x_2' in Fig. 2.5.1 are orthogonal, the direction cosines satisfy the conditions

$$a_{1i}a_{2i} = 0, \tag{2.5.13a}$$
$$a_{1i}a_{1i} = 1, \tag{2.5.14a}$$
$$a_{2i}a_{2i} = 1. \tag{2.5.15a}$$

The reader may verify these requirements from a_{ij} in Eq. (1.2.12). To simultaneously satisfy the conditions given by Eqs. (2.5.12) and (2.5.13), we construct a scalar function f as a superposition of Eqs. (2.5.12) and (2.5.13):

$$f = a_{1i}a_{2i}\lambda_{(i)} - \mu a_{1i}a_{2i} - \mu_1(a_{1i}a_{2i} - 1) - \mu_2(a_{2i}a_{2i} - 1),$$

where μ, μ_1, and μ_2 represent Lagrange multipliers, similar to the case of principal strains discussed earlier. The necessary conditions for an extremum (maximum or minimum) of f with respect to the direction cosines are

$$\frac{\partial f}{\partial a_{1i}} = (\lambda_{(i)} - \mu)a_{2i} - 2\mu_1 a_{1i} = 0, \tag{2.5.14a}$$

$$\frac{\partial f}{\partial a_{2i}} = (\lambda_{(i)} - \mu)a_{1i} - 2\mu_2 a_{2i} = 0. \tag{2.5.14b}$$

Multiply Eq. (2.5.14a) by a_{1i} and satisfy Eqs. (2.5.15a) and (2.5.13b) to give

$$\lambda_{(i)} a_{1i} a_{2i} - 2\mu_1 = 0$$

or

$$\mu_1 = \frac{1}{2}\lambda_{(i)} a_{1i} a_{2i} = \frac{1}{2}\gamma'_{12}.$$

Similarly, from Eqs. (2.5.14b), (2.5.13a), and (2.5.13c),

$$\mu_2 = \frac{1}{2}\lambda_{(i)} a_{1i} a_{2i} = \frac{1}{2}\gamma'_{12}.$$

If we multiply Eq. (2.5.14a) by a_{2i} and enforce the conditions of Eqs. (2.5.13a) and (2.5.13c), we get

$$\mu = \lambda_{(i)} a_{2i} a_{2i} = \gamma'_{22}.$$

Likewise from Eqs. (2.5.14b), (2.5.13a), and (2.5.13b),

$$\mu = \lambda_{(i)} a_{1i} a_{1i} = \gamma'_{11}.$$

The substitution of μ, μ_1, and μ_2 into Eqs. (2.5.14) yields

$$\left[\lambda_{(1)} - \gamma'_{11}\right] a_{21} - \gamma'_{12} a_{11} = 0, \tag{2.5.15a}$$

$$\left[\lambda_{(2)} - \gamma'_{11}\right] a_{22} - \gamma'_{12} a_{12} = 0, \tag{2.5.15b}$$

$$\left[\lambda_{(3)} - \gamma'_{11}\right] a_{23} - \gamma'_{12} a_{13} = 0, \tag{2.5.15c}$$

$$\left[\lambda_{(1)} - \gamma'_{11}\right] a_{11} - \gamma'_{12} a_{21} = 0, \tag{2.5.15d}$$

$$\left[\lambda_{(2)} - \gamma'_{11}\right] a_{12} - \gamma'_{12} a_{22} = 0, \tag{2.5.15e}$$

$$\left[\lambda_{(3)} - \gamma'_{11}\right] a_{13} - \gamma'_{12} a_{23} = 0. \tag{2.5.15f}$$

From Eqs. (2.5.15a) and (2.5.15d), we get

$$a_{21} = \frac{\gamma'_{12}}{\lambda_{(1)} - \gamma'_{11}} a_{11},$$

$$a_{21} = \frac{\lambda_{(1)} - \gamma'_{11}}{\gamma'_{12}} a_{11},$$

which leads to $a_{11}^2 = a_{21}^2$ and $a_{11} = \pm a_{21}$. Similarly, $a_{22}^2 = a_{12}^2$, $a_{21}^2 = a_{22}^2$, and $a_{13}^2 = a_{23}^2 = 0$. Thus we note that

$$a_{11}^2 = a_{12}^2 = a_{21}^2 = a_{22}^2 = B^2.$$

From Eqs. (2.5.13b) and (2.5.13c), we must have

$$2B^2 = 1,$$
$$B = \pm\frac{1}{\sqrt{2}}.$$

To satisfy the conditions in Eq. (2.5.13a), we require that the signs of all B's be the same except for one; for example, we could take

$$a_{11} = a_{21} = a_{22} = \frac{1}{\sqrt{2}},$$
$$a_{12} = -\frac{1}{\sqrt{2}}.$$

Depending on the choice of signs, we obtain either positive or negative shear strains. Note also that angles corresponding to the direction cosines are odd multiples of $\cos^{-1}(1/\sqrt{2}) = 45^\circ$ with x_i' bisecting x_i. The maximum shear strain in the $x_m x_n$ $(m \neq n)$ plane, therefore, is given by

$$\gamma'_{mn} = a_{mi} a_{ni} \lambda_{(i)} \quad (m \neq n), \tag{2.5.16}$$
$$\gamma'_{12} = \frac{1}{\sqrt{2}}\frac{1}{\sqrt{2}}\lambda_{(1)} - \frac{1}{\sqrt{2}}\frac{1}{\sqrt{2}}\lambda_{(2)} = \frac{1}{2}\left(\lambda_{(1)} - \lambda_{(2)}\right).$$

Similar procedures may be followed for the x_1x_3 and x_2x_3 planes. Thus,

$$\gamma'_{13} = \frac{1}{2}\left[\lambda_{(1)} - \lambda_{(3)}\right],$$
$$\gamma'_{23} = \frac{1}{2}\left[\lambda_{(2)} - \lambda_{(3)}\right].$$

These results indicate that the maximum shear strains represent the radii of Mohr circles, as shown in Example 2.5.2.

It is obvious from Eq. (2.5.16) that shear strains are zero (minimum), for all $a_{ij} = 0$ corresponding to the principal planes.

EXAMPLE 2.5.2. MOHR CIRCLE REPRESENTATION. Given the results in Example 2.5.1, plot the Mohr circles (Mohr, 1914) that show the principal strains, normal and shear strains, maximum shear strains, and orientations of planes between the principal plane and the plane on which the strain data are given.

Solution. First verify the results from elementary relations. Principal strains and orientations of planes are given by (refer to undergraduate texts on the mechanics of materials)

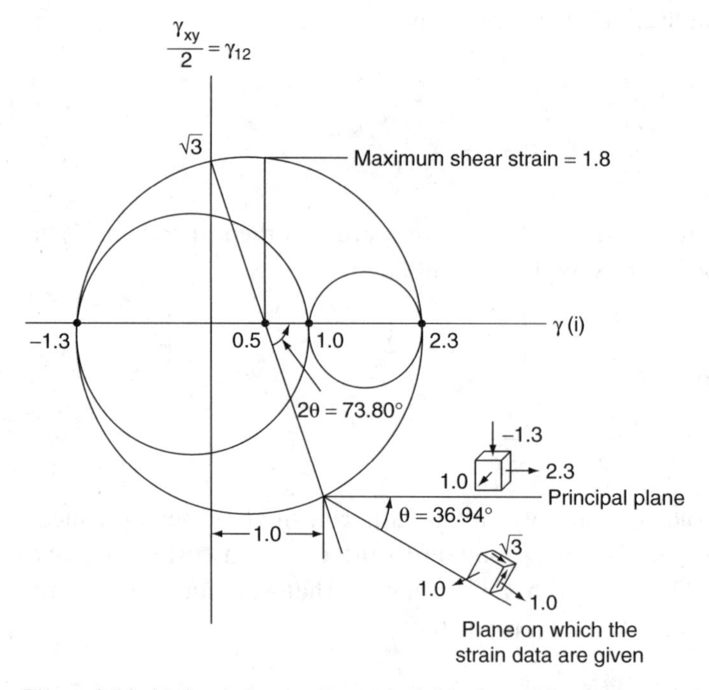

Figure 2.5.2. Mohr circles for Example 2.5.2. Quantities shown on the cube represents strains. Intermediate principal strain is located at the intersection of two inner circles that indicate major and minor principal strains in contact with the exterior enclosing circle. The plane on which the original stain data are given is located at an angle rotated clockwise from the principal plane. Note that the original strain data appear on the exterior circle.

$$\gamma_{p1,2} = \frac{\gamma_{xx} + \gamma_{yy}}{2} \pm \sqrt{\left(\frac{\gamma_{xx} - \gamma_{yy}}{2}\right)^2 + \left(\frac{\gamma_{xy}}{2}\right)^2}$$

$$= \frac{1+0}{2} \pm \sqrt{\left(\frac{1-0}{2}\right)^2 + \left(\frac{2\sqrt{3}}{2}\right)^2} = \begin{cases} 2.3 \\ -1.3 \end{cases},$$

$$\gamma_{p1,3} = \frac{\gamma_{xx} + \gamma_{zz}}{2} \pm \sqrt{\left(\frac{\gamma_{xx} - \gamma_{zz}}{2}\right)^2 + \left(\frac{\gamma_{xz}}{2}\right)^2} = \begin{cases} 1 \\ 1 \end{cases},$$

$$\gamma_{p2,3} = \frac{\gamma_{yy} + \gamma_{zz}}{2} \pm \sqrt{\left(\frac{\gamma_{yy} - \gamma_{zz}}{2}\right)^2 + \left(\frac{\gamma_{yz}}{2}\right)^2} = \begin{cases} 1 \\ 0 \end{cases},$$

$$\tan 2\theta = \frac{\gamma_{xy}/2}{(\gamma_{xx} - \gamma_{yy})/2} = 3.46,$$

$$\theta = 36.94^\circ.$$

The Mohr circle representation is shown in Fig. 2.5.2. If the strain data are fully populated in three dimensions, rotations of planes for which the strain data are given with respect to the planes of principal planes cannot be shown in three-dimensional configurations. Thus, in this case, the usefulness of Mohr circles is limited to two dimensions.

2.6 Dilatational and Deviatoric Properties

Changes in volume before and after deformation in the convective coordinates are referred to as *dilatation*, D, given by

$$D = \frac{d\Omega}{d\Omega_0} = J = \sqrt{|G_{ij}|} = \sqrt{|\delta_{ij} + 2\gamma_{ij}|}. \tag{2.6.1a}$$

We calculate the determinant of the 3×3 matrix and expand the result in infinite series arising from the square root. Neglect all higher-order terms and obtain, for small strains,

$$D \cong 1 + \gamma_{ii}. \tag{2.6.1b}$$

It follows from Eqs. (2.6.1) that the volumetric strain Δ is

$$\Delta = \frac{d\Omega - d\Omega_0}{d\Omega_0} = D - 1 \cong \gamma_{ii},$$

which indicates that only the extensional strains are responsible for volumetric strain if all higher-order terms in the strain tensor are neglected.

The mean volumetric strain γ becomes

$$\gamma = \frac{\Delta}{3} \cong \frac{\gamma_{ii}}{3}. \tag{2.6.2}$$

The total strain is given by the sum of the *deviatoric* (shear) part $\hat{\gamma}_{ij}$ and the dilatational (mean volumetric or spherical) part $\gamma\delta_{ij}$:

$$\gamma_{ij} = \hat{\gamma}_{ij} + \gamma\delta_{ij} = \hat{\gamma}_{ij} + \frac{\Delta}{3}\delta_{ij} \cong \hat{\gamma}_{ij} + \frac{1}{3}\gamma_{kk}\delta_{ij}.$$

Accepting the small strain approximation, we obtain

$$\hat{\gamma}_{ij} = \gamma_{ij} - \frac{1}{3}\gamma_{kk}\delta_{ij}, \tag{2.6.3}$$

where $\hat{\gamma}_{ij}$ is known as the deviatoric strain tensor. The formula for the deviatoric strain tensor given by Eq. (2.6.3) is correct only for small strains, but is not valid for large strains. It is interesting to note that the exact deviatoric strain tensor is of infinite series. In practice, however, the deviatoric strain tensor itself is not used, but its usefulness will be associated indirectly with the deviatoric stress tensor as a failure criterion discussed in Sections 3.5 and 7.4.

The rate-of-deformation tensor in Eulerian coordinates has no geometric constraints of large velocity gradients such as occur in the strain tensor. Thus, for Eulerian coordinates, the deviatoric part of the rate-of-deformation tensor $\hat{d}_{ij}$ is given exactly by

$$\hat{d}_{ij} = d_{ij} - \frac{1}{3}d_{kk}\delta_{ij}. \tag{2.6.4}$$

The viscous flow is founded on the deviatoric rate-of-deformation tensor as given by Eq. (2.6.4), one of the most important concepts in fluid mechanics. This is discussed in Subsection 5.2.2.

Concluding Remarks

As we close this chapter, we note that there are three most important aspects of kinematics distinguishing solid mechanics in Lagrangian coordinates from fluid mechanics in Eulerian coordinates. They are viewed as distinct differences or rather distinct resemblances between solid mechanics and fluid mechanics.

(1) Conservation of mass:

Solids,

$$d\Omega = \frac{\partial z_m}{\partial x_1}\frac{\partial z_n}{\partial x_2}\frac{\partial z_p}{\partial x_3}\epsilon_{mnp} d\Omega_0 = \left|\frac{\partial z_i}{\partial x_j}\right| d\Omega_0 = J d\Omega_0 = \sqrt{|G_{ij}|} d\Omega_0 = \sqrt{G} d\Omega_0,$$

$$m = \int_\Omega \rho d\Omega = \int_\Omega \rho_0 d\Omega_0,$$

$$\int_\Omega (\rho J - \rho_0) d\Omega_0 = 0, \text{ global form,}$$

$$\rho J - \rho_0 = 0, \text{ local form.}$$

Fluids,

$$\frac{D}{Dt}\int_\Omega (\rho J - \rho_0) d\Omega = \int_\Omega \left(\frac{\partial \rho}{\partial t} + \frac{\partial}{\partial z_i}(\rho V_i)\right) d\Omega = 0, \text{ global form,}$$

$$\frac{\partial \rho}{\partial t} + \frac{\partial}{\partial z_i}(\rho V_i) = 0, \text{ local form.}$$

(2) Strain tensor in solids and rate-of-deformation tensor in fluids:

Strain tensor in solids,

$$ds^2 - ds_0^2 = (G_{ij} - \delta_{ij}) dx_i dx_j = 2\gamma_{ij} dx_i dx_j,$$

$$\gamma_{ij} = \frac{1}{2}\left(\frac{\partial u_i}{\partial x_j} + \frac{\partial u_j}{\partial x_i} + \frac{\partial u_m}{\partial x_i}\frac{\partial u_m}{\partial x_j}\right), \text{ for large strains,}$$

$$\gamma_{ij} = \frac{1}{2}\left(\frac{\partial u_i}{\partial x_j} + \frac{\partial u_j}{\partial x_i}\right), \text{ for small strains.}$$

Rate-of-deformation tensor in fluids:

$$\frac{D}{Dt}(ds^2 - ds_0^2) = 2 d_{ij} dz_i dz_j,$$

$$d_{ij} = \frac{1}{2}\left(\frac{\partial V_i}{\partial z_j} + \frac{\partial V_j}{\partial z_i}\right), \text{ valid without restriction.}$$

(3) Deviatoric strain tensor in solids and deviatoric rate-of-deformation tensor in fluids:

Deviatoric strain tensor in solids,

$$\hat{\gamma}_{ij} = \gamma_{ij} - \frac{1}{3}\gamma_{kk}\delta_{ij}, \text{ valid only for small strains.}$$

Deviatoric rate of deformation tensor in fluids,

$$\hat{d}_{ij} = d_{ij} - \frac{1}{3}d_{kk}\delta_{ij}, \text{ valid without restriction.}$$

It is evident that the preceding results, including the conservation of mass in solids and fluids, strain tensors, rate-of-deformation tensors, and deviatoric strain tensors, could not have been calculated without using the convective coordinates. The tangent vector $\mathbf{G}_i$ and the metric tensor G_{ij} are instrumental in these processes. Although the tensor algebra involved in convective coordinates is complicated, the benefits far outweigh the effort. An important role of continuum mechanics is to show, through tensor algebra, analogies, differences, and relationships among the various substances (solids and fluids) in their physical behavior. We shall encounter more of these aspects as we study the forthcoming chapters. In Chap. 6 we discuss the standard curvilinear coordinates in which the tangent vectors and derivatives of tangent vectors are involved in the undeformed body.

PROBLEMS

2.1 Show that the Jacobian is equal to the square root of the determinant of the metric tensor.

2.2 Transform the Lagrangian to Eulerian coordinates to show that the conservation of mass for fluids can be derived from the conservation of mass for solids. Include complete details of the algebra involved.

2.3 Let the motion of a fluid in steady-state flow be given in the Eulerian coordinates by

$$V_1 = kz_1,$$

$$V_2 = -kz_2,$$

$$V_3 = 0.$$

(a) Show that the motions and velocities in Lagrangian coordinates are

$$z_1 = x_1 e^{k(t-t_0)}, \qquad v_1 = kx_1 e^{k(t-t_0)},$$

$$z_2 = x_2 e^{-k(t-t_0)}, \qquad v_2 = -kx_2 e^{-k(t-t_0)},$$

$$z_3 = x_3, \qquad v_3 = 0.$$

(b) Transform the Lagrangian velocities into Eulerian coordinates to prove that the results in part (a) are recovered.

(c) Plot the Eulerian and Lagrangian velocity distributions.

2.4 Derive the Cayley–Hamilton equation using the Lagrange multiplier method and identify the principal strain invariants.

2.5 From the laboratory measurements the following strain tensor components are obtained:

$$\gamma_{ij} = \begin{bmatrix} 2 & 1 & 1 \\ 1 & 2 & 0 \\ 1 & 0 & 1 \end{bmatrix}.$$

Determine

(a) the principal strain invariants,

(b) the principal strains,

(c) the principal direction matrix a_{ij} and verify your results by recalculating the principal strains from principal direction cosines.

2.6 Given $\gamma'_{ij} = a_{ir}a_{js}\gamma_{rs}$, prove that the maximum shear strains are

$$(\gamma'_{ij})_{\max} = \frac{1}{2}[\lambda_{(i)} - \lambda_{(j)}], \quad i \neq j,$$

and that they are oriented 45° from the planes of principal planes.

3 Kinetics

3.1 General

The notion of *stress* originates from the need to quantify internal or external forces distributed, respectively, in a body or along its boundary in equilibrium. Body forces such as gravity act inside the body whereas surface forces act on its bounding surface. Stresses are those forces distributed over an infinitesimal unit area cut out of a body in certain directions or over an infinitesimal unit area on the bounding surface. Stresses may also arise from hydrodynamic pressure and/or velocity gradients in a fluid. Furthermore, changes in temperature in solids or fluids give rise to stresses.

Stresses may be related to strains in solids or rates of deformation in fluids through constitutive laws. Stresses are described in many different ways, depending, for example, on the coordinate systems used, the magnitudes of the strains (in solids) or the velocity gradients (in fluids), or the types of substances involved. The basic groundwork for stresses is developed in this chapter, but related subjects are discussed throughout the remainder of this book. The study of stress-related problems, in general, is referred to as *kinetics* for the body in equilibrium.

Discussions in this chapter include definitions of forces and stresses, conservation laws of linear and angular momentum, coordinate transformations of stresses, and deviatoric stresses. These topics, together with the concept of kinematics discussed in Chap. 2, constitute the basis for the theories of elasticity in Chap. 4 and fluid mechanics in Chap. 5.

In this chapter the stress is defined on the surface of a body with small strains. The stress with large strains is discussed in Chap. 7.

3.2 Forces and Stresses

Consider a body subjected to an external load, as shown in Fig. 3.2.1(a). We examine how the external load influences an interior point [Fig. 3.2.1(b)]. The stress vector $\boldsymbol{\sigma}$ in the undeformed state in Lagrangian coordinates is defined as

$$\boldsymbol{\sigma} = \lim_{\Delta A \to 0} \frac{\Delta \mathbf{F}}{\Delta A} = \frac{d\mathbf{F}}{dA}, \tag{3.2.1}$$

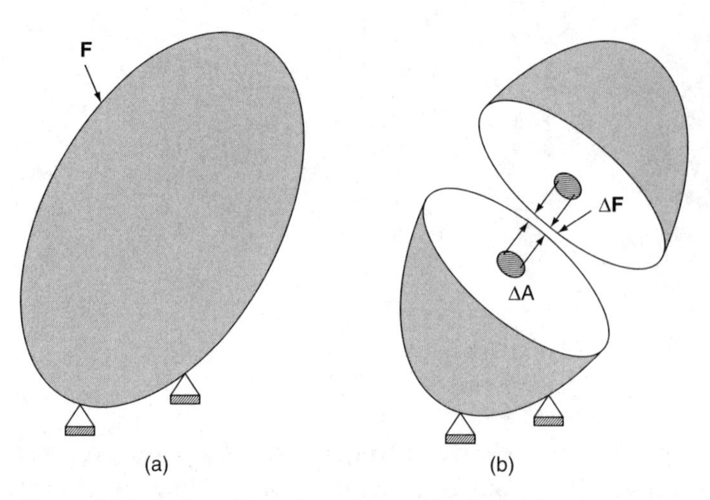

Figure 3.2.1. Body subjected to external load and interior stresses: (a) body under load **F**, (b) interior stresses.

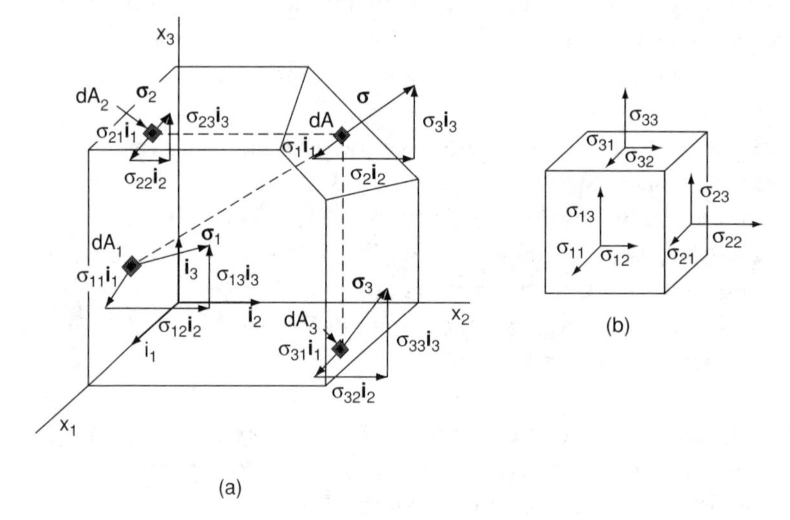

Figure 3.2.2. Undeformed boundary surface and stress components: (a) undeformed boundary surface, (b) components of stress in undeformed coordinates.

where $\Delta\mathbf{F}$ is the incremental internal force activated from the external load $\mathbf{F}$ and ΔA is the incremental area on which $\Delta\mathbf{F}$ is acting.

The differential internal force $d\mathbf{F} = \boldsymbol{\sigma}\, dA$ that acts on the differential area dA can also be said to be acting on an inclined surface dA_i of an infinitesimal body, as shown in Fig. 3.2.2. The infinitesimal force $d\mathbf{F}$ can be written in various forms:

$$\begin{aligned} d\mathbf{F} &= \boldsymbol{\sigma} dA = \sigma_i \mathbf{i}_i dA = \boldsymbol{\sigma}_i dA_i = \sigma_{ij} \mathbf{i}_j dA_i \\ &= (\sigma_{11}\mathbf{i}_1 + \sigma_{12}\mathbf{i}_2 + \sigma_{13}\mathbf{i}_3)\, dA_1 \\ &\quad + (\sigma_{21}\mathbf{i}_1 + \sigma_{22}\mathbf{i}_2 + \sigma_{23}\mathbf{i}_3)\, dA_2 + (\sigma_{31}\mathbf{i}_1 + \sigma_{32}\mathbf{i}_2 + \sigma_{33}\mathbf{i}_3)\, dA_3. \end{aligned} \tag{3.2.2}$$

Here the σ_i's denote components of the stress vector $\boldsymbol{\sigma}$ acting on the inclined surface dA, the $\boldsymbol{\sigma}_i$'s are the resultant stress vectors in the surface dA_i projected from dA on each of the three planes constructed on the Cartesian coordinate axes x_i, and the σ_{ij}'s are the components of the stress on these projected areas, as collected from Eq. (3.2.2):

$$\sigma_{ij} = \begin{bmatrix} \sigma_{11} & \sigma_{12} & \sigma_{13} \\ \sigma_{21} & \sigma_{22} & \sigma_{23} \\ \sigma_{31} & \sigma_{32} & \sigma_{33} \end{bmatrix},$$

where σ_{ij} is a symmetric matrix (this will be proven later). These stress components are shown schematically in Fig. 3.2.2(b). The external force $d\mathbf{F}$ results in activation of $\boldsymbol{\sigma}$ and its components σ_i on dA, three stress vector components σ_i on dA_i, and, finally, a total of nine stress components σ_{ij} on dA_i. Here, σ_{ij} is formally identified as the *stress tensor*. Notice that the first and second indices on σ_{ij} represent, respectively, the planes and directions the stress components are acting upon. The physical and mathematical processes demonstrated in Eq. (3.2.2) identify the existence of the stress tensor, rather than being postulated from intuition, as usually shown in the undergraduate textbooks.

3.3 Basic Conservation Laws

3.3.1 Conservation of Linear Momentum

Linear momentum $\mathbf{L}$ within a volumetric unit $d\Omega$ is defined as the velocity vector summed over the differential mass or as the product of a velocity vector with density summed over the differential volume:

$$\mathbf{L} = \int_m \mathbf{v}\, dm = \int_\Omega \mathbf{v}\rho\, d\Omega.$$

The time rate of change of this linear momentum in Lagrangian coordinates is then defined as the resultant force that represents the conservation of linear momentum:

$$\frac{d\mathbf{L}}{dt} = \frac{d}{dt}\int_\Omega \mathbf{v}\rho\, d\Omega = \int_\Omega \rho\mathbf{F}\, d\Omega + \int_\Gamma \boldsymbol{\sigma}(n)\, d\Gamma. \tag{3.3.1}$$

In Eq. (3.3.1), $\mathbf{F}$ is the body force per unit mass and $\boldsymbol{\sigma}(n)$ is the stress vector normal to the surface:

$$\boldsymbol{\sigma}(n) = \boldsymbol{\sigma}_i n_i = \sigma_{ij}\mathbf{i}_j n_i, \tag{3.3.2a}$$

where each n_i is a component of a vector normal to the surface. It follows from the Green–Gauss theorem that

$$\int_\Gamma \sigma_{ij}\mathbf{i}_j n_i\, d\Gamma = \int_\Omega \frac{\partial \sigma_{ij}}{\partial x_i}\mathbf{i}_j\, d\Omega = \int_\Omega \sigma_{ij,i}\mathbf{i}_j\, d\Omega. \tag{3.3.2b}$$

Here $d\Gamma = n_i d\Gamma_i$ requires $\boldsymbol{\sigma}(n)$ to be the normal component of $\boldsymbol{\sigma}$ in Eq. (3.3.2).

Recall that the conservation of mass in solids is given by

$$\rho\, d\Omega = \rho_0\, d\Omega_0,$$

$$\frac{d}{dt}(\rho_0\, d\Omega_0) = \frac{d}{dt}(\rho\, d\Omega) = 0.$$

With these requirements and from Eqs. (3.3.1) and (3.2.2), we obtain

$$\int_\Omega (\sigma_{ij,i}\mathbf{i}_j + \rho\mathbf{F} - \rho\dot{\mathbf{v}})\, d\Omega = 0 \tag{3.3.3a}$$

or

$$\int_\Omega (\sigma_{ij,i} + \rho F_j - \rho\dot{v}_j)\, i_j\, d\Omega = 0. \tag{3.3.3b}$$

This is known as the global form of *Cauchy's first law of motion*. Here, some authors use direct tensors for derivatives of a stress tensor

$$\sigma_{ij,i}\mathbf{i}_j \rightarrow \nabla\cdot\mathbf{T},$$

where the left-hand side is clearly the vector. The right-hand side quantity $\nabla\cdot\mathbf{T}$, however, is the divergence of the "direct" stress tensor $\mathbf{T}$, equivalent to $\sigma_{ij,i}\,\mathbf{i}_j$. Direct tensor notation without indices is avoided in this book because the summing process is unspecified and the algebra required is unclear to the beginner. Only in Sections 5.2 and 7.2 is the direct tensor used in favor of traditional practice.

For the integral equation, (3.3.3b), to be valid for all arbitrary volumes, it is necessary that the integrand vanish, so that

$$\sigma_{ij,i} + \rho F_j - \rho\dot{v}_j = 0. \tag{3.3.3c}$$

This is the local form of *Cauchy's first law of motion* or, simply, the equation of motion, and is associated with undeformed Cartesian coordinates.

Expand Eq. (3.3.3c) with standard notation, and write

$$\begin{aligned}
&\frac{\partial\sigma_{xx}}{\partial x} + \frac{\partial\sigma_{yx}}{\partial y} + \frac{\partial\sigma_{zx}}{\partial z} + \rho F_x - \rho\ddot{u}_x = 0,\\
&\frac{\partial\sigma_{xy}}{\partial x} + \frac{\partial\sigma_{yy}}{\partial y} + \frac{\partial\sigma_{zy}}{\partial z} + \rho F_y - \rho\ddot{u}_y = 0,\\
&\frac{\partial\sigma_{xz}}{\partial x} + \frac{\partial\sigma_{yz}}{\partial y} + \frac{\partial\sigma_{zz}}{\partial z} + \rho F_z - \rho\ddot{u}_z = 0.
\end{aligned} \tag{3.3.3d}$$

It can be shown that these equations may also be obtained from summing all force components that act on the surfaces and inside of an infinitesimal cube in equilibrium. This process is equivalent to identifying all components of the surface integral (left-hand side) and the domain integral (right-hand side) of Eq. (3.3.2b), an excellent example of the Green–Gauss theorem introduced in Section 1.3. See Problem 3.1 for implementation of this process.

3.3.2 Conservation of Angular Momentum

The time rate of change of angular momentum is given by the cross product of the position vector with the time rate of change of the linear momentum:

$$\frac{d}{dt}\int_{\Omega} \mathbf{r}\times\rho\mathbf{v}\,d\Omega = \int_{\Omega}\mathbf{r}\times\rho\mathbf{F}\,d\Omega + \int_{\Gamma}\mathbf{r}\times\boldsymbol{\sigma}(n)\,d\Gamma, \tag{3.3.4}$$

where

$$\begin{aligned}\int_{\Gamma}\mathbf{r}\times\boldsymbol{\sigma}(n)\,d\Gamma &= \int_{\Gamma}\mathbf{r}\times\sigma_{ij}\mathbf{i}_j n_i\,d\Gamma \\ &= \int_{\Omega}(\mathbf{r}\times\sigma_{ij}\mathbf{i}_j)_{,i}\,d\Omega \\ &= \int_{\Omega}\mathbf{r}\times\sigma_{ij,i}\mathbf{i}_j\,d\Omega + \int_{\Omega}\mathbf{r}_{,i}\times\sigma_{ij}\mathbf{i}_j\,d\Omega.\end{aligned} \tag{3.3.5}$$

Substituting Eq. (3.3.5) into Eq. (3.3.4) yields

$$\int_{\Omega}\mathbf{r}\times(\rho\dot{v}_j - \sigma_{ij,i} - \rho F_j)\mathbf{i}_j\,d\Omega - \int_{\Omega}\mathbf{r}_{,i}\times\sigma_{ij}\mathbf{i}_j\,d\Omega = 0. \tag{3.3.6}$$

If the conservation of linear momentum is to be maintained, then we must have

$$\int_{\Omega}\mathbf{r}_{,i}\times\sigma_{ij}\mathbf{i}_j\,d\Omega = 0.$$

Because the derivative of a position vector on the undeformed state is a unit vector (Fig. 2.2.1),

$$\mathbf{r}_{,i} = (x_k\mathbf{i}_k)_{,i} = \delta_{ki}\mathbf{i}_k = \mathbf{i}_i,$$

we obtain

$$\int_{\Omega}\mathbf{i}_i\times\sigma_{ij}\mathbf{i}_j\,d\Omega = \int_{\Omega}\sigma_{ij}\epsilon_{ijk}\mathbf{i}_k\,d\Omega = 0, \tag{3.3.7a}$$

or the integrand of Eq. (3.3.7a) becomes

$$(\sigma_{23}-\sigma_{32})\,\mathbf{i}_1 + (\sigma_{31}-\sigma_{13})\,\mathbf{i}_2 + (\sigma_{12}-\sigma_{21})\,\mathbf{i}_3 = 0, \tag{3.3.7b}$$

which requires that each term in the parentheses vanish, that is,

$$\sigma_{ij} = \sigma_{ji}. \tag{3.3.7c}$$

This is the consequence of the conservation of angular momentum, which requires the symmetry of the stress tensor. Thus the symmetry of the stress tensor dictates the conservation of linear momentum and angular momentum. The relation given by Eqs. (3.3.7) is also referred to as *Cauchy's second law of motion*. In view of Eq. (3.3.6), it is clear that the conservation of angular momentum is contingent on the conservation of the linear momentum. The symmetry of the stress tensor makes it possible to write Eq. (3.3.3c) in the form

$$\sigma_{ij,j} + \rho F_i - \rho\ddot{u}_i = 0.$$

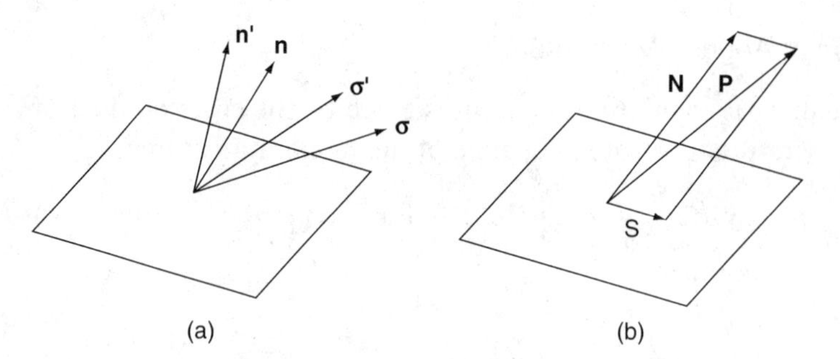

Figure 3.4.1. Coordinate transformations for normal and shear stresses: (a) normal stresses, (b) Shear stresses.

However, for conformity with Eq. (3.3.2b), we maintain the form of Eq. (3.3.3c) for the remainder of this book.

3.4 Coordinate Transformations for Stresses

The coordinate transformations for the stress tensor can be performed in a manner similar to those for the strain tensor. Referring to Fig. 3.4.1, we see that the old and new coordinates x_i and x_i' are related as

$$n_i' = a_{ij} n_j,$$

$$\sigma_i' = \sigma_{ij}' n_j', \tag{3.4.1a}$$

and

$$\sigma_i' = a_{ir}\sigma_r = a_{ir}\sigma_{rs}n_s = a_{ir}\sigma_{rs}a_{js}n_j'. \tag{3.4.1b}$$

Equate Eqs. (3.4.1a) and (3.4.1b) to yield

$$\sigma_{ij}' = a_{ir}a_{js}\sigma_{rs}, \tag{3.4.2}$$

in which the stress tensor σ_{rs} in the old coordinates is transformed into σ_{ij}' in the new coordinates by means of the fourth-order tensor, $a_{ir}a_{js}$. Note that the transformation process for the stress tensor is significantly different from that for the strain tensor as given by Eq. (2.5.3). The stress tensor σ_{ij}, rotated to the principal plane by n_j [Fig. 3.4.1(a)] with $\sigma = \lambda$ introduced as a scalar that corresponds to a principal stress acting on the plane n_i, may be written as

$$\sigma_{ij}n_j = \lambda n_i.$$

This relation can be written in the form

$$(\sigma_{ij} - \lambda\delta_{ij})\, n_j = 0. \tag{3.4.3}$$

We can derive the expression in Eq. (3.4.3) more rigorously by considering a scalar in the principal direction, similar to the case of strains in Section 2.5:

$$\sigma(n) = \sigma_{ij}n_j n_i, \tag{3.4.4a}$$

with the constraint

$$n_i n_i = 1. \tag{3.4.4b}$$

We use the Lagrange multiplier λ to combine Eqs. (3.4.4a) and (3.4.4b) and construct a scalar function

$$f = \sigma_{ij} n_j n_i - \lambda(n_i n_i - 1).$$

We now seek the extrema of f with respect to the direction cosines n_i such that

$$\frac{\partial f}{\partial n_i} = 2(\sigma_{ij} - \lambda\delta_{ij})n_j = 0,$$

which results in Eq. (3.4.3). This is identical in form to the case of strains, Eq. (2.5.5), which leads to the eigenvalue problem,

$$|\sigma_{ij} - \lambda\delta_{ij}| = 0, \tag{3.4.5}$$

from which we obtain the Caley–Hamilton equation of the form

$$-\lambda^3 + J_1\lambda^2 - J_2\lambda + J_3 = 0,$$

where J_1, J_2, and J_3 are called the *principal stress invariants*.

First principal stress invariant:

$$J_1 = \sigma_{ii}. \tag{3.4.6a}$$

Second principal stress invariant:

$$J_2 = \frac{1}{2}(\sigma_{ii}\sigma_{jj} - \sigma_{ij}\sigma_{ij}). \tag{3.4.6b}$$

Third principal stress invariant:

$$J_3 = |\sigma_{ij}|. \tag{3.4.6c}$$

The principal stress invariants are important in plasticity (material nonlinearity) and constitute important failure criteria for plastic materials. This is discussed in Section 7.4. For linear elasticity, however, as in the case of the principal strain invariants discussed in Section 2.5, these principal stress invariants are used only in formulating the eigenvalue problems to determine the principal stresses and principal planes.

The notion that the principal stresses and their directions are identified as eigenvalues and eigenvectors of Eq. (3.4.5), respectively, is similar to our earlier study on principal strains in Section 2.5. The solution procedure is the same as given in Example 2.5.1.

EXAMPLE 3.4.1. For convenience we use the same stress tensor components as those used for the strain. Consider the stress measurement data on a plane rotated from the plane of principal stresses given by

$$\sigma_{ij} = \begin{bmatrix} 1 & \sqrt{3} & 0 \\ \sqrt{3} & 0 & 0 \\ 0 & 0 & 1 \end{bmatrix}.$$

Determine (1) principal stress invariants, (2) principal stresses, (3) principal directions, and (4) draw the Mohr circles for this problem.

Solution. Note that stress components are the same as in Example 2.5.1 for the strain components. Therefore, using the results of Examples 2.5.1 and 2.5.2, we have the following information.

(1) **Principal stress invariants**:

$$J_1 = 2,$$
$$J_2 = -2,$$
$$J_3 = -3.$$

(2) **Principal stresses:**

$$\sigma_{(1)} = \lambda_{(1)} = \frac{1+\sqrt{13}}{2},$$
$$\sigma_{(2)} = \lambda_{(2)} = 1,$$
$$\sigma_{(3)} = \lambda_{(3)} = \frac{1-\sqrt{13}}{2}.$$

(3) **Principal direction cosine matrix:**

$$a_{ij} = \begin{bmatrix} 0.8 & 0.6 & 0 \\ 0 & 0 & 1 \\ 0.6 & -0.8 & 0 \end{bmatrix}.$$

(4) **Mohr circle representation:**

$$\sigma_{p1,2} = \frac{\sigma_{xx}+\sigma_{yy}}{2} \pm \sqrt{\left(\frac{\sigma_{xx}-\sigma_{yy}}{2}\right)^2 + \sigma_{xy}^2} = \begin{cases} 2.3 \\ -1.3 \end{cases},$$

$$\sigma_{p1,3} = \begin{cases} 1 \\ 1 \end{cases},$$

$$\sigma_{p2,3} = \begin{cases} 1 \\ 0 \end{cases},$$

$$\tan 2\theta = \frac{\sigma_{xy}}{(\sigma_{xx}-\sigma_{yy})/2} = 3.46,$$
$$\theta = 36.94^\circ.$$

Note that the only difference here from Example 2.5.2 is the treatment of shear stresses versus shear strains:

$$\sigma_{12} = \sigma_{xy}, \qquad \sigma_{21} = \sigma_{yx},$$
$$\gamma_{12} = \frac{1}{2}\gamma_{xy}, \quad \gamma_{21} = \frac{1}{2}\gamma_{yx}.$$

Thus the Mohr circles will be identical to Fig. 2.5.2 except that the horizontal and vertical axes are to be labeled $\sigma_{(i)}$ and σ_{xy}, respectively.

The shear stresses acting on any plane and the maximum shear stresses are directly associated with the normal stresses on that plane and the principal stresses, respectively. Consider the normal component $N_{(i)}$ and the shear component $S_{(i)}$ of any resultant stress $P_{(i)}$ in Fig. 3.4.1(b):

$$S_{(i)}^2 = P_{(i)}^2 - N_{(i)}^2. \tag{3.4.7a}$$

Because $P_i = \sigma_{ij} n_j$ and $N_i = P_{(i)} n_i$ and σ_{ij} can be transformed into the principal stress $\sigma_{(i)}$, we write

$$P_i = \sigma_{(i)} n_i,$$
$$N_{(i)} = \sigma_{(i)} n_i n_i.$$

Substitution of these into Eq. (3.4.7a) yields

$$S_{(i)}^2 = \sigma_{(i)}^2 n_i n_i - \sigma_{(i)} n_i n_i \sigma_{(j)} n_j n_j. \tag{3.4.7b}$$

Note that the index with parentheses indicates the direction and the requirement that the component not be summed. We are now concerned with a direction n_i for which shear stress is a maximum or minimum, subject to the constraint given in Eq. (3.4.4b). In this respect, we can use the Lagrange multiplier method by constructing a scalar function,

$$f = S_{(i)}^2 + \lambda(n_i n_i - 1),$$

and we find the extremum condition

$$\frac{\partial f}{\partial n_k} = \frac{\partial}{\partial n_k}\left[\sigma_{(i)}^2 n_i n_i - \sigma_{(i)} n_i n_i \sigma_{(j)} n_j n_j + \lambda(n_i n_i - 1)\right] = 0$$

or

$$\left[\sigma_{(i)}^2 - 2\sigma_{(i)} n_j n_j \sigma_{(j)} + \lambda\right] n_i = 0, \tag{3.4.8}$$

which represents the requirement for the shear stresses to be either maximum or minimum.

EXAMPLE 3.4.2. DETERMINATION OF PLANES OF MINIMUM AND MAXIMUM SHEAR STRESSES AND THEIR MAGNITUDES

Solution. The solution involves two cases. Case 1 is for minimum shear stresses and Case 2 is for maximum shear stresses.

Case 1. Minimum shear stresses:

$$n_1 = \pm 1, \quad n_2 = n_3 = 0, \quad \lambda = \sigma_{(1)}^2, \quad S_{(1)} = 0,$$

$$n_2 = \pm 1, \quad n_1 = n_3 = 0, \quad \lambda = \sigma_{(2)}^2, \quad S_{(2)} = 0,$$

$$n_3 = \pm 1, \quad n_1 = n_2 = 0, \quad \lambda = \sigma_{(3)}^2, \quad S_{(3)} = 0.$$

Case 2. Maximum shear stresses:

For $n_1 = 0$, $\quad n_2 = n_3 = \pm\frac{1}{\sqrt{2}}, \quad \lambda = \sigma_{(2)}\sigma_{(3)}, \quad S_{(1)} = \frac{1}{2}[\sigma_{(2)} - \sigma_{(3)}].$

For $n_2 = 0$, $\quad n_1 = n_3 = \pm\frac{1}{\sqrt{2}}, \quad \lambda = \sigma_{(3)}\sigma_{(1)}, \quad S_{(2)} = \frac{1}{2}[\sigma_{(3)} - \sigma_{(1)}].$

For $n_3 = 0$, $\quad n_1 = n_2 = \pm\frac{1}{\sqrt{2}}, \quad \lambda = \sigma_{(1)}\sigma_{(2)}, \quad S_{(3)} = \frac{1}{2}[\sigma_{(1)} - \sigma_{(2)}].$

To verify these results, we first expand Eq. (3.4.8) as follows:

Step 1 $\quad \{\sigma_{(1)}^2 - 2\sigma_{(1)}[\sigma_{(1)}n_1^2 + \sigma_{(2)}n_2^2 + \sigma_{(3)}n_3^2] + \lambda\}n_1 = 0.$

Step 2 $\quad \{\sigma_{(2)}^2 - 2\sigma_{(2)}[\sigma_{(1)}n_1^2 + \sigma_{(2)}n_2^2 + \sigma_{(3)}n_3^2] + \lambda\}n_2 = 0.$

Step 3 $\quad \{\sigma_{(3)}^2 - 2\sigma_3[\sigma_{(1)}n_1^2 + \sigma_{(2)}n_2^2 + \sigma_{(3)}n_3^2] + \lambda\}n_3 = 0.$

The equations must be satisfied together with the following step.

Step 4 $\quad n_1^2 + n_2^2 + n_3^2 = 1.$

For $n_1 = 0$, we have $n_2^2 = 1 - n_3^2$ and, Steps 2 and 3 are as shown in Steps 5 and 6, respectively.

Step 5 $\quad \{\sigma_{(2)}^2 - 2\sigma_{(2)}\,[\sigma_{(2)}(1 - n_3^2) + \sigma_{(3)}n_3^2] + \lambda\}\,(1 - n_3^2)^{1/2} = 0.$

Step 6 $\quad \{\sigma_{(3)}^2 - 2\sigma_{(3)}\,[\sigma_{(2)}(1 - n_3^2) + \sigma_{(3)}n_3^2] + \lambda\}\,n_3 = 0.$

For $n_1 = 0, n_2 = 0$, by using Step 6 with $n_3 = \pm 1$, we have

$$\sigma_{(3)}^2 - 2\sigma_{(3)}^2 + \lambda = 0 \quad \rightarrow \quad \lambda = \sigma_{(3)}^2.$$

Thus, from Eq. (3.4.7b)

$$S_{(i)}^2 = \sigma_{(3)}^2 - \sigma_{(3)}^2 = 0 \quad \rightarrow \quad S_{(i)} = S_{(3)} = 0.$$

For $n_1 = 0, n_3 = 0$, by using Step 5 with $n_2 = \pm 1$, we get

$$\sigma_{(2)}^2 - 2\sigma_{(2)}^2 + \lambda = 0 \quad \rightarrow \quad \lambda = \sigma_{(2)}^2,$$
$$S_{(i)}^2 = \sigma_{(2)}^2 - \sigma_{(2)}^2 = 0 \quad \rightarrow \quad S_{(i)} = S_{(2)} = 0.$$

Similarly, for $n_2 = n_3 = 0, n_1 = \pm 1$, we have

$$\lambda = \sigma_{(1)}^2,$$
$$S_{(i)} = S_{(1)} = 0.$$

Physically, these results indicate that the shear stresses are zero at the principal planes with direction cosines $(\pm 1, 0, 0)$, $(0, \pm 1, 0)$, and $(0, 0, \pm 1)$. To investigate the case of $n_1 = 0, n_2 \neq 0$, and $n_3 \neq 0$, we return to Steps 5 and 6 and write the following steps.

Step 7 $\quad \sigma_{(2)}^2 - 2\sigma_{(2)}^2 \left(1 - n_3^2\right) - 2\sigma_{(2)}\sigma_{(3)} n_3^2 + \lambda = 0.$

Step 8 $\quad \sigma_{(3)}^2 - 2\sigma_{(3)}\sigma_{(2)} \left(1 - n_3^2\right) - 2\sigma_{(3)}^2 - n_3^2 + \lambda = 0.$

Solve Steps 7 and 8 simultaneously to obtain

$$n_3^2 = \frac{1}{2} \frac{(\sigma_{(3)} - \sigma_{(2)})^2}{(\sigma_{(3)} - \sigma_{(2)})^2} = \frac{1}{2},$$
$$n_3 = \pm \frac{1}{\sqrt{2}}.$$

It follows from Step 4 that $n_2 = \pm 1/\sqrt{2}$. Substituting these results into Step 8, we have

$$\sigma_{(3)}^2 - 2\sigma_{(3)}\sigma_{(2)} \left(\frac{1}{2}\right) - 2_{(3)}^2 \left(\frac{1}{2}\right) + \lambda = 0,$$

which gives

$$\lambda = \sigma_{(3)}\sigma_{(2)}.$$

For these direction cosines, from (3.4.7b) we obtain

$$S_{(i)}^2 = \frac{\sigma_{(2)}^2}{2} + \frac{\sigma_{(3)}^2}{2} - \left[\frac{\sigma_{(2)}}{2} + \frac{\sigma_{(3)}}{2}\right]^2 = \frac{1}{4}[\sigma_{(2)} - \sigma_{(3)}]^2.$$

Here we define $S^2_{(i)} = S^2_{(1)}$ and write

$$S_{(1)} = \frac{1}{2}\left[\sigma_{(2)} - \sigma_{(3)}\right].$$

Similarly, for $n_1 \neq 0$, $n_2 = 0$, and $n_3 \neq 0$, and $n_1 \neq 0$, $n_2 \neq 0$, and $n_3 = 0$, respectively, we have

$$n_1 = n_3 = \pm\frac{1}{\sqrt{2}}, \qquad \lambda = \sigma_{(3)}\sigma_{(1)}, \quad S_{(i)} = S_{(2)} = \frac{1}{2}\left[\sigma_{(3)} - \sigma_{(1)}\right];$$

$$n_1 = n_2 = \pm\frac{1}{\sqrt{2}}, \qquad \lambda = \sigma_{(1)}\sigma_{(2)}, \quad S_{(i)} = S_{(3)} = \frac{1}{2}\left[\sigma_{(1)} - \sigma_{(2)}\right].$$

Here, $S_{(1)}$, $S_{(2)}$, and $S_{(3)}$ are called the *maximum shear stresses*. They occur along the planes oriented 45° from the planes of the principal stresses because the direction cosines are $\pm 1/\sqrt{2}$ with respect to the principal planes.

3.5 The Deviatoric Stress Tensor

As was the case in Section 2.8 for the strain tensor, it is often useful to separate the stress tensor into two parts, the deviatoric part $\hat{\sigma}_{ij}$ and the spherical, or hydrostatic, part $1/3\sigma_{kk}\delta_{ij}$, so that

$$\sigma_{ij} = \hat{\sigma}_{ij} + \frac{1}{3}\sigma_{kk}\delta_{ij} \tag{3.5.1}$$

or

$$\hat{\sigma}_{ij} = \sigma_{ij} - \frac{1}{3}\sigma_{kk}\delta_{ij}. \tag{3.5.2}$$

This holds true, irrespective of the geometric constraints of small or large strains. This is in contrast to the deviatoric strains in which only the small strain approximation makes the relation, such as Eq. (2.6.3), valid.

If $i = j$, then we have

$$\hat{\sigma}_{ii} = \sigma_{ii} - \frac{1}{3}\sigma_{kk}\delta_{ii} = \sigma_{ii} - \frac{1}{3}\sigma_{ii}(3) = 0, \tag{3.5.3}$$

On the other hand, if $i \neq j$, then we have $\hat{\sigma}_{ij} = \sigma_{ij}$.

The concept of deviatoric stresses is central to viscous flows in fluid mechanics (Section 5.2) and the failure criteria for plasticity (Section 7.4).

The principal deviatoric stress invariants are particularly important in constitutive theories of inelastic materials (see Section 7.4). To determine these quantities, we begin with the eigenvalue problems by using a process similar to that of Eq. (3.4.5):

$$|\sigma^*_{ij} - \sigma^*\delta_{ij}| = 0$$

or

$$-(\sigma^*)^3 + \hat{J}_1(\sigma^*)^2 - \hat{J}_2\sigma^* + \hat{J}_3 = 0, \tag{3.5.4}$$

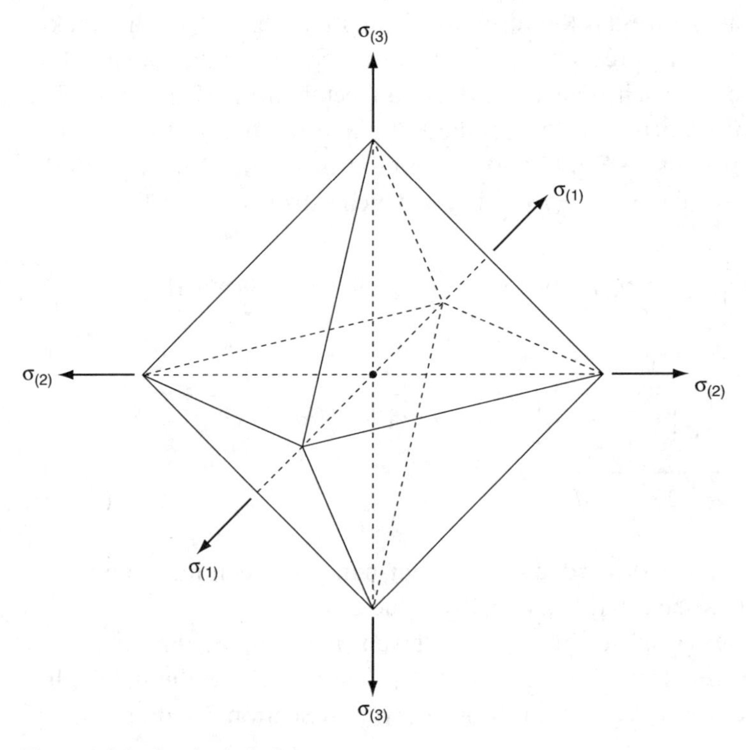

Figure 3.5.1. Octahedral planes.

where the *deviatoric stress invariants* are identified as

$$\hat{J}_1 = \sigma^*_{ii} = 0, \tag{3.5.5}$$

$$\hat{J}_2 = \frac{1}{3}\sigma^2_{kk} - J_2 = \frac{1}{2}\sigma^*_{ij}\sigma^*_{ij}, \tag{3.5.6}$$

$$\hat{J}_3 = J_3 - \frac{1}{3}\sigma_{kk}J_2 + \frac{2}{27}\sigma^3_{kk} = \frac{1}{3}\sigma^*_{ij}\sigma^*_{jk}\sigma^*_{ki}. \tag{3.5.7}$$

Expanding $\hat{J}_2$ further, we have

$$\begin{aligned}\hat{J}_2 &= \frac{1}{2}\left[(\sigma^*_{11})^2 + (\sigma^*_{22})^2 + (\sigma^*_{33})^2\right] + (\sigma_{12})^2 + (\sigma_{23})^2 + \sigma^2_{31} \\ &= \frac{1}{6}\left[(\sigma_{11} - \sigma_{22})^2 + (\sigma_{22} - \sigma_{33})^2 + (\sigma_{33} - \sigma_{11})^2\right] + \sigma^2_{12} + \sigma^2_{23} + \sigma^2_{31} \\ &= \frac{3}{2}\sigma^2_{\text{oct}},\end{aligned} \tag{3.5.8}$$

where σ_{oct} denotes the octahedral shear stress acting on the octahedral plane, as shown in Fig. 3.5.1:

$$\sigma_{\text{oct}} = \frac{1}{3}\sqrt{[(\sigma_{11} - \sigma_{22})^2 + (\sigma_{22} - \sigma_{33})^2 + (\sigma_{33} - \sigma_{11})^2] + 6\left(\sigma^2_{12} + \sigma^2_{23} + \sigma^2_{31}\right)}. \tag{3.5.9}$$

The octahedral shear stress is the resultant shear stress on a plane that makes the same angle with the three principal directions. Such a plane is called an octahedral plane; eight such planes can form an octahedron (Fig. 3.5.1). The direction cosines of a normal to the octahedral plane relative to the principle axes are $1/\sqrt{3}$. Hence, using Eq. (3.5.9), we obtain the octahedral stress as the square root of $S^2_{(i)}$, with $n_1 = n_2 = n_3 = 1/\sqrt{3}$, which satisfies $n_i n_i = 1$:

$$\sigma^2_{oct} = \frac{2}{9}\left[\sigma^2_{(1)} + \sigma^2_{(2)} + \sigma^2_{(3)} - \sigma_{(1)}\sigma_{(2)} - \sigma_{(2)}\sigma_{(3)} - \sigma_{(3)}\sigma_{(1)}\right]$$

or

$$\begin{aligned}\sigma_{oct} &= \frac{1}{3}\sqrt{[\sigma_{(1)} - \sigma_{(2)}]^2 + [\sigma_{(2)} - \sigma_{(3)}]^2 + [\sigma_{(3)} - \sigma_{(1)}]^2} \\ &= \frac{1}{3}\sqrt{2J_1^2 - 6J_2},\end{aligned} \qquad (3.5.10)$$

where the negative sign is discarded on the right-hand side as it has no physical significance. Thus it is seen that Eq. (3.5.10) reduces to (3.5.9).

In 1913, von Mises proposed that, in any given material, yielding occurs at a constant value of the second deviatoric stress invariant $\hat{J}_2$ or the octahedral shear stress. Further discussion of this topic is given in Section 7.4, on plasticity.

Remarks

We have investigated strains in Chap. 2 and stresses in this chapter. With these tools at hand, we are now prepared to explore fundamental equations and the corresponding physical behavior in both solids and fluids. Specifically, we investigate how stresses and strains are related through material properties in constructing theories of constitutive equations and thermomechanical properties for elastic solids in Chap. 4. Similarly, for fluids, we examine stresses and the rate-of-deformation tensors in the derivation of the governing equations of momentum and energy related to various types of flows. These topics are presented in Chap. 5.

PROBLEMS

3.1 Derive Cauchy's first law of motion in Cartesian coordinates and verify this law by summing the forces in all three directions on an infinitesimal cube in equilibrium. *Hint*: This amounts to deriving Eq. (3.3.2b) graphically. Identify all force components in terms of stresses acting on differential areas as they vary from one face to another. For example, in the x direction, there is a force $\sigma_{xx}\,dy\,dz$ on one face and $(\sigma_{xx} + \partial\sigma_{xx}/\partial x\,dx)\,dy\,dz$ on the other plus the inertia forces $\rho\ddot{u}_x\,dx\,dy\,dz$ and body forces $\rho F_x\,dx\,dy\,dz$ acting inside the cube, and similarly for the y and z directions. Sum all force components in all directions for equilibrium.

3.2 Let the components of stress with reference to a set of x_i axes be given by

$$\sigma_{ij} = \begin{bmatrix} 1 & 2 & 1 \\ 2 & 4 & 2 \\ 1 & 2 & 6 \end{bmatrix}.$$

(a) Calculate the principal stress invariants and the principal stress components.

(b) Determine the principal directions and check the correctness of the results by recovering the principal stresses through coordinate transformation.

3.3 Prove that the maximum shear stresses are

$$S_{(1)} = \frac{1}{2}\left[\sigma_{(2)} - \sigma_{(3)}\right],$$

$$S_{(2)} = \frac{1}{2}\left[\sigma_{(3)} - \sigma_{(1)}\right],$$

$$S_{(3)} = \frac{1}{2}\left[\sigma_{(1)} - \sigma_{(2)}\right],$$

and that they act on the planes 45° from the principal planes. Show also that shears are zero on the principal planes.

4 Linear Elasticity

In the previous chapters, we discussed the concepts of strain and stress without focusing on any particular material. In this chapter, we examine a linear elastic materia1. The nonlinear elastic material (nonlinear elasticity) is discussed in Chap. 7. By elastic we mean that the material may be deformed linearly or nonlinearly but will return to its original configuration on release of the applied loads. On the contrary, plastic materials may not return to their original position on the release of applied loads. In another case, the material known as viscoelastic exhibits time-dependent properties. Plasticity and viscoelasticity with such inelastic behavior are treated in Chap. 7.

In this chapter we begin with constitutive equations of linear elastic solids, followed by Navier equations, energy principles, and the thermodynamics of solids (Love, 1934; Truesdell and Toupin, 1960; Little, 1973). Thermomechanically coupled equations of motion and heat conduction are shown to emerge as a result of the thermodynamic principles applied to elastic solids by use of the first and second laws of thermodynamics. Finally, mechanics of fiber composite materials is discussed as a part of linear elasticity.

4.1 Constitutive Equations for Linear Elastic Solids

4.1.1 Three-Dimensional Solids

In the theory of linear elasticity, we are concerned with an ideal material governed by Hooke's law. This law was proposed by Robert Hooke in 1678 in his essay "*Ut tensio sic vis*," stating that "the power of a springy body is in the same proportion as the extension." Thus the stress σ and the strain γ in one dimension are related by Young's modulus E,

$$\sigma = E\gamma,$$

which represents a linear relationship between the stress and the strain. A material that obeys Hooke's law is referred to as a Hookean material.

We may derive a generalized three-dimensional Hooke's law by assuming the existence of an elastic potential W that is invariant with coordinate transformations of the strain tensor γ_{ij}:

$$W = W(\gamma_{ij}). \tag{4.1.1}$$

Expanding this function in Taylor series about $\gamma_{ij} = 0$, we obtain Clapeyron's formula

$$W = W_0 + \frac{\partial W}{\partial \gamma_{ij}}\gamma_{ij} + \frac{1}{2}\frac{\partial^2 W}{\partial \gamma_{ij}\partial \gamma_{km}}\gamma_{ij}\gamma_{km} + \frac{1}{3!}\frac{\partial^3 W}{\partial \gamma_{ij}\partial \gamma_{km}\partial \gamma_{np}}\gamma_{ij}\gamma_{km}\gamma_{np} + \cdots +, \tag{4.1.2a}$$

or, using more convenient notation,

$$W = W_0 + E_{ij}\gamma_{ij} + \frac{1}{2}E_{ijkm}\gamma_{ij}\gamma_{km} + \frac{1}{3!}E_{ijkmnp}\gamma_{ij}\gamma_{km}\gamma_{np} + \cdots +, \tag{4.1.2b}$$

where W_0 is a constant and E_{ij}, E_{ijkm}, E_{ijkmn}, E_{ijkmnp}, etc., called the tensors of elastic moduli, denote tensorial properties required for maintaining the invariant properties of W. Physically, the second term represents the energy that is due to residual stresses, the third term corresponds to the so-called strain energy describing the linear elastic deformations of small strains, and the fourth term is indicative of nonlinear behavior. The rest of the terms indicate highly nonlinear properties.

Consider δW, which is regarded as a small change in energy that is due to a small change in strain $\delta\gamma_{ij}$, related by

$$\delta W = \sigma_{ij}\delta\gamma_{ij}, \tag{4.1.3a}$$

where the symbol δ denotes the *virtual*, *incremental*, or simply small variation. The physical quantity implied in this relation may also be defined from Eq. (4.1.1) as

$$\delta W = \frac{\partial W}{\partial \gamma_{ij}}\delta\gamma_{ij}. \tag{4.1.3b}$$

Subtract Eq. (4.1.3b) from Eq. (4.1.3a) to yield

$$\left(\sigma_{ij} - \frac{\partial W}{\partial \gamma_{ij}}\right)\delta\gamma_{ij} = 0.$$

Because $\delta\gamma_{ij}$ is arbitrary, it is obvious that

$$\sigma_{ij} = \frac{\partial W}{\partial \gamma_{ij}}. \tag{4.1.4a}$$

Substitute Eq. (4.1.2b) into Eq. (4.1.4a) to obtain

$$\sigma_{rs} = \frac{\partial W}{\partial \gamma_{rs}} = \frac{\partial}{\partial \gamma_{rs}}\left(E_{ij}\gamma_{ij} + \frac{1}{2}E_{ijkm}\gamma_{ij}\gamma_{km} + \frac{1}{6}E_{ijkmnp}\gamma_{ij}\gamma_{km}\gamma_{np} + \cdots +\right)$$

$$= E_{ij}\delta_{ir}\delta_{js} + E_{ijkm}\delta_{ir}\delta_{js}\gamma_{km} + \frac{1}{2}E_{ijkmnp}\delta_{ir}\delta_{js}\gamma_{km}\gamma_{np} + \cdots + \qquad (4.1.4b)$$

$$= E_{rs} + E_{rskm}\gamma_{km} + \frac{1}{2}E_{rskmnp}\gamma_{km}\gamma_{np}, \ldots,$$

where $\partial\gamma_{ij}/\partial\gamma_{rs}$ implies a total of 81 terms of derivatives, all resulting in Kronecker deltas. This would have been a difficult task had the power of tensor analysis not been utilized. With the indices r and s now replaced with i and j, respectively, we have

$$\sigma_{ij} = E_{ij} + E_{ijkm}\gamma_{km} + \frac{1}{2}E_{ijkmnp}\gamma_{km}\gamma_{np} + \cdots +, \qquad (4.1.4c)$$

where E_{ij} may be taken as zero at time $t = 0$ for the unstrained condition. It is seen that the stress tensor results from a partial derivative of an elastic potential with respect to the strain tensor. The relation expressed in Eq. (4.1.4c) is regarded as a general form of a constitutive equation for three-dimensional solids. For linear elasticity, the nonlinear terms in Eq. (4.1.4c) are neglected. Thus the generalized Hooke's law for small strains in linear elasticity takes the form, in three dimensions, of

$$\sigma_{ij} = E_{ijkm}\gamma_{km} \qquad (i, j, k, m = 1, 2, 3), \qquad (4.1.5)$$

in which the linear part of the strain tensor for small strains is given by the first two terms on the right-hand side of Eq. (2.3.5),

$$\gamma_{ij} = \frac{1}{2}(u_{i,j} + u_{j,i}).$$

Here E_{ijkm} is a 9× 9 matrix, which has a total of 81 constants corresponding to the 9 stress or strain components ($\sigma_{11}, \sigma_{12}, \sigma_{13}, \sigma_{21}, \sigma_{22}, \sigma_{23}\, \sigma_{31}, \sigma_{32}, \sigma_{33}$, similarly for strains) and must be symmetric, $E_{ijkm} = E_{kmij} = E_{ijmk} = E_{jikm} = E_{jimk}$, because of the symmetry of the stress tensor and strain tensor. Thus from symmetry, E_{ijkm}, now called the fourth-order tensor of linear elastic moduli, has the array of $6 \times 6 = 36$ constants as dictated by Eq. (4.1.5), corresponding to the 6 stress or strain components ($\sigma_{11}, \sigma_{22}, \sigma_{33}, \sigma_{12}, \sigma_{23}, \sigma_{31}$, similarly for strains)

$$E_{ijkm} = \begin{bmatrix} E_{1111} & E_{1122} & E_{1133} & E_{1112} & E_{1123} & E_{1131} \\ & E_{2222} & E_{2233} & E_{2212} & E_{2223} & E_{2231} \\ & & E_{3333} & E_{3312} & E_{3323} & E_{3331} \\ \text{symm.} & & & E_{1212} & E_{1223} & E_{1231} \\ & & & & E_{2323} & E_{2331} \\ & & & & & E_{3131} \end{bmatrix}. \qquad (4.1.6)$$

This indicates that only 21 coefficients are needed to characterize anisotropic Hookean material in general.

By means of coordinate transformation, we are able to relate the material properties in one coordinate system (old) x_i to another coordinate system (new) x'_i. Thus, by substituting Eq. (2.5.3) into the quadratic terms of Eq. (4.1.2), which correspond to the linear elastic behavior, we obtain

$$W = \frac{1}{2} E_{rstu} \gamma_{rs} \gamma_{tu} = \frac{1}{2} E_{rstu} a_{ir} a_{js} a_{kt} a_{mu} \gamma'_{ij} \gamma'_{km}. \tag{4.1.7}$$

On the other hand, the elastic potential for the new coordinates can be written as

$$W = \frac{1}{2} E_{ijkm} \gamma'_{ij} \gamma'_{km}. \tag{4.1.8}$$

It follows from Eqs. (4.1.7) and (4.1.8) that

$$E_{ijkm} = a_{ir} a_{js} a_{kt} a_{mu} E_{rstu}. \tag{4.1.9}$$

This characterizes the relationship of the fourth-order tensor of material constants with the eighth-order tensor $a_{ir} a_{js} a_{kt} a_{mu}$ for coordinate transformations in order to maintain the symmetry of E_{ijkm} associated with anisotropy or isotropy of the material.

Monotropic Material

If the material is symmetric with respect to one plane [Fig. 4.1.1(a)], then the following coordinate transformation will apply:

$$x'_i = a_{ij} x_j,$$

where

$$a_{ij} = \begin{bmatrix} 1 & 0 & 0 \\ 0 & 1 & 0 \\ 0 & 0 & -1 \end{bmatrix}, \tag{4.1.10}$$

in which the negative sign for a_{33} refers to the symmetry of the mirror image with respect to the x_3 plane. Note that $a_{ij} = \delta_{ij}$ would have made both sides of Eq. (4.1.9) remain equal, which is the case of general anisotropy given by Eq. (4.1.6). On substituting Eq. (4.1.10) into Eq. (4.1.9) and expanding all terms with repeated indices, we find that E_{ijkm} is of the form

$$E_{ijkm} = \begin{bmatrix} E_{1111} & E_{1122} & E_{1133} & E_{1112} & 0 & 0 \\ & E_{2222} & E_{2233} & E_{2212} & 0 & 0 \\ & & E_{3333} & E_{3312} & 0 & 0 \\ \text{symm.} & & & E_{1212} & 0 & 0 \\ & & & & E_{2323} & E_{2331} \\ & & & & & E_{3131} \end{bmatrix}. \tag{4.1.11}$$

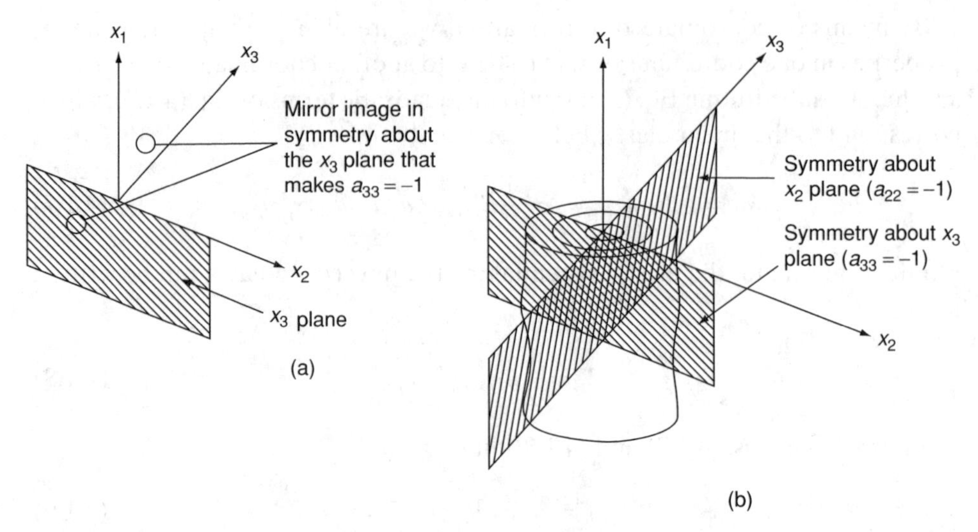

Figure 4.1.1. (a) Monotropic material – x_3 plane of symmetry, no symmetry about x_1 and x_2 planes; (b) orthotropic material – x_2 and x_3 planes of symmetry, no symmetry about x_1 plane.

A material such as the one just characterized is known as a monotropic material. Equation (4.1.11) shows that for monotropic materials there are 13 nonzero coefficients. Note that in Eq. (4.1.11) all terms of E_{ijkm} with the index 3 occurring an odd number of times are set equal to zero. These terms are made equal to zero to ensure the validity of Eq. (4.1.9). For example, $E_{1123} = a_{1r}a_{1s}a_{2t}a_{3u}E_{rstu} = a_{11}a_{11}a_{22}a_{33}E_{1123} = (1)(1)(1)(-1)E_{1123} = -E_{1123}$. Obviously this is not possible, and the only way the relation expressed in Eq. (4.1.9) can remain valid is if E_{1123} vanishes. All other zero entries in Eq. (4.1.11) can be verified in a similar manner.

Orthotropic Material

Material that is symmetric with respect to two planes [Fig. 4.1.1(b)] is called *orthotropic*, such as in wood. For example, if the x_2 and x_3 planes are chosen for symmetry, then

$$a_{ij} = \begin{bmatrix} 1 & 0 & 0 \\ 0 & -1 & 0 \\ 0 & 0 & -1 \end{bmatrix}. \tag{4.1.12}$$

If we proceed in a manner similar to that for the monotropic material, we obtain

$$E_{ijkm} = \begin{bmatrix} E_{1111} & E_{1122} & E_{1133} & 0 & 0 & 0 \\ & E_{2222} & E_{2233} & 0 & 0 & 0 \\ & & E_{3333} & 0 & 0 & 0 \\ \text{symm.} & & & E_{1212} & 0 & 0 \\ & & & & E_{2323} & 0 \\ & & & & & E_{3131} \end{bmatrix}. \tag{4.1.13}$$

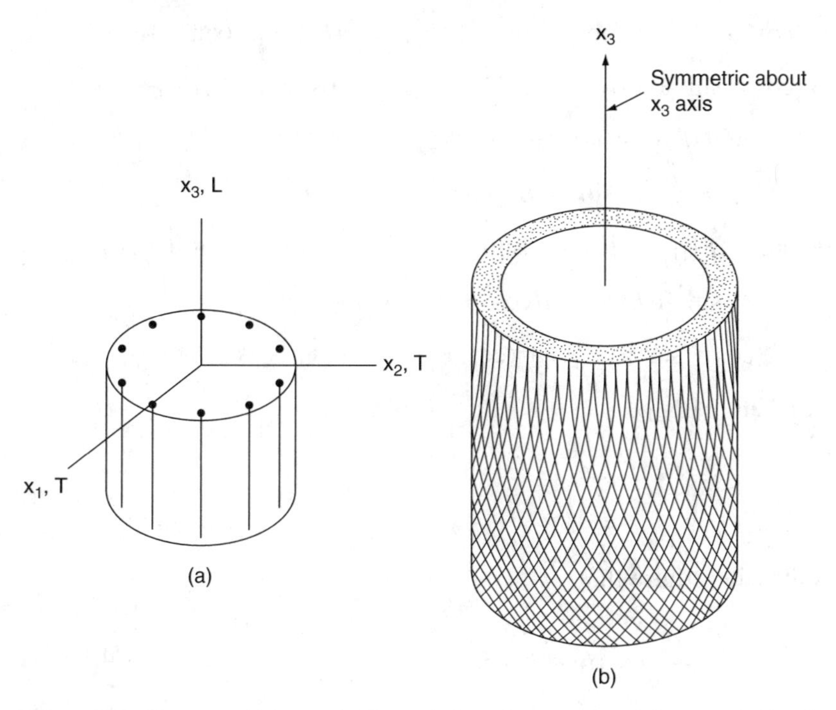

Figure 4.1.2. Transversely isotropic fiber-reinforced composites: (a) properties in all transverse (T) directions are symmetric about the longitudinal (L) axis, (b) fiber composites wrapped symmetrically about the x_3 axis.

Here, all terms of E_{ijkm} with the indices 3 and 2 occurring an odd number of times are again set equal to zero to satisfy Eq. (4.1.9). The proof is left to the reader. Note that, in this case, there are nine nonzero coefficients to be characterized.

Transversely Isotropic Material

A material is *transversely isotropic* if there is a preference for directions normal to all but one of the three axes. If this axis is x_3, then through right-handed counterclockwise rotations about the x_3 axis we must have (see Figs. 1.2.2 and 4.1.2)

$$a_{ij} = \begin{bmatrix} \cos\theta & \sin\theta & 0 \\ -\sin\theta & \cos\theta & 0 \\ 0 & 0 & 1 \end{bmatrix}. \tag{4.1.14}$$

This transformation is required in addition to the two previous transformations required for monotropic and orthotropic materials. Substituting Eq. (4.1.14) into Eq. (4.1.9), and in view of the symmetry of E_{ijkm} (it is convenient to make use of the 9×9 symmetric matrix to identify all symmetric terms; see Problem 4.1), we find

$$
\begin{aligned}
E_{1111} &= (\cos^4\theta)E_{1111} + (\cos^2\theta\sin^2\theta)(2E_{1122} + 4E_{1212}) + (\sin^4\theta)E_{2222},\\
E_{1122} &= (\cos^2\theta\sin^2\theta)E_{1111} + (\cos^4\theta)E_{1122} - 4(\cos^2\theta\sin^2\theta)E_{1212}\\
&\quad + (\sin^4\theta)E_{2211} + (\sin^2\theta\cos^2\theta)E_{2222},\\
E_{1133} &= (\cos^2\theta)E_{1133} + (\sin^2\theta)E_{2233},\\
E_{2222} &= (\sin^4\theta)E_{1111} + (\sin^2\theta\cos^2\theta)(2E_{1122} + 4E_{1212}) + (\cos^4\theta)E_{2222},\\
E_{1212} &= (\cos^2\theta\sin^2\theta)E_{1111} - 2(\cos^2\theta\sin^2\theta)E_{1122},\\
&\quad - 2(\cos^2\theta\sin^2\theta)E_{1212} + (\cos^4\theta)E_{1212} + (\sin^2\theta\cos^2\theta)E_{2222}\\
&\quad + (\sin^4\theta)E_{1212}\\
&\quad \vdots\\
&\quad \text{etc.}
\end{aligned}
$$

These results will require that

$$E_{1111} = E_{2222}, \tag{4.1.15a}$$

$$E_{1133} = E_{2233}, \tag{4.1.15b}$$

$$E_{2323} = E_{3131}, \tag{4.1.15c}$$

$$E_{1212} = \frac{1}{2}(E_{1111} - E_{1122}). \tag{4.1.15d}$$

In view of these results, we obtain

$$
E_{ijkm} = \begin{bmatrix}
E_{1111} & E_{1122} & E_{1133} & 0 & 0 & 0\\
 & E_{2222} & E_{2233} & 0 & 0 & 0\\
 & & E_{3333} & 0 & 0 & 0\\
\text{symm.} & & & \frac{1}{2}(E_{1111} - E_{1122}) & 0 & 0\\
 & & & & E_{2323} & 0\\
 & & & & & E_{3131}
\end{bmatrix}. \tag{4.1.16}
$$

Thus there are five independent coefficients involved in transversely isotropic materials.

Isotropic Material

If the material is symmetric with respect to two planes (x_3 and x_2 planes) and every axis, then the elastic properties are identical in all directions. Material that exhibits such a property is said to be *isotropic* and possesses only two nonzero coefficients. To see this we proceed as follows.

To characterize an isotropic material, we require coordinate transformations with rotations about the x_2 and x_1 axes in addition to all previous coordinate transformations. We begin with the right-handed counterclockwise rotation about the x_2 axis:

$$a_{ij} = \begin{bmatrix} \cos\theta & 0 & -\sin\theta \\ 0 & 1 & 0 \\ \sin\theta & 0 & \cos\theta \end{bmatrix}. \tag{4.1.17}$$

We follow a procedure similar to the case of transversely isotropic material to obtain

$$E_{1111} = E_{3333},$$
$$E_{3131} = \frac{1}{2}(E_{1111} - E_{1133}). \tag{4.1.18}$$

Now, with the last counterclockwise rotation about the x_1 axis given by

$$a_{ij} = \begin{bmatrix} 1 & 0 & 0 \\ 0 & \cos\theta & \sin\theta \\ 0 & -\sin\theta & \cos\theta \end{bmatrix}, \tag{4.1.19}$$

it follows that

$$E_{1122} = E_{1133}, \tag{4.1.20}$$

$$E_{3131} = \frac{1}{2}(E_{3333} - E_{1133}), \tag{4.1.21}$$

$$E_{2323} = \frac{1}{2}(E_{2222} - E_{2233}). \tag{4.1.22}$$

From these and previous results, we have

$$E_{ijkm} = \begin{bmatrix} E_{1111} & E_{1122} & E_{1133} & 0 & 0 & 0 \\ & E_{2222} & E_{2233} & 0 & 0 & 0 \\ & & E_{3333} & 0 & 0 & 0 \\ \text{symm.} & & & a & 0 & 0 \\ & & & & b & 0 \\ & & & & & c \end{bmatrix}, \tag{4.1.23}$$

with $a = \frac{1}{2}(E_{1111} - E_{1122})$, $b = \frac{1}{2}(E_{2222} - E_{2233})$, and $c = \frac{1}{2}(E_{3333} - E_{1133})$.

If we denote $E_{1122} = E_{1133} = E_{2233} = \lambda$ and $E_{1212} = E_{2323} = E_{3131} = \mu$, then from Eqs. (4.1.20)–(4.1.22) we find that $E_{1111} = E_{2222} = E_{3333} = \lambda + 2\mu$. Substitute these definitions into Eq. (4.1.23) to yield

$$E_{ijkm} = \lambda\delta_{ij}\delta_{km} + \mu(\delta_{ik}\delta_{jm} + \delta_{im}\delta_{kj}). \tag{4.1.24}$$

These relations indicate that there are only two constants (λ and μ), known as the Lamé constants, involved in isotropic material.

We recognize that the expression for an isotropic material characterized by means of energy invariants through an elastic potential function and, finally, by the fourth-order tensor, Eq. (4.1.24), coincides with the mathematical expression for the fourth-order isotropic tensor, as in the second-order isotropic tensor, the Kronecker delta, shown in Eq. (1.2.17a), Chap. 1.

On substituting Eq. (4.1.24) into Eq. (4.1.5), we obtain

$$\sigma_{ij} = [\lambda\delta_{ij}\delta_{km} + \mu(\delta_{ik}\delta_{jm} + \delta_{im}\delta_{jk})]\gamma_{km} \tag{4.1.25a}$$

or

$$\sigma_{ij} = \lambda\gamma_{kk}\delta_{ij} + 2\mu\gamma_{ij}. \tag{4.1.25b}$$

This represents the constitutive relationship of stress tensor and strain tensor. Had the substance been fluids rather than solids the derivation would have been the same. For this reason, the same expression is used for fluid mechanics with the strain tensor γ_{ij} replaced with the rate-of-deformation tensor d_{ij} and the stress tensor σ_{ij} for solids replaced with the viscous stress tensor τ_{ij} for fluids, as is shown in Subsection 5.2.2, Chap. 5:

$$\tau_{ij} = \lambda d_{kk}\delta_{ij} + 2\mu d_{ij}. \tag{4.1.25c}$$

To solve for the strain tensor in terms of the stress tensor, we proceed as follows. For $i = j$, Eq. (4.1.25b) takes the form

$$\sigma_{ii} = (3\lambda + 2\mu)\gamma_{ii}. \tag{4.1.26}$$

Then we substitute Eq. (4.1.26) into (4.1.25b) and solve for γ_{ij}:

$$\gamma_{ij} = \frac{1}{2\mu}\sigma_{ij} - \frac{\lambda\delta_{ij}}{2\mu(3\lambda + 2\mu)}\sigma_{kk}, \tag{4.1.27}$$

where the Lamé constants λ and μ, known respectively as the dilatational modulus and shear modulus, are given by

$$\lambda = \frac{\nu E}{(1+\nu)(1-2\nu)}, \tag{4.1.28a}$$

$$\mu = \frac{E}{2(1+\nu)}, \tag{4.1.28b}$$

where E and ν are Young's modulus and Poisson's ratio, respectively. It is possible to derive these Lamé constants λ and μ deductively, as guided by laboratory measurement data with E and ν. Toward this end, let us consider a simple case of an axial loading of a bar in tension or compression in the x_1 direction. Then we have

$$\sigma_{11} = \sigma \text{ with all other } \sigma_{ij} = 0.$$

In view of Eq. (4.1.27), all strain components are written as

$$\gamma_{11} = \frac{\lambda + \mu}{\mu(3\lambda + 2\mu)}\sigma,$$

$$\gamma_{22} = \gamma_{33} = -\frac{\lambda}{2\mu(3\lambda + 2\mu)}\sigma,$$

$$\gamma_{12} = \gamma_{23} = \gamma_{13} = 0.$$

From the elementary relations,

$$\gamma_{11} = \frac{\sigma}{E},$$

$$\nu = -\frac{\gamma_{22}}{\gamma_{11}} = -\frac{\gamma_{33}}{\gamma_{11}},$$

it follows that

$$\frac{1}{E} = \frac{\lambda + \mu}{\mu(3\lambda + 2\mu)},$$

$$\nu = \frac{\lambda}{2(\lambda + \mu)}.$$

Solving these two equations for the Lamé constants, we obtain Eqs. (4.1.28a) and (4.1.28b).

Let us now expand Eqs. (4.1.25b) and (4.1.27) to obtain

$$\begin{bmatrix} \sigma_{11} \\ \sigma_{22} \\ \sigma_{33} \\ \sigma_{12} \\ \sigma_{23} \\ \sigma_{31} \end{bmatrix} = a \begin{bmatrix} 1 & b & b & 0 & 0 & 0 \\ b & 1 & b & 0 & 0 & 0 \\ b & b & 1 & 0 & 0 & 0 \\ 0 & 0 & 0 & c & 0 & 0 \\ 0 & 0 & 0 & 0 & c & 0 \\ 0 & 0 & 0 & 0 & 0 & c \end{bmatrix} \begin{bmatrix} \gamma_{11} \\ \gamma_{22} \\ \gamma_{33} \\ 2\gamma_{12} \\ 2\gamma_{23} \\ 2\gamma_{31} \end{bmatrix}, \tag{4.1.29a}$$

with

$$a = \frac{E(1-\nu)}{(1+\nu)(1-2\nu)},$$

$$b = \frac{\nu}{1-\nu},$$

$$c = \frac{1-2\nu}{2(1-\nu)},$$

$$\begin{bmatrix} \gamma_{11} \\ \gamma_{22} \\ \gamma_{33} \\ \gamma_{12} \\ \gamma_{23} \\ \gamma_{31} \end{bmatrix} = \frac{1}{E} \begin{bmatrix} 1 & -\nu & -\nu & 0 & 0 & 0 \\ -\nu & 1 & -\nu & 0 & 0 & 0 \\ -\nu & -\nu & 1 & 0 & 0 & 0 \\ 0 & 0 & 0 & 1+\nu & 0 & 0 \\ 0 & 0 & 0 & 0 & 1+\nu & 0 \\ 0 & 0 & 0 & 0 & 0 & 1+\nu \end{bmatrix} \begin{bmatrix} \sigma_{11} \\ \sigma_{22} \\ \sigma_{33} \\ \sigma_{12} \\ \sigma_{23} \\ \sigma_{31} \end{bmatrix}. \tag{4.1.29b}$$

Here, the 6×6 matrices in Eqs. (4.1.29a) and (4.1.29b) are called the elasticity matrix $[E]$ and the compliance matrix $[C]$, respectively, with $[E] = [C]^{-1}$. The reader will recall that these results are introduced in elementary mechanics courses simply as stress–strain relations, without the rigorous derivations presented here.

It is interesting to note that, in terms of deviatoric strains and stresses, the expressions given by Eqs. (4.1.25) and (4.1.27) can be shown to be of the form

$$\hat{\sigma}_{ij} = 2\mu\hat{\gamma}_{ij},$$

which indicates that deviatoric stresses are related to deviatoric strains only through the shear modulus μ, with no dilatational modulus λ being involved.

Consider now the mean hydrostatic pressure defined as

$$p = \frac{-\sigma_{kk}}{3} = -\left(\lambda + \frac{2}{3}\mu\right)\gamma_{kk} = -\kappa\gamma_{kk}, \quad (4.1.30)$$

where the negative sign indicates the mean hydrostatic pressure in compression. The quantity κ is known as the *bulk modulus*:

$$\kappa = \lambda + \frac{2}{3}\mu = \frac{E}{3(1-2\nu)}. \quad (4.1.31)$$

Substituting Eqs. (4.1.28a) and (4.1.28b) into Eq. (4.1.30) gives the expression for mean pressure in the form

$$p = \frac{-E}{3(1-2\nu)}\gamma_{kk}.$$

An incompressible material has the bulk modulus $\kappa \to \infty$ in Eq. (4.1.31), which arises when $\nu = \frac{1}{2}$. This is the condition when $\gamma_{kk} = 0$ in Eq. (4.1.30). For a *stable material*, we must have $\kappa > 0$, $\mu > 0$, and $E > 0$, because these constants are involved in the elastic potential function, and positive work is required to cause any deformation. Therefore the Poisson ratio must be in the range $-1 \leq \nu \leq 1/2$, with $\nu = 0$ representing a one-dimensional deformation. Physically, no material is likely to have the Poisson ratio less than zero, and thus the range of the Poisson ratio is $0 \leq \nu \leq 1/2$.

4.1.2 Plane Stress and Plane Strain

In many engineering applications, three-dimensional problems may be idealized as two-dimensional, or plane, problems. If one of the dimensions is small in comparison with the other dimensions, then the stress in the direction of the small dimension (assume that will be x_3) is negligible. The state of stress in this case is called *plane stress* [Fig. 4.1.3(a)]. Thus we may simply ignore the small dimension and perform our analysis for the two-dimensional plane of the larger dimensions only. On the other hand, if one of the dimensions is extremely large in comparison with the other two dimensions (again, assume x_3), then it is possible

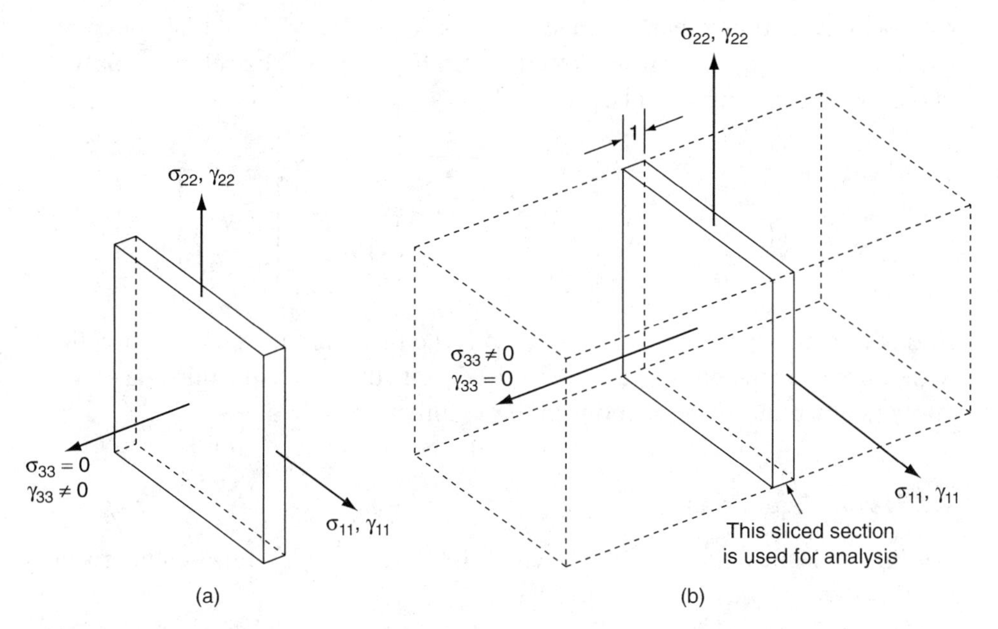

Figure 4.1.3. Plane problems with applied loads remaining in plane: (a) plane stress thickness in the x_3 direction is small in comparison with other directions ($\sigma_{33} = 0$, $\gamma_{33} \neq 0$), (b) plane strain thickness in the x_3 direction is large in comparison with other directions so that we may use a plane section with unit thickness in the 3–3 direction for the analysis by setting $\gamma_{33} = 0$ and $\sigma_{33} \neq 0$.

that the strain in this direction will be negligible. This condition is referred to as *plane strain* [Fig. 4.1.3(b)]. We may then be able to perform two-dimensional analysis on a sliced plane with unit thickness along the x_3 axis.

For plane stress, the stress–strain relation given by Eq. (4.1.25b) is modified by the conditions

$$\sigma_{33} = \sigma_{31} = \sigma_{32} = 0.$$

Thus,

$$\sigma_{33} = 0 = \lambda(\gamma_{11} + \gamma_{22} + \gamma_{33}) + 2\mu\gamma_{33},$$

from which

$$\gamma_{33} = \frac{-\lambda}{\lambda + 2\mu}(\gamma_{11} + \gamma_{22}).$$

The substitution of this last equality into Eq. (4.1.25b) gives

$$\begin{bmatrix} \sigma_{11} \\ \sigma_{22} \\ \sigma_{12} \end{bmatrix} = \frac{E}{1 - \nu^2} \begin{bmatrix} 1 & \nu & 0 \\ \nu & 1 & 0 \\ 0 & 0 & \dfrac{1 - \nu}{2} \end{bmatrix} \begin{bmatrix} \gamma_{11} \\ \gamma_{22} \\ 2\gamma_{12} \end{bmatrix}. \tag{4.1.32}$$

Note that the total shear strain is seen as $\gamma_{xy} = 2\gamma_{12}$. We obtain the stress–strain relation for plane strain directly from Eq. (4.1.29a) by retaining only the two-dimensional components:

$$\begin{bmatrix} \sigma_{11} \\ \sigma_{22} \\ \sigma_{12} \end{bmatrix} = \frac{E(1-\nu)}{(1+\nu)(1-2\nu)} \begin{bmatrix} 1 & \dfrac{\nu}{1-\nu} & 0 \\ \dfrac{\nu}{1-\nu} & 1 & 0 \\ 0 & 0 & \dfrac{1-2\nu}{2(1-\nu)} \end{bmatrix} \begin{bmatrix} \gamma_{11} \\ \gamma_{22} \\ 2\gamma_{12} \end{bmatrix}. \tag{4.1.33}$$

In this case, the structure to be analyzed is the plane in the x_1 and x_2 directions, with a unit dimension in the x_3 direction. Thus the results of this analysis can apply to any other plane sections in the x_3 direction.

4.2 Navier Equations

For linear elasticity, we invoke Cauchy's first law of motion for the undeformed coordinates given by Eq. (3.3.3c),

$$\sigma_{ij,i} + \rho F_j - \rho \ddot{u}_j = 0, \tag{4.2.1}$$

subject to the following boundary conditions: The Dirichlet or essential boundary condition is

$$u = \bar{u} \quad \text{on} \quad \Gamma_1,$$

and the Neumann or natural boundary condition is

$$\sigma_{ij} n_i = \bar{S}_j \quad \text{on} \quad \Gamma_2.$$

By virtue of the constitutive equations that result from Eq. (4.1.25b) and the exchange of indices between i and j, we arrive at the so-called Navier equation:

$$\mu u_{i,jj} + (\lambda + \mu) u_{j,ji} + \rho F_i - \rho \ddot{u}_i = 0 \tag{4.2.2a}$$

or

$$\mu \nabla^2 \mathbf{u} + (\lambda + \mu) \nabla (\nabla \cdot \mathbf{u}) + \rho \mathbf{F} - \rho \ddot{\mathbf{u}} = 0. \tag{4.2.2b}$$

In the absence of inertial forces, we have

$$\mu u_{i,jj} + (\lambda + \mu) u_{j,ji} + \rho F_i = 0. \tag{4.2.3}$$

For two-dimensional problems with plane strain conditions where $\gamma_{33} = \gamma_{13} = \gamma_{23} = 0$, we have

$$\sigma_{\alpha\beta} = \lambda \gamma_{\eta\eta} \delta_{\alpha\beta} + 2\mu \gamma_{\alpha\beta},$$

with $\alpha, \beta, \eta = 1, 2$ and $\sigma_{33} = \lambda \gamma_{\eta\eta}$. The Navier equation assumes the form

$$\mu \left(u_{\alpha,\beta\beta} + \frac{1}{1-2\nu} u_{\beta,\beta\alpha} \right) + \rho F_\alpha = 0. \tag{4.2.4}$$

For plane stress conditions, however, we have

$$\sigma_{33} = \sigma_{13} = \sigma_{23} = 0,$$

$$\gamma_{33} = -\frac{\lambda}{2\mu + \lambda}(\gamma_{11} + \gamma_{22}) = -\frac{\nu}{1 - \nu}\gamma_{\eta\eta}.$$

Therefore the Navier equation for plane stress takes the form

$$\mu\left(u_{\alpha,\beta\beta} + \frac{1 + \nu}{1 - \nu}u_{\beta,\beta\alpha}\right) + \rho F_{\alpha} = 0. \tag{4.2.5}$$

Solutions of the Navier equations – Eqs. (4.2.2) and (4.2.3) for three-dimensional problems, Eq. (4.2.4) for plane strain, and Eq. (4.2.5) for plane stress – have been the subject of extensive research for more than a century. Various energy principles that facilitate solution procedures may be employed. In the presence of heat energy, the Navier equations must be modified for thermomechanical coupling, which gives rise to the concept of entropy. These and other topics are presented in the following sections.

4.3 Energy Principles

Problems of elasticity are intimately related to the energy principles or variational principles that define equilibrium conditions (Lanczos, 1949; Mikhlin, 1964; Finlayson, 1970; and Oden and Reddy, 1976). This section highlights the principles of virtual work and total potential energy. The concept of energy principles is of paramount importance in the solution of elasticity problems in engineering. We discuss these subjects and associated boundary conditions.

4.3.1 Variational Principle, Virtual Work, and Total Potential Energy

We may derive the variational principle for three-dimensional linear elasticity by constructing the orthogonal projection of the residual space of governing differential equations on the subspace spanned by δu_j, known as virtual displacements. Denote the residual of the steady-state case of Eq. (4.2.1) as

$$R_j = \sigma_{ij,i} + \rho F_j$$

to construct an inner product, such that the virtual work δI summed over the domain will vanish:

$$\delta I = (R_j, \delta u_j) = \int_{\Omega} R_j \delta u_j \, d\Omega = 0 \tag{4.3.1a}$$

or

$$\delta I = \int_{\Omega} (\sigma_{ij,i} + \rho F_j)\delta u_j \, d\Omega = 0. \tag{4.3.1b}$$

Physically, this implies the condition of equilibrium, stating that the virtual work or infinitesimal energy deviating from equilibrium must vanish in order to maintain stability, known as the stable equilibrium. It should be pointed out that the notation δ has a dual role: physical and mathematical. Physically it represents virtual or infinitesimal; mathematically it means differential or derivative. In Eqs. (4.3.1) δ is meant to be physical.

The integration of Eq. (4.3.1b) by parts or use of the Green–Gauss theorem yields

$$\delta I = \int_\Gamma \sigma_{ij} n_i \delta u_j \, d\Gamma - \int_\Omega \sigma_{ij} \delta u_{j,i} \, d\Omega + \int_\Omega \rho F_j \delta u_j \, d\Omega = 0 \qquad (4.3.2a)$$

or

$$\delta I = \int_\Gamma \sigma_{ij} n_i \delta u_j \, d\Gamma - \int_\Omega E_{ijkm} \gamma_{km} \delta u_{j,i} \, d\Omega + \int_\Omega \rho F_j \delta u_j \, d\Omega = 0, \qquad (4.3.2b)$$

where σ_{ij}, when integrated and now residing in the boundary surface, is no longer the variable but becomes the constant acting as the surface traction $\sigma_{ij} n_i$ as applied at the boundary, with n_i representing the components of a vector normal to the boundary surface. Because σ_{ij} and E_{ijkm} are symmetric, the displacement gradient u_{km} can be written as a sum of the strain tensor γ_{km} (linear part only) and rotational or spin tensor $\omega_{km} = \frac{1}{2}(u_{k,m} - u_{m,k})$:

$$u_{k,m} = \gamma_{km} + \omega_{km} = \frac{1}{2}(u_{k,m} + u_{m,k}) + \frac{1}{2}(u_{k,m} - u_{m,k}), \qquad (4.3.3a)$$

so that

$$E_{ijkm} u_{k,m} = E_{ijkm}(\gamma_{km} + \omega_{km}) = E_{ijkm} \gamma_{km}, \qquad (4.3.3b)$$

where $E_{ijkm}\omega_{km} = 0$. Note that the proof of Eq. (4.3.3b) can be shown alternatively by use of the definition of the linear part of γ_{km} that is due to the symmetry of E_{ijkm},

$$E_{ijkm} \gamma_{km} = E_{ijkm} \frac{1}{2}(u_{k,m} + u_{m,k}) = E_{ijkm} u_{k,m}, \qquad (4.3.3c)$$

which can be proved without introducing the spin tensor simply because E_{ijkm} is symmetric. Note that $\gamma_{km} \neq u_{k,m}$ whereas $E_{ijkm}\gamma_{km} = E_{ijkm} u_{k,m}$ only if E_{ijkm} is symmetric. The reader is encouraged to prove this point.

It follows from Eqs. (4.3.2b) and (4.3.3c), and changing the signs, that

$$\delta I = \int_\Omega E_{ijkm} u_{k,m} \delta u_{j,i} \, d\Omega - \int_\Omega \rho F_j \delta u_j \, d\Omega - \int_\Gamma \sigma_{ij} n_i \delta u_j \, d\Gamma = 0 \qquad (4.3.4a)$$

or

$$\delta I = \delta \left(\int_\Omega \frac{1}{2} E_{ijkm} u_{k,m} u_{j,i} \, d\Omega - \int_\Omega \rho F_j u_j \, d\Omega - \int_\Gamma \sigma_{ij} n_i u_j \, d\Gamma \right) = 0, \qquad (4.3.4b)$$

where δ operates this time as a differential d (one of the roles of δ) such that the first term of Eq. (4.3.4b) can be shown:

$$
\begin{aligned}
\delta \int_\Omega \frac{1}{2} E_{ijkm} u_{k,m} u_{j,i}\, d\Omega &= \int_\Omega \frac{1}{2} E_{ijkm} \left(\frac{\partial u_{k,m}}{\partial u_{r,s}} \delta u_{r,s} u_{j,i} + u_{k,m} \frac{\partial u_{j,i}}{\partial u_{r,s}} \delta u_{r,s} \right) d\Omega \\
&= \int_\Omega \frac{1}{2} E_{ijkm} (\delta_{kr}\delta_{ms}\delta u_{r,s} u_{j,i} + u_{k,m}\delta_{jr}\delta_{is}\delta u_{r,s})\, d\Omega \\
&= \int_\Omega \frac{1}{2} (E_{ijrs} u_{j,i} \delta u_{r,s} + E_{ijkm} u_{k,m} \delta u_{j,i})\, d\Omega \\
&= \int_\Omega E_{ijkm} u_{k,m} \delta u_{j,i}\, d\Omega,
\end{aligned}
\tag{4.3.5}
$$

which produces the first term in Eq. (4.3.4a). The last step is due to the symmetry of E_{ijrs} and the exchange of indices $r, s \to i, j$ and $i, j \to k, m$. We recognize that $\partial u_{k,m}/\partial u_{r,s}$ implies a total of 81 terms of derivatives, differentiations performed similarly as in $\partial \gamma_{ij}/\partial \gamma_{rs}$, Eq. (4.1.4b). Also we note that ρF_j (constant body force) and $\sigma_{ij} n_i$ (constant boundary data) are independent of variation and only u_j is being subjected to variation. The stationary condition is a requirement of equilibrium, and it dictates that the quantity I must be equated to the integrals inside the parentheses of the right-hand side of Eqs. (4.3.4), so that

$$
I = \int_\Omega \frac{1}{2} \sigma_{ij} \gamma_{ij}\, d\Omega - \int_\Omega \rho F_j u_j\, d\Omega - \int_\Gamma \sigma_{ij} n_i u_j\, d\Gamma. \tag{4.3.6}
$$

This is known as the variational energy or the total potential energy, defined as

$$
I = U - W.
$$

Here, U is referred to as the strain energy,

$$
U = \int_\Omega \frac{1}{2} \sigma_{ij} \gamma_{ij}\, d\Omega, \tag{4.3.7}
$$

and W is the external work

$$
W = W(b) + W(s),
$$

where $W(b)$ and $W(s)$ are the external works that are due to body forces and surface tractions, respectively:

$$
W(b) = \int_\Omega \rho F_j u_j\, d\Omega, \tag{4.3.8a}
$$

$$
W(s) = \int_\Gamma \sigma_{ij} n_i u_j\, d\Gamma. \tag{4.3.9a}
$$

The variational energy expressed in the integral form, Eq. (4.3.6), is referred to as the *variational principle*. The integrand $\frac{1}{2}\sigma_{ij}\gamma_{ij}$ in Eq. (4.3.6), which contains derivatives (both σ_{ij} and γ_{ij} are functions of displacement gradients), is called the *variational functional*. The word "functional" is used to distinguish this term from a "function" that does not contain derivatives. The reader may notice that Eq. (4.3.6) is introduced in other textbooks merely as the statement for the variational principle. The derivation given here clarifies the origin. The process of differentiations involved in Kronecker deltas in Eq. (4.3.5) shows this origin. The usefulness of the variational principle is well known, particularly for the solutions of differential equations. One such application is demonstrated in Example 4.3.2.

4.3.2 Boundary Conditions

To verify the existence of boundary conditions required for a given physical problem and to derive formally the explicit forms of boundary conditions in three-dimensional elastic solids, we invoke the concept of invariant energy, as briefly discussed in Chap. 1. The representation given in Eqs. (4.3.1) is manifest for the notion of invariant energy, and the results of integration by parts given by Eq. (4.3.2) reveal the existence of the Neumann (natural) boundary condition, which is physically identified as the surface traction $\sigma_{ij}n_i$. The boundary integral $\int_\Gamma \sigma_{ij}n_i u_j\, d\Gamma$ represents the energy required for maintaining such boundary conditions. It is important to realize that the derivation of variational energy produces only the natural boundary conditions. Thus Dirichlet (essential) boundary conditions must be specified additionally for the solution of boundary-value problems.

If we intend simply to examine the existence and explicit forms of boundary conditions in general, there is no need to seek the virtual work. Rather, it is sufficient to work with energy created because of actual deformations. Thus we write

$$(R_j, u_j) = \int_\Omega (\sigma_{ij,i} + \rho F_j) u_j \, d\Omega.$$

Integrating by parts, we obtain

$$\begin{aligned}
(R_j, u_j) &= \int_\Gamma \sigma_{ij} n_i u_j \, d\Gamma - \int_\Omega E_{ijkm} u_{k,m} u_{j,i} \, d\Omega + \int_\Omega \rho F_j u_j \, d\Omega \\
&= \int_\Gamma \sigma_{ij} n_i u_j \, d\Gamma - \int_\Gamma E_{ijkm} u_k u_{j,i} \, d\Gamma \\
&\quad + \int_\Omega E_{ijkm} u_k u_{j,im} \, d\Omega + \int_\Omega \rho F_j u_j \, d\Omega \\
&= \int_\Gamma (\sigma_{ij} n_i u_j - u_k n_m E_{ijkm} u_{j,i}) \, d\Gamma \\
&\quad + \int_\Omega u_k E_{ijkm} u_{j,im} \, d\Omega + \int_\Omega \rho F_j u_j \, d\Omega.
\end{aligned} \tag{4.3.9a}$$

Here the boundary conditions include the Dirichlet (essential) boundary data u_k in addition to the Neumann boundary data $\sigma_{ij}n_i$.

For the one-dimensional case with a bar of length 1, cross-section area $A = 1$, $\sigma_{11} = \sigma$, $u_1 = u$, and $F_1 = F$, we have

$$(R, u) = \int_0^1 \left(\frac{\partial \sigma}{\partial x} + \rho F \right) u\, dx = \sigma u \,|_0^1 - \int_0^1 \sigma \frac{\partial u}{\partial x}\, dx + \int_0^1 \rho F u\, dx$$

$$= \sigma u \,|_0^1 - uE\frac{\partial u}{\partial x} \,|_0^1 + \int_0^1 uE\frac{\partial^2 u}{\partial x^2}\, dx + \int_0^1 \rho F u\, dx. \qquad (4.3.9b)$$

If we consider a bar fixed at $x = 0$ that is undergoing extensional deformation because of its own weight, we set $u = 0$ at $x = 0$ and $\sigma = 0$ at $x = 1$.

Note that each term on the right-hand side in Eq. (4.3.9b) represents energy mobilized to maintain equilibrium.

Neumann boundary condition:
$(\sigma) u \,|_0^1$ – energy required for maintaining the stress-free boundary condition $\sigma = 0$ at $x = 1$.

Dirichlet boundary condition:
$(u)E\frac{\partial u}{\partial x} \,|_0^1 = (u)\sigma \,|_0^1$ – energy required for maintaining the boundary condition $u = 0$ at $x = 0$.

The quantities inside the parentheses imply boundary data. The last two terms on the right-hand side of Eq. (4.3.9b) may be combined such that

$$\int_0^1 \left(E\frac{\partial^2 u}{\partial x^2} + \rho F \right) u\, dx = \int_0^1 \left(\frac{\partial \sigma}{\partial x} + \rho F \right) u\, dx.$$

Thus, mathematically, the right- and left-hand sides of Eq. (4.3.9b) are the same when $\sigma = 0$ and $u = 0$ for free boundaries.

EXAMPLE 4.3.1. Derive the variational principle for a rod (with length 1 and cross-sectional area A) fixed at the top undergoing deformation because of its own weight.

Solution. This is a one-dimensional problem; thus the governing equation is as follows:

Step 1 $\quad \dfrac{\partial \sigma_{11}}{\partial x_1} + \rho F_1 = 0.$

Step 2 $\quad \sigma_{11} = \sigma_x = E\dfrac{\partial u}{\partial x}.$

Substituting Step 2 into Step 1 yields

$$R = E\frac{\partial^2 u}{\partial x^2} + \rho F_x.$$

Thus, the virtual work becomes

$$\delta I = (R, \delta u) = \iiint \left(E\frac{\partial^2 u}{\partial x^2} + \rho F_x \right) \delta u \, dx\, dy\, dz$$
$$= A \int_0^1 \left(E\frac{\partial^2 u}{\partial x^2} + \rho F_x \right) \delta u \, dx$$
$$= AE\frac{\partial u}{\partial x}\delta u \,|_0^1 - \int_0^1 A \left(E\frac{\partial u}{\partial x}\frac{\partial \delta u}{\partial x} - \rho F_x \delta u \right) dx$$
$$= \delta \left\{ AE\frac{\partial u}{\partial x} u \,|_0^1 - A \int_0^1 \left[\frac{E}{2}\left(\frac{\partial u}{\partial x}\right)^2 - \rho F_x \delta u \right] dx \right\} = 0.$$

It follows that the variational principle or the total potential energy is

$$I = U - W,$$

where, with $A = 1$,

$$U = \int_0^1 \frac{1}{2} E \left(\frac{\partial u}{\partial x}\right)^2 dx,$$
$$W = \int_0^1 \rho F_x u \, dx + E\frac{\partial u}{\partial x} u \,|_0^1.$$

Note that U is the strain energy and W is the external energy consisting of the dead load that is due to the weight density (ρF_x) and the surface traction

$$E\frac{\partial u}{\partial x}\,|_0^1,$$

which vanishes at $x = 1$. This implies that the Neumann boundary condition,

$$\left(E\frac{\partial u}{\partial x} \right),$$

at $x = 1$ is zero (stress is zero at the end of the rod). Note also that the Dirichlet boundary condition (u) at $x = 0$ (the fixed end) is zero.

EXAMPLE 4.3.2. Reconsider the problem in Example 4.3.1 and determine the displacement by minimizing the total potential energy (the variational principle).

Solution. **Total potential energy**

Step 1 $$I = \int_0^L \frac{1}{2} E \left(\frac{\partial u}{\partial x}\right)^2 dx - \int_0^L \rho F_x u \, dx - E\frac{\partial u}{\partial x} u \,|_0^1.$$

Neumann boundary condition

Step 2 $$\frac{\partial u}{\partial x} = 0 \text{ at } x = L.$$

Dirichlet boundary condition

Step 3 $\quad u = 0$ at $x = 0$.

A functional representation for u that satisfies these conditions may be given by the so-called Rayleigh–Ritz (Ritz, 1909) approximations:

Step 4a $\quad u = c_i\phi_i(x) = c_1\phi_1 + c_2\phi_2 + \cdots + .$

In this case, choose a function that will satisfy both boundary conditions in Steps 2 and 3.

Step 4b $\quad u = c_1(x^2 - 2xL).$

Equilibrium is ensured when the total potential energy assumes a minimum.

Step 5 $\quad \delta I = \dfrac{\partial I}{\partial c_i}\delta c_i = 0 \quad \text{or} \quad \dfrac{\partial I}{\partial c_i} = 0.$

Substitute Step 4b into Step 1 and, subsequently, into Step 5 to yield Step 6.

Step 6 $\quad \dfrac{\partial I}{\partial c_1} = \displaystyle\int_0^L [E(2x - 2L)^2 c_1 - f(x^2 - 2xL)]\, dx = 0,$

where $f = \rho F_x$. Solving for c_1 from Step 6 gives $c_1 = -\frac{f}{2E}$. Thus substitution of this constant into Step 4b leads to

$$u = -\frac{f}{2E}(x^2 - 2xL).$$

This agrees with the exact solution, which may be obtained by integration of $E(\partial^2 u/\partial x^2) + f = 0$, subject to the given boundary conditions.

EXAMPLE 4.3.3. Derive the variational principle for the biharmonic equation

$$\frac{\partial^4\psi}{\partial x^4} + 2\frac{\partial^4\psi}{\partial x^2\partial y^2} + \frac{\partial^4\psi}{\partial y^4} = 0,$$

where ψ is the transverse displacement in plate bending.

Solution

$$\delta I = \int_\Omega \left(\frac{\partial^4\psi}{\partial x^4} + 2\frac{\partial^4\psi}{\partial x^2\partial y^2} + \frac{\partial^4\psi}{\partial y^4}\right)\delta\psi\, dx\, dy = 0.$$

Here it is cautioned that the cross-derivative term must be split so that integration by parts can be carried out separately with respect to x and y for each term of the right-hand side [see Eq. (1.3.11)]. Note that this process is automatically taken care of if the tensorial approach is followed:

$$
\begin{aligned}
\delta I &= \int_\Omega \psi_{,iijj}\delta\psi\, d\Omega = \int_\Gamma \psi_{,iij}\delta\psi\, n_j d\Gamma - \int_\Omega \psi_{,iij}\delta\psi_{,j}\, d\Omega \\
&= \int_\Gamma \psi_{,iij}\delta\psi\, n_j d\Gamma - \int_\Gamma \psi_{,ii}\delta\psi_{,j}\, n_j d\Gamma + \int_\Omega \psi_{,ii}\delta\psi_{,jj}\, d\Omega \\
&= \delta\left[\int_\Gamma (\psi_{,iij}\psi - \psi_{,ii}\psi_{,j}) n_j\, d\Gamma + \frac{1}{2}\int_\Omega \psi_{,ii}\psi_{,jj}\, d\Omega\right] = 0,
\end{aligned}
$$

where $i, j = 1, 2$. Note that $\psi_{,iij}$ and $\psi_{,ii}$ in the boundary integral represent the Neumann boundary condition data. Further integrations by parts down to the zeroth-order derivative of ψ would have revealed the existence of Dirichlet boundary conditions [see Eq. (1.3.11)], but integration by parts for the derivation of the variational principle ends at the completion of the Neumann boundary data. Thus the variational principle is obtained as

$$
I = \frac{1}{2}\int_\Omega \psi_{,ii}\psi_{,jj}\, d\Omega + \int_\Gamma (\psi_{,iij}\psi - \psi_{,ii}\psi_{,j}) n_j\, d\Gamma.
$$

Writing explicitly, we have

$$
\begin{aligned}
I = &\frac{1}{2}\int_\Omega \left[\left(\frac{\partial^2\psi}{\partial x^2}\right)^2 + \left(\frac{\partial^2\psi}{\partial y^2}\right)^2 + 2\frac{\partial^2\psi}{\partial x^2}\frac{\partial^2\psi}{\partial y^2}\right] d\Omega \\
&+ \int_\Gamma \left(\frac{\partial^3\psi}{\partial x^3}\psi n_1 + \frac{\partial^3\psi}{\partial x^2 \partial y}\psi n_2 + \frac{\partial^3\psi}{\partial y^2 \partial x}\psi n_1 + \frac{\partial^3\psi}{\partial y^3}\psi n_2 \right. \\
&\left. - \frac{\partial^2\psi}{\partial x^2}\frac{\partial\psi}{\partial x} n_1 - \frac{\partial^2\psi}{\partial x^2}\frac{\partial\psi}{\partial y} n_2 - \frac{\partial^2\psi}{\partial y^2}\frac{\partial\psi}{\partial x} n_1 - \frac{\partial^2\psi}{\partial y^2}\frac{\partial\psi}{\partial y} n_2\right) d\Gamma.
\end{aligned}
$$

For two-dimensional problems, the boundary integrals are line integrals that consist of third- and second-order derivatives of ψ, which represent boundary shears and moments, respectively. Once again, Dirichlet boundary conditions are the first- and zeroth-order derivatives, which indicate boundary slopes and displacements, respectively, but do not appear in the derivation of the variational principle. The variational principle for the Laplace equation ($\psi_{ii} = 0$) and Poisson equation ($\psi_{,ii} = f$) can be derived similarly, resulting in simpler forms.

4.4 Thermodynamics of Solids

4.4.1 General

If a body is heated, strains and stresses develop. Conversely, if a body is strained rapidly, then heat is generated inside the body. The system undergoing these processes can be characterized by various functions called *state variables*. One basic state variable is temperature, which abstracts the degree of hot and cold. It is known that no system can be cooled below a certain limit. If the greatest lower bound is assigned the value zero, then the temperature is said to be absolute. The absolute temperature θ for the scale of measurement therefore has the property

$$\theta \geq 0.$$

The absolute temperature θ is also defined as

$$\theta = T_0 + T,$$

where T_0 is the reference temperature and T is the temperature change from T_0. The heat energy supplied to a body arises from sources within the body and from the heat flux on its surfaces given by

$$Q = \int_\Omega \rho\, r d\Omega \pm \int_\Gamma \mathbf{q} \cdot \mathbf{n}\, d\Gamma, \tag{4.4.1}$$

in which r is the heat supply per unit mass, $\mathbf{q}$ is the heat flux vector per unit surface area, and $\mathbf{n}$ is the unit vector normal to the surface. The plus or minus sign is determined from the choice of positive or negative slope for the heat flux with the minus sign indicating heat flowing from hot to cold. This is governed by the Fourier's heat conduction law, $\mathbf{q} = \pm K \Delta T$, with K being the thermal conductivity.

When we consider the thermal environment and the mechanical loading, our concern is to derive governing equations that will couple displacements with temperature so that their combined influence can be determined. If deformations produce heat and if the temperature rise leads to deformations (expansion), and if these actions are simultaneous, then what will be the consequence? To this end, we invoke the first and second laws of thermodynamics in the subsections that follow.

4.4.2 The First Law of Thermodynamics

We consider a motion governed by the functions $z_i, \sigma_{ij}, F_i, \epsilon, q_i, r$, and η of particle $\mathbf{x}$ and time t in a thermoelastic *body*. We define the functions in the Lagrangian coordinates,

$$z_i = z_i(\mathbf{x}, t), \tag{4.4.2}$$

as the components of the spatial position at the material point $\mathbf{x}$ at time t that characterize the *motion*. Here we assume that components of the functions z_i are continuously differentiable. The stress tensor σ_{ij} and the body force per unit mass F_i are defined in terms of the Cartesian coordinates, ϵ is the internal energy density per unit mass, q_i are the components of the heat flux vector, r represents the heat supply per unit mass, and η is the entropy per unit mass.

The first law of thermodynamics may be stated as follows: The time rate of change of kinetic energy and internal energy in a body is equal to the external energies including the rate of work done on the body plus the changes in all other energies, such as heat, magnetic, electrical, and chemical, per unit time, or

$$\frac{\partial K}{\partial t} + \frac{\partial U}{\partial t} = M + \sum E_i. \tag{4.4.3}$$

In the absence of the external energies other than those that are due to mechanical power and heat, we write

$$\frac{\partial K}{\partial t} + \frac{\partial U}{\partial t} = M + Q, \tag{4.4.4}$$

where the individual variables are defined as follows.

Kinetic energy

$$K = \frac{1}{2}\int_\Omega \rho \mathbf{v} \cdot \mathbf{v}\, d\Omega. \tag{4.4.5}$$

Internal energy

$$U = \int_\Omega \rho \epsilon \, d\Omega. \tag{4.4.6}$$

Mechanical energy

$$M = \int_\Omega \rho \mathbf{F} \cdot \mathbf{v}\, d\Omega + \int_\Gamma \boldsymbol{\sigma}(n) \cdot \mathbf{v}\, d\Gamma. \tag{4.4.7}$$

Heat energy

$$Q = \text{Eq. (4.4.1)}.$$

Note that the surface traction energy in Eq. (4.4.7) can be transformed to the domain integral by means of the Green–Gauss theorem discussed in Section 1.3:

$$\begin{aligned}\int_\Gamma \boldsymbol{\sigma}(n) \cdot \mathbf{v}\, d\Gamma &= \int_\Gamma \sigma_{ij} \mathbf{i}_j n_i \cdot v_m \mathbf{i}_m \, d\Gamma = \int_\Gamma \sigma_{ij} v_j n_i \, d\Gamma = \int_\Omega (\sigma_{ij} v_j)_{,i} \, d\Omega \\ &= \int_\Omega (\sigma_{ij,i} v_j + \sigma_{ij} v_{j,i})\, d\Omega,\end{aligned} \tag{4.4.8}$$

where

$$\sigma_{ij} v_{j,i} = \sigma_{ij} \dot{\gamma}_{ij}, \tag{4.4.9}$$

which is true only if the stress and strain tensors are symmetric, realizing that $v_{j,i} \neq \dot{\gamma}_{ij}$. Substituting Eq. (4.4.1) and Eqs. (4.4.5)–(4.4.9) into Eq. (4.4.4) yields

$$\int_\Omega (\rho \dot{v}_j - \rho F_j - \sigma_{ij,i}) v_j \, d\Omega + \int_\Omega (\rho \dot{\epsilon} - \sigma_{ij} \dot{\gamma}_{ij} - q_{i,i} - \rho r)\, d\Omega = 0, \tag{4.4.10}$$

where we used the relation given in Eq. (2.2.22a) for the time derivative of $\rho\, d\Omega$ for the conservation of mass. Thus, for Eq. (4.4.10) to be valid for all arbitrary components of v_j and $d\Omega$, it is necessary that the first integrand vanish, so that

$$\sigma_{ij,i} + \rho F_j - \rho \dot{v}_j = 0. \tag{4.4.11}$$

This leads to

$$\rho \dot{\epsilon} - \sigma_{ij}\dot{\gamma}_{ij} - q_{i,i} - \rho\, r = 0. \tag{4.4.12}$$

Equations (4.4.11) and (4.4.12) are the local forms of Cauchy's first law of motion (or the momentum equation) and the energy equation, respectively. Returning to Eq. (4.4.10), we can see that the conservation of momentum (the vanishing of the first integral) is prerequisite to the conservation of energy (the vanishing of the second integral). We conclude that the first law of thermodynamics provides conservation of momentum as well as conservation of energy. Note, however, that the energy equation given by Eq. (4.4.12) is not useful, because the internal energy density ϵ cannot be defined in terms of quantities available in solid mechanics. It can be related only through *entropy* and *Helmholtz free energy*, which are to be characterized by the second law of thermodynamics. We discuss this subject in the following subsection.

4.4.3 The Second Law of Thermodynamics

The second law of thermodynamics is based on the concept of entropy associated with irreversible thermodynamic processes. The entropy is regarded as a measure of change of energy dissipation with respect to temperature or the disorder of the thermodynamic system (Onsager, 1931; Eckart, 1940; Biot, 1956; Boley and Weiner, 1960; Truesdell and Toupin, 1960; Prigogin, 1961; Truesdell and Noll, 1965).

We define entropy S as an additive continuous function,

$$S = \int_\Omega \rho \eta \, d\Omega,$$

where η is the entropy density per unit mass. Furthermore, the total entropy production B is defined as

$$B = \dot{S} - \int_\Gamma \left(\frac{\mathbf{q}}{\theta}\right) \cdot \mathbf{n}\, d\Gamma - \int_\Omega \left(\frac{\rho r}{\theta}\right) d\Omega \geq 0. \tag{4.4.13}$$

This expression is referred to as the second law of thermodynamics in solids, which states that the total entropy production is always greater than or equal to zero. This is also known as the *Clausius–Duhem inequality*. Using the Green-Gauss theorem, we may rewrite Eq. (4.4.13) as

$$\int_\Omega \left[\rho \dot{\eta} - \nabla \cdot \left(\frac{\mathbf{q}}{\theta}\right) - \left(\frac{\rho r}{\theta}\right)\right] d\Omega \geq 0$$

or

$$\theta\rho\dot{\eta} - \nabla \cdot \mathbf{q} + \frac{1}{\theta}\mathbf{q} \cdot \nabla\theta - \rho r \geq 0, \tag{4.4.14}$$

which is the local form of the Clausius–Duhem inequality.

The quantity $(-\theta\eta)$ is the heat energy that is due to entropy as related to temperature, with the negative sign indicating that compressive reaction results from thermal expansion (temperature rise) in a body. The sum of internal energy (ϵ) and heat energy $(-\theta\eta)$ is known as Helmholtz free energy (Helmholtz, 1858):

$$\Phi = \epsilon - \theta\eta. \tag{4.4.15}$$

Substituting Eq. (4.4.15) into Eq. (4.4.12), we obtain

$$\rho\dot{\Phi} = \sigma_{ij}\dot{\gamma}_{ij} - \rho\dot{\theta}\eta - D, \tag{4.4.16}$$

where D is defined as internal dissipation:

$$D = \theta\rho\dot{\eta} - q_{i,i} - \rho r. \tag{4.4.17}$$

In view of Eqs. (4.4.17) and (4.4.14), it follows that

$$D + \frac{1}{\theta}q_i\theta_{,i} \geq 0, \tag{4.4.18}$$

which is called the general dissipation inequality. For an irreversible process, we find

$$D > 0. \tag{4.4.19a}$$

From the physical ground, it is always true that

$$\frac{1}{\theta}q_i\theta_{,i} \geq 0. \tag{4.4.19b}$$

This is because, if the Fourier heat conduction law is used for the heat flux (the temperature gradient), then $q_i\theta_{,i}$ is quadratic in the temperature gradient, always rendering this product positive. Thus Eq. (4.4.18) remains valid without the term $\frac{1}{\theta}q_i\theta_{,i}$ participating in Eq. (4.4.18).

For the reversible process, $D = 0$, Eq. (4.4.17) represents the energy equation replacing Eq. (4.4.12),

$$\rho\theta\dot{\eta} - q_{i,i} - \rho r = 0, \tag{4.4.20}$$

in which the entropy η appears, replacing the internal energy density ϵ. The next step is to invoke the constitutive theory for thermodynamics of solids in order to replace entropy with temperature.

4.4.4 Constitutive Theory for Thermodynamics of Solids

To characterize the thermal deformations effectively, the constitutive theory for linear elasticity, discussed in Section 4.1, must be reinforced with the theory of thermodynamics of solids (Truesdell and Toupin, 1960). Toward this end, it is necessary that the constitutive theory be developed in such a way that these governing equations are not violated. This requirement is called *physical admissibility*. This and other rules for constitutive equations are subsequently summarized (Gurtin, 1963; Truesdell and Noll, 1965).

Rule 1: Physical admissibility
All constitutive equations must be consistent with the basic physical laws of conservation of mass, conservation of momentum, conservation of energy, and the Clausius–Duhem inequality.

Rule 2: Determinism
The values of the constitutive variables $(\sigma_{ij}, q_i, \eta, \Phi)$ at a material point $\mathbf{x}$ of a body at time t are determined by the histories of the motion and the temperature of all points of the body:

$$\Phi = \hat{\Phi}\left[z_i(x_i', t'),\ \theta(x_i', t'),\ \nabla\theta(x_i', t')\right], \tag{4.4.21a}$$

$$\sigma_{ij} = \hat{\sigma}_{ij}\left[z_i(x_i', t'),\ \theta(x_i', t'),\ \nabla\theta(x_i', t')\right], \tag{4.4.21b}$$

$$\eta = \hat{\eta}\left[z_i(x_i', t'),\ \theta(x_i', t'),\ \nabla\theta(x_i', t')\right], \tag{4.4.21c}$$

$$q_i = \hat{q}_i\left[z_i(x_i', t'),\ \theta(x_i', t'),\ \nabla\theta(x_i', t')\right], \tag{4.4.21d}$$

where $z_i(x_i', t')$, $\theta(x_j', t')$, and $\nabla\theta(x_i', t')$ are the histories of motion and temperature with $t' \leq t$ and x_i' corresponding to t'. Here we assume that all constitutive functionals depend on the same set of independent variables. This leads to the principle of equipresence.

Rule 3: Equipresence
At the outset, a quantity that appears as an independent variable in one constitutive equation should appear likewise in all constitutive equations unless ruled out by physical laws.

Rule 4: Local action (the principle of neighborhood)
This rule imposes certain restrictions on the smoothness of the constitutive functionals in the neighborhood of a material point $\mathbf{x}$. The dependent constitutive variables at $\mathbf{x}$ are not appreciably affected by the values of the independent variables at material points distant from $\mathbf{x}$.

Rule 5: Material objectivity (material frame indifference)
The response of a material is independent of the spatial reference frame established to describe it. This implies that the constitutive equations are invariant under observer transformations.

Rule 6: Material symmetry
The constitutive functionals for all ideal materials possess some type of isotropy group. Thus constitutive equations must be form invariant with

respect to a group of unimodular transformations of the material's frame of reference.

Applications of these rules are demonstrated in the following subsection. Additional discussion of this topic appears in Chap. 5 for Newtonian fluids and Chap. 7 for nonlinear continuum.

4.4.5 Thermomechanically Coupled Equations of Motion and Heat Conduction

We recast Eqs. (4.4.21), Rule 2 of the constitutive equations, in the form corresponding to a thermoelastic body:

$$\Phi = \hat{\Phi}(\gamma_{ij}, \theta, \theta_{,i}), \tag{4.4.22a}$$

$$\sigma_{ij} = \hat{\sigma}(\gamma_{ij}, \theta, \theta_{,i}), \tag{4.4.22b}$$

$$\eta = \hat{\eta}(\gamma_{ij}, \theta, \theta_{,i}), \tag{4.4.22c}$$

$$q_i = \hat{q}(\gamma_{ij}, \theta, \theta_{,i}). \tag{4.4.22d}$$

Here the motion z_i at $t' = t$ is characterized by strains γ_{ij}. It is well known that strains and temperatures affect the free energy, but the dependency of temperature gradients on free energy remains to be verified. Note also that all six rules of constitutive theories are fulfilled in the first and second laws of thermodynamics, although the material histories or memory effects are not considered in Eqs. (4.4.22).

It follows from Eq. (4.4.22a) that the time rate of change of free energy takes the form

$$\rho\dot{\Phi} = \rho\frac{\partial\Phi}{\partial\gamma_{ij}}\dot{\gamma}_{ij} + \rho\frac{\partial\Phi}{\partial\theta}\dot{\theta} + \rho\frac{\partial\Phi}{\partial\theta_{,i}}\dot{\theta}_{,i}.$$

Substitute the preceding relation into Eq. (4.4.16) to yield

$$\left(\sigma_{ij} - \rho\frac{\partial\Phi}{\partial\gamma_{ij}}\right)\dot{\gamma}_{ij} - \left(\rho\eta + \rho\frac{\partial\Phi}{\partial\theta}\right)\dot{\theta} - \rho\frac{\partial\Phi}{\partial\theta_{,i}}\dot{\theta}_{,i} - D = 0. \tag{4.4.23}$$

For a reversible process ($D = 0$) and for all arbitrary values of $\dot{\gamma}_{ij}$, $\dot{\theta}$, and $\dot{\theta}_{,i}$, it is necessary that the following relations hold:

$$\sigma_{ij} = \rho\frac{\partial\Phi}{\partial\gamma_{ij}}, \tag{4.4.24}$$

$$\rho\eta = -\rho\frac{\partial\Phi}{\partial\theta}, \tag{4.4.25}$$

$$\rho\frac{\partial\Phi}{\partial\theta_{,i}} = 0. \tag{4.4.26}$$

Equation (4.4.26) implies that the free energy must be independent of the temperature gradient. Thus, we may revise Eqs. (4.4.22):

$$\Phi = \hat{\Phi}(\gamma_{ij}, \theta), \tag{4.4.27a}$$

$$\sigma_{ij} = \hat{\sigma}(\gamma_{ij}, \theta), \tag{4.4.27b}$$

$$\eta = \hat{\eta}(\gamma_{ij}, \theta), \tag{4.4.27c}$$

$$q_i = \hat{q}(\gamma_{ij}, \theta). \tag{4.4.27d}$$

We see that the principle of equipresence (Rule 3, Subsection 4.4.4) is to serve merely as a guide. The final form of the constitutive equation must be dictated by physical observation.

The constitutive equations, (4.4.26) and (4.4.27), suggest that, had a form of free energy been known, it would have been possible to determine the stress tensor and entropy. To this end, we may expand Eq. (4.4.27a) into a Taylor series and retain only the second-order terms (squares and product of variables) in a manner similar to that of Eqs. (4.1.1) and (4.4.2). In the present case, however, we have two arguments, γ_{ij} and θ. Thus the free energy per unit volume may be written in the form

$$\rho\Phi = \frac{1}{2}E_{ijkm}\gamma_{ij}\gamma_{km} - \frac{1}{2}\frac{c}{T_0}\theta^2 - \beta_{ij}\theta\gamma_{ij}, \tag{4.4.28}$$

where E_{ijkm} and β_{ij} are the elastic modulus tensor and the thermoelastic modulus tensor, respectively, and c is the heat capacity ($c = \rho c_v$, where c_v is the specific heat at constant volume). The negative signs for the last two terms on the right-hand side of Eq. (4.4.28) indicate that a temperature rise leads to a compressive reaction on the restrained body. The second term signifies the thermal energy that is due to temperature, and the last term is the energy that is due to the coupling effect between temperature and mechanical deformations. Note that Eq. (4.4.28) is equivalent to Eq. (4.1.2b), Clapeyron's formula, as extended to thermoelastic solids. On substitution of Eq. (4.4.28) into Eqs. (4.4.24) and (4.4.25), we obtain explicit forms of the constitutive equations for the stress tensor and entropy,

$$\sigma_{ij} = E_{ijkm}\gamma_{km} - \beta_{ij}T, \tag{4.4.29a}$$

$$\rho\eta = \frac{c}{T_0}T + \beta_{ij}\gamma_{ij}, \tag{4.4.29b}$$

with $\theta = T$ because the initial stress and entropy contributed by T_0 are zero.

Recall that the validity of constitutive relations expressed in Eqs. (4.4.24) and (4.4.25) is based on reversibility. For an irreversible process ($D > 0$), however, it is necessary to provide adequate mathematical models for the internal dissipation.

For example, we may assume that $D = D_{ijkm}\gamma_{ij}\dot{\gamma}_{km}$, with D_{ijkm} being the fourth-order tensor of damping constants as used in the theory of viscoelasticity [see Eq. (7.5.6) for more complicated cases]. The constitutive equation for heat flux may be given by the *Fourier heat conduction law*,

$$q_i = \pm k_{ij} T_{,j}, \tag{4.4.30}$$

where k_{ij} is the *thermal conductivity tensor* for a general anisotropic material. Note that the plus or minus sign follows from Eq. (4.4.1).

For isotropic materials with small strains, the constitutive laws given by Eqs. (4.4.29a), (4.4.29b), and (4.4.30) are written as follows:

$$\beta_{ij} = \beta\delta_{ij}, \tag{4.4.31}$$

$$k_{ij} = k\delta_{ij}, \tag{4.4.32}$$

$$\sigma_{ij} = E_{ijkm}\gamma_{km} - \beta\delta_{ij} T, \tag{4.4.33}$$

$$\rho\eta = \frac{c}{T_0} T + \beta\gamma_{ii}, \tag{4.4.34}$$

$$q_i = kT_{,i}, \text{ [as chosen by the plus sign in Eq. (4.4.1)]}, \tag{4.4.35}$$

where E_{ijkm} is as given by Eq. (4.1.24), k is the isotropic thermal conductivity, and the isotropic thermoelastic constant β assumes the form

$$\beta = \frac{E\alpha}{1-2v} \quad \text{for three-dimensional and plane strain cases}, \tag{4.4.36}$$

$$\beta = \frac{E\alpha}{1-v} \text{ for plane stress}, \tag{4.4.37}$$

with α being the coefficient of thermal expansion. Note that we can derive the thermoelastic constant β by setting Eq. (4.1.26) equal to the thermal stress $\sigma_{ii} = \sigma_T = \beta T$ and by defining the thermal strain as $\gamma_{ii} = \gamma_T = \alpha T$:

$$\sigma_T = \beta T = \sigma_{ii} = (3\lambda + 2\mu)\gamma_{ii} = (3\lambda + 2\mu)\alpha T,$$

which gives

$$\beta = (3\lambda + 2\mu)\alpha = \frac{E\alpha}{1-2v}.$$

For the case of plane stress, it follows from Eq. (4.1.32) that

$$\sigma_T = \beta T = \frac{E}{1-v^2}(1+v)\gamma_T = \frac{E\alpha T}{1-v}.$$

Thus,

$$\beta = \frac{E\alpha}{1-v}.$$

With these preliminaries, the equations of motion and heat conduction for an isotropic material can be derived for the reversible process. We can now write the momentum equation (4.4.11) and energy equation (4.4.20), respectively, as

$$\sigma_{ij,i} + \rho F_j - \rho \ddot{u}_j = 0, \tag{4.4.38a}$$

$$\frac{\theta}{T_0} c\dot{T} + \theta\beta\dot{u}_{i,i} - kT_{,ii} - \rho r = 0. \tag{4.4.38b}$$

More explicitly, we obtain the thermomechanically coupled Navier equations or equations of motion

$$\mu u_{i,jj} + (\lambda + \mu)\, u_{j,ji} - \alpha(3\lambda + 2\mu)\, T_{,i} + \rho F_i - \rho \ddot{u}_i = 0 \tag{4.4.39a}$$

or

$$\mu \nabla^2 \mathbf{u} + (\lambda + \mu)\, \nabla(\nabla \cdot \mathbf{u}) - \alpha(3\lambda + 2\mu)\, \nabla T + \rho \mathbf{F} - \rho \frac{\partial^2 \mathbf{u}}{\partial t^2} = 0 \tag{4.4.39b}$$

and the thermomechanically coupled heat conduction equation

$$c\dot{T} + T_0 \alpha(3\lambda + 2\mu)\, \dot{u}_{i,i} - kT_{,ii} - \rho r = 0 \tag{4.4.40a}$$

or

$$c\frac{\partial T}{\partial t} + T_0 \alpha(3\lambda + 2\mu) \frac{\partial}{\partial t} \nabla \cdot \mathbf{u} - k\nabla^2 T - \rho r = 0, \tag{4.4.40b}$$

where, for the transient state ($T \ll T_0$), the absolute temperature is set equal to the reference temperature ($\theta \cong T_0$), which is the common assumption accepted in the field of heat conduction. Equations (4.4.39) and (4.4.40) are known as thermomechanically coupled equations of motion and heat conduction for small strains. Both sets of equations involve displacements and temperature, thus requiring the simultaneous solution with the effects of both displacement and temperature fields taken into account.

It is important to realize that the presence of temperature terms in the momentum equations and displacement terms in the energy equation is the consequence of the second law of thermodynamics through the concepts of entropy, internal energy density, and Helmholtz free energy, as well as the first law of thermodynamics. In Chap. 5, similar approaches are followed for the derivation of the Navier–Stokes system of equations for fluid mechanics.

It should be noted that in traditional thermoelasticity thermomechanical coupling by means of the second law of thermodynamics is not considered. See for example Nowacki (1962) and Parkus (1962, 1968).

Boundary Conditions

The equations of motion and heat conduction are subject to the following boundary conditions:

$$u = \hat{u} \text{ on } \Gamma_1, \tag{4.4.41a}$$

$$\sigma_{ij} n_i = \hat{S}_j \text{ on } \Gamma_2, \tag{4.4.41b}$$

$$T = \hat{T} \text{ on } \Gamma_3, \tag{4.4.41c}$$

$$k\frac{\partial T}{\partial n} = kT_{,i} n_i = -\hat{q} - \bar{\alpha}(T - T') \text{ on } \Gamma_4. \tag{4.4.41d}$$

As defined in Subsection 4.3.4, the values of variables prescribed on boundaries such as u_i and T are the Dirichlet boundary conditions, whereas the gradients normal to the surface, such as $\sigma_{ij} n_i$ and $kT_{,i}\, n_i$, are the Neumann boundary conditions. However, the right-hand side of Eq. (4.4.41d) indicates the mixture of the value of a variable (T, T') and the heat flux (the temperature gradient q). This is sometimes referred to as the Cauchy or Robin boundary condition, where $\hat{q}$ is the heat flux on the boundary's surface, $\bar{\alpha}$ is the heat transfer (film) coefficient, and T' denotes the ambient temperature.

The existence and explicit forms of boundary conditions for the equations of equilibrium were shown in Subsection 4.3.4. We follow the identical procedure for heat conduction. However, in boundary-value problems, only the second derivative term is involved in producing the boundary conditions. Therefore it follows that

$$-\int_\Omega kT_{,ii} T \, d\Omega = -\int_\Gamma kT_{,i} n_i T \, d\Gamma + \int_\Omega kT_{,i} T_{,i} \, d\Omega. \tag{4.4.42}$$

Further integration by parts is to verify the existence of Dirichlet boundary conditions. The Neumann boundary conditions $kT_{,i}\, n_i$, as indicated by the first term on the right-hand side of Eq. (4.4.42), may consist of surface heat flux $\hat{q}$ and additional heat flux generated by the temperature difference between the surface temperature T and the ambient temperatures T' associated with heat transfer coefficient $\bar{\alpha}$, as shown in (4.4.41d).

Remarks

We note that the equations of motion (4.4.39) are of a mixed hyperbolic–elliptic nature, whereas the heat conduction equation (4.4.40) is a parabolic partial differential equation. Heat conduction in some materials (helium, for example) exhibits hyperbolic behavior, a phenomenon known as *second sound* (Ackerman and Berman, 1966) in which the Fourier heat conduction law, Eq. (4.4.36), must be modified. The details on second sound are beyond the scope of this book.

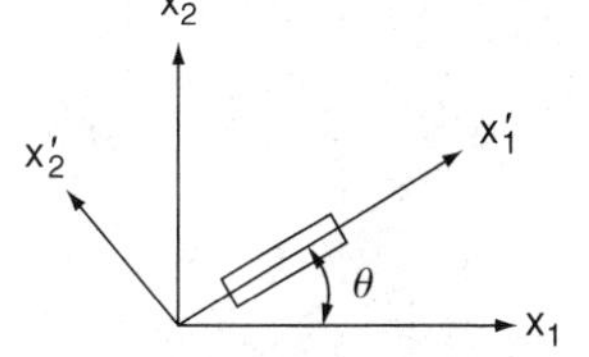

Figure 4.5.1. Coordinate transformations for a fiber-composite solid.

4.5 Fiber Composites

4.5.1 General

In Section 4.1 a general approach was presented that led to constitutive equations for anisotropic material. Subsequently, an isotropic solid was chosen in Section 4.2 to derive the governing equations for linear elasticity. Modern technology, in aerospace industries in particular, has prompted the use of composite materials with metal or glass fibers. Composite materials consist of two or more materials bonded together to exhibit desirable strength–weight properties that cannot be attained by a single individual material alone. The resulting behavior of composite materials is extremely complicated on a microscopic scale. In the design of fiber-composite materials, the concept of micromechanics may be used although, in general, the macromechanics approach will be adequate for most engineering problems.

4.5.2 Coordinate Transformations for Transversely Isotropic Materials

The fiber composites may be orthotropic, as characterized by Eq. (4.1.12), or transversely isotropic (Eq. 4.1.14), as shown in Fig. 4.1.2. Suppose that the fiber is oriented in the direction x_1' (see Fig. 4.5.1), then the coordinate transformation for a simple fiber-composite solid is governed by Eqs. (4.1.16) and (2.5.3) so that the strain components with the x_1' axis coinciding with the longitudinal direction of the fiber are written as

$$\begin{bmatrix} \gamma_{11}' \\ \gamma_{22}' \\ \gamma_{33}' \\ \gamma_{12}' \\ \gamma_{23}' \\ \gamma_{31}' \end{bmatrix} = \begin{bmatrix} \hat{c}^2 & \hat{s}^2 & 0 & 2\hat{c}\hat{s} & 0 & 0 \\ \hat{s}^2 & \hat{c}^2 & 0 & -2\hat{c}\hat{s} & 0 & 0 \\ 0 & 0 & 1 & 0 & 0 & 0 \\ -\hat{c}\hat{s} & \hat{s}\hat{c} & 0 & \hat{c}^2 - \hat{s}^2 & 0 & 0 \\ 0 & 0 & 0 & 0 & \hat{c} & -\hat{s} \\ 0 & 0 & 0 & 0 & \hat{s} & \hat{c} \end{bmatrix} \begin{bmatrix} \gamma_{11} \\ \gamma_{22} \\ \gamma_{33} \\ \gamma_{12} \\ \gamma_{23} \\ \gamma_{31} \end{bmatrix}, \qquad (4.5.1)$$

with $\hat{c} = \cos\theta$, $\hat{s} = \sin\theta$.

Suppose that the fiber-reinforced cylinder shown in Fig. 4.5.2(a) has the local coordinates $x_1' = r$, $x_2' = T$, and $x_3' = L$, and the global coordinates $x_1 = r$, $x_2 = z$,

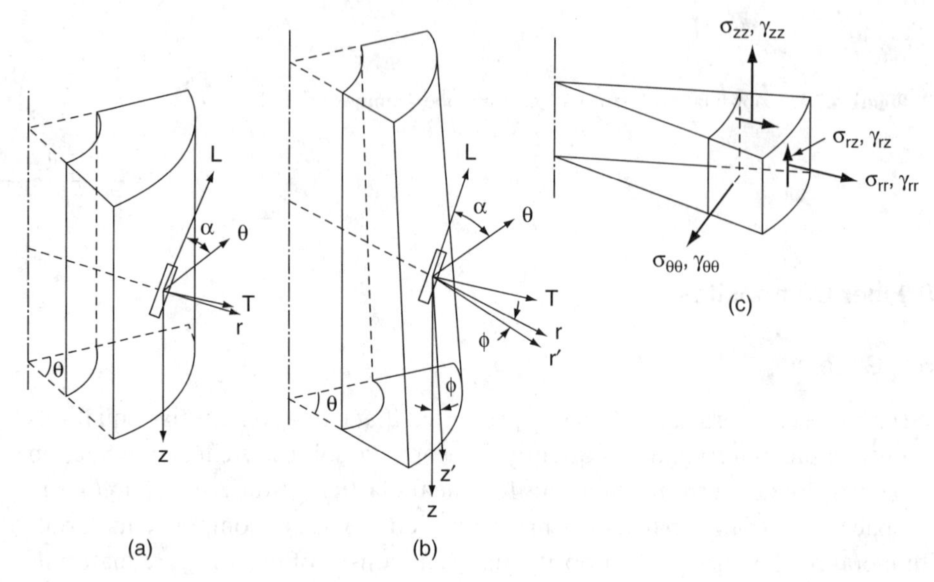

Figure 4.5.2. (a) Coordinate transformations for axisymmetric cylindrical fiber composites, (b) general axisymmetric body for transversely isotropic materials, and (c) stress and strain components.

and $x_3 = \theta$. The rotation about the axis $x_1' = x_1 = r$ is represented by

$$\begin{bmatrix} x_1' \\ x_2' \\ x_3' \end{bmatrix} = \begin{bmatrix} 1 & 0 & 0 \\ 0 & c & s \\ 0 & -s & c \end{bmatrix} \begin{bmatrix} x_1 \\ x_2 \\ x_3 \end{bmatrix}, \tag{4.5.2}$$

where $c = \cos\alpha$ and $s = \sin\alpha$. The relation between global strains ($\gamma_{rr}, \gamma_{zz}, \gamma_{\theta\theta}, \mathit{and}\, \gamma_{rz}$) and local strains ($\gamma_{rr}, \gamma_{TT}, \gamma_{LL}, \gamma_{rT}, \gamma_{rL}, \mathit{and}\, \gamma_{TL}$) is given by Eqs. (2.5.3) and (4.5.2) as [see Fig. 4.5.2(c)]

$$\begin{bmatrix} \gamma_{rr} \\ \gamma_{TT} \\ \gamma_{LL} \\ \gamma_{rT} \\ \gamma_{rL} \\ \gamma_{TL} \end{bmatrix} = \begin{bmatrix} 1 & 0 & 0 & 0 \\ 0 & c^2 & s^2 & 0 \\ 0 & s^2 & c^2 & 0 \\ 0 & 0 & 0 & 0 \\ 0 & 0 & 0 & -s \\ 0 & -2cs & 2cs & 0 \end{bmatrix} \begin{bmatrix} \gamma_{rr} \\ \gamma_{zz} \\ \gamma_{\theta\theta} \\ \gamma_{rz} \end{bmatrix}. \tag{4.5.3}$$

Note that the tensorial shear components have been transformed into total shear components:

$$(\gamma_{ij})_{\text{total}} = 2\,(\gamma_{ij})_{\text{tensor}} \quad \text{if } i \neq j.$$

Let us now consider an axisymmetric body with the generator inclined at an angle ϕ from the vertical axis, as shown in Fig. 4.5.2(b). The local and global coordinates are labeled $x_i' = (r', z')$ and $x_i = (r, z)$. Consider the intermediate coordinates x_i^* related by the global coordinates $x_1 = r$, $x_2 = z$, and $x_3 = \theta$,

$$\begin{bmatrix} x_1^* \\ x_2^* \\ x_3^* \end{bmatrix} = \begin{bmatrix} \bar{c} & \bar{s} & 0 \\ -\bar{s} & \bar{c} & 0 \\ 0 & 0 & 1 \end{bmatrix} \begin{bmatrix} x_1 \\ x_2 \\ x_3 \end{bmatrix}, \tag{4.5.4}$$

with $\bar{c} = \cos\phi$ and $\bar{s} = \sin\phi$. The intermediate coordinates are also related by the local coordinates $x_1' = r, x_2' = T, x_3' = L$:

$$\begin{bmatrix} x_1' \\ x_2' \\ x_3' \end{bmatrix} = \begin{bmatrix} 1 & 0 & 0 \\ 0 & c & s \\ 0 & -s & c \end{bmatrix} \begin{bmatrix} x_1^* \\ x_2^* \\ x_3^* \end{bmatrix}. \tag{4.5.5}$$

With rotations about the axis $x_3 = \theta = x_3^*$ and subsequently about the axis $x_1' = r = x_1^*$, we obtain the transformation for the x_i and x_i' coordinates:

$$\begin{aligned} \begin{bmatrix} x_1' \\ x_2' \\ x_3' \end{bmatrix} &= \begin{bmatrix} 1 & 0 & 0 \\ 0 & c & s \\ 0 & -s & c \end{bmatrix} \begin{bmatrix} \bar{c} & \bar{s} & 0 \\ -\bar{s} & \bar{c} & 0 \\ 0 & 0 & 1 \end{bmatrix} \begin{bmatrix} x_1 \\ x_2 \\ x_3 \end{bmatrix} \\ &= \begin{bmatrix} \bar{c} & \bar{s} & 0 \\ -c\bar{s} & c\bar{s} & s \\ s\bar{s} & -s\bar{c} & c \end{bmatrix} \begin{bmatrix} x_1 \\ x_2 \\ x_3 \end{bmatrix}. \end{aligned} \tag{4.5.6}$$

It follows from Eqs. (2.5.3) and (4.5.6) that the relation between the local and global strains for the general axisymmetric body is given by

$$\begin{bmatrix} \gamma_{rr} \\ \gamma_{TT} \\ \gamma_{LL} \\ \gamma_{rT} \\ \gamma_{rL} \\ \gamma_{TL} \end{bmatrix} = \begin{bmatrix} \bar{c}^2 & \bar{s}^2 & 0 & \bar{c}\bar{s} \\ c^2\bar{s}^2 & c^2\bar{s}^2 & s^2 & c^2\bar{c}\bar{s} \\ s^2\bar{s}^2 & s^2\bar{c}^2 & c^2 & -s^2\bar{s}\bar{c} \\ -2c\bar{c}\bar{s} & 2s\bar{c}\bar{s} & 0 & c\bar{c}^2 - c\bar{s}^2 \\ 2s\bar{c}\bar{s} & -2s\bar{c}\bar{s} & 0 & s\bar{s}^2 - s\bar{c}^2 \\ -2cs\bar{s}^2 & -2ss\bar{c}^2 & 2sc & 2cs\bar{c}\bar{s} \end{bmatrix} \begin{bmatrix} \gamma_{rr} \\ \gamma_{zz} \\ \gamma_{\theta\theta} \\ \gamma_{rz} \end{bmatrix}, \tag{4.5.7}$$

in which the tensor shear strains have been converted to the total shear strain components.

4.5.3 Characterization of Material Constants for Orthotropic and Transversely Isotropic Composites

To determine explicit forms of elastic constants for an orthotropic material, we define the Poisson ratio as

$$\nu_{(ij)} = -\frac{\gamma_{(ij)}}{\gamma_{(ii)}}. \tag{4.5.8}$$

Here the index i represents the direction in which the stress is applied, and j denotes the direction transverse to i. The strain–stress relations are of the form

$$\gamma_{ij} = C_{ijkm}\sigma_{km}, \tag{4.5.9}$$

where the compliance C_{ijkm} is given by

$$C_{ijkm} = \begin{bmatrix} \frac{1}{E_1} & \frac{-\nu_{21}}{E_2} & \frac{-\nu_{31}}{E_3} & 0 & 0 & 0 \\ & \frac{1}{E_2} & \frac{-\nu_{32}}{E_3} & 0 & 0 & 0 \\ \text{symm.} & & \frac{1}{E_3} & 0 & 0 & 0 \\ & & & \frac{1}{\mu_{12}} & 0 & 0 \\ & & & & \frac{1}{\mu_{23}} & 0 \\ & & & & & \frac{1}{\mu_{31}} \end{bmatrix}. \tag{4.5.10}$$

Note that $E_{(i)}$ is Young's modulus in the direction $x_{(i)}$ and $\mu_{(ij)}$ is the shear modulus in the plane (ij). Symmetry of compliance requires that

$$\frac{\nu_{(ij)}}{E_{(i)}} = \frac{\nu_{(ji)}}{E_{(j)}}.$$

Furthermore, an inverse of compliance matrix equation (4.5.10) is equal to the elastic matrix:

$$E_{ijkm} = (C_{ijkm})^{-1}. \tag{4.5.11}$$

Thus we obtain

$$\begin{aligned}
E_{1111} &= \frac{1 - \nu_{23}\nu_{32}}{E_2 E_3 J}, \\
E_{1122} &= \frac{\nu_{21} + \nu_{31}\nu_{23}}{E_2 E_3 J} = \frac{\nu_{12} + \nu_{32}\nu_{13}}{E_1 E_3 J}, \\
E_{1133} &= \frac{\nu_{31} + \nu_{21}\nu_{32}}{E_2 E_3 J} = \frac{\nu_{13} + \nu_{12}\nu_{23}}{E_1 E_2 J}, \\
E_{2222} &= \frac{1 - \nu_{13}\nu_{31}}{E_1 E_3 J}, \\
E_{2233} &= \frac{\nu_{32} + \nu_{12}\nu_{31}}{E_1 E_3 J} = \frac{\nu_{23} + \nu_{21}\nu_{13}}{E_1 E_2 J}, \\
E_{3333} &= \frac{1 - \nu_{12}\nu_{21}}{E_1 E_2 J}, \\
E_{1212} &= \mu_{12}, \quad E_{2323} = \mu_{23}, \quad E_{3131} = \mu_{31},
\end{aligned} \tag{4.5.12}$$

where

$$J = \frac{1 - \nu_{12}\nu_{21} - \nu_{23}\nu_{32} - \nu_{31}\nu_{13} - 2\nu_{21}\nu_{32}\nu_{13}}{E_1 E_2 E_3}.$$

For transversely isotropic solids (see Fig. 4.1.2), the results are simplified by Eqs. (4.1.15). Let

$$E_1 = E_2 = E_T, \quad E_3 = E_L, \quad \mu_{12} = G_T, \quad \mu_{23} = \mu_{31} = G_{TL},$$
$$\nu_{21} = \nu_{12} = \nu_{IT}, \quad \nu_{31} = \nu_{13} = \nu_{32} = \nu_{23} = \nu_{TL};$$

then we have

$$\begin{aligned}
E_{1111} &= E_{2222} = E_T\left(E_L - E_T \nu_{TL}^2\right)k,\\
E_{1122} &= E_T\left(E_L \nu_{TT} + E_T \nu_{TL}^2\right)k,\\
E_{1133} &= E_{2233} = E_T E_L \nu_{TL}(1 + \nu_{TT})k,\\
E_{3333} &= E_L^2\left(1 - \nu_{TT}^2\right)k,\\
E_{1212} &= G_T,\\
E_{2323} &= E_{3131} = G_{TL},
\end{aligned} \tag{4.5.13}$$

where

$$k = \left[E_L\left(1 - \nu_{TT}^2\right) - 2E_T \nu_{TL}^2 (1 + \nu_{TT})\right]^{-1}.$$

For the case of a plane stress, components of the fourth-order tensor of elastic constants are simplified to

$$\begin{aligned}
E_{1111} &= \frac{E_1}{1 - \nu_{12}\nu_{21}},\\
E_{1122} &= \frac{\nu_{12}E_2}{1 - \nu_{12}\nu_{21}} = \frac{\nu_{21}E_1}{1 - \nu_{12}\nu_{21}},\\
E_{2222} &= \frac{E_2}{1 - \nu_{12}\nu_{21}},\\
E_{1212} &= G_{12}.
\end{aligned} \tag{4.5.14}$$

It was shown how coordinate transformations between the strains of local and global coordinates for a simple composite solid, axisymmetric cylinders, and general axisymmetric bodies can be performed. The transformed strain components are then substituted into the generalized Hooke's law for composites:

$$\sigma_{ij} = E_{ijkm}\gamma'_{km}. \tag{4.5.15}$$

Here γ'_{km} is computed from Eqs. (4.5.1), (4.5.3), and (4.5.7) for a simple composite solid, axisymmetric cylinders, and general axisymmetric bodies, respectively. Furthermore, the fourth-order tensor E_{ijkm} takes the forms given in Eqs. (4.5.12), (4.5.13), and (4.5.14) for orthotropic materials, general transversely isotropic fiber-reinforced solids, and the plane stress case of transversely isotropic materials, respectively.

Once the anisotropic materials of the type given by Eq. (4.5.15) are characterized, it is then a simple matter to substitute Eq. (4.5.15) into Cauchy's first law of motion or the Navier equations for solutions of partial differential equations.

PROBLEMS

4.1 Show all components of σ_{ij} (9×1 column), E_{ijkm} (9×9 matrix), and γ_{km} (9×1 column) of Eq. (4.1.5) in a standard matrix form. Expand all components of the shear stresses σ_{12} and σ_{21} in terms of the components associated with E_{ijkm} and γ_{km} and identify symmetric terms of E_{ijkm}. The results of these and similar exercises will be useful for Problem 4.2.

4.2 Derive the constitutive equation or stress–strain relation for a linear isotropic material, Eq. (4.1.25b). Start from the concept of an elastic potential, the generalized Hooke's law, and perform all coordinate transformations. Demonstrate all intermediate results for various anisotropic materials.

4.3 From laboratory test data for the Young's modulus E and the Poisson ratio ν, show that the Lamé constants λ and μ may be derived in the form

$$\lambda = \frac{\nu E}{(1+\nu)(1-2\nu)}, \tag{4.5.16}$$

$$\mu = \frac{E}{2(1+\nu)}. \tag{4.5.17}$$

4.4 From the stress tensor derived in Problem 4.2, write all components of the stress and strain tensors in matrix form in terms of Young's modulus and Poisson's ratio.

4.5 Show that the deviatoric stresses are related to deviatoric strains through only the shear modulus.

4.6 Derive the stress–strain relation for plane stress and plane strain.

4.7 If the plane strain data consist of $\gamma_{11} = 0.02$, $\gamma_{12} = \gamma_{21} = 0.01$, and $\gamma_{22} = 0.03$, what are the normal and shear stresses along the x_1' axis rotated $\theta = 30^\circ$ counterclockwise about the x_3 axis? The material constants are $E = 3 \times 10^6$ psi and $\nu = 0.3$.

4.8 Derive the Navier equations, (4.2.2), (4.2.4), and (4.2.5), and expand the results into three-dimensional, plane strain, and plane stress problems in terms of standard variables, x, y, z, u, v, and w.

4.9 Derive the variational principle (total potential energy) from Cauchy's first law of motion in three dimensions.

4.10 Repeat Problem 4.9 for one dimension and show that the strain energy per unit volume is the area of the triangle under the stress–strain curve.

4.11 Derive the equations of motion and heat conduction for small strains, using the first and second laws of thermodynamics. Expand the results into three-dimensional, plane strain, and plane stress problems in terms of standard variables, x, y, z, u, v, and w.

4.12 Prove Eqs. (4.4.36) and (4.4.37).

4.13 Derive all elements of the fourth-order tensor of material constants for transversely isotropic materials.

4.14 Verify the coordinate transformations required between the local and global coordinates for (a) simple composite solids and (b) general axisymmetric composite bodies.

5 Newtonian Fluid Mechanics

5.1 Introduction

In contrast to the study of solid mechanics, in which we were preoccupied with deformed geometries, our main concern in fluid mechanics is to view fluid particles in motion as continua and to determine the velocities at any given point in space (Navier, 1821; Lamb, 1879; Lamb, 1959; Milne-Thompson, 1960; Robertson, 1965). For this purpose we invoke the Eulerian coordinates whose properties were detailed in Subsection 2.2.2. Most fluids, whether gases or liquids, are called *Newtonian* fluids, in which the stress tensor is linearly proportional to the velocity gradients. On the other hand, certain chemical fluids, polymers, rheologically complex fluids, and suspensions, among others, are referred to as *non-Newtonian* fluids, in which the stress tensor is nonlinearly proportional to the velocity gradients. Non-Newtonian fluid mechanics is discussed in Section 7.2.

The subject of Newtonian fluids is generally referred to as fluid mechanics, which encompasses diverse topics, such as the motion of airplanes and missiles through the atmosphere, satellites through the outer atmosphere, submarines and ships through water, the flow of liquids and gases through ducts, the transfer of heat and mass by fluid motion, propagation of sound through gases and liquids, the study of ocean waves and tides, the study of air masses in the atmosphere, astrophysical, geophysical, and meteorological problems, and reacting fluids.

Fluids, like all matter, are made up of molecules. Thus the properties of fluid motion, such as those observed, may be studied on the basis of the mechanics of the molecules that compose the fluid. Although such a procedure may be feasible in principle, it will be a formidable task to achieve solutions of practical problems. As mentioned in Chap. 1, we are generally not interested in the details of the mechanics of the molecules. We wish to establish relations between various macroscopically observable quantities that pertain to a fluid at rest or in motion. Such observable properties are mean values in space and time that we obtain by taking the average over a sufficiently large volume that contains a large number of molecules and over a sufficiently long time compared with a certain time related to the mechanics of the molecules. For instance, at normal temperature and pressure, a 10^{-2} cc volume of air will contain about 2.7×10^{7} molecules, which is a large number. This being so, it is a reasonable approximation to regard a fluid, whether at rest or in motion, as a continuous distribution of matter. We then speak of the fluid as a continuous medium or as a continuum.

The properties of density, pressure, and temperature at a point in a liquid or a gas in static equilibrium are well known. At any point, it is the magnitude of the normal stress acting on an elemental plane area passing through that point. In a fluid at rest, only normal stresses occur and, in general, they are "compressive" in nature. When only normal stresses occur, it is easy to show that, at any point, they should be equal in all directions.

The fact that only normal stresses occur for a fluid in static equilibrium contrasts with what happens to a solid in static equilibrium in which both normal and shear stresses prevail. The absence of shear stresses in a state of static equilibrium distinguishes a fluid from a solid and may be considered the property that defines a fluid.

All fluids undergo changes in volume under changes of pressure and temperature. The ability to change the volume of a mass of fluid is known as *compressibility*. It is well known that gases are more easily compressed than liquids. When a fixed mass of a fluid undergoes changes in volume, its density also changes, a phenomenon also known as compressibility. Under normal circumstances, the density changes in liquids that are due to pressure changes are small, whereas density changes that all due to temperature differences can be substantial. If the temperature differences are sufficiently small, the density changes in a liquid are almost nil and the liquid may then be regarded as an incompressible fluid. An incompressible fluid is one whose elements undergo no changes in volume or density. Compressibility is easy to notice in gases. In certain circumstances, however, the changes in volume or density of an element of a gas may be negligibly small. In such a case, as a reasonable approximation, the gas may be regarded as an incompressible fluid.

It is a matter of experience that even smoothly shaped bodies moving with a constant velocity through an otherwise undisturbed fluid encounter a resistance to their motion. Similarly, a fluid flowing through a pipe offers resistance. Thus, for a fluid in motion, shear and normal stresses occur on any elemental plane passing through a point in the fluid. The phenomenon of internal friction or *viscosity* is associated with these shear stresses. Such stresses give rise to a resistance to the nonuniform motion of a fluid, which results in *velocity gradients*. Shear stresses are, in general, linearly proportional to the velocity gradients through *viscosity constants*. A fluid of this type is Newtonian, as defined earlier.

When a fluid in static equilibrium is heated nonuniformly, heat may be transferred (without causing motion of the fluid) from points at which the temperature is high to those at which it is low by what is called *thermal conduction*. Observations show that under usual circumstances the *heat flux*, which is the amount of heat transferred across the surface element per unit time per unit area in the direction of the normal to the element, is proportional to the spatial rate of change of the temperature at that point in the direction of the normal. The heat flow occurs in the direction of decreasing temperature. Also, the static equilibrium of a fluid in which the temperature is not constant is unstable unless certain conditions are fulfilled. Instability of this kind leads to the appearance of convective currents in the fluid, which tend to mix the fluid in such a way that the temperature is equalized, a phenomenon known as *thermal convection*. Heat transfer also occurs in the form of *radiation* through participating or nonparticipating media. A process during which no heat is transferred to or from a system is known as an *adiabatic process*.

At this stage, it is convenient to introduce the concept of a perfect gas, which is the simplest working fluid in thermodynamics and hence is useful in the detailed study of thermodynamic processes. Measurements of the thermal properties of gases show that for low densities the thermal equation of state approaches the same form for all gases, namely

$$pV = RT \tag{5.1.1}$$

or

$$p = \rho RT, \tag{5.1.2}$$

where V is the specific volume $(1/\rho)$ and R is the specific gas constant. Equation (5.1.1) or (5.1.2) represents the *equation of state* for a perfect (ideal) gas, known as a *calorically perfect* gas in which specific heats are constant. In practice, most fluids have been observed to follow the perfect gas law given by Eq. (5.1.1) or (5.1.2). In general, pressure is a function of both density and temperature. If pressure is a function of only density, independent of temperature, then the fluid is *baratropic*.

In what follows, various subjects in fluid mechanics and heat transfer are presented. We begin in Section 5.2 with the theory of constitutive equations for the stress tensor. Some historical developments are briefly reviewed. Section 5.3 gives the derivations of the most general forms of the governing equations in fluid mechanics by use of the first and second laws of thermodynamics. On the basis of these general forms, explicit equations for compressible viscous flow are derived for Cartesian coordinates in Section 5.4. We discuss an ideal flow in terms of velocity potential and stream functions and the Bernoulli equation in Section 5.5.

Applied subjects in fluid mechanics include rotational flows and vorticity transport equations, turbulence, boundary-layer, convective heat transfer, high-speed aerodynamics, acoustics, and reacting flows. These topics are presented in Sections 5.6–5.12.

5.2 Constitutive Equations

The basic principles for the development of constitutive theories for fluids are the same as those for solids. We proceed in accordance with the rules of physical admissibility, determinism, equipresence, local action, material objectivity, and material symmetry, as discussed in Chap. 4. First we review historical developments and examine transformation laws of stress tensors and the relations between the stresses and velocity gradients, thus arriving at the results analogous to those obtained for solids and providing the basis for the most rigorous governing equations in compressible viscous flows.

5.2.1 Historical Review

To begin, a brief historical review of fluid mechanics is in order. Archimedes studied fluid behavior as early as 250 B.C.E., but it was not until the 18th century that Daniel Bernoulli described a simple constitutive law of inviscid flow,

$$\sigma_{ij} = -p\delta_{ij}. \tag{5.2.1}$$

However, this law was inadequate for some fluid flows, as noted by Newton, Cauchy, Navier, Saint-Venant, Stokes, and others. In 1821, for example, Navier proposed that additional terms proportional to velocity gradients ($V_{i,j}$) must be added to Eq. (5.2.1). In 1831 Poisson assumed such terms to be of the form

$$\lambda d_{kk}\delta_{ij} + 2\mu d_{ij}, \tag{5.2.2}$$

which is identical in form to the stress tensor derived for an isotropic solid [Eqs. (4.1.25)], with d_{ij} being the rate-of-deformation tensor [Eq. (2.4.4)]:

$$d_{ij} = \frac{1}{2}(V_{i,j} + V_{j,i}) \tag{5.2.3}$$

Recall that, in Chap. 2, we distinguished $V_{i,j} = \partial V_i/\partial z_j$ described in the Eulerain coordinates from $u_{i,j} = \partial u_i/\partial x_j$ in the Lagrangian coordinates. This restriction is no longer necessary as our discussion is limited to fluid mechanics throughout this chapter. Thus, following the tradition, we adopt the notation $V_{i,j} = \partial V_i/\partial x_j$, with the spatial coordinates traditionally designated by x_i instead of z_i.

The idea in Eq. (5.2.2) was suggested by Newton in 1687. Subsequently, in 1845, Stokes proposed that, in general, the following relation should hold:

$$3\lambda + 2\mu = 0, \tag{5.2.4}$$

as we shall derive in Eq. (5.2.27). However, this assertion was disputed by others during the early part of the 20th century on the basis of experimental data on some special types of fluids that are dependent on higher orders of velocity gradients [see Lamb (1879) for the classical fluid mechanics of the 17th–19th centuries]. Reiner (1945) studied a general theory of fluids using the principle of objectivity in which the fluid flow is assumed to be dependent on an isotropic function of the rate-of-deformation tensor d_{ij}. Reiner's theory was later criticized by Rivlin (1948), and, subsequently, Rivlin and Eriksen (1955) offered a more general theory in which the time rate of change of d_{ij} of higher orders is added to Reiner's theory. This work was later modified by Green and Rivlin (1957), who included a dependence on the total history of deformation. Coleman and Noll (1959) and Noll (1965) proposed a theory of simple fluids in which they considered the stress tensors to be dependent on only the history of deformation. A rather complete and systematic development of this theory is given by Truesdell and Noll (1965). These topics belong to non-Newtonian fluid mechanics, discussed in Section 7.2.

5.2.2 The Stress Tensor

The stress tensor for fluids incapable of sustaining shear stresses is given by Eq. (5.2.1). When thermodynamic considerations are taken into account, we have

$$\sigma_{ij} = -\pi(\rho)\delta_{ij}, \tag{5.2.5}$$

where $\pi(\rho)$ is the thermodynamic pressure. To generalize Eq. (5.2.5) and to allow shears, Stokes, in 1845, proposed that

$$\sigma_{ij} = -\pi\delta_{ij} + f_{ij}(\mathbf{d}), \tag{5.2.6}$$

where $f_{ij}(\mathbf{d})$ indicates a function of the rate-of-deformation tensor d_{ij}, with $\mathbf{d}$ representing the direct tensor notation for d_{ij}. If f_{ij} is linear, as in Eq. (5.2.3), then the fluid is said to be Newtonian.

One of the most widely used constitutive equations for viscous fluids is known as the Reiner–Rivlin equation,

$$\sigma_{ij} = -\pi\delta_{ij} + f_{ij}, \tag{5.2.7}$$

with

$$f_{ij} = \alpha_0\delta_{ij} + \alpha_1 d_{ij} + \alpha_2 d_{ik}d_{kj} \tag{5.2.8}$$

where α_0, α_1, and α_2 are functions of the invariants of d_{ij}:

$$I_1 = d_{ii} = d(1) + d(2) + d(3), \tag{5.2.9a}$$

$$\begin{aligned} I_2 &= \frac{1}{2}(d_{ii}d_{jj} - d_{ij}d_{ij}) \\ &= d(1)d(2) + d(2)d(3) + d(3)d(1), \end{aligned} \tag{5.2.9b}$$

$$I_3 = |d_{ij}| = d(1)d(2)d(3), \tag{5.2.9c}$$

which are in a form similar to the invariants of the strain tensor γ_{ij} discussed in Section 2.7.

The function $f_{ij}(\mathbf{d})$ in Eq. (5.2.6) may be written in a general form as

$$f_{ij} = f_{ij}(d_{ij}, d_{ik}d_{kj}, d_{ik}d_{km}d_{mj}, \ldots,). \tag{5.2.10}$$

It is possible for f_{ij} to be dependent on the past motion, as in certain polymer solutions. Fluids for which this is true are able to "remember" their preceding configurations. On the other hand, if we neglect all higher orders of d_{ij}, then we get

$$f_{ij} = f_{ij}(d_{ij}), \tag{5.2.11a}$$

$$f_{ij} = f_{ij}\left(\frac{\partial V_i}{\partial x_j}\right). \tag{5.2.11b}$$

A fluid that obeys this relation, known as a Newtonian fluid, is unable to remember its preceding motion, although it may be able to remember its immediately preceding state of deformation history. With this in mind, we write $\partial V_i/\partial x_j$ as a sum of symmetric and antisymmetric parts [similarly as in Eq. (4.3.3a)]:

$$\frac{\partial V_i}{\partial x_j} = \frac{1}{2}\left(\frac{\partial V_i}{\partial x_j} + \frac{\partial V_j}{\partial x_i}\right) + \frac{1}{2}\left(\frac{\partial V_i}{\partial x_j} - \frac{\partial V_j}{\partial x_i}\right)$$

or

$$\frac{\partial V_i}{\partial x_j} = d_{ij} + w_{ij}, \tag{5.2.12}$$

where w_{ij}, known as the rotational (spin) tensor (per unit time), is given by

$$w_{ij} = \frac{1}{2}\left(\frac{\partial V_i}{\partial x_j} - \frac{\partial V_j}{\partial x_i}\right) \tag{5.2.13}$$

and is antisymmetric, representing an angular motion. Thus Eq. (5.2.11) may be revised for a new coordinate configuration as

$$\bar{f}_{ij} = f_{ij}(d_{ij}, w_{ij}). \tag{5.2.14}$$

By virtue of the principle of material objectivity (frame indifference), as discussed in Subsection 4.4.4, the orthogonal transformation is, from Eq. (2.5.3),

$$\bar{f}_{ij} = a_{ir}a_{js}f_{rs}, \tag{5.2.15}$$

and similarly,

$$\bar{d}_{ij} = a_{ir}a_{js}d_{rs}. \tag{5.2.16}$$

However, from

$$a_{ir}a_{jr} = \delta_{ij} \tag{5.2.17}$$

and its time derivative

$$\frac{d}{dt}(a_{ir}a_{jr}) = \dot{a}_{ir}a_{jr} + a_{ir}\dot{a}_{jr} = 0,$$

we have

$$\begin{aligned}\bar{w}_{ij} &= \frac{1}{2}(\dot{a}_{ir}a_{jr} - a_{ir}\dot{a}_{jr}) + a_{ir}a_{js}w_{rs} \\ &= \dot{a}_{ir}a_{jr} + a_{ir}a_{js}w_{rs} \\ &= a_{ir}a_{js}(w_{rs} + \dot{a}_{rm}a_{sm}).\end{aligned} \tag{5.2.18}$$

This indicates that w_{rs} is not frame indifferent because orthogonal transformation is not available, as seen by the antisymmetry of $\dot{a}_{rm}a_{sm}$. In view of Eqs. (5.2.14) and (5.2.18), we have

$$a_{ir}a_{js}f_{rs} = f_{ij}[a_{ir}a_{js}d_{rs}, a_{ir}a_{js}(w_{rs} + \dot{a}_{rm}a_{sm})]. \tag{5.2.19}$$

Let us now consider the case of $a_{ir} = \delta_{ir}, \dot{a}_{ij} \neq 0$ (nonzero angular velocity). Then Eq. (5.2.19) becomes

$$f_{ij} = f_{ij}(d_{ij}, w_{ij} + \dot{a}_{ij}). \tag{5.2.20}$$

Consequently, this requires that angular motion vanish if the frame indifference is to be recovered and if the stress is to be independent of rigid-body motion, as dictated by Eq. (5.2.11a) or by the relation

$$a_{ir}a_{js}f_{rs} = f_{ij}(a_{ir}a_{js}d_{rs}). \tag{5.2.21}$$

Equation (5.2.21) requires that the fluid be isotropic, as deduced from Eqs. (5.2.15) and (5.2.16).

The explicit form of f_{ij} is thus written in terms of an isotropic tensor, E_{ijkm}, derived in Section 4.1,

$$f_{ij} = E_{ijkm}d_{km} = [\lambda\delta_{ij}\delta_{km} + \mu(\delta_{ik}\delta_{jm} + \delta_{im}\delta_{jk})]\, d_{km}$$

or

$$f_{ij} = \lambda d_{kk}\delta_{ij} + 2\mu d_{ij}. \tag{5.2.22}$$

Here, f_{ij} may be referred to as *excess stress tensor* or shear stress tensor, denoted as τ_{ij}, which is contributed by velocity gradients, leading to the postulate suggested by Poisson (1831) given in Eq. (5.2.2):

$$\tau_{ij} = \lambda d_{kk}\delta_{ij} + 2\mu d_{ij}. \tag{5.2.23}$$

Note that this remarkable expression for the "isotropic" stress tensor was derived from the most general concept of invariant energy (elastic potential) for originally "anisotropic" solids in Section 4.1 in terms of the strain tensor rather than of the deformation-rate tensor [see Eqs. (4.1.25a) and (4.1.25b)]. We could have obtained the expression given by Eq. (5.2.23) by using a similar approach and taking the substantial time derivative for the strain tensor. Such a derivation is seldom mentioned in the fluid mechanics literature. Instead, Eq. (5.2.23) is accepted merely as a definition.

In fluid mechanics, λ and μ are called the dilatational viscosity constant and the shear viscosity constant, respectively, which we referred to as Lamé constants in elastic solids in Chap. 4. To establish the relation between these two constants, we invoke Stokes hypothesis. Toward this end, we set $i = j$ in Eqs. (5.2.7) and (5.2.23):

$$\sigma_{ii} = -3\pi(\rho) + 3\lambda d_{ii} + 2\mu d_{ii}. \tag{5.2.24}$$

We denote the mean pressure as [(see Eq. 4.1.30)]

$$-p = \frac{1}{3}\sigma_{ii}, \tag{5.2.25}$$

with the negative sign indicating the mean hydrodynamic pressure in compression. Now, combining Eqs. (5.2.24) and (5.2.25), we obtain

$$\pi(\rho) - p = \left(\lambda + \frac{2}{3}\mu\right) d_{ii}. \tag{5.2.26}$$

According to Stokes, writing in 1845, the thermodynamic pressure $\pi(\rho)$ is approximately equal to the mean pressure p, which is true for most types of liquids and gases. For all arbitrary values of d_{ii}, this leads to

$$\lambda + \frac{2}{3}\mu = 0, \tag{5.2.27}$$

which agrees with Eq. (5.2.4). Thus the dilatational viscosity constant can be eliminated in Eq. (5.2.23) by use of the relation

$$\lambda = -\frac{2}{3}\mu. \tag{5.2.28}$$

where μ is known as the dynamic viscosity.

Substituting Eq. (5.2.28) and the relation $\pi(\rho) = p$ into Eq. (5.2.23) gives the total stress tensor,

$$\sigma_{ij} = -p\delta_{ij} + \tau_{ij}, \tag{5.2.29}$$

where τ_{ij}, the shear stress tensor, takes the form

$$\tau_{ij} = 2\mu\left(d_{ij} - \frac{1}{3}d_{kk}\delta_{ij}\right). \tag{5.2.30}$$

Thus

$$\sigma_{ij} = -p\delta_{ij} + 2\mu\hat{d}_{ij}, \tag{5.2.31}$$

where

$$\hat{d}_{ij} = d_{ij} - \frac{1}{3}d_{kk}\delta_{ij}, \tag{5.2.32}$$

which is the deviatoric part of the rate-of-deformation tensor, already mentioned in Eq. (2.6.4), similar to the deviatoric strain tensor of Eq. (2.6.3). This implies that τ_{ij} is due only to a deviatoric part of the rate-of-deformation tensor. Thus the shear stress in fluid mechanics may be called the deviatoric part of the total stress. Fluid mechanics governed by the stress tensor given by Eq. (5.2.31) is known as Newtonian fluid mechanics.

In certain fluids dilatational aspects may be taken into account. To this end, let us reexamine the general form of f_{ij} in Eq. (5.2.10) or Eq. (5.2.8) to consider three scalar invariants of $\mathbf{d}$ in the form

$$f[d(1), d(2), d(3)] = \alpha_0 + \alpha_1 d(1) + \alpha_2 d^2(1), \tag{5.2.33a}$$

$$f[d(2), d(3), d(1)] = \alpha_0 + \alpha_1 d(2) + \alpha_2 d^2(2), \tag{5.2.33b}$$

$$f[d(3), d(1), d(2)] = \alpha_0 + \alpha_1 d(3) + \alpha_2 d^2(3), \tag{5.2.33c}$$

where α_0, α_1, and α_2 are functions of three scalar invariants of $\mathbf{d}$ and are symmetric functions of $d(1)$, $d(2)$, and $d(3)$. Thus we may choose

$$\alpha_0 = \alpha_0(I_1, I_2, I_3), \tag{5.2.34a}$$

$$\alpha_1 = \alpha_1(I_1, I_2, I_3), \tag{5.2.34b}$$

$$\alpha_2 = \alpha_2(I_1, I_2, I_3), \tag{5.2.34c}$$

where the invariants I_1, I_2, and I_3 are given by Eq. (5.2.9). Therefore, Eqs. (5.2.33) may be written as

$$\begin{aligned} \text{diag}\,[\mathbf{f}(1), \mathbf{f}(2), \mathbf{f}(3)] = \alpha_0 \mathbf{I} + \alpha_1\, \text{diag}\,[d(1), d(2), d(3)] \\ + \alpha_2 \text{diag}[d^2(1), d^2(2), d^2(3)] \end{aligned} \tag{5.2.35}$$

or

$$\mathbf{f} = \alpha_0 \mathbf{I} + \alpha_1 \mathbf{d} + \alpha_2 \mathbf{d}^2, \tag{5.2.36}$$

which is the direct tensor representation of Eq. (5.2.8). In general, for compressible flows, the stress tensor is a function of density ρ and temperature T in addition to the rate-of-deformation tensor:

$$\mathbf{f} = \hat{\mathbf{f}}(\mathbf{d}, \rho, T). \tag{5.2.37}$$

On the other hand, for incompressible flows, the stress is independent of density:

$$\mathbf{f} = \hat{\mathbf{f}}(\mathbf{d}, T) \tag{5.2.38}$$

Furthermore, α_0, α_1, and α_2 are independent of I_1 and the density for incompressible flows. Note that the absence of ρ and T in Eq. (5.2.11) should not have affected the derivation of Eq. (5.2.37), because ρ and T are frame independent.

The classical Newtonian fluid results from Eq. (5.2.36) by setting $\alpha_2 = 0$ and

$$\alpha_0 = \hat{\alpha}_0 I_1 = \hat{\alpha}_0 d_{ii},$$

$$\alpha_1 = 2\mu.$$

These modifications lead to

$$f_{ij} = \hat{\alpha}_0 d_{kk} \delta_{ij} + 2\mu d_{ij} \tag{5.2.39}$$

or

$$f_{ij} = \hat{\alpha} d_{kk}\delta_{ij} + 2\mu\left(d_{ij} - \frac{1}{3}d_{kk}\delta_{ij}\right), \tag{5.2.40}$$

where

$$\hat{\alpha} = \hat{\alpha}_0 + \frac{2}{3}\mu.$$

This is known as the *coefficient of bulk viscosity*. In most of the fluids considered in engineering, however, the bulk viscosity is negligible. Thus it follows that the stress tensor contributed by shears assumes the form

$$f_{ij} = \tau_{ij} = 2\mu\left(d_{ij} - \frac{1}{3}d_{kk}\delta_{ij}\right). \tag{5.2.41}$$

This corresponds to the shear stress for the Newtonian fluids given by Eq. (5.2.30).

It is seen that the stress tensor in Newtonian fluids consists of hydrodynamic pressure and excess stress due to velocity gradients or the deviatoric part of the rate-of-deformation tensor associated with the shear viscosity constant μ $(\frac{\text{N-s}}{\text{m}^2})$. This constant is known as the dynamic viscosity. Sometimes the kinematic viscosity, defined as $\nu = \mu/\rho(\frac{\text{m}^2}{\text{s}})$, is used for convenience in fluid mechanics and heat transfer problems.

5.3 The Navier–Stokes System of Equations

5.3.1 The First Law of Thermodynamics

It was shown in Section 3.3 that the conservation of linear momentum leads to the equations of motion or Navier equations for solids. It was also shown in Section 4.4 that the first law of thermodynamics yields the conservation of mass, momentum, and energy for solids. Exactly the same approaches may be pursued for fluid mechanics, except that here we must use Eulerian coordinates, as indicated in Subsection 2.2.2.

Just as in Eq. (4.4.4) for the Lagrangian coordinate system, the first law of thermodynamics for the Eulerian coordinate system is written with the exception that substantial derivatives now replace the ordinary time derivatives:

$$\frac{DK}{Dt} + \frac{DU}{Dt} = M + Q. \tag{5.3.1}$$

We define the following terms.

Kinetic energy

$$K = \frac{1}{2}\int_\Omega \rho \mathbf{V}\cdot\mathbf{V}\,d\Omega = \frac{1}{2}\int_\Omega \rho V_i V_i\,d\Omega. \tag{5.3.2}$$

Internal energy

$$U = \int_\Omega \rho \epsilon \, d\Omega, \tag{5.3.3}$$

where ϵ is the internal energy density. For an ideal gas, we have

$$\epsilon = c_v T \tag{5.3.4a}$$

or

$$\epsilon = H - \frac{p}{\rho}, \tag{5.3.4b}$$

where $H = c_p T$ is the enthalpy, and c_v and c_p are the specific heats at constant volume and constant pressure, respectively.

Mechanical power

$$M = \int_\Omega \rho \mathbf{F} \cdot \mathbf{V} \, d\Omega + \int_\Gamma \boldsymbol{\sigma}(n) \cdot \mathbf{V} \, d\Gamma = \int_\Omega \rho F_i V_i \, d\Omega + \int_\Gamma \sigma_{ij} V_j n_i \, d\Gamma, \tag{5.3.5}$$

where $\mathbf{F}$ is the body force and $\boldsymbol{\sigma}(n)$ is the surface traction normal to the boundary surface given by

$$\boldsymbol{\sigma}(n) = \sigma_{ij} n_i \mathbf{i}_j. \tag{5.3.6}$$

Heat energy

$$Q = \int_\Omega \rho r \, d\Omega \pm \int_\Gamma \mathbf{q} \cdot \mathbf{n} \, d\Gamma = \int_\Omega \rho r \, d\Omega \pm \int_\Gamma q_i n_i \, d\Gamma, \tag{5.3.7}$$

where r is the heat supply and $\mathbf{q}$ is the heat flux, as defined in Subsection 4.4.2. The sign convention for heat flux is the same as in the case of solids.

The most significant part of the Eulerian formulation is the substantial derivative of $\rho d\Omega$, as shown in Eq. (2.2.29):

$$\int_\Omega \frac{D}{Dt}(\rho \, d\Omega) = \int_\Omega \left[\frac{\partial \rho}{\partial t} + \nabla \cdot (\rho \mathbf{V}) \right] d\Omega. \tag{5.3.8}$$

Thus the substantial derivative of the kinetic energy takes the form

$$\begin{aligned} \frac{D}{Dt} \int_\Omega \frac{1}{2} \rho \mathbf{V} \cdot \mathbf{V} \, d\Omega &= \int_\Omega \frac{1}{2} \mathbf{V} \cdot \mathbf{V} \frac{D}{Dt}(\rho \, d\Omega) + \int_\Omega \rho \frac{D}{Dt} \left(\frac{1}{2} \mathbf{V} \cdot \mathbf{V} \right) d\Omega \\ &= \int_\Omega \frac{1}{2} V_i V_i \left[\frac{\partial \rho}{\partial t} + (\rho V_i)_{,i} \right] d\Omega + \int_\Omega \left(\rho \frac{\partial V_j}{\partial t} + \rho V_{j,i} V_i \right) V_j \, d\Omega. \end{aligned} \tag{5.3.9}$$

Similarly, the substantial derivative of the internal energy is

$$\begin{aligned} \frac{D}{Dt} \int_\Omega \rho \epsilon d\Omega &= \int_\Omega \epsilon \frac{D}{Dt}(\rho \, d\Omega) + \int_\Omega \rho \frac{D\epsilon}{Dt} \, d\Omega \\ &= \int_\Omega \epsilon \left[\frac{\partial \rho}{\partial t} + (\rho V_i)_{,i} \right] d\Omega + \int_\Omega \rho \left(\frac{\partial \epsilon}{\partial t} + \epsilon_{,i} V_i \right) d\Omega. \end{aligned} \tag{5.3.10}$$

The surface integral becomes

$$\int_{\Gamma} \sigma_{ij} V_j n_i \, d\Gamma = \int_{\Omega} (\sigma_{ij} V_j)_{,i} \, d\Omega = \int_{\Omega} (\sigma_{ij,i} V_j + \sigma_{ij} V_{j,i}) \, d\Omega. \qquad (5.3.11)$$

Substitute Eqs. (5.3.2)–(5.3.11) into Eq. (5.3.1) and adjust the indices to obtain

$$\int_{\Omega} E \left[\frac{\partial \rho}{\partial t} + (\rho V_i)_{,i} \right] d\Omega + \int_{\Omega} V_j \left(\rho \frac{\partial V_j}{\partial t} + \rho V_{j,i} V_i - \sigma_{ij,i} - \rho F_j \right) d\Omega$$
$$+ \int_{\Omega} \left(\rho \frac{\partial \epsilon}{\partial t} + \rho \epsilon_{,i} V_i - \sigma_{ij} V_{j,i} - q_{i,i} - \rho r \right) d\Omega = 0, \qquad (5.3.12)$$

where E is the total (stagnation) energy,

$$E = \epsilon + \frac{1}{2} V_i V_i.$$

Note that Eq. (5.3.12) is the consequence of the first law of thermodynamics, referred to as the FLT (first law of thermodynamics) equation. At this point, we observe that, for Eq. (5.3.12) to vanish, all integrands must vanish independently such that, for all arbitrary values of E, V_j, and $d\Omega$, this requirement leads to

$$\frac{\partial \rho}{\partial t} + (\rho V_i)_{,i} = 0, \qquad (5.3.13)$$

$$\rho \frac{\partial V_j}{\partial t} + \rho V_{j,i} V_i - \sigma_{ij,i} - \rho F_j = 0, \qquad (5.3.14)$$

$$\rho \frac{\partial \epsilon}{\partial t} + \rho \epsilon_{,i} V_i - \sigma_{ij} V_{j,i} - q_{i,i} - \rho r = 0. \qquad (5.3.15)$$

These three equations represent the equations of continuity, momentum, and energy, respectively. Equations (5.3.13)–(5.3.15) are known as the nonconservation form of the Navier–Stokes system of equations. The momentum equations alone are often called the Navier–Stokes equations, corresponding to the Navier equations for solid mechanics. Note that Eq. (5.3.13) was derived as the expression for the conservation of mass in Eq. (2.2.28). It is seen in Eq. (5.3.12) that the conservation of energy depends on the conservation of mass and momentum. Actually, as will be demonstrated in Subsection 5.3.2, the conservation of mass is the prerequisite to the conservation of momentum, and the conservation of both mass and momentum is the prerequisite to the conservation of energy.

The heat flux q_i consists of conductive and radiative parts,

$$q_i = q_i^{(c)} + q_i^{(r)}, \qquad (5.3.16)$$

in which the conductive heat flux $q_i^{(c)}$ may be given by the Fourier heat conduction law, as in Eq. (4.4.35), and the radiative heat flux $q_i^{(r)}$ is a complicated integral expression given in terms of various radiative parameters,

$$q_i^{(r)} = \int_0^\infty q_{\lambda i}^{(r)} \, d\lambda = \int_0^\infty \int_{4\pi} I_\lambda n_i \, d\omega d\lambda \tag{5.3.17}$$

where λ is the wavelength, I_λ is the spectral radiation intensity, n_i are the components of the vector normal to the surface, and ω is the *solid angle*. For further details on radiative heat transfer, see Sparrow and Cess (1966) and Chung and Kim (1984).

Substituting Eq. (5.3.4b) into Eq. (5.3.15) yields

$$\rho c_p \frac{\partial T}{\partial t} + \rho c_p T_{,i} V_i - \frac{\partial p}{\partial t} - p_{,i} V_i - V_{i,i} p - \sigma_{ij} V_{j,i} - q_{i,i} - \rho r = 0. \tag{5.3.18}$$

Note that in Eq. (5.3.18) the continuity equation, (5.3.13), is utilized to arrive at

$$\frac{1}{\rho}\left(\frac{\partial \rho}{\partial t} + \rho_{,i} V_i\right) = -V_{i,i}.$$

We have now obtained three differential equations (continuity, momentum, and energy), but there are four dependent variables (velocity, pressure, density, and temperature) to be determined. To this end, we require an additional equation, called the equation of state:

$$f(\rho, p, T) = 0. \tag{5.3.19}$$

For an ideal (perfect) gas, the equation of state assumes the form given in Eq. (5.1.2):

$$p = \rho R T. \tag{5.3.20}$$

Equations (5.3.13)–(5.3.20) represent the most general Cartesian form of governing equations in fluid mechanics. The specific forms of the stress tensor σ_{ij} and the equation of state will distinguish one type of *fluid* from another (i.e., viscous or inviscid, compressible or incompressible, ideal or real gases). Furthermore, we must distinguish one type of *flow* from another, based on flow speeds (subsonic, transonic, supersonic, hypersonic), flow patterns (laminar or turbulent, rotational or irrotational, with or without boundary layers, oscillatory or nonoscillatory), and the influence of body forces (natural or forced convection in heat transfer). Thus the subject of fluid mechanics is divided into various specialized areas to deal with the various types of "fluids" and "flows," which require each specialty to employ its own version of the governing equations.

The fluid mechanics equations given in Eqs. (5.3.13)–(5.3.15) are the consequence of the first law of thermodynamics. However, the most significant factor involved in this process is the substantial derivative of Jacobian [Eq. (2.2.30)] containing the metric tensor G_{ij} leading to the conservation of mass. The validity of entire fluid mechanics equations is solely dependent on the geometrical deformation process depicted in Fig. 2.2.2.

5.3.2 The Conservation Form of the Navier–Stokes System of Equations and Control-Volume–Control-Surface Equations

The Navier–Stokes system of equations derived from the first law of thermodynamics in Subsection 5.3.1 may be recast in a *conservation form*. For simplicity, let us consider a two-dimensional case as follows.

Continuity

$$\frac{\partial \rho}{\partial t} + \frac{\partial}{\partial x}(\rho u) + \frac{\partial}{\partial y}(\rho v) = 0. \tag{5.3.21}$$

Momentum

$$\frac{\partial}{\partial t}(\rho u) + \frac{\partial}{\partial x}(\rho u^2) + \frac{\partial}{\partial y}(\rho v u) = -\frac{\partial p}{\partial x} + \frac{\partial \tau_{xx}}{\partial x} + \frac{\partial \tau_{yx}}{\partial y} + \rho F_x, \tag{5.3.22a}$$

$$\frac{\partial}{\partial t}(\rho v) + \frac{\partial}{\partial x}(\rho u v) + \frac{\partial}{\partial y}(\rho v^2) = -\frac{\partial p}{\partial x} + \frac{\partial \tau_{xy}}{\partial x} + \frac{\partial \tau_{yy}}{\partial y} + \rho F_y. \tag{5.3.22b}$$

Energy

$$\frac{\partial}{\partial t}(\rho E) + \frac{\partial}{\partial x}(\rho E u) + \frac{\partial}{\partial y}(\rho E v) = -\frac{\partial}{\partial x}(pu) - \frac{\partial}{\partial y}(pv) + \frac{\partial}{\partial x}(\tau_{xx} u + \tau_{xy} v)$$
$$+ \frac{\partial}{\partial y}(\tau_{yx} u + \tau_{yy} v) + \frac{\partial q_x}{\partial x} + \frac{\partial q_y}{\partial y} + \rho r + \rho F_x u + \rho F_y v. \tag{5.3.23}$$

We obtain these results as a consequence of the conservation of mass, momentum, and energy in the control volumes and on the control surfaces, as shown in Fig. 5.3.1, by summing all components that enter and exit through the inflow and outflow boundaries.

It is now a simple matter to combine Eqs. (5.3.21)–(5.3.23) into a compact form in a three-dimensional geometry:

$$\frac{\partial \mathbf{U}}{\partial t} + \frac{\partial \mathbf{F}_i}{\partial x_i} + \frac{\partial \mathbf{G}_i}{\partial x_i} = \mathbf{B}, \tag{5.3.24}$$

which is referred to as the CNS equation (the conservation form of the Navier–Stokes system of equations). Here, $\mathbf{U}$, $\mathbf{F}_i$, $\mathbf{G}_i$, and $\mathbf{B}$ are the conservation flow

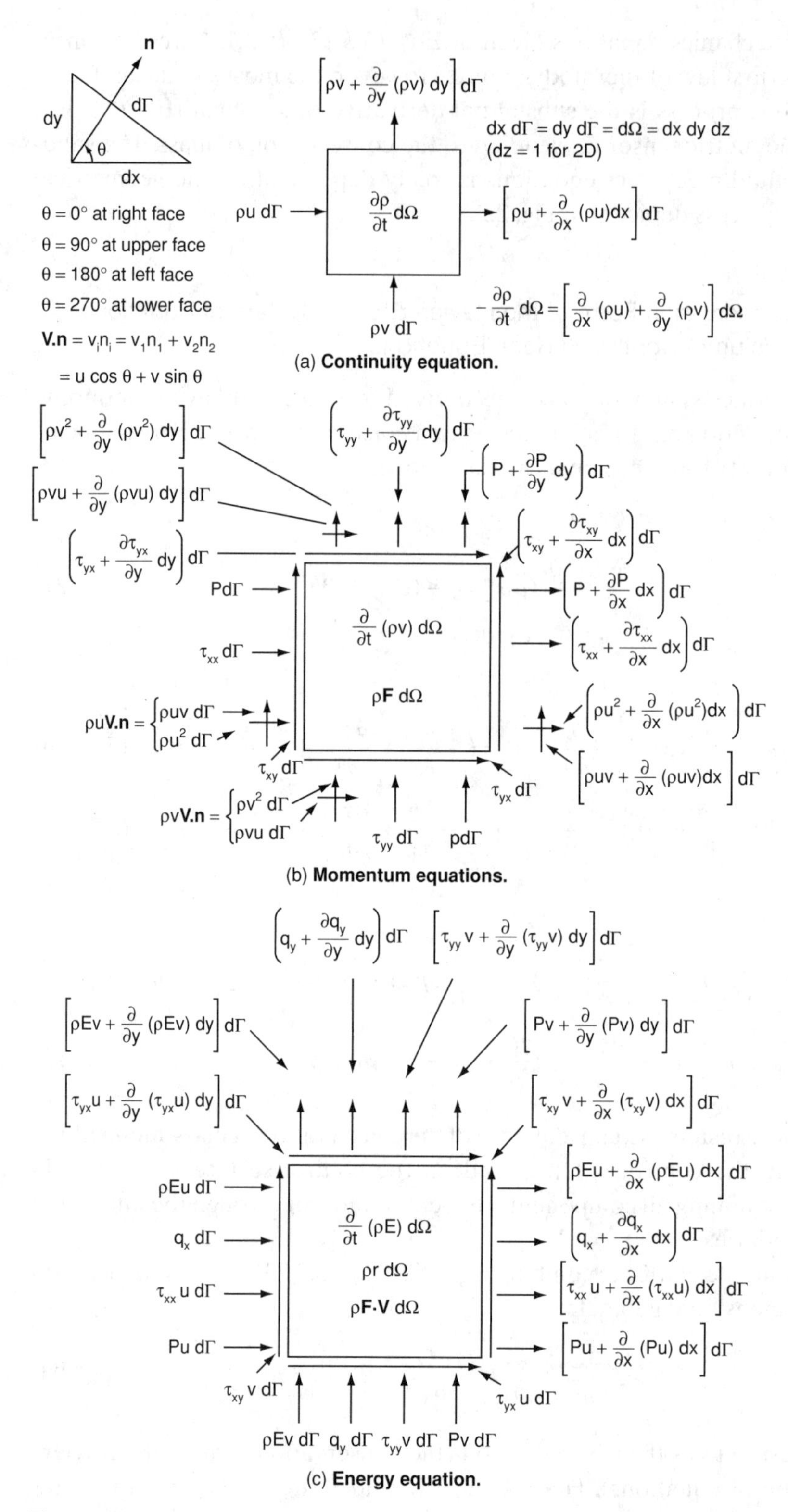

Figure 5.3.1. Free-body diagrams for Navier–Stokes system of equations drawn from Eqs. (5.3.31)

variables, convection flux variables, diffusion flux variables, and source terms, respectively:

$$\mathbf{U} = \begin{bmatrix} \rho \\ \rho V_j \\ \rho E \end{bmatrix}, \tag{5.3.25a}$$

$$\mathbf{F}_i = \begin{bmatrix} \rho V_i \\ \rho V_i V_j + p\delta_{ij} \\ \rho E V_i + p V_i \end{bmatrix}, \tag{5.3.25b}$$

$$\mathbf{G}_i = \begin{bmatrix} 0 \\ -\tau_{ij} \\ -\tau_{ij} V_j - q_i \end{bmatrix}, \tag{5.3.25c}$$

$$\mathbf{B} = \begin{bmatrix} 0 \\ \rho F_j \\ \rho r + \rho F_j V_j \end{bmatrix}, \tag{5.3.25d}$$

$$\tau_{ij} = \mu\left(V_{i,j} + V_{j,i} - \frac{2}{3} V_{k,k}\delta_{ij}\right), \tag{5.3.26}$$

$$p = (\gamma - 1)\rho\left(E - \frac{1}{2} V_j V_j\right), \tag{5.3.27a}$$

$$T = \frac{1}{c_v}\left(E - \frac{1}{2} V_j V_j\right). \tag{5.3.27b}$$

where all variables inside the brackets in Eqs. (5.3.25) are referred to as primitive variables. The Navier-Stokes system of equations without the diffusion flux terms ($\mathbf{G}_i$) is known as the *Euler equations*:

$$\frac{\partial \mathbf{U}}{\partial t} + \frac{\partial \mathbf{F}_i}{\partial x_i} = \mathbf{B}. \tag{5.3.28}$$

Let us now examine the consequence of the CNS given by Eq. (5.3.24). To this end, we substitute Eqs. (5.3.25a)–(5.3.25d) into Eq. (5.3.24) and perform the required differentiations. As a result, we obtain three separate equations:

$$\frac{\partial \rho}{\partial t} + (\rho V_i)_{,i} = 0, \tag{5.3.29a}$$

$$V_j\left[\frac{\partial \rho}{\partial t} + (\rho V_i)_{,i}\right] + \rho\frac{\partial V_j}{\partial t} + \rho V_i V_{j,i} + p_{,j} - \tau_{ij,i} - \rho F_j = 0, \tag{5.3.29b}$$

$$E\left[\frac{\partial \rho}{\partial t} + (\rho V_i)_{,i}\right] + V_j\left(\rho\frac{\partial V_j}{\partial t} + \rho V_i V_{j,i} + p_{,j} - \tau_{ij,i} - \rho F_j\right) + \rho\frac{\partial \epsilon}{\partial t}$$
$$+ \rho\epsilon_{,i} V + p V_{i,i} - \tau_{ij,i} V_{j,i} - q_{i,i} - \rho r = 0. \tag{5.3.29c}$$

A glance at Eq. (5.3.29b) indicates that conservation of mass is a prerequisite to the conservation of momentum, as given in Eq. (5.3.14). Similarly, Eq. (5.3.29c) states that conservation of both mass and momentum is a prerequisite to the conservation of energy, as given in Eq. (5.3.15). Thus the conservation form, Eq. (5.3.24), is equivalent to the first law of thermodynamics on differentiation, resulting in the nonconservation form of the Navier–Stokes system of equations given by Eqs (5.3.13)–(5.3.15). Indeed, the third equation, (5.3.29c), is identical to the integrands of Eq. (5.3.12) as derived from the first law of thermodynamics, which led to the nonconservation form of the Navier–Stokes system of equations.

If the conservation form (Eq. 5.3.24) is integrated over the domain,

$$\int_{\Omega}\left(\frac{\partial \mathbf{U}}{\partial t}+\frac{\partial \mathbf{F}_i}{\partial x_i}+\frac{\partial \mathbf{G}_i}{\partial x_i}-\mathbf{B}\right)d\Omega=0, \tag{5.3.30}$$

we obtain

$$\int_{\Omega}\left(\frac{\partial \mathbf{U}}{\partial t}-\mathbf{B}\right)d\Omega+\int_{\Gamma}(\mathbf{F}_i+\mathbf{G}_i)\,n_i\,d\Gamma=0 \tag{5.3.31a}$$

or in a discrete form for control volumes (indicated by the subscript cv) and control surfaces (indicated by the subscript cs),

$$\sum_{\text{cv}}\left(\frac{\partial \mathbf{U}}{\partial t}-\mathbf{B}\right)d\Omega+\sum_{\text{cs}}(\mathbf{F}_i+\mathbf{G}_i)\,n_i\,d\Gamma=0, \tag{5.3.31b}$$

which are referred to as the CVS equations. Note that these equations lend themselves to a convenient numerical solution scheme because the boundary surface integrals in Eqs. (5.3.31) can be evaluated numerically on boundaries. They can be computed along the exterior boundary surfaces and along the interior control surfaces of a control volume for the computational grid points, as in *finite-volume methods* (Chung, 2002).

Note that for one-dimensional problems the surface integrals in Eq. (5.3.31a) play a major role in evaluating the simple physical phenomena governing the conservation of mass, momentum, and energy. Furthermore, both the volume and surface integrals in Eq. (5.3.31a) can be decomposed into all relevant components on free-body diagrams, as shown in Fig. 5.3.1 for two dimensions. To see this, the surface integrals [Eq. (5.3.31a)] and the corresponding discrete forms [Eq. (5.3.31b)] are written as follows:

Continuity

$$\int_{\Gamma}\rho V_i n_i\,d\Gamma=a,$$

$$\sum_{\text{cs}}\rho V_i n_i\,d\Gamma=a. \tag{5.3.32a}$$

Momentum

$$\int_{\Gamma}(\rho V_i V_j+p\delta_{ij}-\tau_{ij})n_i\,d\Gamma=b_j,$$

$$\sum_{\text{cs}}(\rho V_i V_j+p\delta_{ij}-\tau_{ij})n_i\,d\Gamma=b_j. \tag{5.3.32b}$$

Energy

$$\int_{\Gamma} (\rho E V_i + p V_i - \tau_{ij} V_j - q_i) n_i \, d\Gamma = c,$$

$$\sum_{\mathrm{cs}} (\rho E V_i + p V_i - \tau_{ij} V_j - q_i) n_i \, d\Gamma = c. \tag{5.3.32c}$$

Expansion of these equations for a two-dimensional domain ($i, j = 1, 2$) represents all components that appear on the left-hand and bottom control surfaces (upstream sides) in Fig. 5.3.1. The changes (differentials) of these quantities across the segments dx and dy, which are represented by the integrands of Eq. (5.3.31a), appear on the right-hand and top control surfaces (downstream sides). The time derivative and sources terms are shown inside the square. Although all of these components may be intuitively predicted, as shown in traditional textbooks, it is seen here that they actually arise from the integration of the CNS equations, as verified by the first law of thermodynamics. Note that the direction cosines n_i ($n_1 = \cos\theta, n_2 = \sin\theta$, with θ measured from the x axis counterclockwise to the vector normal to the control surfaces) are used in determining the signs of all components in Fig. 5.3.1 and Eqs. (5.3.21)–(5.3.23). It is emphasized that the conservation of mass, momentum, and energy as these components move across the control volume is maintained mathematically with correct signs through appropriate direction cosines, as they appear in Eqs. (5.3.32). Thus the arrows shown in Fig. 5.3.1 merely indicate the direction of flow and are not indicative of signs to be used in determining the balance or equilibrium. The collection of equations that arise from Eqs. (5.3.31) in line with all components in Fig. 5.3.1 constitutes the CVS equations. It is interesting to note that subtraction of upstream components from downstream components results in the recovery of the CNS equations, Eq. (5.3.30). The reader is encouraged to prove this process, implying that it is the reverse process of integration with Eqs. (5.3.31) converted back to Eq. (5.3.30). In fact the conventional derivation of the Navier–Stokes system of equations presented in other textbooks is based on the determination of the free-body diagrams first as shown in Fig. (5.3.1) and performing exactly the reverse process to the approaches taken in this book. In this process, the guidance such as Eqs. (5.3.32a) is not available and must be constructed from intuition. Such intuition would be difficult, if not impossible, for the beginner particularly for the case of energy equation. Figure 5.3.1 can be revised for the three-dimensional geometry by showing all components of Eqs. (5.3.32) in three dimensions, as called for in Problem 5.3. The free-body sketches for this problem can be used to reconstruct Eq. (5.3.30) and eventually Eqs. (5.3.1), the first law of thermodynamics, in precisely the reverse order (Chung, 1996).

The importance of a thorough understanding of the relationships among the FLT equation, (5.3.12), CNS equations, (5.3.24), and CVS equations, (5.3.31), cannot be overemphasized. Modifications and simplifications of these equations will be discussed in the remainder of this chapter, as we relate them to various types of fluid flows.

5.3.3 Initial and Boundary Conditions

The basic approach in determining the boundary conditions is to classify the partial differential equations into elliptic, parabolic, and hyperbolic forms. Each form dictates a specific type of initial and boundary condition.

For simplicity, let us consider the elliptic, parabolic, and hyperbolic equations in the form

$$\phi_{,ii} = 0, \text{ elliptic;} \tag{5.3.33}$$

$$\frac{\partial \phi}{\partial t} - \phi_{,ii} = 0, \text{ parabolic;} \tag{5.3.34}$$

$$\frac{\partial^2 \phi}{\partial t^2} - \phi_{,ii} = 0, \text{ hyperbolic.} \tag{5.3.35}$$

In general, initial and boundary conditions are written as

$$\alpha(x_i, t) + \beta(x_i, t)\phi_{,i} n_i = \gamma(x_i, t). \tag{5.3.36}$$

Elliptic equations are associated with the Dirichlet condition ($\beta = 0$) or Neumann boundary conditions ($\alpha = 0$) in a closed region. On the other hand, parabolic equations are associated with the Dirichlet condition ($\beta = 0$) or the Neumann condition ($\alpha = 0$) in an open region. For hyperbolic equations, however, either Dirichlet or Neumann conditions with $\partial\phi/\partial t$ should be specified in an open region. For the case of the first-order hyperbolic (Euler) equation, the outflow boundary conditions are unspecified. For mixed elliptic and parabolic equations it is often required to provide the so-called Cauchy (or Robin) condition ($\alpha \neq 0, \beta \neq 0$), such as in convective heat transfer boundary conditions.

The Navier–Stokes system of equations, in general, is considered as the mixed elliptic, parabolic, and hyperbolic equations if all terms in Eq. (5.3.30) are included. The Neumann boundary conditions are given by

$$\int_\Gamma (\mathbf{F}_i + \mathbf{G}_i) n_i \, d\Gamma = \mathbf{g}, \tag{5.3.37}$$

and explicit components are given in Eqs. (5.3.32).

Correct specifications of Dirichlet (essential) boundary conditions and Neumann (natural) boundary conditions render the solution of the Navier–Stokes system of equations as "well posed." Further details are found in Strikweda (1977), Gustafsson and Sundström (1978), Dutt (1988), and Chung (2002).

5.3.4 The Second Law of Thermodynamics

Recall that we used the second law of thermodynamics to derive the heat conduction equation for solids in Section 4.4. This is because the only way the internal energy density can be characterized in thermomechanically loaded solids is to use the Helmholtz free energy in terms of entropy, which calls for the second law of thermodynamics. On the other hand, in fluid mechanics, the internal energy can be expressed by enthalpy [Eq. (5.3.4b)] rather than by entropy and Helmholtz free energy. For this reason, the second law of thermodynamics is not required for heat transfer involved in the energy equation. However, in acoustic oscillations in which thermal expansion of fluids occurs, such as in sound emission and absorption, it is necessary that the second law of thermodynamics be invoked. Toward this end, a brief introduction to thermodynamic relations in gases is in order.

The first law of thermodynamics in gases states that the heat added to the system δQ in an infinitesimal process is

$$\delta Q = d\epsilon + p\,dV, \tag{5.3.38}$$

where V is the specific volume $V = 1/\rho$ in nonreacting gases. Here δ implies different processes for Q. This equation can also be written in terms of the enthalpy H per unit mass,

$$H = \epsilon + pV. \tag{5.3.39}$$

Thus, in view of Eqs. (5.3.38) and (5.3.39),

$$\delta Q = dH - V\,dp. \tag{5.3.40}$$

The second law of thermodynamics in gases is stated in a different form from the case of the thermodynamics of solids discussed in Section 4.4:

$$d\eta \geq \frac{\delta Q}{T}, \tag{5.3.41}$$

where η is the specific entropy previously defined in the thermodynamics of solids (the more commonly used notation for entropy in thermodynamics in traditional texts is S), and the equality and inequality signs denote reversible and irreversible processes, respectively. For the reversible process, from Eq. (5.3.40) or (5.3.38) we have

$$T\,d\eta = \delta Q = dH - V\,dp \tag{5.3.42a}$$

or

$$T\,d\eta = \delta Q = d\epsilon + p\,dV. \tag{5.3.42b}$$

Note that these results confirm the existence of *Gibbs energy* in the form

$$G = H - T\eta,$$

which is related to the enthalpy and entropy.

To compute entropy changes in a variety of processes, it is necessary to have equations in which the changes are expressed in terms of measurable quantities.

This requires the use of some relationships between partial derivatives of thermodynamic properties. These relationships are known as Maxwell's thermodynamic relations. To this end, we begin by expanding $\eta = \eta(T, p)$:

$$d\eta = \left(\frac{\partial \eta}{\partial T}\right)_p dT + \left(\frac{\partial \eta}{\partial p}\right)_T dp, \tag{5.3.43}$$

where the subscripts p and T indicate that pressure and temperature are, respectively, held constant for the appropriate partial derivatives. The following Maxwell's equations are obtained from Eqs. (5.3.38)–(5.3.43):

$$\left(\frac{\partial \eta}{\partial p}\right)_T = -\left(\frac{\partial V}{\partial T}\right)_p, \tag{5.3.44}$$

$$\left(\frac{\partial \eta}{\partial V}\right)_T = \left(\frac{\partial p}{\partial T}\right)_V, \tag{5.3.45}$$

$$\left(\frac{\partial \eta}{\partial V}\right)_p = \left(\frac{\partial p}{\partial T}\right)_\eta, \tag{5.3.46}$$

$$\left(\frac{\partial \eta}{\partial p}\right)_V = -\left(\frac{\partial V}{\partial T}\right)_p. \tag{5.3.47}$$

Additional properties associated with Maxwell's equations are

$$\left(\frac{\partial \eta}{\partial T}\right)_p = \frac{c_p}{T}, \tag{5.3.48}$$

$$c_p = c_v + p\left(\frac{\partial V}{\partial T}\right)_p = c_v + R, \tag{5.3.49}$$

where c_p and c_v are the specific heats at constant pressure and constant volume, respectively:

$$c_p = \left(\frac{\partial H}{\partial T}\right)_p,$$

$$c_v = \left(\frac{\partial \epsilon}{\partial T}\right)_v.$$

In view of Eqs. (5.3.43), (5.3.44), and (5.3.48), we obtain

$$d\eta = \frac{c_p}{T} dT - \left(\frac{\partial V}{\partial T}\right)_p dp \tag{5.3.50}$$

or

$$d\eta = \frac{c_p}{T} dT - V\alpha\, dp, \tag{5.3.51}$$

where α is the coefficient of thermal expansion defined as

$$\alpha = \frac{1}{V}\left(\frac{\partial V}{\partial T}\right)_p = -\frac{1}{\rho}\left(\frac{\partial \rho}{\partial T}\right)_p. \tag{5.3.52}$$

Similarly, using $\eta = \eta(T, V)$, we obtain

$$T\,d\eta = c_v\,dT + T\left(\frac{\partial p}{\partial T}\right)_v dV. \tag{5.3.53}$$

Let us now return to the energy equation, (5.3.15), derived in Subsection 5.3.1:

$$\rho\frac{D\epsilon}{Dt} = \sigma_{ij}V_{j,i} + q_{i,i} + \rho r. \tag{5.3.54}$$

Also, we rewrite relation (5.3.41) in the form

$$T\frac{D\eta}{Dt} = \frac{D\epsilon}{Dt} + p\frac{D}{Dt}\left(\frac{1}{\rho}\right) = \frac{D\epsilon}{Dt} + \frac{p}{\rho}V_{i,i}, \tag{5.3.55}$$

with

$$p\frac{D}{Dt}\left(\frac{1}{\rho}\right) = -\frac{p}{\rho^2}\left[\frac{\partial\rho}{\partial t} + (\mathbf{V}\cdot\nabla)\rho\right] = \frac{p}{\rho}\mathbf{V}\cdot\nabla. \tag{5.3.56}$$

It follows from Eq. (5.3.51) that

$$T\frac{D\eta}{Dt} = c_p\frac{DT}{Dt} - \frac{\alpha T}{\rho}\frac{Dp}{Dt}. \tag{5.3.57}$$

Equating Eqs. (5.3.55) and (5.3.57) yields

$$\rho c_p\frac{DT}{Dt} - \alpha T\frac{Dp}{Dt} - pV_{i,i} - \sigma_{ij}V_{j,i} - \nabla\cdot\mathbf{q} - \rho r = 0 \tag{5.3.58}$$

or

$$\rho c_p\frac{\partial T}{\partial t} + \rho c_p T_{,i}V_i - \alpha T\frac{\partial p}{\partial t} - \alpha T p_{,i}V_i - \tau_{ij}V_{j,i} - q_{i,i} - \rho r = 0, \tag{5.3.59}$$

where τ_{ij} is given by Eq. (5.3.26). Here, we notice the interesting similarity of the term

$$-\alpha T\frac{\partial p}{\partial t}$$

associated with the thermal expansion coefficient α, to the counterpart

$$\alpha T_0(3\lambda + 2\mu)\frac{\partial}{\partial t}(u_{i,i})$$

in Eqs. (4.4.40) for solid mechanics. The difference in sign stems from the fact that, in solids, thermal stresses act against the mechanical resistance, whereas in fluids such restraint is absent. Applications of Eq. (5.3.59) can be made for the problems of sound-wave emission or radiation and absorption from fluctuating forces in acoustics.

The second law of thermodynamics is also invoked when shock waves are curved, in which enthalpy and entropy gradients normal to the streamlines are nonzero. The flow is rotational behind the shock waves. This subject is discussed in Section 5.10.

If specific heats c_p and c_v are not constant but are dependent on temperature, then such a gas is known as *thermally perfect*. A *real gas* arises when it is dependent on both temperature and pressure. These topics are beyond the scope of this text.

In concluding this section on the Navier–Stokes system of equations, we have witnessed that the derivation of the Navier–Stokes system of equations in fluid mechanics is similar to that of the Navier equations for solid mechanics. The first and second laws of thermodynamics constitute the basis of both solid mechanics and fluid mechanics. It was shown that the deformation of solids in convective Lagrangian coordinates leads to the conservation of mass not only for solids but also for fluids through transformation to Eulerian coordinates with substantial derivatives as demonstrated by Eq. (2.2.29), (2.2.30), or (5.3.8). Without the power of tensor analysis, these consequences would not have been made possible. Note that the expression given by Eq. (5.3.8) lays the foundation for fluid mechanics. Recall that it all began with the combined convective Lagrangian and Eulerian coordinate system with the Jacobian connecting the large deformation and the initial configuration as shown in Fig. 2.2.2 and Eqs. (2.2.15).

5.4 Compressible Viscous Flow

Having established the stress tensor for Newtonian fluids in Eq. (5.2.29) and the conservation equations for mass, momentum, and energy in Eqs. (5.3.13)–(5.3.15), we now proceed to the governing equations for compressible viscous fluids in the nonconservation form.

Continuity

$$\frac{\partial \rho}{\partial t} + \nabla \cdot (\rho \mathbf{V}) = 0. \tag{5.4.1}$$

Momentum

$$\rho \frac{\partial \mathbf{V}}{\partial t} + \rho(\mathbf{V} \cdot \nabla)\mathbf{V} + \nabla p - \mu \left[\nabla^2 \mathbf{V} + \frac{1}{3}\nabla(\nabla \cdot \mathbf{V}) \right] - \rho \mathbf{F} = 0. \tag{5.4.2}$$

Energy

$$\rho \frac{\partial \epsilon}{\partial t} + \rho(V \cdot \nabla)\epsilon + p\nabla \cdot \mathbf{V} + \mu \left(\frac{2}{3} V_{i,i} V_{j,j} - V_{i,j} V_{j,i} - V_{i,j} V_{i,j} \right)$$
$$- k\nabla^2 T - \rho r = 0. \tag{5.4.3}$$

To these, the equation of state for perfect gas given in Eq. (5.1.2) must be added.

The momentum and energy equations may be written explicitly in terms of index notation.

Momentum

$$\rho \frac{\partial V_j}{\partial t} + \rho V_{j,i} V_i + p_{,j} - \mu \left(V_{j,ii} + \frac{1}{3} V_{i,ij} \right) - \rho F_j = 0. \tag{5.4.4}$$

Energy

$$\rho c_v \frac{\partial T}{\partial t} + \rho c_v T_{,i} V_i + p V_{i,i} + \mu \left(\frac{2}{3} V_{i,i} V_{j,j} - V_{i,j} V_{j,i} - V_{j,i} V_{j,i} \right)$$
$$- k T_{,ii} - \rho r = 0 \tag{5.4.5a}$$

or

$$\rho c_p \frac{\partial T}{\partial t} + \rho c_p T_{,i} V_i - \frac{\partial p}{\partial t} - p_{,i} V_i + \mu \left(\frac{2}{3} V_{i,i} V_{j,j} - V_{i,j} V_{j,i} - V_{j,i} V_{j,i} \right)$$
$$- k T_{,ii} - \rho r = 0. \tag{5.4.5b}$$

Here the dynamic viscosity μ and the coefficient of thermal conductivity k are, in general, given as functions of temperature for compressible flows. For two-dimensional flow ($x_1 = x, x_2 = y, V_1 = u, V_2 = v$), we may write in traditional forms as follows:

Continuity

$$\frac{\partial \rho}{\partial t} + \rho \left(\frac{\partial u}{\partial x} + \frac{\partial v}{\partial y} \right) + u \frac{\partial \rho}{\partial x} + v \frac{\partial \rho}{\partial y} = 0. \tag{5.4.6}$$

Momentum

$$\rho \frac{\partial u}{\partial t} + \rho \left(u \frac{\partial u}{\partial x} + v \frac{\partial u}{\partial y} \right) + \frac{\partial p}{\partial x} - \mu \left(\frac{4}{3} \frac{\partial^2 u}{\partial x^2} + \frac{\partial^2 u}{\partial y^2} + \frac{1}{3} \frac{\partial^2 v}{\partial x \partial y} \right) - \rho F_x = 0, \tag{5.4.7a}$$

$$\rho \frac{\partial v}{\partial t} + \rho \left(u \frac{\partial v}{\partial x} + v \frac{\partial v}{\partial y} \right) + \frac{\partial p}{\partial y} - \mu \left(\frac{\partial^2 v}{\partial x^2} + \frac{4}{3} \frac{\partial^2 v}{\partial y^2} + \frac{1}{3} \frac{\partial^2 u}{\partial y \partial x} \right) - \rho F_y = 0. \tag{5.4.7b}$$

Energy

$$\rho c_v \frac{\partial T}{\partial t} + \rho c_v \left(u \frac{\partial T}{\partial x} + v \frac{\partial T}{\partial y} \right) + p \left(\frac{\partial u}{\partial x} + \frac{\partial v}{\partial y} \right)$$
$$- \mu \left\{ \frac{4}{3} \left[\left(\frac{\partial u}{\partial x} \right)^2 + \left(\frac{\partial v}{\partial y} \right)^2 - \frac{\partial u}{\partial x} \frac{\partial v}{\partial y} \right] + 2 \frac{\partial u}{\partial y} \frac{\partial v}{\partial x} + \left(\frac{\partial u}{\partial y} \right)^2 + \left(\frac{\partial v}{\partial x} \right)^2 \right\}$$
$$- k \left(\frac{\partial^2 T}{\partial x^2} + \frac{\partial^2 T}{\partial y^2} \right) - \rho r = 0, \tag{5.4.8a}$$

or

$$\rho c_p \frac{\partial T}{\partial t} + \rho c_p \left(u \frac{\partial T}{\partial x} + v \frac{\partial T}{\partial y} \right) - \frac{\partial p}{\partial t} - u \frac{\partial p}{\partial x} - v \frac{\partial p}{\partial y}$$
$$- \mu \left\{ \frac{4}{3} \left[\left(\frac{\partial u}{\partial x} \right)^2 + \left(\frac{\partial v}{\partial y} \right)^2 - \frac{\partial u}{\partial x} \frac{\partial v}{\partial y} \right] + 2 \frac{\partial u}{\partial y} \frac{\partial v}{\partial x} + \left(\frac{\partial u}{\partial y} \right)^2 + \left(\frac{\partial v}{\partial x} \right)^2 \right\}$$
$$- k \left(\frac{\partial^2 T}{\partial x^2} + \frac{\partial^2 T}{\partial y^2} \right) - \rho r = 0, \tag{5.4.8b}$$

where the dependent variables (u, v, ρ, p, T) are called primitive. Thus the non-conservation form of the Navier-Stokes system of equations is sometimes called the primitive variable formulation.

A glance at these equations indicates that there are nonlinear terms in the momentum and energy equations. They arise from $(\mathbf{V}\cdot\nabla)\mathbf{V}$ and $(\mathbf{V}\cdot\nabla)T$, which are known as *convective terms*. The presence of these terms renders the solution difficult both analytically and numerically. These terms represent complicated physical phenomena, such as shock waves, turbulence, and convective heat transfer. Furthermore, the terms generated from $\tau_{ij}V_{j,i}$, called *thermoviscous dissipation*, are also nonlinear. The nonlinearity of these terms is not as troublesome as that of the convective terms, but it is still the source of difficulty in analytical and numerical solutions.

In summary, we have discussed a complete set of governing equations for compressible viscous flow in Cartesian coordinates. We may obtain equations for other types of flow by simplifying and modifying the equations for compressible viscous flow. We discuss these topics in the sections that follow.

5.5 Ideal Flow

5.5.1 General

An ideal flow is inviscid, incompressible, and irrotational, characterized by the vanishing of the divergence and curl of the velocity vector, respectively:

Condition of incompressibility

$$\nabla\cdot\mathbf{V}=0. \tag{5.5.1}$$

Condition of irrotationality

$$\boldsymbol{\omega}=\nabla\times\mathbf{V}=0, \tag{5.5.2}$$

where $\boldsymbol{\omega}$ is the vorticity vector.

Recall that $\nabla\cdot\mathbf{V}=0$ indicates the conservation of mass for incompressible flow. Note that density must remain constant in space and time in this case. On the other hand, $\boldsymbol{\omega}=\nabla\times\mathbf{V}=0$ implies irrotationality or that the vorticity vector $\boldsymbol{\omega}$ is zero.

The solution of Eqs. (5.5.1) and (5.5.2) is facilitated by the scalar functions called *velocity potential function* ϕ and *stream function* ψ such that, for two dimensions,

$$\mathbf{V}=\nabla\phi=\phi_{,i}\mathbf{i}_i \quad (i=1,2) \tag{5.5.3}$$

or

$$\mathbf{V}=\epsilon_{ij}\psi_{,j}\mathbf{i}_i \quad (i,j=1,2), \tag{5.5.4}$$

where ϵ_{ij} is the second-order permutation symbol defined by $\epsilon_{12}=1$, $\epsilon_{21}=-1$, with all other ϵ_{ij} equal to zero. We prove relation (5.5.4) in Subsection 5.5.2. Substituting Eqs. (5.5.3) and (5.5.4) into Eqs. (5.5.1) and (5.5.2), respectively, we have

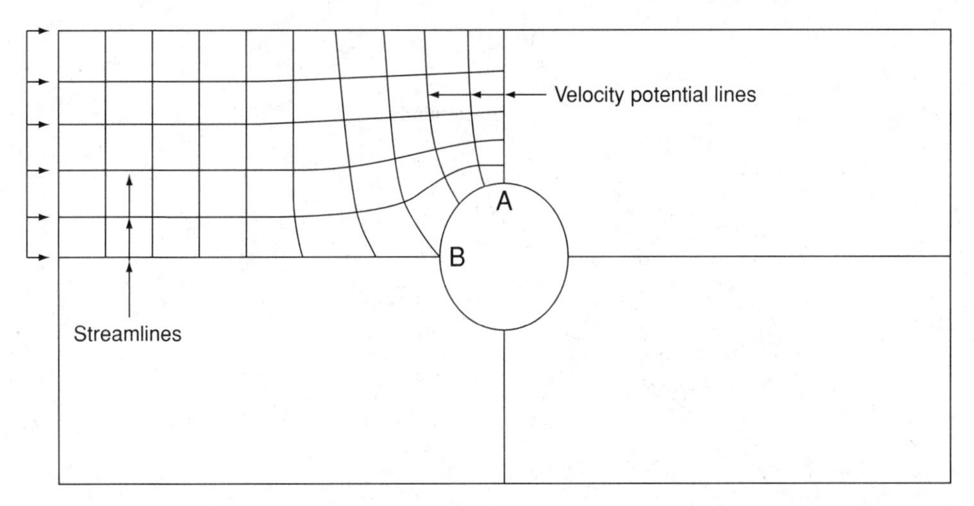

Figure 5.5.1. Ideal flow over a cylinder. Uniform inflow velocity, symmetric about horizontal and vertical center lines.

$$\nabla \cdot \mathbf{V} = \nabla^2 \phi = \phi_{,ii} = 0 \quad (i = 1, 2), \tag{5.5.5}$$

$$\begin{aligned} \boldsymbol{\omega} = \nabla \times \mathbf{V} &= V_{j,i} \epsilon_{ijk} \mathbf{i}_k = \psi_{,mi} \epsilon_{jm} \epsilon_{ijk} \mathbf{i}_k = (V_{2,1} - V_{1,2}) \mathbf{i}_3 \\ &= -(\psi_{,11} + \psi_{,22}) \mathbf{i}_3 = \omega_3 \mathbf{i}_3 = 0 \quad (i, j = 1, 2), \end{aligned}$$

indicating that only one vorticity component ($\omega_3 = \omega$) arises perpendicular to the $x - y$ plane.

$$\omega_3 = \omega = -\nabla^2 \psi = -\psi_{,ii} = 0 \quad (i = 1,\ 2). \tag{5.5.6}$$

Equations (5.5.5) and (5.5.6) are the Laplace equations, written as

$$\nabla^2 \phi = \frac{\partial^2 \phi}{\partial x^2} + \frac{\partial^2 \phi}{\partial y^2} = 0,$$

$$\nabla^2 \psi = \frac{\partial^2 \psi}{\partial x^2} + \frac{\partial^2 \psi}{\partial y^2} = 0.$$

The solution of Laplace equations $\nabla^2 \phi = 0$ and $\nabla^2 \psi = 0$ will provide data for potential lines and streamlines, respectively. These lines must intersect orthogonally, i.e., $\nabla \phi \cdot \nabla \psi = 0$, everywhere in the domain; the proof of this is left to the reader. An example of ideal flow over a cylinder (infinite in the direction perpendicular to the plane) is shown in Fig. 5.5.1, with points A and B indicating the positions of highest velocity and zero velocity (stagnation), respectively. In the absence of viscosity (friction) and vorticity (rotation), the flowfield is symmetrical about the horizontal and vertical lines drawn intersecting the center of the cylinder, with ψ and ϕ lines forming perfect squares, known as "flow net," as $\Delta x \to 0$ and $\Delta y \to 0$. The regions of high velocity and low velocity are indicated by smaller squares and larger squares, respectively. The symmetric

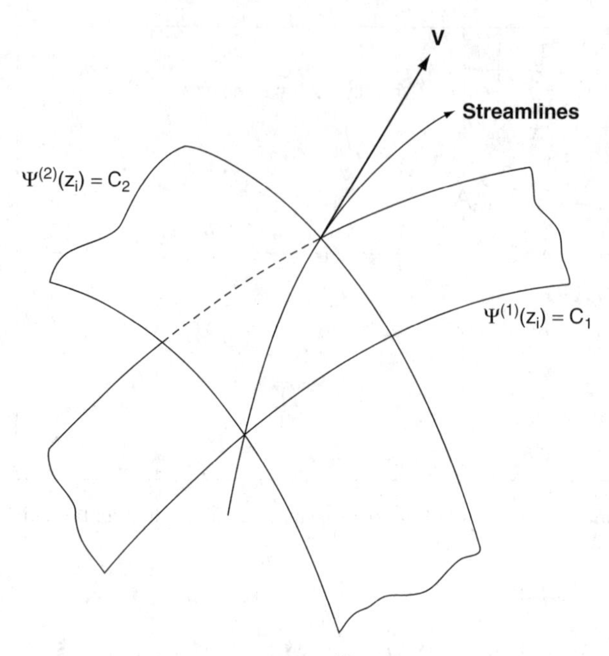

Figure 5.5.2. Stream functions $[\psi^{(1)}, \psi^{(2)}]$ on surfaces C_1 and C_2 and streamlines parallel to **V**.

flow pattern will no longer be maintained if viscosity becomes significant. Separation of streamlines, boundary layers, vorticities, and even turbulence may dominate the flowfield. In this case the preceding Laplace equations are no longer valid and we must use the standard fluid mechanics equations discussed in Section 5.4.

5.5.2 Existence of the Stream Function in Two Dimensions

The relation between the velocity vector and stream function given by Eq. (5.5.4) is proved in this section. To this end, let $d\mathbf{R} \equiv d\mathbf{x}$ be an element of the streamline passing through any point and let $\mathbf{V}$ denote the velocity vector at this point (see Fig. 5.5.2) such that $d\mathbf{x}$ is parallel to $\mathbf{V}$:

$$d\mathbf{x} \times \mathbf{V} = 0 \tag{5.5.7}$$

or

$$dx_i\mathbf{i}_i \times V_j\mathbf{i}_j = dx_i V_j \epsilon_{ijk}\mathbf{i}_k = 0.$$

On expansion, we obtain

$$V_3 dx_2 - V_2 dx_3 = 0, \tag{5.5.8a}$$

$$V_1 dx_3 - V_3 dx_1 = 0, \tag{5.5.8b}$$

$$V_2 dx_1 - V_1 dx_2 = 0, \tag{5.5.8c}$$

which may be written as

$$\frac{dx_1}{V_1} = \frac{dx_2}{V_2} = \frac{dx_3}{V_3}. \tag{5.5.9}$$

Equation (5.5.8) or (5.5.9) is the Cartesian form of the set of differential equations for a streamline. Because only two of the equations can be independent in equation set (5.5.8), we may write

$$C_{mj}(x_k)dx_j = 0 \quad (m = 1, 2 \quad j = 1, 2, 3). \tag{5.5.10}$$

Equations of this kind, (5.5.8)–(5.5.10), are known as *Pfaffian differential equations* (Sneddon, 1957), and represent a streamline [see Eq. (2.2.33)]. The solution of Eq. (5.5.10) is given by two independent functions:

$$\psi^{(1)}(x_i) = c_1,$$
$$\psi^{(2)}(x_i) = c_2.$$

This indicates that a line in space, such as a streamline, may be described as the curve of the intersection of two surfaces such that

$$\mathbf{V} \cdot \nabla\psi^{(1)} = 0, \tag{5.5.10a}$$

$$\mathbf{V} \cdot \nabla\psi^{(2)} = 0, \tag{5.5.10b}$$

or

$$\mathbf{V} = \nabla\psi^{(1)} \times \nabla\psi^{(2)}. \tag{5.5.12}$$

Equation (5.5.12) implies that $\mathbf{V}$ is parallel to the cross product $\nabla\psi^{(1)} \times \nabla\psi^{(2)}$ (see Fig. 5.5.2).

For two-dimensional problems ($V_3 = 0$), the streamline equation takes the form

$$\frac{dx_1}{V_1} = \frac{dx_2}{V_2} = \frac{dx_3}{0}. \tag{5.5.13}$$

This requires that $dx_3 = 0$ and $x_3 =$ constant, suggesting that

$$\psi^{(1)} = \psi(x_1, x_2), \tag{5.5.14a}$$

$$\psi^{(2)} = x_3. \tag{5.5.14b}$$

We expand Eq. (5.5.12) to get

$$\begin{aligned}\mathbf{V} &= \psi^{(1)}_{,i}\psi^{(2)}_{,j}\epsilon_{ijk}\mathbf{i}_k \\ &= (\psi^{(1)}_{,2}\psi^{(2)}_{,3} - \psi^{(1)}_{,3}\psi^{(2)}_{,2})\mathbf{i}_1 + (\psi^{(1)}_{,3}\psi^{(2)}_{,1} - \psi^{(1)}_{,1}\psi^{(2)}_{,3})\mathbf{i}_2 \\ &\quad + (\psi^{(1)}_{,1}\psi^{(2)}_{,1} - \psi^{(1)}_{,2}\psi^{(2)}_{,1})\mathbf{i}_3.\end{aligned} \tag{5.5.15}$$

Then we substitute Eqs. (5.5.14) into Eq. (5.5.15), which gives

$$\mathbf{V} = V_1\mathbf{i}_1 + V_2\mathbf{i}_2 = \psi^{(1)}_{,2}\psi^{(2)}_{,3}\mathbf{i}_1 - \psi^{(1)}_{,1}\psi^{(2)}_{,3}\mathbf{i}_2$$

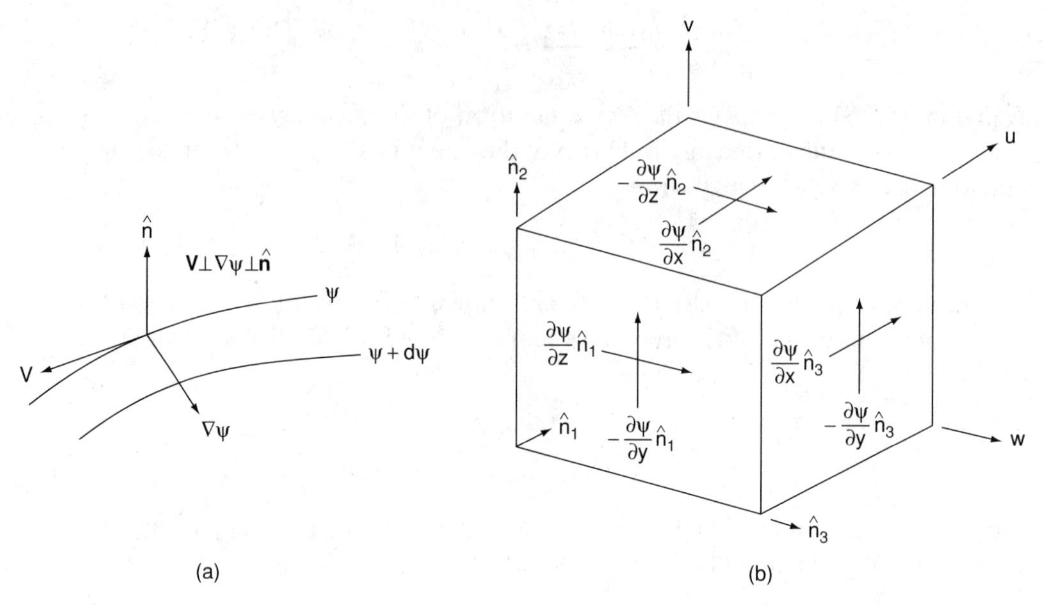

Figure 5.5.3. General descriptions of three-dimensional stream functions.

or, in view of Eq. (5.5.14),

$$\mathbf{V} = V_1\mathbf{i}_1 + V_2\mathbf{i}_2 = \psi_{,2}\mathbf{i}_1 - \psi_{,1}\mathbf{i}_2 = \epsilon_{ij}\psi_{,j}\mathbf{i}_i \quad (i, j = 1, 2),$$

which is in agreement with Eq. (5.5.4). Using the standard notation, we obtain

$$V_1 = \frac{\partial\psi}{\partial y},$$

$$V_2 = -\frac{\partial\psi}{\partial x}. \tag{5.5.16}$$

This result satisfies Eq. (5.5.1):

$$\nabla\cdot\mathbf{V} = \epsilon_{ij}\psi_{,ji} = 0,$$

indicating that the velocity–stream-function relationship is in conformance with the conservation of mass.

5.5.3 Existence of the Stream Function in Three Dimensions

The most general approach to derive the relationship between the velocity and the stream function in three-dimensional flow is to consider that the velocity $\mathbf{V}$ is parallel or tangent to the streamline identified by a stream function ψ (thus $\nabla\psi$ is perpendicular to $\mathbf{V}$) and that $\mathbf{V}$ is equal to the cross product of $\nabla\psi$ and the unit normal vector $\hat{\mathbf{n}}$ (perpendicular to the plane of $\nabla\psi$ and $\mathbf{V}$), as shown in Fig. 5.5.3(a):

$$\mathbf{V} = \nabla\psi \times \hat{\mathbf{n}} = \epsilon_{ijk}\psi_{,j}\hat{n}_k\mathbf{i}_i \quad (i, j, k = 1, 2, 3), \tag{5.5.17}$$

with each component given by

$$V_1 = \frac{\partial \psi}{\partial y}\hat{n}_3 - \frac{\partial \psi}{\partial z}\hat{n}_2,$$

$$V_2 = \frac{\partial \psi}{\partial z}\hat{n}_1 - \frac{\partial \psi}{\partial x}\hat{n}_3,$$

$$V_3 = \frac{\partial \psi}{\partial x}\hat{n}_2 - \frac{\partial \psi}{\partial y}\hat{n}_1, \qquad (5.5.18)$$

where derivatives of normal unit vectors vanish regardless of changes in directions. We may identify the stream function described on each surface designated by $\hat{n}_i$ as [see Fig. 5.5.3(b)]

$$\Psi_1 = \psi\hat{n}_1,$$

$$\Psi_2 = \psi\hat{n}_2,$$

$$\Psi_3 = \psi\hat{n}_3. \qquad (5.5.19)$$

These definitions suggest that there exist three independent stream-function components, with the three-dimensional stream-function vector defined as

$$\boldsymbol{\Psi} = \Psi_i\mathbf{i}_i \quad (i = 1, 2, 3),$$

so that

$$\mathbf{V} = \nabla \times \boldsymbol{\Psi} = \epsilon_{ijk}\Psi_{k,j}\mathbf{i}_i, \qquad (5.5.20)$$

with $\Psi_k = \psi\hat{n}_k$, as given by Eqs. (5.5.19). Equations (5.5.17) and (5.5.20) satisfy the conservation of mass in steady state and reduce automatically to the two-dimensional incompressible flow in terms of a single stream function $\psi = \psi_3$:

$$V_1 = u = \frac{\partial \psi}{\partial y}\hat{n}_3 = \frac{\partial \psi}{\partial y} \quad (\hat{n}_3 = 1 \text{ on } \hat{n}_3 \text{ surface}),$$

$$V_2 = v = -\frac{\partial \psi}{\partial x}\hat{n}_3 = -\frac{\partial \psi}{\partial x} \quad (\hat{n}_3 = 1 \text{ on } \hat{n}_3 \text{ surface}),$$

which confirms the two-dimensional definition given by Eq. (5.5.4). The deduction from Eqs. (5.5.17)–(5.5.20), as depicted in Fig. 5.5.3(b), has an important physical significance in which a scalar stream function ψ is extended to the three-dimensional stream-function vector $\boldsymbol{\Psi}$.

As a result of the preceding assessments, the vorticity vector assumes the form [see Eq. (1.2.27)]

$$\boldsymbol{\omega} = \nabla \times \mathbf{V} = \nabla \times (\nabla \times \boldsymbol{\Psi}) = \nabla(\nabla \cdot \boldsymbol{\Psi}) - \nabla^2\boldsymbol{\Psi} = -\nabla^2\boldsymbol{\Psi}, \qquad (5.5.21)$$

where $\nabla \cdot \boldsymbol{\Psi} = 0$ arises simply from the geometrical property $\nabla\psi \cdot \hat{\mathbf{n}} = 0$. Thus an irrotational ideal flow in three dimensions can be given by $\nabla^2\boldsymbol{\Psi} = 0$ in terms of the three-dimensional stream-function vector.

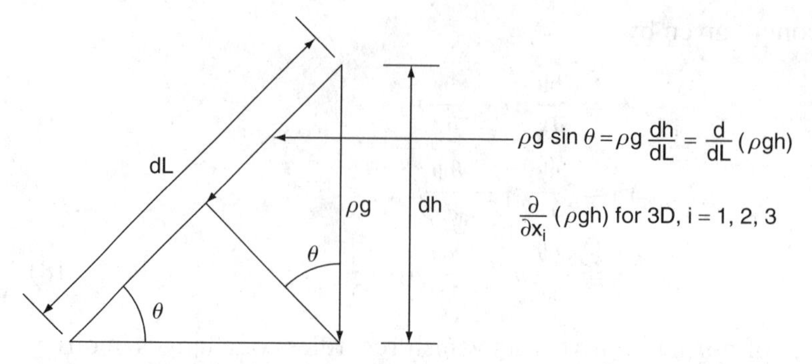

Figure 5.5.4. Gravity contributions to Bernoulli equation.

The foregoing analyses allow us to formulate a three-dimensional analysis in terms of three stream-function components through the vorticity transport equation discussed in Section 5.6.

The notion of the three-dimensional stream-function vector described is in contrast to the so-called three-dimensional vector potential concept in which the velocity vector is assumed to be the sum of the irrotational part and the rotational (solenoidal) parts,

$$\mathbf{V} = \nabla\phi + \nabla \times \mathbf{A},$$

where ϕ is the scalar potential and $\mathbf{A}$ is the vector potential, different from the three-dimensional stream-function vector $\boldsymbol{\Psi}$, described previously. Unfortunately, the vector potential approach to dealing with three-dimensional flow encounters unnecessary complications in the solution of the resulting differential equations, particularly with regard to the specification of boundary conditions (Elshabka and Chung, 1999).

5.5.4 The Bernoulli Equation

An ideal flow may also be treated by means of the *Bernoulli equation*. To derive this equation we may begin with momentum equations for incompressible and irrotational flow ($\nabla \cdot \mathbf{V} = 0$, and $\boldsymbol{\omega} = \nabla \times \mathbf{V} = 0$):

$$\frac{\partial \mathbf{V}}{\partial t} + (\mathbf{V} \cdot \nabla)\mathbf{V} + \frac{1}{\rho}(\nabla p - \rho \mathbf{F}) - \nu \nabla^2 \mathbf{V} = 0, \tag{5.5.22}$$

where ν is the kinematic viscosity.

The body force may be written in the form (see Fig. 5.5.4)

$$\rho \boldsymbol{F} = -\nabla(\rho g h), \tag{5.5.23}$$

where g is the gravitational acceleration and h is the height. It follows from Eq. (1.2.27) that

$$\nabla^2 \mathbf{V} = 0$$

for incompressible flow ($\nabla^2 \cdot \mathbf{V} = 0$) and irrotational flow ($\nabla \times \mathbf{V} = 0$), indicating that viscosity plays no role in Eq. (5.5.22) even if the flow is indeed viscous. Thus Eq. (5.5.22) results in the form known as the Euler equation, given by

$$\frac{\partial \mathbf{V}}{\partial t} + (\mathbf{V} \cdot \nabla)\mathbf{V} = -\nabla \left(\frac{p}{\rho} + gh \right). \tag{5.5.24}$$

Recall that we may write the convective term by using the vector identity (see Problem 1.7):

$$(\mathbf{V} \cdot \nabla)\mathbf{V} = \nabla \left(\frac{V^2}{2} \right) - \mathbf{V} \times \boldsymbol{\omega}, \tag{5.5.25}$$

where

$$V^2 = \mathbf{V} \cdot \mathbf{V}.$$

It follows that for steady motion Eq. (5.5.24) takes the form

$$\mathbf{V} \times \boldsymbol{\omega} = \nabla \left(\frac{p}{\rho} + \frac{V^2}{2} + gh \right). \tag{5.5.26}$$

For irrotational flow, or $\boldsymbol{\omega} = 0$, we have

$$\nabla \left(\frac{p}{\rho} + \frac{V^2}{2} + gh \right) = 0. \tag{5.5.27}$$

We integrate this equation to obtain for any direction x_1, x_2, or x_3 in three dimensions,

$$\frac{p}{\rho} + \frac{V^2}{2} + gh = C, \tag{5.5.28}$$

where C is the constant of integration. This is known as the Bernoulli equation. This may be generalized to include an unsteady motion in the form

$$\frac{\partial \mathbf{V}}{\partial t} + \nabla \left(\frac{p}{\rho} + \frac{V^2}{2} + gh \right) = 0$$

or

$$\frac{\partial \phi}{\partial t} + \frac{p}{\rho} + \frac{V^2}{2} + gh = C(t), \tag{5.5.29}$$

where the integration constant $C(t)$ is a function of time.

5.6 Rotational Flow

5.6.1 General

For fluids in rotational motion, the curl of the velocity produces the vorticity vector $\boldsymbol{\omega}$:

$$\boldsymbol{\omega} = 2\mathbf{w} = \nabla \times \mathbf{V} = \epsilon_{ijk} V_{j,i} \mathbf{i}_k, \tag{5.6.1}$$

where **w** is the rotation vector. For two-dimensional flow, we have

$$\omega_3 = V_{2,1} - V_{1,2} \tag{5.6.2}$$

or we may write

$$\omega = \omega_3 = \frac{\partial v}{\partial x} - \frac{\partial u}{\partial y}.$$

This defines the vorticity in two-dimensional flow (in the xy plane) whose direction is oriented toward the z axis.

Substituting the definition of the stream function, Eq. (5.5.16), we obtain the Poisson equation of the form

$$\frac{\partial^2 \psi}{\partial x^2} + \frac{\partial^2 \psi}{\partial y^2} = -\omega. \tag{5.6.3}$$

The vorticity normal to a surface element ds, $\boldsymbol{\omega} \cdot \mathbf{n}\, ds$, is called the *circulation*. The circulation around any closed curve that contains a given set of fluid particles remains constant with time as long as the external force field remains conservative (or constant). This phenomenon is known as the *Kelvin circulation theorem*. It follows from the Kelvin circulation theorem that a vortex line can neither begin nor end in the fluid; hence it appears as a closed loop or ends on a boundary.

Vortex motions cannot be formed in an ideal fluid, but regions of localized vorticity are often produced in portions of real fluids, thus affecting the fluid motion. The action of viscosity is, in general, a prerequisite to the appearance of vorticity. The regions in which rotational motion and circulation exist are relatively small, and the effect of vortices of fluid motion is felt through the velocities that they induce throughout the rest of the flowfield.

In certain real fluid flows, we encounter flow patterns that may be idealized as perfect fluids that contain infinite or semi-infinite rows of vortices, called *vortex streets*. The occurrence of vortex streets is usually associated with the formation of wakes behind bodies, which would lead to a kinematic instability of a form that might be conducive to possible rearrangements of vortices. In this regard, von Kármán showed that only one particular arrangement of alternating vortices is stable; this is known as the *Kármán vortex sheet*.

In what follows we discuss the vorticity transport equation coupled with either the velocity or the stream function and the associated boundary conditions.

5.6.2 The Vorticity Transport Equation

To obtain the governing equations for incompressible vortex flow, we take the curl of the momentum equation, (5.4.2), with ($\nabla \cdot \mathbf{V} = 0$):

$$\nabla \times \left[\frac{\partial \mathbf{V}}{\partial t} + (\mathbf{V} \cdot \nabla)\mathbf{V} + \frac{1}{\rho}\nabla p - \nu \nabla^2 \mathbf{V} - \mathbf{F} \right] = 0. \tag{5.6.4}$$

It follows from Eq. (1.2.25) that

$$\nabla \times (\mathbf{V} \times \boldsymbol{\omega}) = (V_{i,j}\omega_j + V_i\omega_{j,j} - V_{j,j}\omega_i - V_j\omega_{i,j})\mathbf{i}_i,$$

where

$$V_{i,j}\omega_j\mathbf{i}_i = (\boldsymbol{\omega} \cdot \nabla)\mathbf{V} = [(\nabla \times \mathbf{V}) \cdot \nabla]\mathbf{V} = \epsilon_{ijk}V_{j,i}V_{m,k}\mathbf{i}_m = 0$$

for two-dimensional flow. Furthermore, $V_{j,j}\omega_i = 0$ for incompressible flow, and because the velocity is the continuous function, we obtain

$$V_i\omega_{j,j}\mathbf{i}_i = \mathbf{V}(\nabla \cdot \boldsymbol{\omega}) = \mathbf{V}[\nabla \cdot (\nabla \times \mathbf{V})] = \mathbf{V}(\epsilon_{ijk}V_{j,ik}) = 0,$$

where $\partial^2 V_1/\partial x_1 \partial x_2 = \partial^2 V_1/\partial x_2 \partial x_1$, etc. Note also that $\nabla \times \mathbf{F} = 0$ for the conservative field or $\nabla \times \nabla(gh) = 0$ and $\nabla \times \nabla p = 0$ from the definition of vector algebra. With these observations, it follows from Eq. (5.6.4) that, in two dimensions ($\omega_3 = \omega$),

$$\frac{\partial \omega}{\partial t} + (\mathbf{V} \cdot \nabla)\omega - \nu\nabla^2\omega = 0. \tag{5.6.5}$$

This is known as the *vorticity transport equation* for two-dimensional incompressible flow.

In terms of stream function, the vorticity transport equation for two dimensions becomes

$$\frac{\partial \omega}{\partial t} + \omega_{,i}\epsilon_{ij}\psi_{,j} - \nu\omega_{,jj} = 0 \quad (i, j = 1, 2). \tag{5.6.6}$$

Because there are two unknowns (ω, ψ), we require an additional equation, namely, the stream-function Poisson equation, (5.6.3). Equations (5.6.6) and (5.6.3) should be solved simultaneously for ω and ψ.

To calculate the pressure, we take the divergence of the momentum equation,

$$\nabla \cdot \left[\frac{\partial \mathbf{V}}{\partial t} + (\mathbf{V} \cdot \nabla)\mathbf{V} + \frac{1}{\rho}\nabla p - \nu\nabla^2\mathbf{V} - \mathbf{F}\right] = 0,$$

in which the first and last terms vanish because of incompressibility, and $\nabla \cdot \mathbf{F} = 0$ for the conservative field. Thus we obtain the pressure Poisson equation:

$$\nabla^2 p = -B, \tag{5.6.7}$$

where

$$B = \rho\nabla \cdot [(\mathbf{V} \cdot \nabla)\mathbf{V}] = \rho(V_{i,j}V_j)_{,i}. \tag{5.6.8}$$

We expand Eq. (5.6.8) and use the incompressibility condition in two dimensions to obtain

$$\nabla^2 p = 2\rho\left(\frac{\partial u}{\partial x}\frac{\partial v}{\partial y} - \frac{\partial u}{\partial y}\frac{\partial v}{\partial x}\right). \tag{5.6.9}$$

Once the velocity field is calculated, then Eq. (5.6.9) can be used to determine the pressure.

An alternative approach to rotational flow analysis is to rewrite Eq. (5.6.6) by eliminating the vorticity through Eq. (5.5.4) to arrive at

$$\nabla^4 \psi = R, \tag{5.6.10}$$

where

$$R = \frac{1}{\nu}\left(\frac{\partial^3 \psi}{\partial x \partial y^2}\frac{\partial \psi}{\partial y} - \frac{\partial^3 \psi}{\partial x^2 \partial y}\frac{\partial \psi}{\partial x} + \frac{\partial^3 \psi}{\partial x^3}\frac{\partial \psi}{\partial y} - \frac{\partial^3 \psi}{\partial y^3}\frac{\partial \psi}{\partial x}\right),$$

in which a steady state is assumed. This approach has the advantage over Eq. (5.6.6) by involving only one variable. Recall that the boundary conditions that arise in Eq. (5.6.10) were discussed extensively in Section 1.3 and Subsection 4.3.2.

The standard boundary conditions to be specified for Eqs. (5.6.6) and (5.6.3) include $\psi = \hat{\psi}$ on Γ_1, $\omega = \hat{\omega}$ on Γ_2, $\psi_{,i} n_i = 0$ on Γ_3, and $\omega_{,i} n_i = 0$ on Γ_4. These boundary conditions are to be chosen depending on the governing equations used for the solution.

For three-dimensional incompressible flows, the two-dimensional vorticity transport equation must be modified by use of the definition of a stream-function vector with three stream-function vector components, as shown in Eq. (5.5.19). The vorticity transport equation now has an extra term, $(\boldsymbol{\omega} \cdot \nabla)\mathbf{V}$, for three dimensions:

$$\frac{\partial \boldsymbol{\omega}}{\partial t} + (\mathbf{V} \cdot \nabla)\boldsymbol{\omega} - (\boldsymbol{\omega} \cdot \nabla)\mathbf{V} = \nu \nabla^2 \boldsymbol{\omega}. \tag{5.6.11}$$

It follows from Eqs. (5.5.21) and (5.6.11) that

$$\frac{\partial}{\partial t}\nabla^2 \boldsymbol{\Psi} + (\nabla \times \boldsymbol{\Psi} \cdot \nabla)\nabla^2 \boldsymbol{\Psi} - (\nabla^2 \boldsymbol{\Psi} \cdot \nabla)(\nabla \times \boldsymbol{\Psi}) = \nu \nabla^4 \boldsymbol{\Psi}. \tag{5.6.12}$$

Once the components of $\boldsymbol{\Psi}$ are calculated, then the velocity and vorticity components and pressure are determined from Eqs. (5.5.20), (5.5.21), and (5.6.7), respectively. The numerical solution of Eq. (5.6.12) is shown in Elshabka and Chung (1999).

For compressible flows we must include $\nabla \times \frac{1}{\rho}\nabla p$ or $-\frac{1}{\rho^2}\nabla \rho \times \nabla p$ on the right-hand side of (5.6.11) and (5.6.12). This will then require that the pressure Poisson equation be solved simultaneously.

5.7 Turbulence

5.7.1 Time and Ensemble Averages

It is well known that the motion of fluids may occur in irregular fluctuations, resulting in *mixing*, *eddying*, or both. Such motions are called *turbulent flows* (Hinz, 1975; Schlichting, 1979b). When the Reynolds number $\mathrm{Re} = u_\infty L/v$ (where u_∞ is the free-stream velocity and L is the characteristic length) is increased, internal flows and boundary layers around solid bodies change from

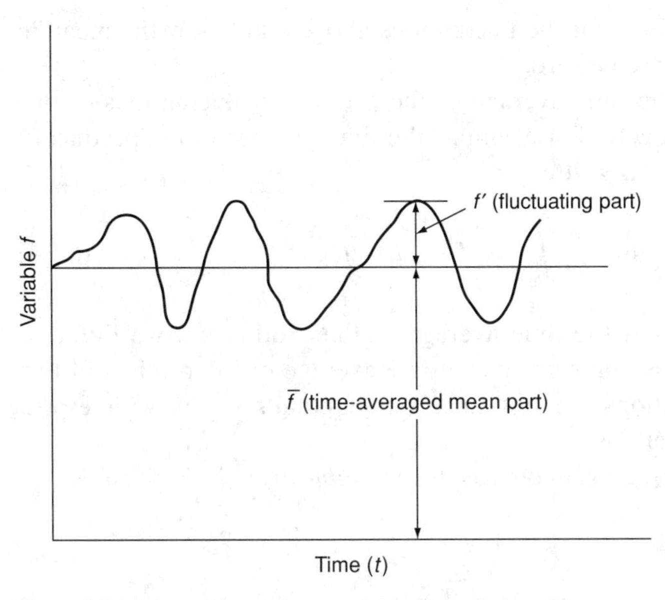

Figure 5.7.1. Fluctuations of any variable f in turbulent flow.

laminar to turbulent. Such a transition is influenced by geometries, pressure gradients, suction, compressibility, and heat transfer. The exact mathematical treatment of turbulence is hopelessly complex. In turbulence, the motion appears as if the viscosity were increased tremendously. At large Reynolds numbers, there exists a continuous transport of energy from the main flow into large eddies. The velocity and pressure at a fixed point in space do not remain constant with time, but undergo large, irregular fluctuations of high frequency. The flow is then assumed to consist of a mean motion and an eddying motion. Therefore, in the analysis of turbulent flow, we assume that the instantaneous value of any variable f for the motion is represented as the sum of the time-averaged mean part $\bar{f}$ and the eddying (fluctuating) part f':

$$f = \bar{f} + f'. \tag{5.7.1}$$

This relationship is shown in Fig. 5.7.1. The *time average* is given by

$$\bar{f} = \lim_{\Delta t \to \infty} \frac{1}{\Delta t} \int_{t_0}^{t_0+\Delta t} f dt, \tag{5.7.2}$$

where Δt is a very small time interval, but large in comparison with the *time scale* of the turbulence and sufficiently small in comparison with the period of slow variations of the averaged quantities in the flowfield. It is apparent from the definition in Eq. (5.7.2) that the time average of the fluctuation is zero,

$$\bar{f}' = \lim_{\Delta t \to \infty} \frac{1}{\Delta t} \int_{t_0}^{t_0+\Delta t} f' dt = 0, \tag{5.7.3}$$

which conforms to the fact that the fluctuations above and below the mean in Fig. 5.7.1 are summed up to be zero.

On the other hand, the time average of the product of fluctuations of two different variables is not zero. For example, the time average of the product of f' and another variable g' is written as

$$\lim_{\Delta t \to \infty} \frac{1}{\Delta t} \int_{t_0}^{t_0+\Delta t} f'g'\, dt = \overline{f'g'}, \tag{5.7.4}$$

where the overbar denotes the time average of the product of two different fluctuation variables. This implies that the time average of the product of two different variable fluctuations does not cancel out, but tends to grow with respect to time, known as correlations.

Another means of averaging, known as the *ensemble average*, is given by

$$\bar{f} = \frac{1}{N} \sum_{k=1}^{N} f_k, \tag{5.7.5}$$

where f_k denotes the values of N identical measurements. For quasi-steady systems, i.e., for $\partial f/\partial t = 0$ and $\partial \bar{f}/\partial t = 0$, it may be assumed that the two average systems lead to the same result.

5.7.2 Correlations

For compressible viscous flows, the variables involved in turbulence are the velocity, pressure, temperature, and density, consisting of the mean values and fluctuations as follows:

$$V_i = \bar{V}_i + V_i', \tag{5.7.6a}$$

$$p = \bar{p} + p', \tag{5.7.6b}$$

$$T = \bar{T} + T', \tag{5.7.6c}$$

$$\rho = \bar{\rho} + \rho'. \tag{5.7.6d}$$

For simplicity, consider an isothermal flow governed by the conservation form of the Navier–Stokes system of equations as follows:

Continuity

$$\frac{\partial \rho}{\partial t} + (\rho V_i)_{,i} = 0. \tag{5.7.7}$$

Momentum

$$\frac{\partial}{\partial t}(\rho V_i) + (\rho V_i V_j)_{,j} = -p_{,i} + \left[\mu\left(V_{i,j} + \frac{1}{3} V_{j,i}\right)\right]_{,j}. \tag{5.7.8}$$

Substituting Eq. (5.7.6) into Eqs. (5.7.7) and (5.7.8), taking time averages, and satisfying the conservation of mass lead to

$$\frac{\partial \bar{\rho}}{\partial t} + \left(\bar{\rho}\,\bar{V}_i + \overline{\rho' V_i'}\right)_{,i} = 0, \tag{5.7.9}$$

$$\rho \frac{\partial \bar{V}_i}{\partial t} + \bar{\rho}\,\bar{V}_{i,j}\bar{V}_j = -\bar{p}_{,i} + \mu\left(\bar{V}_{i,jj} + \frac{1}{3}\bar{V}_{j,ji}\right) + \Phi_i, \tag{5.7.10}$$

where

$$\Phi_i = -\left[\frac{\partial}{\partial t}(\overline{\rho' V_i'}) + (\bar{\rho}\,\overline{V_i' V_j'})_{,j} + (\overline{\rho' V'}_j \bar{V}_i)_{,j} + (\overline{\rho' V_i'}\,\bar{V}_j)_{,j} + (\overline{\rho' V_i' V_j'})_{,j}\right], \tag{5.7.11}$$

in which the last four terms of the variables within the brackets in Eq. (5.7.11) are known as turbulent (or Reynolds) stresses. These additional terms in Eq. (5.7.10), which originated from the convective terms in Eq. (5.7.8), are responsible for turbulent flow.

For incompressible flow, Eq. (5.7.10) is simplified to read

$$\rho\left(\frac{\partial \bar{V}_i}{\partial t} + \bar{V}_{i,j}\bar{V}_j\right) = -\bar{p}_{,i} + (\mu \bar{V}_{i,j} - \rho\,\overline{V_i' V_j'})_{,j}. \tag{5.7.12}$$

The total stress tensor may be written as

$$\sigma_{ij} = -\bar{p}\delta_{ij} + 2\mu \bar{d}_{ij} - \rho\,\overline{V_i' V_j'}. \tag{5.7.13}$$

The turbulent (Reynolds) stress, $-\rho\,\overline{V_i' V_j'}$, comes from the convective terms, and it is only logical that the convective motion results in turbulence. It is important to realize that the momentum equations for turbulent flows with the turbulent stress appearing as shown in Eq. (5.7.12) arise only through the conservation form, not through the nonconservation form. In 1877 Boussinesq proposed the concept of *eddy viscosity* $\hat{\mu}$, such that

$$\sigma_{ij} = -\bar{p}\delta_{ij} + 2(\mu + \hat{\mu})\bar{d}_{ij}. \tag{5.7.14}$$

By equating this with Eq. (5.7.13), we get

$$-\rho\,\overline{V_i' V_j'} = 2\hat{\mu}\bar{d}_{ij} = \hat{\mu}(\bar{V}_{i,j} + \bar{V}_{j,i}). \tag{5.7.15}$$

If we assume that $i = j$ in Eq. (5.7.15), then

$$-\rho\,\overline{V_i' V_i'} = 2\hat{\mu}\bar{V}_{i,i}. \tag{5.7.16}$$

The right-hand side vanishes for incompressible flow, whereas the left-hand side is not zero unless turbulence is absent. This is clearly a contradiction.

Another possibility is to assume that the eddy viscosity can be expressed in terms of the second- or fourth-order tensor:

$$-\rho\,\overline{V_i' V_j'} = \hat{\mu}_{ik}\bar{d}_{kj} \tag{5.7.17}$$

or

$$-\rho\,\overline{V_i'V_j'} = \hat{\mu}_{ijkm}\bar{d}_{km}. \tag{5.7.18}$$

However, these relations are not invariant with respect to a rotation of the coordinate system with a constant angular velocity, and so they violate the principle of material objectivity (see Subsection 4.4.4).

In searching for acceptable forms of representation for eddy viscosity, let us examine the equation

$$-\rho\,\overline{V_i'V_j'} = \hat{\mu}_0\delta_{ij} + \hat{\mu}_1\bar{d}_{ij} + \hat{\mu}_2\bar{d}_{ik}\bar{d}_{kj}, \tag{5.7.19}$$

where the scalars $\hat{\mu}_0$, $\hat{\mu}_1$, and $\hat{\mu}_2$ are functions of the principal invariants of $\bar{d}_{ij}$, much as the scalars were defined for the constitutive equation of Reiner–Rivlin fluids [see Eqs. (5.2.7) and (5.2.8)]. Unfortunately, the model suggested in Eq. (5.7.19) has not been supported by experimental evidence, despite the fact that Eq. (5.7.19) meets the requirement for material objectivity (frame indifference).

The turbulence process may be thought of as a memory effect of the sort that occurs in viscoelasticity, giving rise to an integral of the form

$$-\rho\,\overline{V_i'V_j'} = \bar{p}_t\delta_{ij} + A\bar{p}_t\int_0^\infty M(\tau)\left[\frac{\partial V_i}{\partial x_j}(t-\tau) + \frac{\partial V_j}{\partial x_i}(t-\tau)\right]d\tau, \tag{5.7.20}$$

where $\bar{p}_t$ is the average turbulence pressure, defined as

$$\bar{p}_t = \frac{1}{3}\rho\,\bar{V}_k'\bar{V}_k', \tag{5.7.21}$$

and $M(\tau)$ is an exponential-type memory function characteristic of turbulent flow (Hinze, 1975). Although this approach is appealing from the theoretical viewpoint, it is not likely to contribute to practical applications.

The turbulence correlation for the energy equation can be obtained in a similar manner:

$$\frac{\partial\bar{T}}{\partial t} + \bar{V}_i\bar{T}_{,i} = (\hat{\alpha}\bar{T}_{,i} - \overline{V_i'T'})_{,i}, \tag{5.7.22}$$

where $\hat{\alpha} = k/\rho c_p$ is the thermal diffusivity. Also, the second-order correlation $-\overline{V_i'T'}$ between the turbulent fluctuations of the velocity and the temperature may be given by

$$-\overline{V_i'T'} = (\alpha^*\bar{T})_{,i}, \tag{5.7.23}$$

where α^* is the eddy-transport coefficient. If this coefficient is chosen to be of second order, then

$$-\overline{V_i'T'} = (\alpha_{ij}^*\bar{T})_{,j}. \tag{5.7.24}$$

Along with these correlation data, we need to examine the notion of isotropic and homogeneous properties, and kinetic and dissipative energies in turbulence, which we discuss in the subsections that follow.

5.7.3 Isotropic and Homogeneous Turbulence

The turbulence shear stress

$$\tau_{ij}^{*} = -\rho \overline{V_i' V_j'}$$

has a normal component ($i = j$) and a shear component ($i \neq j$). Let us consider turbulence stress in a direction normal to an arbitrary surface of a fluid element in incompressible flows,

$$\tau_{(i)}^{*} = \tau_{ij}^{*} n_j. \tag{5.7.25}$$

The energy required for this fluid element in the normal direction is

$$E_{(n)} = \tau_{ij}^{*} n_i n_j = \rho \overline{V_i' V_j'} n_i n_j, \tag{5.7.26}$$

because

$$n_i n_j = \frac{1}{3}\delta_{ij},$$

$$E_{(n)} = \frac{1}{3}\rho \overline{V_i' V_i'} = \frac{2}{3}K^{*}, \tag{5.7.27}$$

where K^{*} is the turbulent kinetic energy:

$$K^{*} = \frac{1}{2}\rho \overline{V_i' V_i'}. \tag{5.7.28}$$

If the turbulence does not show a "preference" for any particular direction, i.e., if $V_1'^2 = V_2'^2 = V_3'^2 = u'^2$, then

$$K^{*} = \frac{3}{2}\rho u'^2 = \frac{3}{2}E_{(n)}. \tag{5.7.29}$$

Equations (5.7.26)–(5.7.29) imply that the surface of this fluid element becomes a sphere and they hold with respect to any coordinate system. They are invariant under rotation of the coordinate system. Such a flowfield is said to show *isotropic turbulence*, in which its statistical features have no preference for any direction. Here, no average shear stress can occur, the gradient of the mean velocity is zero, and the mean velocity is constant throughout the field. If the mean velocity has a gradient, then the turbulence is *anisotropic*, as it is for example in wall turbulence, where the average shear stress must be taken into account.

In general, viscosity effects result in conversion of the kinetic energy of the flow into heat. Other effects of viscosity make the turbulence more homogeneous. If the turbulence has the same structure throughout the flowfield or is invariant under translation of coordinate systems, it is said to be *homogeneous turbulence*. Otherwise, the turbulence is inhomogeneous.

5.7.4 Transport Equations of Reynolds Stresses, Turbulent Kinetic Energy, and Dissipation Rates

The equation of motion for an incompressible fluid with a steady mean flow reads

$$\frac{\partial}{\partial t}(\bar{V}_i + V_i') + (\bar{V}_k + V_k')(\bar{V}_i + V_i')_{,k} = -\frac{1}{\rho}(\bar{p} + p')_{,i} + \nu(\bar{V}_i + V_i')_{,kk}. \quad (5.7.30)$$

We obtain the so-called Reynolds stress transport equation by (1) subtracting Eq. (5.7.12) from Eq. (5.7.30), (2) multiplying the result by V_j', (3) repeating (1) and (2) by reversing the preceding i and j, (4) adding the resulting equations obtained in (2) and (3), and (5) taking the time average of the final equation obtained in (4):

$$\frac{D}{Dt}(\overline{V_i'V_j'}) = \underbrace{-(\overline{V_i'V_k'}\bar{V}_{j,k} + \overline{V_j'V_k'}\bar{V}_{i,k})}_{\text{I}} \underbrace{- 2\nu\overline{V_{i,k}'V_{j,k}'}}_{\text{II}} + \underbrace{\frac{1}{\rho}\overline{p'(V_{i,j}' + V_{j,i}')}}_{\text{III}}$$

$$\underbrace{-\left[(\overline{V_i'V_j'V_k'}) + \frac{1}{\rho}(\overline{p'V_i'\delta_{jk} + p'V_j'\delta_{ik}}) - \nu(\overline{V_i'V_j'})_{,k}\right]_{,k}}_{\text{IV}}, \quad (5.7.31)$$

where the quantities designated by I, II, III, and IV are identified as the production, destruction, pressure strain, and diffusive transport, respectively.

We obtain the differential equation for turbulent kinetic energy per unit mass by setting $j = i$ in Eq. (5.7.31):

$$\frac{DK}{Dt} = -\overline{V_i'V_j'}\,\overline{V}_{i,j} - \epsilon - \left(\frac{1}{2}\overline{V'_iV_i'V_j'} - \nu K_{,j} + \frac{1}{\rho}\overline{p'V_j'}\right)_{,j} = 0, \quad (5.7.32)$$

where K and ϵ are the turbulent kinetic energy per unit mass and its dissipation rate, respectively:

$$K = \frac{1}{2}\overline{V_i'V_i'}, \quad (5.7.33)$$

$$\epsilon = \nu\overline{V_{i,j}'V_{i,j}'}. \quad (5.7.34)$$

We can derive the differential equation for the dissipation rates of the turbulent kinetic energy by differentiating the momentum equation with respect to x_k, multiplying throughout by $2\nu V_{i,k}$, and time averaging the resulting equation:

$$\frac{D\epsilon}{Dt} = -2\nu(\overline{V_{i,k}'V_{j,k}'} + \overline{V_{k,i}'V_{k,j}'})\bar{V}_{i,j} - 2\nu\overline{V_j'V_{i,k}'}\,\bar{V}_{i,jk}$$

$$-2\nu\overline{V_{i,j}'V_{i,k}'V_{j,k}'} - 2\nu^2\overline{V_{i,jk}'V_{i,jk}'}$$

$$-\left(\upsilon\overline{V_j'V_{i,k}'V_{i,k}'} + \frac{2\psi}{\rho}\overline{V_{j,k}'p_{,k}'} - \psi\,\epsilon_{,j}\right)_{,j}. \quad (5.7.35)$$

From Eqs. (5.7.32) and (5.7.35), the following observations are made. (1) The turbulent shear stress, $\overline{V_i'V_j'}$ $(i \neq j)$, is produced only if the main flow is not uniform; (2) the pressure velocity-gradient correlations tend to decrease the nonisotropy by equalizing the three turbulence velocity components and by decreasing the turbulent shear stress; (3) this tendency to isotropy is greater in the smaller-scale range of turbulence; and (4) the viscosity effect, through dissipation, increases with increasing intensity of turbulence and contributes to the damping of the greater-intensity components at a higher rate than the smaller, the effect being greater in the higher-wave-number range, although such a process, because of viscosity, is rather slow. Note that the transport equations for the Reynolds stress, turbulent kinetic energy, and dissipation rates contain many unknowns of multiple velocity products. They must be modeled empirically in terms of known quantities.

Discussions carried out so far indicate that we have introduced additional unknowns, namely, the product terms with overbars in Eq. (5.7.1). To solve the Navier–Stokes system, we then require additional equations, with the number of equations matching the number of unknowns. To this end, we are able to use the transport equations of the Reynolds stress [Eq. (5.7.31)], turbulent kinetic energy [Eq. (5.7.32)], and dissipation rates [Eq. (5.7.35)]. Such a process is known as *turbulence closure*. Although we do not delve into this subject, it is informative to examine the earliest and simplest method of turbulence closure in the next subsection.

5.7.5 Phenomenological Theories

In earlier theories, such as the Prandtl mixing-length theory, the diffusion action of turbulence is considered to result in a constant, known as the eddy viscosity or eddy heat conductivity. This implies that the local gradient of the mean property may be used together with such constants. To this end, in a way that was analogous to the reasoning in the kinetic theory of gases, Prandtl (1904) assumed that the eddy viscosity was equal to the product of the *mixing length* and some suitable velocity. If we consider a plane flow along with its mean velocity in the x direction and negligible x-direction velocity gradients (boundary layer), then the Reynolds stress originally proposed by Boussinesq reads

$$-\overline{\rho u'v'}, = \hat{\mu}\frac{d\bar{u}}{dy}, \tag{5.7.36}$$

where $\hat{\mu}$ is the dynamic eddy viscosity,

$$\hat{\mu} = \rho L^2 \frac{d\bar{u}}{dy}, \tag{5.7.37}$$

and L is the Prandtl mixing length,

$$L = \kappa y, \tag{5.7.38}$$

where $\kappa = 0.4$ is an empirical dimensionless constant, as proposed by von Kármán. This is considered the simplest phenomenological model in which the u velocity is zero at the wall and parabolic u velocity distributions are assumed along the y direction known as the boundary layer, as discussed in Section 5.8. Here we require no additional equation because no additional variable is introduced, and so the model is known as a *zero-equation model.*

One of the most popular *two-equation models* is known as the $K-\epsilon$ model, in which the incompressible flow transport equations for K and ϵ are given by

$$\frac{\partial}{\partial t}(\bar{\rho}K) + (\bar{\rho}KV_i')_{,i} = (\mu_k K_{,i})_{,i} + (\bar{\tau}_{ij}V_j')_{,i} - \bar{\rho}\epsilon, \tag{5.7.39a}$$

$$\frac{\partial}{\partial t}(\bar{\rho}\epsilon) + (\bar{\rho}\epsilon V_i')_{,i} = (\mu_\epsilon \epsilon_{,i})_{,i} + c_{\epsilon 1}(\bar{\tau}_{ij}V_j')_{,i} - c_{\epsilon 2}\bar{\rho}\frac{\epsilon^2}{K}, \tag{5.7.39b}$$

with

$$\bar{\tau}_{ij} = -\rho\overline{V_i'V_j'} = \hat{\mu}\left(\bar{V}_{i,j} + \bar{V}_{j,i} - \frac{2}{3}\bar{V}_{k,k}\delta_{ij}\right) - \frac{2}{3}\bar{\rho}K\delta_{ij},$$

$$\hat{\mu} = \bar{\rho}c_\mu\frac{K^2}{\epsilon}, \quad \mu_k = \mu + \frac{\hat{\mu}}{\sigma_k}, \quad \mu_\epsilon = \mu + \frac{\hat{\mu}}{\sigma_\epsilon},$$

$$c_\mu = 0.09, \quad C_{\epsilon 1} = 1.45 - 1.55, \quad C_{\epsilon 2} = 1.92 \sim 2.0,$$

$$\sigma_k = 1, \quad \sigma_\epsilon = 1.3.$$

Very complicated models can be developed that take into account many variables. Various empirical assumptions have been advanced, but no phenomenological models developed to date are considered perfect. It is not likely that we will ever achieve a complete understanding of the mechanism of turbulence through phenomenological models, because its physical nature is extremely complicated.

For further details concerning the subject of turbulence, the reader should consult specialized books such as those by Hinze (1975) for theory and Hanjalic and Launder (1976) for numerical modeling, in which the unknowns in Eqs. (5.7.31), (5.7.32), and (5.7.35) are replaced with explicit empirical models. Details of the various turbulence models are beyond the scope of this book.

5.7.6 Other Approaches to Turbulence

In turbulence, large eddies interact strongly with the mean flow. Small eddies are created mainly by nonlinear interactions among large eddies. Most transport of mass, momentum, energy, and concentration is due to large eddies. Small eddies dissipate fluctuations of these quantities, but affect the mean properties only slightly. Large eddies are anisotropic, whereas small eddies are nearly isotropic.

In view of these properties, large structures in the flow are computed explicitly and small ones are modeled. This approach is known as large-eddy simulation (LES) (Ferziger, 1977). In this case, variables are space averaged rather than time averaged, so that the large-scale field is defined as

$$\bar{f}(\mathbf{x}) = \int_\Omega G(\mathbf{x}, \mathbf{x}') f(\mathbf{x}') d\Omega, \tag{5.7.40}$$

where $G(\mathbf{x}, \mathbf{x}')$ is a filter function. To this end, we define a subgrid scale velocity V_i' as

$$V_i' = V_i - \bar{V}_i. \tag{5.7.41}$$

Thus the filtered continuity and momentum equations can be written as

$$\frac{\partial \bar{V}_i}{\partial x_i} = 0, \tag{5.7.42a}$$

$$\frac{\partial \bar{V}_i}{\partial t} + \frac{\partial}{\partial x_j}(\overline{V_i V_j}) = -\frac{1}{\rho}\frac{\partial \bar{p}}{\partial x_i} + \nu \frac{\partial^2 \bar{V}_i}{\partial x_j \partial x_j}, \tag{5.7.42b}$$

where the overbar means a space-averaged quantity, and

$$\overline{V_i V_j} = \overline{\bar{V}_i \bar{V}_j} + \overline{V_i' \bar{V}_j} + \overline{\bar{V}_i V_j'} + \overline{V_i' V_j'}. \tag{5.7.43}$$

The last three terms contain the subgrid scale velocity and therefore must be treated by modeling. The first term depends on only the large-scale velocities $\bar{V}_i$ and can be calculated explicitly. This is similar to the use of Fourier transform in view of the space average given by (5.7.40).

We may compute turbulence without resorting to any turbulence models, an approach known as direct numerical simulation (DNS) [see Ferziger (1983) and Huser and Biringin (1993), among others]. Here, turbulence microscales are resolved by use of refined discretization of the domain.

In summary, the DNS approach is the most accurate but also the most expensive. All other methods involve modeling approximations, and therefore inaccuracy that is due to modeling errors is unavoidable.

5.8 The Boundary Layer

5.8.1 Laminar Boundary Layer

When a fluid is in contact with the surface of a solid body, the flow experiences a marked velocity change in a region from the immediate vicinity of the wall to some distance from the wall, forming a so-called *boundary layer*, an idea first proposed by Prandtl in 1904 (Schlicting, 1979a). Although viscosity plays a major role within the boundary layer, it has little influence in the flow outside the boundary layer, as pictured in Fig. 5.8.1. Boundary layers prevail for both compressible and incompressible flows. There are two types of boundary layers: wall shear boundary layers and free shear boundary layers (two free streams of different velocities). The concept of boundary layers is subsequently illustrated with the wall boundary layers for two-dimensional incompressible flows.

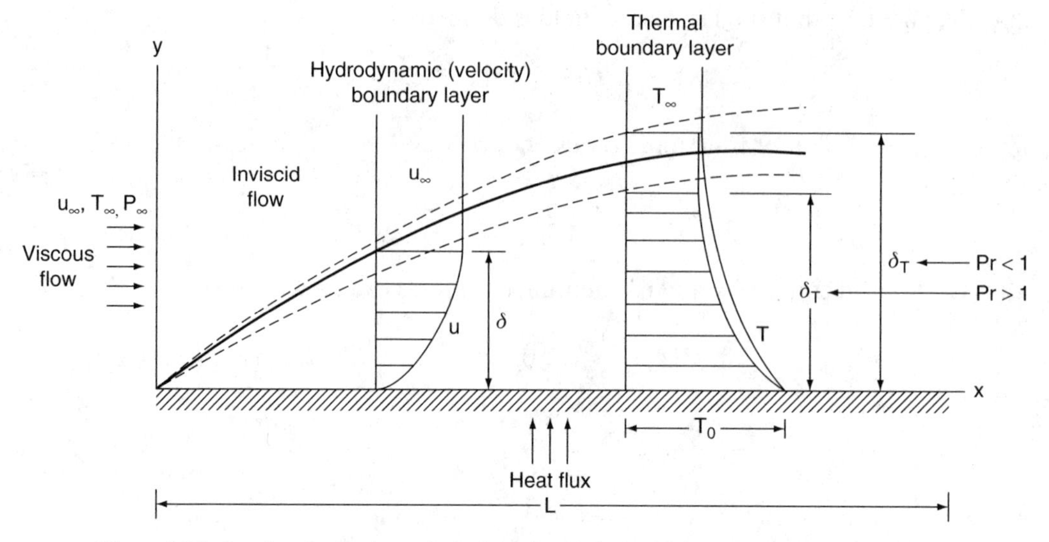

Figure 5.8.1. Laminar hydrodynamic (velocity) and thermal boundary layers along a flat wall. $\delta_T = \delta$ for Pr $= 1$, $\delta_T > \delta$ for Pr > 1, $\delta_T < \delta$ for Pr < 1.

In the study of the wall shear boundary layer, we consider a very thin layer in the immediate neighborhood of the wall, where the velocity gradient normal to the wall is large, contributing to high shear stresses, with the horizontal velocity above the wall varying parabolically until it reaches the free-stream velocity. No large-velocity gradients occur away from the surface of the body or channel, and the flow becomes frictionless (known as potential) in the free stream. This is the basic assumption for the boundary-layer theory. In addition, the following boundary-layer assumptions are made. Denoting δ and L (Fig. 5.8.1) as the boundary-layer thickness and the characteristic length, respectively, we assume that

(1) $\dfrac{\delta}{L} \ll 1,$

(2) $\dfrac{\partial u}{\partial y} \approx \dfrac{u}{\delta},$

(3) $\dfrac{\partial u}{\partial x} \approx \dfrac{u}{L}.$

Using the conservation of mass, $[(\partial u/\partial x) + (\partial v/\partial y)] = 0$, $\partial v/\partial y$ must be of the same order of magnitude as $\partial u/\partial x$ in order to conserve mass so that

(4) $\dfrac{\partial v}{\partial y} \approx \dfrac{u}{L}.$

Integrating (4), we obtain

(5) $v \approx \dfrac{u\delta}{L},$

(6) $\dfrac{\partial v}{\partial x} \approx \dfrac{u\delta}{L^2}.$

Using these boundary-layer assumptions and neglecting the terms of negligibly small magnitude, the two-dimensional steady-state momentum equations [see Eqs. (5.4.7)] for the incompressible flow are revised for the boundary layer as follows:

The x momentum

$$u\frac{\partial u}{\partial x}+v\frac{\partial v}{\partial y}=-\frac{1}{\rho}\frac{\partial p}{\partial x}+\nu\frac{\partial^2 u}{\partial y^2}. \tag{5.8.1}$$

The y momentum

$$\frac{1}{\rho}\frac{\partial p}{\partial y}=0, \tag{5.8.2}$$

with the y momentum simply indicating that the pressure is constant along the y direction. Thus the y momentum is to be discarded from analysis.

From Eqs. (5.4.8) the boundary-layer equation for energy may be derived similarly:

$$u\frac{\partial T}{\partial x}+v\frac{\partial T}{\partial y}=\hat{\alpha}\frac{\partial T}{\partial y}+\frac{\mu}{\rho c_p}\left(\frac{\partial u}{\partial y}\right)^2. \tag{5.8.3}$$

The boundary-layer thickness is proportional to the square root of the kinematic viscosity of the fluid. To see this for the boundary layer on a flat plate, using the momentum equation, we write

$$\rho u\frac{\partial u}{\partial x}\approx\mu\frac{\partial^2 u}{\partial y^2} \tag{5.8.4}$$

or

$$\rho\frac{u_\infty^2}{L}\approx\mu\frac{u_\infty}{\delta^2}, \tag{5.8.5}$$

where u_∞ is the free-stream velocity and L is the plate characteristic length. This gives

$$\delta\sim\sqrt{\frac{\mu L}{\rho u_\infty}}=\sqrt{\frac{\nu L}{u_\infty}}. \tag{5.8.6}$$

Blasius (1908) showed that, for laminar flow,

$$\delta=5\sqrt{\frac{\nu L}{u_\infty}} \tag{5.8.7}$$

or

$$\delta=\frac{5L}{\sqrt{\mathrm{Re}}}, \tag{5.8.8}$$

where $\mathrm{Re}=u_\infty L/\nu$ is the Reynolds number. It is seen that, as the Reynolds number becomes infinity, the boundary-layer thickness vanishes. In general, the boundary-layer thickness is represented by

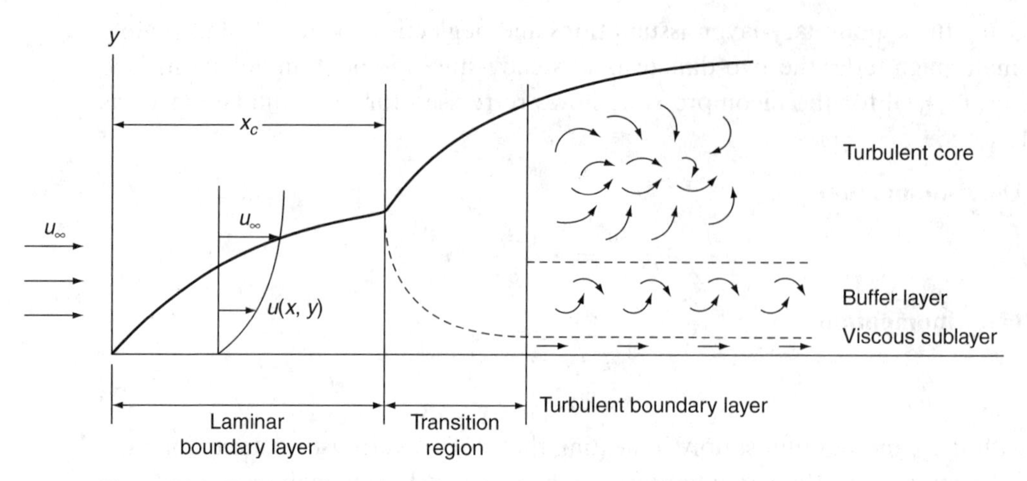

Figure 5.8.2. Turbulent boundary layers for flow over a flat plate.

$$\Delta^2 = \left(\frac{\delta}{L}\right)^2 \approx \frac{1}{\text{Re}} \ll 1. \tag{5.8.9}$$

Similarly, for the thermal boundary layer δ_T (see Fig. 5.8.1), using the energy equation, we write

$$\rho c_p u \frac{\partial T}{\partial x} \approx k \frac{\partial^2 T}{\partial y^2} \tag{5.8.10}$$

or

$$\frac{u}{L} \approx \frac{k}{\rho c_p} \frac{1}{\delta_T^2},$$

$$\Delta_T^2 = \left(\frac{\delta_T}{L}\right)^2 \approx \frac{1}{\text{RePr}} \ll 1, \tag{5.8.11}$$

in which Pr is the Prandtl number, defined as $\text{Pr} = (c_p \mu / k)$. It is clear from these definitions that $\delta_T = \delta$ for $\text{Pr} = 1$, $\delta_T < \delta$ for $\text{Pr} > 1$, and $\delta_T > \delta$ for $\text{Pr} < 1$.

Note that, for gases, Pr is of the order of unity; hence, the velocity and thermal boundary layers coincide. For liquids, the Prandtl number ranges between 10 and 1000; thus the thermal boundary-layer thickness is smaller than the thickness of the velocity (or hydrodynamic) boundary layer. For liquid metals, Pr varies from about 0.003 to 0.03, and the thermal boundary layer is much thicker than the velocity boundary layer.

5.8.2 Turbulent Boundary Layer

As shown in Fig. 5.8.2, starting from the leading edge of the plate, the laminar boundary layer continues to develop until some critical distance x_c is reached, beyond which small disturbances start and grow inside the boundary layer and the transition from laminar to turbulent boundary layer takes place, with the transition region characterized by unstable fluid motions. This critical distance is determined from the relation $x_c = \mu \mathrm{Re}/\rho u_\infty$. Usually, turbulence may occur at $5 \times 10^5 < \mathrm{Re} < 5 \times 10^6$ for a flat plate, but as low as 3000 for a circular tube. In the boundary layer next to the wall, the flow is characterized by a very thin viscous-flow region called the *viscous sublayer*. Adjacent to the sublayer is a highly turbulent region known as the *buffer layer*, in which the mean axial velocity increases rapidly with the distance from the wall. The buffer layer is then followed by the *turbulence core*, in which there is relatively lower-intensity and larger-scale turbulence, with the velocity changing slightly with the distance from the wall.

We can obtain the governing equations for two-dimensional, turbulent, thermal-hydrodynamic boundary-layer flow by substituting Eqs. (5.7.6) into the conservation forms of laminar boundary-layer equations, taking time averages, and using the boundary-layer assumptions as follows:

Continuity

$$\frac{\partial \bar{u}}{\partial x} + \frac{\partial \bar{v}}{\partial y} = 0. \tag{5.8.12}$$

The *x* momentum

$$\bar{u}\frac{\partial \bar{u}}{\partial x} + \bar{v}\frac{\partial \bar{u}}{\partial y} = -\frac{1}{\rho}\frac{d\bar{p}}{dx} + \frac{\partial}{\partial y}\left(\upsilon\frac{\partial \bar{u}}{\partial y} - \overline{u'v'}\right). \tag{5.8.13}$$

Energy

$$\bar{u}\frac{\partial \bar{T}}{\partial x} + \bar{v}\frac{\partial \bar{T}}{\partial y} = \frac{\partial}{\partial y}\left(\hat{\alpha}\frac{\partial \bar{T}}{\partial y} - \overline{T'v'}\right) + \frac{\mu}{\rho c_p}\left(\frac{\partial \bar{u}}{\partial y}\right)^2, \tag{5.8.14}$$

where $\hat{\alpha} = k/\rho c_p$ is the thermal diffusivity. Once again the turbulent stress $\overline{u'v'}$ and the turbulent heat flux $\overline{T'v'}$ require suitable correlations (turbulence models), as discussed in Section 5.7. Also note that the turbulent boundary-layer equations can be obtained only through the CNS equations, not through the nonconservation form. For a large Reynolds number, turbulence is essentially a three-dimensional structure and the simplified two-dimensional analysis is inadequate.

5.9 Convective Heat Transfer

5.9.1 General

The subject of heat transfer is often treated as an independent area of study in engineering (Bejan, 1984; Eckert and Drake, 1987). However, the governing equations derived in Section 5.3 have already suggested that heat transfer arises as a part of the first, the second, or both, laws of thermodynamics. To see this, we return to Eqs. (5.3.16)–(5.3.18), and note that heat transfer consists of conduction,

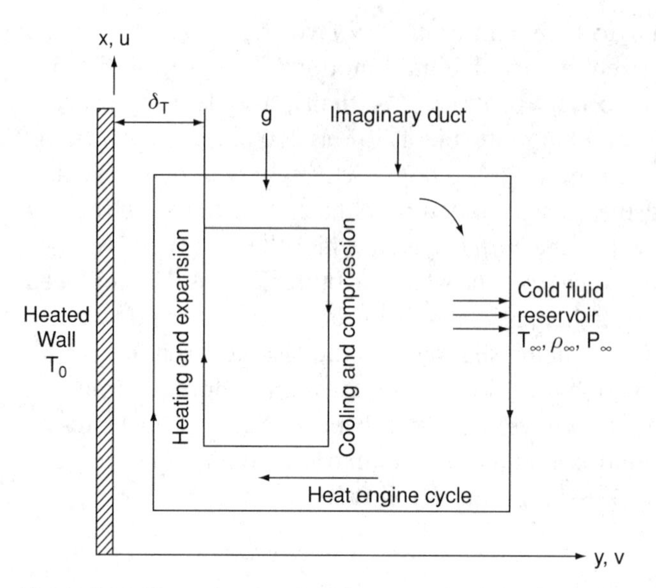

Figure 5.9.1. Natural convection along a vertical wall and analogy of a heat engine responsible for driving the flow.

convection, and radiation. Conductive heat transfer is the simplest form of heat transfer and is characterized by Fourier's heat conduction law. Convective heat transfer is the study of heat transport processes directly affected by the flow of fluids. Radiative heat transfer is the most complicated mode of heat transfer and is characterized by the radiative heat flux given by Eq. (5.3.17), which may or may not participate with the flow media in general. Radiative heat transfer is beyond the scope of this text.

Convection is either of two types: natural (free) or forced. Natural convection is due to of the differences in temperature in space. In a fluid this activates motion because of gravity in the momentum equation, resulting in condensation or vaporization (evaporation). On the other hand, forced convection arises in the energy equation when external disturbances are introduced into the domain by force. We discuss appropriate governing equations for the forced and natural convection in the next subsection.

5.9.2 Forced and Natural Convection

We can solve forced convection problems by using the energy equation together with the continuity and momentum equations with the usual boundary conditions. A simple example of natural convection is a heated body immersed in a cold reservoir; it acts as a heat engine that drives the fluid (see Fig. 5.9.1). Thus in natural convection the heat engine that drives the flow is built into the flow itself, whereas in forced convection the heat engine that drives the flow is external.

Consider the governing equations of steady incompressible viscous flow for the forced convection:

$$V_{i,i} = 0, \tag{5.9.1}$$

$$\rho V_{i,j} V_j = \rho F_i - p_{,i} + \mu V_{i,jj}, \tag{5.9.2}$$

$$T_{,i} V_i = \hat{\alpha} T_{,ii}. \tag{5.9.3}$$

The two-dimensional boundary-layer equations for the flow system for the natural convection as shown in Fig. 5.9.1

$$\rho \left(u \frac{\partial u}{\partial x} + v \frac{\partial v}{\partial y} \right) = -\rho g - \frac{\partial p}{\partial x} + \mu \frac{\partial^2 u}{\partial y^2}, \tag{5.9.4}$$

$$u \frac{\partial T}{\partial x} + v \frac{\partial T}{\partial y} = \hat{\alpha} \frac{\partial^2 T}{\partial y^2}, \tag{5.9.5}$$

where $F_1 = F_x$ in Eq. (5.9.2) is set equal to gravity g. Note that $\partial p/\partial x = dp_\infty/dx$ is equivalent to the gravity force $-\rho_\infty g$ in the reservoir so that

$$\rho \left(u \frac{\partial u}{\partial x} + \nu \frac{\partial u}{\partial y} \right) = \mu \frac{\partial^2 u}{\partial y^2} + (\rho_\infty - \rho) g. \tag{5.9.6}$$

Define the thermal expansion coefficient as

$$\alpha = -\frac{1}{\rho} \left(\frac{\partial \rho}{\partial T} \right)_p \tag{5.9.7}$$

and invoke the Boussinesq (1877) approximation as

$$\alpha \approx -\frac{1}{\rho} \frac{\rho - \rho_\infty}{T - T_\infty} \tag{5.9.8}$$

so that we may write the momentum equation in the form

$$u \frac{\partial u}{\partial x} + v \frac{\partial u}{\partial y} = \nu \frac{\partial^2 u}{\partial y^2} + \alpha g (T - T_\infty). \tag{5.9.9}$$

Here the term $\alpha g(T - T_\infty)$ represents the driving force that is due to the body force and the equivalent pressure gradient at infinity in Eq. (5.9.4). This implies that, if forced convection is combined with natural convection, we must add another pressure-gradient term that accounts for the contribution from additional driving forces applied externally to the boundaries. Thus the momentum equation, (5.9.9), is revised to read

$$u \frac{\partial u}{\partial x} + v \frac{\partial u}{\partial y} = -\frac{1}{\rho} \frac{\partial p}{\partial x} + \nu \frac{\partial^2 u}{\partial y^2} + \alpha g (T - T_\infty). \tag{5.9.10}$$

Typical examples of boundary layers for natural convection are condensation and vaporization (or evaporation), as depicted in Fig. 5.9.2. The governing equation for both condensation and vaporization is given by Eq. (5.9.9), with boundary conditions as evident from Fig. 5.9.2.

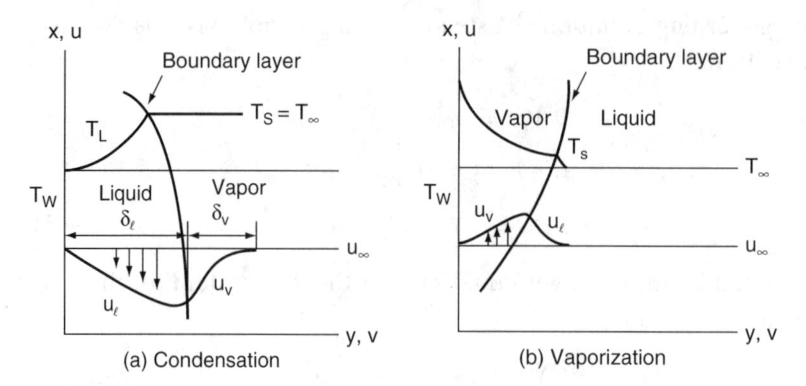

Figure 5.9.2. Condensation and vaporization.

For a combined system of natural and forced convective heat transfer, it is necessary that momentum equation (5.9.10) be combined with the equations of continuity and energy for simultaneous solutions. In general, convective heat transfer with natural and forced convection will require the solution of the CNS equations given by Eq. (5.3.30) with the body force replaced with the Boussinesq approximation. With an adequate numerical scheme and appropriate boundary conditions, the boundary-layer flow previously described can be realized without use of the simplified boundary-layer equations.

5.10 High-Speed Aerodynamics

5.10.1 General

In high-speed flows a convenient measure of speed is a quantity called the Mach number, M, defined as the ratio of the flow velocity u to the speed of sound a:

$$M = \frac{u}{a}.$$

When a Mach number is less than approximately 0.3 the flow is considered, in general, to be incompressible (see Problem 5.18). However, for $M \geq 0.3$ the flow becomes compressible, and density changes are significant. For Mach numbers near unity (transonic: $0.8 \leq M \leq 1.2$) shock waves may develop and persist throughout supersonic ($M > 1$) and hypersonic ($M \geq 5$) ranges.

In this section, we explore the nature of the governing equation for transonic flows, its mathematical implications, and the computational difficulties closely associated with the complicated physical phenomenon of shock waves.

5.10.2 Full Potential Equation

Consider the steady compressible flow governed by the Euler equations:

Continuity

$$(\rho V_i)_{,i} = 0. \tag{5.10.1}$$

Momentum

$$V_{i,j}V_j + \frac{1}{\rho}p_{,i} = 0. \tag{5.10.2}$$

Defining the speed of sound as $a = (\partial p/\partial\rho)^{1/2}$, we obtain from Eq. (5.10.1)

$$\frac{1}{a^2}p_{,i}V_i + \rho V_{i,i} = 0. \tag{5.10.3}$$

Multiplying Eq. (5.10.2) by V_i, we obtain

$$V_i V_{i,j} V_j + \frac{1}{\rho}p_{,i}V_i = 0. \tag{5.10.4}$$

Combining Eqs. (5.10.3) and (5.10.4) yields

$$V_i V_j V_{i,j} - a^2 V_{i,i} = 0. \tag{5.10.5}$$

The expansion of Eq. (5.10.5) in two dimensions gives

$$(1 - M_1^2)V_{1,1} + (1 - M_2^2)V_{2,2} - \frac{V_1 V_2}{a^2}(V_{1,2} + V_{2,1}) = 0, \tag{5.10.6}$$

where $M_1 = V_1/a$ and $M_2 = V_2/a$ are known as local Mach numbers in the x_1 and x_2 directions, respectively. Rearranging Eq. (5.10.6), leads to

$$(1 - M_1^2)V_{1,1} + (1 - M_2^2)V_{2,2} - \frac{2V_1 V_2}{a^2}V_{1,2} = E, \tag{5.10.7}$$

with E, referred to as the sonic vorticity,

$$E = (V_{2,1} - V_{1,2})\frac{V_1 V_2}{a^2}. \tag{5.10.8}$$

For $E = 0$, Eq. (5.10.7) is known as the full potential equation for irrotational flow. If $E \neq 0$, then Eq. (5.10.7) represents the full potential equations with non-vanishing vorticity.

For a general case, the sonic vorticity E may be evaluated from the thermodynamic point of view. To this end, we begin with the definitions of total enthalpy $\hat{H}$ and entropy η as follows:

$$\hat{H} = H + \frac{1}{2}V_j V_j, \tag{5.10.9}$$

$$\eta = c_v \ln \frac{p}{\rho^\gamma}. \tag{5.10.10}$$

Along the streamline, the velocity of sound is

$$a = \left(\sqrt{\frac{\partial p}{\partial \rho}}\right)_{\eta = \text{constant}} = \sqrt{\frac{\gamma p}{\rho}} = \sqrt{\gamma RT}. \tag{5.10.11}$$

For an ideal gas, it follows that

$$\frac{p}{\rho^\gamma} = \frac{p_0}{\rho_0^\gamma}\exp\frac{(\eta - \eta_0)}{c_v} = C\exp\frac{\eta}{c_v}, \tag{5.10.12}$$

with the subscript 0 indicating the initial condition, whereas the function C remains constant along each streamline. By virtue of Eqs. (5.10.10)–(5.10.12) and from $p = f(\rho, \eta)$, we obtain

$$p_{,i} = a^2\rho_{,i} + \frac{\rho a^2}{c_p}\eta_{,i}. \tag{5.10.13}$$

Recall that Eq. (5.10.2) may be rewritten in the form

$$\frac{1}{2}(V_jV_j)_{,i} - \epsilon_{ijk}\epsilon_{kmn}V_{n,m}V_j + \frac{1}{\rho}p_{,i} = 0. \tag{5.10.14}$$

Using the definition of enthalpy,

$$H = c_pT = \frac{\gamma}{\gamma - 1}\frac{p}{\rho}, \tag{5.10.15}$$

and in view of Eq. (5.10.13), we rewrite Eq. (5.10.14) as

$$T\eta_{,i} + \epsilon_{ijk}\epsilon_{kmn}V_{n,m}V_j - \left(H + \frac{1}{2}V_jV_j\right)_{,i} = 0 \tag{5.10.16a}$$

or

$$\mathbf{V} \times (\nabla \times \mathbf{V}) = \mathbf{V} \times \boldsymbol{\omega} = \nabla\hat{H} - T\nabla\eta, \tag{5.10.16b}$$

where $\hat{H}$ is the total enthalpy. This is known as Crocco's equation. It implies that, if entropy and enthalpy gradients are nonvanishing across the streamlines, then there exists a nonvanishing vorticity, a condition of nonisentropic and nonadiabatic flow.

Note that we can also derive Eq. (5.10.16) by combining the reversible process of the second law of thermodynamics and the momentum equation, respectively:

$$T\nabla\eta = \nabla H - \frac{1}{\rho}\nabla p,$$

$$\nabla\left(\frac{1}{2}\mathbf{V}\cdot\mathbf{V}\right) - \mathbf{V} \times \boldsymbol{\omega} = -\tfrac{1}{\rho}\nabla p,$$

from which Eq. (5.10.16) can be obtained.

For the direction normal to the streamline, Eq. (5.10.16) takes the form

$$T\eta_{,i}n_i + \epsilon_{ijk}\epsilon_{kmn}V_{n,m}V_jn_i - \hat{H}_{,i}n_i = 0 \tag{5.10.17}$$

or

$$T\frac{\partial\eta}{\partial n} - \frac{\partial\hat{H}}{\partial n} + \mathbf{V} \times (\nabla \times \mathbf{V}) \cdot \mathbf{n} = 0, \tag{5.10.18}$$

where, in two dimensions,

$$\mathbf{V} \times (\nabla \times \mathbf{V}) \cdot \mathbf{n} = \epsilon_{ijk}\epsilon_{kmn}V_{n,m}V_jn_i = V^*(V_{2,1} - V_{1,2}),$$

with

$$V^* = V_2n_1 - V_1n_2. \tag{5.10.19}$$

Hence the sonic vorticity is defined in terms of the enthalpy and entropy gradients:

$$E = \frac{1}{V^*}\left(\frac{\partial \hat{H}}{\partial n} - \frac{a^2}{\gamma R}\frac{\partial \eta}{\partial n}\right)\frac{V_1 V_2}{a^2}. \tag{5.10.20}$$

Behind the interacting shock waves of different strengths a slipstream (contact surface, vortex sheet) develops, across which (normal to the slipstream) entropy is discontinuous, resulting in rotational flow with $E \neq 0$.

In terms of the velocity potential function, from Eq. (5.10.5), we obtain

$$\phi_{,ii} - \frac{1}{a^2}\phi_{,i}\phi_{,j}\phi_{,ij} = 0 \tag{5.10.21a}$$

or

$$(1 - M_x^2)\frac{\partial^2 \phi}{\partial x^2} + (1 - M_y^2)\frac{\partial^2 \phi}{\partial y^2} - \frac{2}{a^2}\frac{\partial \phi}{\partial x}\frac{\partial \phi}{\partial y}\frac{\partial^2 \phi}{\partial x \partial y} = 0. \tag{5.10.21b}$$

It is important to note that the same result is obtained from Eq. (5.10.7) with $E = 0$, implying that the velocity potential representation leads to irrotational flow. For incompressible flows the sound speed is infinite and it is seen that Eq. (5.10.21a) or (5.10.21b) becomes a Laplace equation. The boundary equations on the surface of the body are

$$\phi = g(x, y) \quad \text{on} \quad \Gamma_1, \tag{5.10.22}$$

$$\phi_{,i}n_i = 0 \quad \text{on} \quad \Gamma_2. \tag{5.10.23}$$

Because the solution of full potential equations (5.10.21) represents a major computational effort, simplifications known as *small-perturbation* approximations are made in many engineering applications. We assume that a body such as an airfoil is so thin that disturbances (in the y direction) are limited to the immediate vicinity of the body and thus do not propagate far from it. In this case, Eq. (5.10.21b) is simplified to the small-perturbation potential equation

$$(1 - M_\infty^2)\frac{\partial^2 \phi'}{\partial x^2} + (1 - M_\infty^2)\frac{\partial^2 \phi'}{\partial y^2} = M_\infty^2\left(\frac{1+\gamma}{u_\infty}\right)\frac{\partial \phi'}{\partial x}\frac{\partial^2 \phi'}{\partial x^2}, \tag{5.10.24}$$

where M_∞ and u_∞ refer, respectively, to the Mach number and the velocity in the free-stream region, and ϕ' is the perturbation velocity potential defined as

$$\phi' = \phi - u_\infty x.$$

A linearized small-perturbation equation arises when the right-hand side of Eq. (5.10.24) is neglected. Further details are found in Liepmann and Roshko (1957).

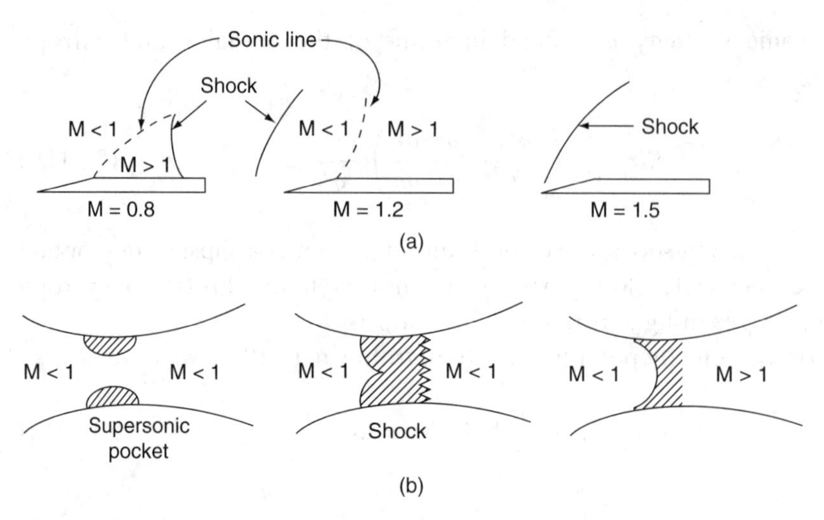

Figure 5.10.1. External and internal flows with various speed ranges: (a) flow around a wedge, (b) nozzle flow. The dashed curves represent $M = 1$, shaded area, supersonic.

5.10.3 Shock Waves

An interesting feature of Eq. (5.10.24) is that, depending on the magnitude of the Mach number, mathematical properties of the governing equation are altered in such a way that the following conditions arise:

for $M_\infty < 1$ (*subsonic speed*), elliptic partial differential equation;
for $M_\infty = 1$ (*transonic*), parabolic partial differential equation;
for $M_\infty > 1$ (*supersonic*), hyperbolic partial differential equation.

Note that the full potential of Eqs. (5.10.21) is of the mixed type in which the local Mach numbers M_x and M_y vary spatially within the flowfield. Shock-wave discontinuities usually occur as the Mach number approaches unity. Shock-wave structures for various geometries and Mach numbers are illustrated in Fig. 5.10.1.

The occurrence of shock waves is a nonisentropic process. If geometrical changes of flow domain (compression corners, etc.) are encountered in high-speed flows (with the Mach number higher than approximately 0.8), then sudden changes of streamline directions are accompanied by an increase in entropy, causing all flow variables to change abruptly. The conservation of mass, momentum, and energy is enforced in this process by the presence of shock waves, which allows the streamlines to bend nonisentropically.

The pressure discontinuity associated with shock waves is the most critical design information for high-speed vehicles. To this end, the pressure coefficient C_p is a convenient means to measure pressure variations over the body:

$$C_p = \frac{p - p_\infty}{\frac{1}{2}\rho_\infty u_\infty^2}. \tag{5.10.25}$$

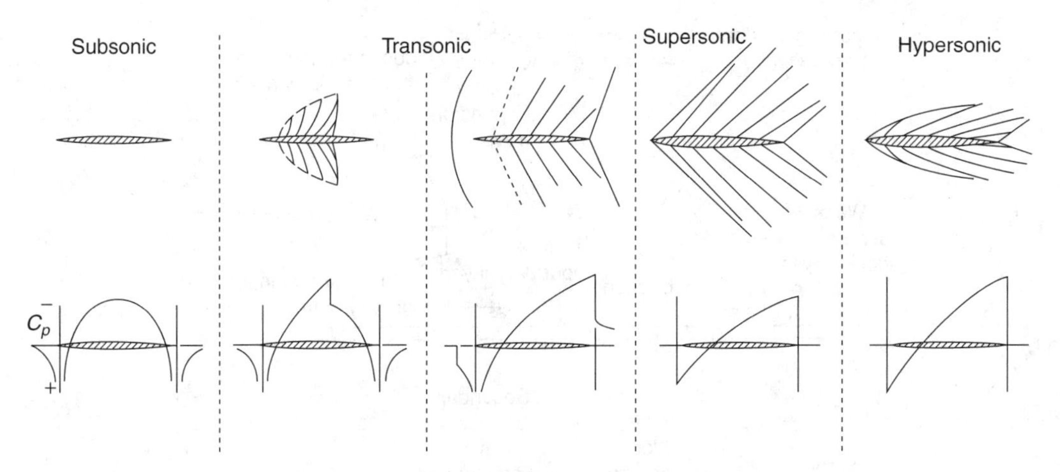

Figure 5.10.2. Surface pressure distributions for a typical airfoil.

For compressible flow, we may modify Eq. (5.10.25) as

$$C_p = \frac{2}{\gamma M_\infty^2}\left(\frac{p}{p_\infty} - 1\right) \tag{5.10.26}$$

whereas, for incompressible flow, Eq. (5.10.25) assumes the form

$$C_p = 1 - \left(\frac{V}{u_\infty}\right)^2, \tag{5.10.27}$$

with $V = (\mathbf{V} \cdot \mathbf{V})^{1/2}$.

A typical pressure coefficient distribution is shown in Fig. 5.10.2. Note that the negative and positive signs imply that $p < p_\infty$ and $p > p_\infty$, respectively. The precise form of C_p depends on the relationship between p and p_∞ for various types of flows and approximations, such as perturbations and linearizations.

The supersonic flow with Mach numbers higher than about 5 is referred to as hypersonic flow. For Mach numbers above about 2, the inviscid flow theory breaks down because the flow around a body is affected by viscous boundary layers. The flow may even become turbulent, leading to one of the most difficult and complicated fluid mechanics problems – shock-wave turbulent boundary-layer interactions, as shown in Fig. 5.10.3. The potential equation is no longer valid, and we must invoke the CNS equation, (5.3.24),

$$\frac{\partial \mathbf{U}}{\partial \mathbf{t}} + \frac{\partial \mathbf{F}_i}{\partial x_i} + \frac{\partial \mathbf{G}_i}{\partial x_i} = \mathbf{B},$$

together with additional equations involved in turbulence, as given in Section 5.7. In Fig. 5.10.3, the incoming incident shock wave intersects the laminar boundary layer upstream, raising it concave upward, producing weak compression shock waves directed away from the boundary layer, which coalesce into a strong

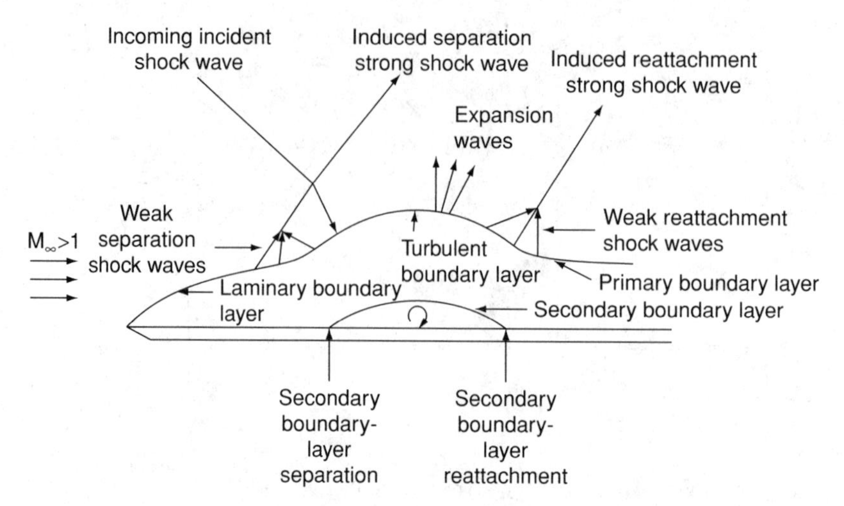

Figure 5.10.3. Boundary-layer interactions with an incoming incident shock wave, resulting in weak separation and reattachment shock waves and induced strong shock waves, expansion waves, and secondary boundary layer.

separation shock wave. The laminar boundary layer changes into a transition to turbulence and subsequently to a fully turbulent boundary-layer flow as it curves concave downward, producing expansion waves outward. The turbulent boundary layer then curves concave upward again, resulting in weak compression shock waves and coalescing into a strong reattachment shock wave. The secondary boundary layer develops below the primary boundary layer and above the wall corresponding to the separation and reattachment shock waves, producing highly viscous and turbulent eddy motions, and resulting in extremely high wall turbulent stresses and temperature. Above the primary boundary layer the flow is inviscid and compressible whereas it is viscous and incompressible below the primary boundary layer. Across the primary boundary layer, complex physical interactions occur between inviscid and viscous actions and between compressibility and incompressibility. Because analytical solutions are not available it is necessary that an accurate numerical scheme for the solution of the CNS equations be used for determining these interactions.

In summary, the nonlinear terms of governing equations (5.10.21), which are responsible for shock discontinuities, originate from the convective terms of Eqs. (5.10.2). This is similar in spirit to the turbulent (Reynolds) stress arising from the convective terms of Eq. (5.7.10). Thus we conclude that the convection of flow leads to both turbulence and shock-wave discontinuities. This is in contrast to solid mechanics, in which nonlinearity does not arise unless strains are large, which is analogous to non-Newtonian fluids in which stress is nonlinearly proportional to velocity gradients. We have explored the fluid mechanics problems only to the extent that tensor analysis plays an important role in the derivation of various forms of the governing equations for fluids.

5.11 Acoustics

5.11.1 General

So far we have discussed the standard problems involved in continuum mechanics, including solid mechanics, fluid mechanics, and heat transfer. The subject of acoustics is also regarded as an application area of continuum mechanics. Wave oscillations take place in both solids and fluids; they include sound waves or acoustic waves through solid materials, in a pressurized chamber filled with gases or liquids, or combustion waves in gases (Rayleigh, 1945; Pierce, 1956; Temkin; 1981). This section focuses on acoustic waves in fluids only in order to limit the scope of our discussion. Waves propagating in neither solids nor combustion are included.

The governing equations for acoustics are based on the conservation laws of fluid mechanics. The prerequisite for a sound wave to be transmitted is the presence of a compressible medium. In the absence of viscosity, the following conservation equations are invoked.

Continuity

$$\frac{D\rho}{Dt} + \rho\nabla\cdot\mathbf{V} = 0. \tag{5.11.1}$$

Momentum

$$\rho\frac{D\mathbf{V}}{Dt} + \nabla p = 0. \tag{5.11.2}$$

Energy

$$\frac{D\eta}{Dt} = 0. \tag{5.11.3}$$

For the isentropic process, the changes of pressure are given by

$$dp = \left(\frac{\partial p}{\partial\rho}\right)_{\eta} d\rho + \left(\frac{\partial p}{\partial\eta}\right)_{\rho} d\eta = a^2 d\rho, \tag{5.11.4}$$

with the speed of sound being defined as

$$a = \left(\frac{\partial p}{\partial\rho}\right)_{\eta_0}^{1/2}.$$

Thus, from Eq. (5.11.4), we obtain

$$\frac{Dp}{Dt} - a^2\frac{D\rho}{Dt} = 0.$$

In view of the continuity equation,

$$\frac{Dp}{Dt} + \rho a^2\nabla\cdot\mathbf{V} = 0. \tag{5.11.5}$$

Consider the state of equilibrium slightly disturbed so that

$$\rho = \rho_0(\mathbf{x}) + \rho'(\mathbf{x}, t), \tag{5.11.6a}$$

$$p = p_0(\mathbf{x}) + p'(\mathbf{x}, t), \tag{5.11.6b}$$

where the symbols 0 and prime denote equilibrium and disturbance, respectively. The spatial representation $\mathbf{x}$ refers to the Eulerian description in this section. Substituting Eq. (5.11.6) into Eq. (5.11.5) gives

$$\frac{\partial p'}{\partial t} + \rho_0(\mathbf{x}) a_0^2(\mathbf{x}) \nabla \cdot \mathbf{V} = 0, \tag{5.11.7}$$

where $a_0^2 = (\partial p/\partial \rho)_{\eta_0, \rho_0}$ and $(\mathbf{V} \cdot \nabla) p_0$ is neglected. This can be justified by the fact that, for the characteristic velocity U, length L, and pressure gradient $\nabla p_0 = \rho_0 g$, the ratio

$$|(V \cdot \nabla) p_0| / \rho_0 a_0^2 \, |\nabla \cdot V| \approx \frac{L}{a_0^2/g} \approx L/12 \text{ km}$$

in air at standard temperature and pressure may be negligibly small, because $L \ll 12$ km in a practical system. Similarly, from momentum equation (5.11.2), we have

$$\rho_0 \frac{\partial \mathbf{V}}{\partial t} + \nabla p' = 0. \tag{5.11.8}$$

Now, taking the time derivative of Eq. (5.11.7) and the spatial derivative of Eq. (5.11.8) leads to, respectively,

$$\frac{\partial^2 p'}{\partial t^2} + \rho_0(\mathbf{x}) a_0^2(\mathbf{x}) \nabla \cdot \frac{\partial \mathbf{V}}{\partial t} = 0, \tag{5.11.9a}$$

$$\nabla \cdot \frac{\partial \mathbf{V}}{\partial t} + \frac{1}{\rho_0} \nabla^2 p' = \frac{1}{\rho_0^2} \nabla p' \cdot \nabla \rho_0 \cong 0. \tag{5.11.9b}$$

We combine Eqs. (5.11.9a) and (5.11.9b), which gives

$$\frac{\partial^2 p'}{\partial t^2} - a_0^2(\mathbf{x}) \nabla^2 p' = 0. \tag{5.11.10}$$

This is the wave equation (or acoustic equation) for a fluid whose equilibrium state is not uniform. It is the same as the classical wave equation, except that the speed of sound a_0 is a function of position.

For many engineering problems, the linearized continuity and momentum equations can be used to derive the wave equations, assuming that the flow is uniform or that

$$\frac{\partial p'}{\partial t} + \rho_0 a_0^2 \nabla \cdot \mathbf{V} = 0, \tag{5.11.11}$$

$$\rho_0 \frac{\partial \mathbf{V}}{\partial t} + \nabla p' = 0. \tag{5.11.12}$$

We introduce the velocity potential ϕ such that $\mathbf{V} = \nabla\phi$, which is substituted into Eq. (5.11.12); hence we have

$$\nabla\left(\rho_0 \frac{\partial \phi}{\partial t} + p'\right) = 0,$$

$$\rho_0 \frac{\partial \phi}{\partial t} + p' = c(t),$$

where the constant of integration $c(t)$ may be set equal to zero without loss of generality. Thus,

$$p' = -\rho_0 \frac{\partial \phi}{\partial t}.$$

Now we substitute this into Eq. (5.11.11) to get

$$\frac{\partial^2 \phi}{\partial t^2} - a_0^2 \nabla^2 \phi = 0. \tag{5.11.13}$$

Similarly, taking the gradient of Eq. (5.11.11) and substituting in Eq. (5.11.12) yields

$$\frac{\partial^2 \mathbf{V}}{\partial t^2} - a_0^2 \nabla^2 \mathbf{V} = 0. \tag{5.11.14}$$

Equations (5.11.10), (5.11.13), and (5.11.14) are the wave equations in terms of pressure, velocity potential, and velocity vector, respectively. The choice depends on convenience as determined by the available boundary conditions and other aspects of the particular engineering application.

5.11.2 Monochromatic Waves

A *monochromatic wave* is one in which the pressure and velocity depend on time only through periodic functions of time of a single (circular) frequency ω. The velocity potential may be written as the real part ($\Re$) of the oscillatory behavior:

$$\Phi(\mathbf{x}, t) = \Re\{\Phi(\mathbf{x}) e^{-i\omega t}\}. \tag{5.11.15}$$

Substituting Eq. (5.11.15) into Eq. (5.11.13) gives

$$\nabla^2 \Phi + k^2 \Phi = 0, \tag{5.11.16}$$

where $k = \omega / a_0$ is the wave number. This is known as the Helmholtz equation, which is now independent of time.

Let us consider a *plane wave*, which is defined as a wave that impinges on a plane in a one-, two-, or three-dimensional geometry. For simplicity, we examine a plane wave in one direction:

$$\frac{d^2 \Phi}{dx^2} + k^2 \Phi = 0. \tag{5.11.17}$$

The general solution of this equation can be written as

$$\Phi = Ae^{ikx} + Be^{-ikx}. \tag{5.11.18}$$

We substitute this into Eq. (5.11.15) to get

$$\Phi(x, t) = Ae^{i(kx-\omega t)} + Be^{-i(kx-\omega t)}. \tag{5.11.19}$$

Setting $B = 0$, we arrive at the *positive-going wave*, defined as

$$\Phi = \tilde{A}\cos(kx - \omega t + \beta), \tag{5.11.20}$$

where $A = |A|\, e^{i\beta}$ and $\tilde{A} = |A|$ are called the *amplitudes* of the wave and the quantity

$$\psi = kx - \omega t + \beta \tag{5.11.21}$$

is known as the *phase*. Note that the phase is constant for an observer moving with velocity *dx/dt*, so that

$$\frac{d\psi}{dt} = k\frac{dx}{dt} - \omega = 0.$$

Thus

$$\frac{dx}{dt} = \frac{\omega}{k} = c_f,$$

where c_f is the phase velocity and in this simple case is equal to the speed of sound a_0. Because the acoustic variables in the wave are sinusoidal functions of time with period T, we have

$$\cos(kx - \omega t + \beta) = \cos\left[kx - \omega(t + T) + \beta\right], \tag{5.11.22}$$

which requires that $\omega = 2\pi/T$. Similarly, the spatial period, called the *wavelength* λ, may be introduced in the form

$$\cos kx = \cos\left[k(x + \lambda)\right]. \tag{5.11.23}$$

This leads to $k = 2\pi/\lambda$ and $a_0 = \lambda/T = f\lambda$, where $f = 1/T$ is the frequency.

Let us now consider a plane wave in three dimensions propagating along some axis $s(\mathbf{x})$ such that $\mathbf{x} = s\mathbf{n}$, $s = \mathbf{n}\cdot\mathbf{x}$, and

$$\Phi = Ae^{ik\mathbf{n}\cdot\mathbf{x}}. \tag{5.11.24}$$

We define a wave vector $\mathbf{k}$ as

$$\mathbf{k} = k\mathbf{n} = \frac{\omega}{a_0}\mathbf{n}.$$

Then,

$$\Phi = Ae^{i\mathbf{k}\cdot\mathbf{x}}$$

or

$$\Phi = Ae^{ik_jx_j}.$$

Thus

$$\frac{\partial \Phi}{\partial x_m} = Aik_j \frac{\partial x_j}{\partial x_m} e^{ik_n x_n} = Aik_m e^{ik_n x_n} = ik_m \Phi.$$

This leads to

$$\frac{\partial^2 \Phi}{\partial x_m \partial x_m} + k^2 \Phi = 0, \tag{5.11.25}$$

in which the following identity has been used:

$$k_j k_j = \mathbf{k} \cdot \mathbf{k} = \frac{\omega}{a_0} \mathbf{n} \cdot \frac{\omega}{a_0} \mathbf{n} = \left(\frac{\omega}{a_0} \right)^2 = k^2.$$

Therefore we conclude that the plane wave in three dimensions assumes the solution

$$\Phi(\mathbf{x}, t) = Ae^{i(k_j x_j - \omega t)} + Be^{-i(k_j x_j + \omega t)}. \tag{5.11.26}$$

This solution is useful in problems that deal with plane waves over boundaries that are neither parallel nor perpendicular to the direction of propagation.

The acoustic pressure p' may be written as

$$p'(\mathbf{x}, t) = \hat{p}'(\mathbf{x}) e^{-i\omega t}, \tag{5.11.27}$$

whereas the velocity vector $\mathbf{V}$ is expressed in the form

$$\mathbf{V} = \hat{\mathbf{V}}(\mathbf{x}) e^{-i\omega t}. \tag{5.11.28}$$

From the momentum equation,

$$\frac{\partial \mathbf{V}}{\partial t} = -\frac{1}{\rho_0} \nabla p'. \tag{5.11.29}$$

For plane waves,

$$\hat{p}' = p_0 e^{ik_j x_j},$$

$$\hat{p}'_{,j} = ik_j p_0 e^{ik_n x_n} = ik_j \hat{p}'.$$

Thus, from Eq. (5.11.29) together with Eqs. (5.11.27) and (5.11.28), we obtain

$$\hat{\mathbf{V}}'(\mathbf{x}) = \frac{1}{i\rho_0 \omega} \nabla \hat{p}'(\mathbf{x}) \tag{5.11.30}$$

or

$$\hat{V}'_j = \frac{1}{i\rho_0 \omega} \hat{p}'_{,j} = \frac{k_j \hat{p}'}{\rho_0 \omega}. \tag{5.11.31}$$

For the one-dimensional case,

$$\hat{u} = \frac{1}{i\rho_0 \omega} \frac{\partial \hat{p}'}{\partial x} = \frac{k \hat{p}'}{\rho_0 \omega}. \tag{5.11.32}$$

Let us now examine the incident wave in the positive x direction:

$$p' = \bar{A} e^{-i(kx - \omega t)}, \tag{5.11.33}$$

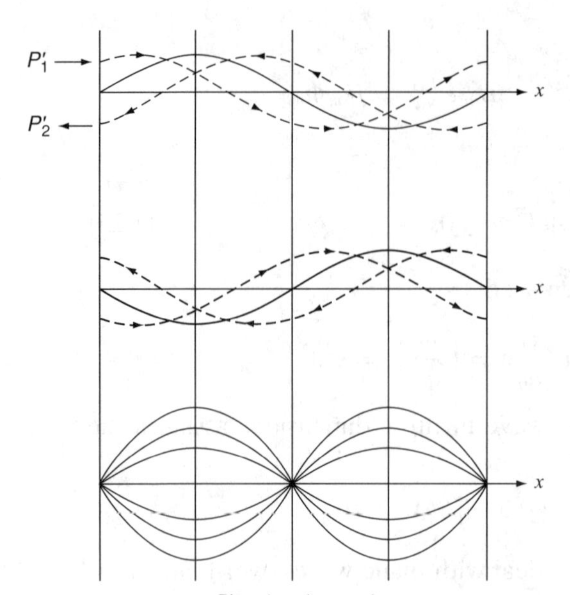

—— Traveling wave: $P_1' = A\cos(\omega t - \kappa x)$
$P_2' = A\cos[\omega t - \kappa(x - l)] = A\cos(\omega t - \kappa x)$

- - - - Standing wave: $P' = P_1' + P_2' = 2A\cos\omega t\cos\kappa x$

Figure 5.11.1. Standing wave. Constant harmonic oscillation at any point x equivalent to superposition of two traveling waves, positive and negative, at any given time, may grow or decay as a function of time only.

and let the rigid reflector be placed at $x = 0$. The solution of the wave equation for the space in front of the reflector is given by

$$p' = \bar{A}e^{i(\omega t - kx)} + \bar{B}e^{i(\omega t + kx)}. \tag{5.11.34}$$

Thus the sound field in front of the reflector consists of the incident wave and a second wave that travels from the reflector to $-\infty$. Obviously this wave can be interpreted as the wave reflected at the rigid reflector, where

$$\begin{aligned} u'\big|_{x=0} = 0 &= \frac{1}{i\omega\rho_0}\left(\frac{\partial p'}{\partial x}\right)_{x=0} = \frac{1}{ik\rho_0 a_0}(-ik\bar{A} + ik\bar{B})e^{i\omega t} \\ &= \frac{1}{\rho_0 a_0}(\bar{A} + \bar{B})e^{i\omega t}, \end{aligned} \tag{5.11.35}$$

which suggests that $\bar{B} = \bar{A}$ and

$$\bar{R} = \frac{\bar{B}}{\bar{A}} = 1, \tag{5.11.36}$$

where $\bar{R}$ is the reflection coefficient for the pressure. The resultant pressure,

$$p' = \bar{A}e^{i\omega t}(e^{-ikx} + e^{ikx}) = 2\bar{A}\cos(kx)e^{i\omega t}, \tag{5.11.37}$$

represents a *standing wave* of amplitude $2\bar{A}\cos(kx)$ and frequency ω. The standing wave is a constant harmonic oscillation at any point x equivalent to the superposition of two *traveling waves*, positive and negative, at any given time, which may grow or decay as a function of time only. The relationship between traveling and standing waves is illustrated in Fig. 5.11.1. The real part of the solution in Eq. (5.11.37) reads

$$p' = 2A\cos(kx)\cos(\omega t + \phi_0). \tag{5.11.38}$$

We obtain the complex particle velocity from Eq. (5.11.35) by setting $\bar{B} = \bar{A}$ so that

$$u' = \frac{1}{\rho_0 a_0}(\bar{A}e^{-ikx} - Ae^{ikx})e^{i\omega t}$$

or

$$u' = -\frac{2i\bar{A}}{\rho_0 a_0}\sin(kx)e^{i\omega t}. \tag{5.11.39}$$

The real part of the solution of the particle velocity is

$$\begin{aligned} u' &= -\frac{2\bar{A}}{\rho_0 a_0}\sin(kx)\sin(\omega t + \phi_0) \\ &= \frac{2\bar{A}}{\rho_0 a_0}\sin(kx)\sin\left(\omega t + \phi_0 - \frac{\pi}{2}\right). \end{aligned} \tag{5.11.40}$$

The particle velocity vanishes at the reflector and lags the pressure in phase by 90°. Note that the pressure is reflected without phase change at a rigid surface. Because the reflected wave travels in the direction with the opposite sign of the incident wave at the reflector, the resultant particle velocity at the reflector is zero.

In acoustics there are three different modes of wave propagation: pressure-mode acoustics, vorticity-mode acoustics, and entropy-mode acoustics. These topics are described in the following subsections.

5.11.3 Pressure-Mode Acoustics

A prominent example of pressure-mode acoustics is the idea of Kirchhoff's formula surrounding the region of a nonlinear flowfield and acoustic sources by a closed surface. In the domain inside the surface, a nonlinear aerodynamic computation is carried out that provides the pressure distribution on the surface as well as its time history. Outside this surface, the acoustic disturbance satisfies the stationary wave, Eq. (5.11.10). To determine $p(\mathbf{x}, t)$, consider the homogeneous Helmholtz equation, (5.11.16), whose solution is the Green's function $G(\mathbf{y}, \mathbf{x}; \omega)$,

$$p(x, \omega) = \int_S \left[G(\mathbf{x}, \mathbf{y}; \omega)\frac{\partial p}{\partial y}(\mathbf{y}, \omega) - p(\mathbf{y}, \omega)\frac{\partial G}{\partial y_j}(\mathbf{x}, \mathbf{y}; \omega)\right] n_j dS(\mathbf{y}), \tag{5.11.41}$$

when n_j is the unit normal on S into the fluid and $\mathbf{y}$ denotes the integration variable for $\mathbf{x}$. Making use of the convolution theorem, we can write the time-domain solution as

$$p(x,t) = \int_S \left[-G(\mathbf{x}, \mathbf{y}; t-\tau) \frac{\partial p}{\partial y_j}(\mathbf{y}, \tau) d\tau + p(\mathbf{y}, \tau) \frac{\partial G}{\partial y_j}(\mathbf{x}, \mathbf{y}; t-\tau) \right] n_j dS(\mathbf{y}), \tag{5.11.42}$$

where the retarded time integration is taken over $(-\infty, \infty)$. The linearized momentum equation gives

$$\rho_0 \frac{\partial V_j}{\partial \tau} = -\frac{\partial p}{\partial y_j}$$

in the absence of the body forces, which leads to

$$p(x,t) = \int_S \left[G(\mathbf{x}, \mathbf{y}; t-\tau) \rho_0 \frac{\partial V_j}{\partial \tau}(\mathbf{y}, \tau) d\tau + p(\mathbf{y}, \tau) \frac{\partial G}{\partial y_j}(\mathbf{x}, \mathbf{y}; t-\tau) \right] n_j dS(\mathbf{y}) d\tau. \tag{5.11.43}$$

Using the free-space Green's function given by

$$G(\mathbf{x}, \mathbf{y}; t-\tau) = \frac{1}{4\pi |\mathbf{x}-\mathbf{y}|} \delta(t-\tau - |\mathbf{x}-\mathbf{y}|/a_0),$$

we find that Eq. (5.11.43) becomes

$$p(x,t) = \frac{\rho_0}{4\pi} \frac{\partial}{\partial t} \int_S \frac{V_n(\mathbf{y}, t - |\mathbf{x}-\mathbf{y}|/a_0)}{(\mathbf{x}-\mathbf{y})} dS(y) - \frac{1}{4\pi} \frac{\partial}{\partial x_j} \int_S \frac{p(\mathbf{y}, t - |\mathbf{x}-\mathbf{y}|/a_0)}{|\mathbf{x}-\mathbf{y}|} n_j dS(y). \tag{5.11.44}$$

Equation (5.11.44) can be reduced to the following form:

$$4\pi p(x,t) = \int_S \left[\frac{p}{r^2} \frac{\partial r}{\partial \mathbf{n}} - \frac{1}{r} \frac{\partial p}{\partial \mathbf{n}} + \frac{1}{a_0 r} \frac{\partial r}{\partial \mathbf{n}} \frac{\partial p}{\partial \tau} \right], \tag{5.11.45}$$

where $|\mathbf{x}-\mathbf{y}| = r$ is the distance between the observer and the source $\partial r/\partial \mathbf{n} = \cos\theta$, where θ is the angle between the normal vector and the radial direction, and $\mathbf{n}$ is the outward normal vector. In (5.11.45), the bracket notation [ ] is used to denote the retarded time.

5.11.4 Vorticity-Mode Acoustics

The sound generated by vorticity in an unbounded fluid is generally referred to as aerodynamic sound (Lighthill, 1952, 1954). Most fluid flows of engineering interest are unsteady in nature, of high Reynolds number, and turbulent. These flows are known to generate noise; that is, turbulent boundary layers, jets, and shear layers. Although the acoustic radiation is a very small by-product of the fluid motion, it is becoming an important part of the flow solutions.

The theory of aerodynamic sound was developed by Lighthill (1952), who rewrote the Navier–Stokes equations into an exact inhomogeneous wave equation whose source terms are important only within the turbulent region. Furthermore, at low Mach numbers, the sound generation and subsequent propagation can be decoupled from the fluid motion. The momentum equation for an ideal, stationary fluid of density ρ_0 and sound speed a_0 subject to the externally applied stress T_{ij} is

$$\frac{\partial(\rho V_i)}{\partial t} + \frac{\partial\left[a_0^2(\rho - \rho_0)\right]}{\partial x_i} = -\frac{\partial T_{ij}}{\partial x_j}. \tag{5.11.46}$$

Using the continuity equation to eliminate (ρV_i) results in the well-known Lighthill acoustic analogy equation:

$$\left(\frac{1}{a_0^2}\frac{\partial^2}{\partial t^2} - \nabla^2\right)\left[a_0^2(\rho - \rho_0)\right] = \frac{\partial^2 T_{ij}}{\partial x_i \partial x_j}. \tag{5.11.47}$$

Momentum transfer is produced solely by the pressure. Here T_{ij} is the Lighthill stress tensor given by

$$T_{ij} = \rho V_i V_j + \left[(p - p_0) - a_0^2(\rho - \rho_0)\right]\delta_{ij} - \tau_{ij}. \tag{5.11.48}$$

The solution of (5.11.47) requires determination of the Lighthill stress tensor given by (5.11.48). When the mean density and sound speed are uniform, the variation in ρ produced by low Mach number, high Reynolds number, and velocity fluctuations are of order $\rho_0 M^2$ and $\rho V_i V_j \approx \rho_0 V_i V_j$ with a relative error $\sim O(M^2) \ll 1$.

Similarly, we have

$$p - p_0 - a_0^2(\rho - \rho_0) \approx (p - p_0)\left(1 - a_0^2/a^2\right) \sim O\left(\rho_0 V^2 M^2\right).$$

Therefore $T_{ij} \approx \rho_0 V_i V_j$, and when viscous stresses are neglected, the solution to Lighthill equations can be written as

$$p(\mathbf{x}, \mathbf{t}) \approx \frac{\partial^2}{\partial x_i \partial x_j}\int \frac{\rho_0 V_i V_j(\mathbf{y}, t - |\mathbf{x} - \mathbf{y}|/a_0)}{4\pi(\mathbf{x} - \mathbf{y})} d^3\mathbf{y}$$

$$\approx \frac{x_i x_j}{4\pi a_0^2 |\mathbf{x}^3|}\frac{\partial^2}{\partial t^2}\int \rho_0 V_i V_j(\mathbf{y}, t - |\mathbf{x} - \mathbf{y}|/a_0) d^3\mathbf{y} \quad |x| \to \infty, \tag{5.11.49}$$

where $p(\mathbf{x}, t) = a_0^2(\rho - \rho_0)$ is the perturbation pressure in the far field.

When Lighthill's acoustic analogy is used in flows with moving boundaries, moving sources, or in turbulent shear layers separating a quiescent medium from a high-speed flow, it is necessary to introduce control surfaces. These surfaces can coincide with existing physical surfaces or correspond to a convenient interface between fluid regions of widely differing mean properties. Suitable boundary conditions are applied on these surfaces.

Let $f(\mathbf{x}, t)$ be an indicator function that vanishes on the surface S and satisfies $f(\mathbf{x}, t) > 0$ in the fluid where Lighthill's equation is to be solved, and $f(\mathbf{x}, t) < 0$

elsewhere. Multiply (5.11.46) by $H(f)$ and rearrange into the form

$$\frac{\partial}{\partial t}[\rho V_i H(f)] + \frac{\partial}{\partial x_i}\left[H(f)a_0^2(\rho - \rho_0)\right]$$
$$= -\frac{\partial}{\partial x_j}[H(f)T_{ij}] + [\rho V_i(V_j - \bar{V}_j] + [p - p_0)\delta_{ij} - \tau_{ij}]\frac{\partial H}{\partial x_j}(f). \quad (5.11.50)$$

A similar process can be applied to the continuity equation to obtain

$$\frac{\partial}{\partial t}\left\{H(f)(\rho - \rho_0) + \frac{\partial}{\partial x_i}[H(f)\rho V_i]\right\} = [\rho(V_i - \bar{V}_i) + \rho_0 \bar{V}_i]\frac{\partial H}{\partial x_i}(f). \quad (5.11.51)$$

Elimination of $H\rho V_i$ between the two preceding equations leads to the well-known Ffowcs-Williams–Hawkings equation:

$$\left(\frac{1}{a_0^2}\frac{\partial^2}{\partial t^2} - \nabla^2\right)\left[H(f)a_0^2(\rho - \rho_0)\right]$$
$$= \frac{\partial^2(H(f)T_{ij})}{\partial x_i \partial x_j} - \frac{\partial}{\partial x_i}\left\{\left[\rho V_i(V_j - \bar{V}_j) + (p - p_0)\delta_{ij} - \tau_{ij}\right]\frac{\partial H}{\partial x_j}(f)\right\}$$
$$+ \frac{\partial}{\partial t}\left\{[\rho(V_j - V_j) + \rho_0 \bar{V}_j]\frac{\partial H}{\partial x_j}(f)\right\}. \quad (5.11.52)$$

This equation is valid throughout the whole space. By using the Green's function we can write a formal outgoing wave solution. Written in an integral form, the Ffowcs–William-Hawking equation (1969) is

$$H(f)a_0^2(\rho - \rho_0) = \frac{\partial^2}{\partial x_i \partial x_j}\int_{V(\tau)} [T_{ij}]\frac{d^3\mathbf{y}}{4\pi|\mathbf{x} - \mathbf{y}|}$$
$$- \frac{\partial}{\partial x_i}\oint_{S(\tau)} [\rho V_i(V_j - \bar{V}_j) + P'_{ij}]\,\frac{dS_j(\mathbf{y})}{4\pi|\mathbf{x} - \mathbf{y}|}$$
$$+ \frac{\partial}{\partial t}\oint_{S(\tau)} [\rho(V_j - \bar{V}_j) + \rho_0 \bar{V}_j]\,\frac{dS_j(\mathbf{y})}{4\pi|\mathbf{x} - \mathbf{y}|}, \quad (5.11.53)$$

where $P'_{ij} = (p - p_0)\delta_{ij} - \tau_{ij}$ and the square bracket denotes the retarded time ($\tau = t - |\mathbf{x} - \mathbf{y}|/a_0$). The surface integrals indicate a monopole and a dipole source. When $\bar{V}$ is zero in (5.11.50) (i.e., stationary surface), the generalized Kirchhoff formula is recovered.

5.11.5 Entropy-Mode Acoustics

If temperature gradients are high, the fluctuation components of temperature can be very large, leading to entropy waves. In this case, we invoke the first and second laws of thermodynamics. At equilibrium in a continuous flowfield domain, the combined first and second laws of thermodynamics and Maxwell's relations (Subsection 5.3.4) can be written as

$$T\frac{D\eta}{Dt} = \frac{DT}{Dt} + p\frac{D}{Dt}\left(\frac{1}{\rho}\right), \tag{5.11.54}$$

$$T\frac{D\eta}{Dt} = \frac{DH}{Dt} - \frac{1}{\rho}\frac{Dp}{Dt}, \tag{5.11.55}$$

or

$$T\frac{D\eta}{Dt} = c_p\frac{DT}{Dt} - \frac{\alpha T}{\rho}\frac{Dp}{Dt}, \tag{5.11.56}$$

where η is the specific entropy and α is the thermal expansion coefficient:

$$\alpha = -\frac{1}{\rho}\frac{\partial\rho}{\partial T}$$

It can be shown that the equations of momentum, energy, and vorticity transport for three-dimensional compressible flows are of the following forms:

Momentum

$$\frac{\partial \mathbf{V}}{\partial t} + \nabla\hat{H} - \mathbf{V}\times\boldsymbol{\omega} = T\nabla\eta + \nu\left[\nabla^2\mathbf{V} + \frac{1}{3}\nabla(\nabla\cdot\mathbf{V})\right], \tag{5.11.57}$$

where $\hat{H}$ denotes the total enthalpy.

Energy

$$\rho c_p\frac{DT}{Dt} - \alpha T\frac{Dp}{Dt} - \tau_{ij}V_{j,i} - k^2\nabla T = 0 \quad \text{for sound emission,} \tag{5.11.58a}$$

$$\rho T\frac{D\eta}{Dt} = \tau_{ij}V_{j,i} - k^2\nabla T = 0 \quad \text{for sound absorption.} \tag{5.11.59a}$$

Continuity

$$\frac{1}{\rho a^2}\frac{Dp}{Dt} + \nabla\cdot\mathbf{V} = \frac{\alpha T}{c_p}\frac{D\eta}{Dt}. \tag{5.11.59}$$

Vorticity transport

$$\frac{\partial\boldsymbol{\omega}}{\partial t} + (\mathbf{V}\cdot\nabla)\boldsymbol{\omega} + \boldsymbol{\omega}\nabla\cdot\mathbf{V} - (\boldsymbol{\omega}\cdot\nabla)\mathbf{V} = \nabla T\times\nabla\eta + \nu\nabla^2\boldsymbol{\omega}' \tag{5.11.60}$$

Taking a time derivative of (5.11.58) and combining with (5.11.57), we obtain the acoustic analogy equation that may be used for determining unstable entropy waves:

$$\frac{\partial}{\partial t}\left(\frac{1}{\rho a^2}\frac{\partial p}{\partial t}\right) - \boldsymbol{\nabla}\cdot\left(\frac{1}{\rho}\boldsymbol{\nabla}p\right) = (V_iV_j)_{,ij} + \frac{\partial}{\partial t}\left(\frac{\alpha T}{c_p}\frac{D\eta}{Dt}\right). \tag{5.11.61}$$

Although unstable entropy waves can be calculated from (5.11.61), it is more convenient to use a form in which the entropy term is replaced with thermodynamic relationships. This approach, known as the entropy-controlled instability (ECI) method, is intended to include pressure and vorticity modes as well as the entropy mode where all three modes can be combined, as detailed in Yoon and Chung (1994).

5.12 Reacting Flows

5.12.1 General

This section considers the various conservation equations as they would apply in a system of reacting chemical species for which macroscopic viewpoints are assumed to be valid. Thus our discussion is limited to a reacting fluid that can be treated as a continuum in which the ideal gas law still holds. In such a situation, the continuum theory presented in the previous sections for Newtonian fluids may be extended to construct the conservation equations for multicomponent reacting systems.

With reference to the index notation we have used so far, there is a conflict in describing chemical species denoted by subscripts with the components of a vector represented by indices. The subscripts for chemical species, of course, do not follow the tensorial rules, and the reader is cautioned to distinguish between the chemical species and tensorial indices as they appear side by side together within a term.

In reacting fluids we are dealing with mass concentrations, molar concentrations, mass fractions, and mole fractions of various species, defined as follows:

(1) *Mass concentration* ρ_i: the mass of species i per unit volume of solution.
(2) *Molar concentration* $C_i = \rho_i / W_i$: the number of moles of species i per unit volume, where W_i is the molecular weight.
(3) *Mass fraction* $Y_i = \rho_i / \rho$: the mass concentration of species i divided by the total mass density of the solution, $\sum_{i=1}^{N} Y_i = 1$, where N is the total number of species.
(4) *Mole fraction* $X_i = C_i / C$: the molar concentration of species i divided by the total molar density of the solution.

Here the subscript i represents the species rather than the tensor index. If tensor indices are used, they are clearly distinguished from the subscripts for the species.

The gaseous state under ordinary conditions (i.e., the air we breathe) is extremely dilute, with the molecules separated by large distances compared with their molecular diameters. Thus, in general, the ideal gas law holds that

$$p = \rho R T, \qquad (5.12.1)$$

where R is the specific gas constant and has the units J kg^{-1}K^{-1}. It is also known that at a constant temperature and in the limit as p approaches zero, the ratio

p/ρ is equal to the inverse molecular weight ratio:

$$\frac{\left(\frac{p}{\rho}\right)_A}{\left(\frac{p}{\rho}\right)_B} = \frac{R_A}{R_B} = \frac{W_A}{W_B}.$$

Therefore we may write

$$R^0 = R_A W_A = R_B W_B,$$

in which R^0 is the universal gas constant for all substances with the value 8.31434 J mol^{-1}K^{-1}. The ideal gas law then becomes

$$p = \frac{1}{V} R^0 T = \frac{\rho}{W} R^0 T,$$

where $V = W/\rho$, the volume of one mole or 6.02283×10^{23} molecules of the gas. The volume of one mole of an ideal gas at $p = 101{,}325$ Pa (or 1 atm) and $T = 273.15$ K (0 °C) is 0.022414 m^3.

For a mixture of thermally perfect gases, the equation of state is expressed in various ways:

$$p = R^0 T \sum_{i=1}^{N} C_i = \rho R^0 T \sum_{i=1}^{N} \frac{Y_i}{W_i} = R^0 T \frac{\sum_{i=1}^{N} N_i}{V}, \tag{5.12.2}$$

where N_i is the number of moles of species i.

The various chemical species in a diffusing mixture have different velocities $\mathbf{v}_i$ measured with respect to stationary coordinate axes. Thus, for a mixture of N species, the local mass-average velocity $\mathbf{v}$ is defined as

$$\mathbf{v} = \sum_{i=1}^{N} \rho_i \mathbf{v}_i \bigg/ \sum_{i=1}^{N} \rho_i. \tag{5.12.3a}$$

The difference between the local mass-average velocity $\mathbf{v}$ and the velocity of the ith species $\mathbf{v}_i$ is known as the diffusion velocity $\mathbf{V}_i$ (Hirschfelder, Curtis, and Bird, 1954):

$$\mathbf{V}_i = \mathbf{v}_i - \mathbf{v}. \tag{5.12.3b}$$

A diffusion velocity is zero if $\mathbf{v}_i$ is equal to $\mathbf{v}$, in which case the motion of component i coincides with the local motion of the fluid stream. Here, the lowercase $\mathbf{v}$ is to be taken as the Eulerian mass-average velocity (recall the earlier use of $\mathbf{v}$ for the Lagrangian velocity).

Similarly, a local molar-average velocity $\mathbf{v}^*$ may be defined as

$$\mathbf{v}^* = \sum_{i=1}^{N} C_i \mathbf{v}_i \bigg/ \sum_{i=1}^{N} C_i. \tag{5.12.4a}$$

Thus the molar diffusion velocity becomes

$$\mathbf{V}_i^* = \mathbf{v}_i - \mathbf{v}^*. \tag{5.12.4b}$$

The mass flux or molar flux of species i is a vector quantity that denotes the mass or number of moles of species i that passes through a unit area per unit time relative to stationary coordinates:

$$\dot{\mathbf{m}}_i = \rho_i \mathbf{v}_i \quad \text{(mass flux)}, \tag{5.12.5a}$$

$$\dot{\mathbf{n}}_i = C_i \mathbf{v}_i \quad \text{(molar flux)}. \tag{5.12.5b}$$

The relative mass flux and relative molar flux are

$$\mathbf{J}_i = \rho(\mathbf{v}_i - \mathbf{v}) = \rho_i \mathbf{V}_i, \tag{5.12.6a}$$

$$\mathbf{J}_i^* = C_i(\mathbf{v}_i - \mathbf{v}^*) = C_i \mathbf{V}_i^*. \tag{5.12.6b}$$

The mass diffusivity $D_{AB} = D_{BA}$ in a binary system (a system with species A and B) is related to these quantities by *Fick's first law*:

$$\mathbf{J}_A = \rho_A \mathbf{V}_A = -\rho D_{AB} \nabla Y_A \quad \text{(mass diffusion flux)}, \tag{5.12.7a}$$

$$\mathbf{J}_A^* = C_A \mathbf{V}_A^* = -C D_{AB} \nabla X_A \quad \text{(molar diffusion flux)}. \tag{5.12.7b}$$

In general, the reaction mechanism for multistep reversible reactions is written as

$$\sum_{i=1}^{N} \nu'_{i,k} M_i \leftrightarrow \sum_{i=1}^{N} \nu''_{i,k} M_i \quad (k = 1, 2, \ldots, M), \tag{5.12.8}$$

in which $\nu_{i,k}$ is the stoichiometric coefficient of species i for reaction step k, the prime and double primes represent the reactant and product, respectively, and M_i is a stand-in for the chemical formula for species i. The rate of production of species i for reaction step k is

$$\omega_i = (\nu''_{i,k} - \nu'_{i,k})\omega'_k \quad (i = 1, 2, \ldots, N;\ k = 1, 2, \ldots, M),$$

with the net forward rate of the kth reaction defined as

$$\omega'_k = K_{f,k} \prod_{i=1}^{N} C_i^{\nu'_{i,k}} - K_{b,k} \prod_{i=1}^{N} C_i^{\nu''_{i,k}} \quad (k = 1, 2, \ldots, M),$$

where $K_{f,k}$ and $K_{b,k}$ denote the specific reaction-rate constants for the forward and backward reactions, respectively, and Π represents a cumulative product.

The specific reaction-rate constant is usually evaluated by the Arrhenius law,

$$K_{f,k}, K_{b,k} = B_k T^{\alpha_k} \exp\left(-\frac{E_k}{R^0 T}\right),$$

where B_k is the frequency factor and α_k is the constant. Thus the reaction rate is written as

$$\omega_i = W_i \sum_{k=1}^{M} (\nu''_{i,k} - \nu'_{i,k}) B_k T^{\alpha_k} \exp\left(-\frac{E_k}{R^0 T}\right) \prod_{j=1}^{N} \left(\frac{X_j p}{R^0 T}\right)^{\nu'_{j,k}}. \tag{5.12.9}$$

With these preliminaries, we are now prepared to develop the equations for the conservation of mass, momentum, and energy in reacting flows.

5.12.2 Conservation of Mass for Mixture and Species

Consider the continuity equation for component A in a binary mixture with a chemical reaction at a rate $\omega_A(\mathrm{kg}^{-3}\,\mathrm{s}^{-1})$, known as the *mass rate of production of species A*:

$$\frac{\partial \rho_A}{\partial t} + \nabla \cdot (\rho_A \mathbf{v}_A) = \omega_A. \tag{5.12.10}$$

Similarly, the equation of continuity for component B is

$$\frac{\partial \rho_B}{\partial t} + \nabla \cdot (\rho_B \mathbf{v}_B) = \omega_B. \tag{5.12.11}$$

Add Eqs. (5.12.10) and (5.12.11) to give

$$\frac{\partial \rho}{\partial t} + \nabla \cdot (\rho \mathbf{v}) = 0. \tag{5.12.12}$$

Equation (5.12.12) results from the laws of conservation of mass, $\omega_A + \omega_B = 0$, $\rho_A + \rho_B = \rho$, and $\rho_A \mathbf{v}_A + \rho_B \mathbf{v}_B = \rho \mathbf{v}$.

In terms of molar units, the continuity equation takes the form

$$\frac{\partial C_A}{\partial t} + \nabla \cdot (C_A \mathbf{v}_A) = \bar{\omega}_A, \tag{5.12.13}$$

where $\bar{\omega}_A$ is the molar rate of production of A per unit volume. It follows from Eqs. (5.12.3), (5.12.5b), (5.12.6b), and (5.12.10) that

$$\frac{\partial \rho_A}{\partial t} + \nabla \cdot (\rho_A \mathbf{v}) = \nabla \cdot (\rho D_{AB} \nabla Y_A) + \omega_A. \tag{5.12.14}$$

Similarly, from Eqs. (5.12.4a), (5.12.5b), (5.12.6b), and (5.12.13), we obtain

$$\frac{\partial C_A}{\partial t} + \nabla \cdot (C_A \mathbf{v}^*) = \nabla \cdot (C D_{AB} \nabla X_A) + \bar{\omega}_A. \tag{5.12.15}$$

Note that, if chemical reactions are absent and all velocities vanish, then

$$\frac{\partial C_A}{\partial t} = D_{AB} \nabla^2 C_A, \tag{5.12.16}$$

which is called *Fick's second law* of diffusion and is valid in solids or stationary nonreacting fluids.

In view of Eqs. (5.12.2) and (5.12.10) the continuity equation for a multicomponent system becomes

$$\frac{\partial}{\partial t}(\rho Y_i) + \nabla \cdot [\rho Y_i (\mathbf{v} + \mathbf{V}_i)] = \omega_i, \tag{5.12.17}$$

where we have used the relation $\rho_i = \rho Y_i$. Carrying out the differentiations in Eq. (5.12.17) and using Eq. (5.12.12) lead to

$$\rho \frac{\partial Y_i}{\partial t} + \rho (\mathbf{v} \cdot \nabla) Y_i + \nabla \cdot (\rho Y_i \mathbf{V}_i) = \omega_i \quad (i = 1, 2, \ldots, N), \tag{5.12.18}$$

which indicates the existence of N equations. Thus the addition of these equations gives the continuity equation for the mixture. It is now obvious that any one of these N equations may be replaced with the continuity equation for the mixture in any given problem, indicating that only $N-1$ of the Y_i's are independent.

On substitution of the relation from Eq. (5.12.7b) and expression (5.12.8) into Eq. (5.12.18), the conservation of mass equation for Y_i with $D_i = D$ takes the form

$$\rho \frac{\partial Y_i}{\partial t} + \rho(\mathbf{v} \cdot \nabla) Y_i + \nabla \cdot (\rho D \nabla Y_i) = \omega_i, \tag{5.12.19}$$

where the reaction rate ω_i is determined by the phenomenological chemical kinetic expression that will be presented in Subsection 5.12.6. Further discussion of diffusion velocities is given in Subsection 5.12.5.

5.12.3 Conservation of Momentum

For a reacting fluid mixture that involves a species k ($k = 1, 2, \ldots, N$), the body forces ρF_i acting on species k in direction i will contribute to the rate of change of the momentum:

$$\rho F_i = \rho \sum_{k=1}^{N} Y_k f_{ki}. \tag{5.12.20}$$

These body forces are now added to the forces that are due to velocity gradients so that, by using Eq. (5.3.17), we can write the momentum equations in the form

$$\rho \frac{\partial \mathbf{v}}{\partial t} + \rho(\mathbf{v} \cdot \nabla)\mathbf{v} = -\nabla p + \mu \left[\nabla^2 \mathbf{v} + \frac{1}{3} \nabla(\nabla \cdot \mathbf{v}) \right] + \rho \sum_{k=1}^{N} Y_k f_k, \tag{5.12.21}$$

in which the bulk modulus is neglected. If Eq. (5.12.21) is written in terms of tensorial components, then we have

$$\rho \frac{\partial v_i}{\partial t} + \rho \frac{\partial v_i}{\partial x_j} v_j = -\frac{\partial p}{\partial x_i} + \mu \left(\frac{\partial^2 v_i}{\partial x_j \partial x_j} + \frac{1}{3} \frac{\partial^2 v_j}{\partial x_j \partial x_i} \right) + \rho \sum_{k=1}^{N} Y_k f_{ki}, \tag{5.12.22}$$

where the indices i, j refer to the directions in Eulerian coordinates and the subscript $k = 1, 2, \ldots, N$ denotes the species.

5.12.4 Conservation of Energy

In deriving the equation for the conservation of energy for a multicomponent system, we must be aware that the heat flux is contributed by the temperature gradient and radiation and by an additional heat flux that is due to diffusion velocities given by

$$\rho \sum_{i=1}^{N} H_i Y_i \mathbf{V}_i,$$

where H_i is the average enthalpy (per unit mass) associated with the species i,

$$H_i = H_i^0 + \int_{T^0}^{T} c_{p_i} dT \quad (i = 1, \ldots, N), \tag{5.12.23}$$

and where H_i^0 is the standard heat of formation per unit mass for species i at reference temperature T^0. The heat flux is further augmented by the "thermal-diffusion" effect, called the *Soret effect*, which represents the effect of temperature gradients on diffusion. On the other hand, the reciprocal "diffusive-thermal" effect, called the *Dufour effect*, which indicates the effect of mass concentration gradients on the energy, also gives rise to the heat flux. These effects are given by

$$R^0 T \sum_{i=1}^{N} \sum_{j=1}^{N} \left(\frac{X_j \alpha_i}{W_i D_{ij}} \right) (\mathbf{V}_i - \mathbf{V}_j),$$

where D_{ij} is the mass diffusivity and α_i is the thermal diffusivity. Thus the total heat flux is of the form

$$\mathbf{q} = \mathbf{q}^{(c)} + \mathbf{q}^{(R)} + \rho \sum_{i=1}^{N} H_i Y_i \mathbf{V}_i + R^0 T \sum_{i=1}^{N} \sum_{j=1}^{N} \left(\frac{X_j \alpha_i}{W_i D_{ij}} \right) (\mathbf{V}_i - \mathbf{V}_j), \tag{5.12.24}$$

in which the heat flux contributions consist of condition (the first term), radiation (the second term), chemical diffusion (the third term), and Soret and Dufour effects (the fourth term).

Let us now consider energy equation (5.3.18),

$$\rho \frac{\partial \epsilon}{\partial t} + \rho (\mathbf{v} \cdot \nabla) \epsilon = \sigma_{ij} \frac{\partial v_j}{\partial x_i} - \nabla \cdot \mathbf{q}, \tag{5.12.25}$$

where the stress tensor σ_{ij} is as given by Eq. (5.2.29) and ϵ is the specific internal energy, defined as

$$\epsilon = H - \frac{p}{\rho} = \sum_{i=1}^{N} H_i Y_i - \frac{p}{\rho}. \tag{5.12.26}$$

Here, the sign of $\nabla \cdot \mathbf{q}$ is the reverse of what it was in Eq. (5.3.18) because the definition for $\mathbf{q}$ has the opposite sign in this case from what it had in Eq. (4.4.36).

In general, the Soret and Dufour effects and the radiative heat flux are negligible. Thus it follows from (5.12.19) and (5.12.25) that the energy equation becomes

$$\begin{aligned} &\rho c_p \frac{\partial T}{\partial t} + \rho c_p (\mathbf{v} \cdot \nabla) T - \frac{\partial p}{\partial t} - (\mathbf{v} \cdot \nabla) p - p \nabla \cdot \mathbf{v} \\ &- \sigma_{ij} \frac{\partial v_j}{\partial x_i} - k \nabla^2 T + \sum_{k=1}^{N} (\rho Y_k v_k \cdot \nabla) H_k = -H_k^0 \omega_k. \end{aligned} \tag{5.12.27}$$

A more popular form of the energy equation results from the so-called Shvab–Zel'dovich formulation, as presented in Subsection 5.12.6.

5.12.5 Physical Derivation of Multicomponent Diffusion

In Subsection 5.12.4 it was observed that the diffusion velocity is related to Fick's first law. The exact method of finding the diffusion velocity, however, can be introduced from the kinetic theory of gases (Hirschfelder, Curtis, and Bird, 1954). The three-dimensional dynamical problem of binary collision between two particles with masses m_i and m_j is found to be mathematically equivalent to a one-body problem in a plane with the reduced mass

$$\mu_{ij} = \frac{m_i m_j}{m_i + m_j}, \tag{5.12.28}$$

which is related to the sum $\mathbf{S}_i$ of the contributions that are due to collision and body forces:

$$\mathbf{S}_i = \sum_{j=1}^{N} \mu_{ij} Z_{ij}(\mathbf{V}_j - \mathbf{V}_i) \sum_{j=1}^{N} \rho Y_i Y_f \mathbf{f}_i \quad (i = 1, 2, \ldots, N), \tag{5.12.29}$$

where $\sum_{j=1}^{N} Y_j = 1$ and Z_{ij} is the total number of collisions per unit volume per second between the molecules of types i and j.

The partial pressure p_i of species i represents physically the momentum of neolecules of type i transported per second across a surface of unit area, traveling with the mass-average velocity of the fluid. Thus the quantity $\mathbf{S}_i$ may also be expressed by the relation

$$\mathbf{S}_i = \rho_i \frac{D\mathbf{v}}{Dt} + \nabla p_i = \rho Y_i \frac{D\mathbf{v}}{Dt} + \nabla p_i \quad (i = 1, 2, \ldots, N). \tag{5.12.30}$$

Let us now introduce Dalton's law of partial pressure,

$$p_i = X_i p \quad (i = 1, 2, \ldots, N), \tag{5.12.31}$$

from which

$$\nabla p_i = p \nabla X_i + X_i \nabla p. \tag{5.12.32}$$

If the viscous forces are neglected, the momentum equation is of the form

$$\rho \frac{D\mathbf{v}}{Dt} = -\nabla p + \rho \sum_{j=1}^{N} Y_i \mathbf{f}_j. \tag{5.12.33}$$

It follows from Eqs. (5.12.29)–(5.12.33) that

$$\nabla X_i = \sum_{j=1}^{N} \frac{\mu_{ij} Z_{ij}}{p} (\mathbf{V}_j - \mathbf{V}_i) + (Y_i - X_i) \frac{\nabla p}{p} + \frac{\rho}{p} \sum_{j=1}^{N} Y_i Y_j (\mathbf{f}_i - \mathbf{f}_j). \tag{5.12.34}$$

The multicomponent diffusion equation derived more rigorously from the kinetic theory of gases leads to

$$\nabla X_i = \sum_{j=1}^{N} \frac{X_i X_j}{D_{ij}} (\mathbf{V}_j - \mathbf{V}_i) + (Y_i - X_i)\frac{\nabla p}{p} + \frac{\rho}{p}\sum_{j=1}^{N} Y_i Y_j (\mathbf{f}_i - \mathbf{f}_j)$$

$$+ \sum_{j=1}^{N} \frac{X_i X_j}{\rho D_{ij}} \left(\frac{\alpha_j}{Y_j} - \frac{\alpha_i}{Y_i} \right) \frac{\nabla T}{T} \quad (i = 1, 2, \ldots, N), \tag{5.12.35}$$

where α_j is the thermal-diffusion coefficient of species j and

$$D_{ij} = \frac{X_i X_j p}{\mu_{ij} Z_{ij}}. \tag{5.12.36}$$

If pressure-gradient diffusion, body forces, and thermal-gradient diffusion (the Soret and Dufour effects) are negligible, then

$$\nabla X_i = \sum_{j=1}^{N} \frac{X_i X_j}{D_{ij}} (\mathbf{V}_j - \mathbf{V}_i). \tag{5.12.37}$$

When the binary-diffusion coefficients of all pairs of species are equal, diffusion equation (5.12.37) reduces to

$$D \nabla X_i = X_i \sum_{j=1}^{N} X_j V_j - X_i V_i \quad (i = 1, 2, \ldots, N). \tag{5.12.38}$$

We multiply Eq. (5.12.38) by Y_i / X_i and sum over i to obtain

$$\sum_{i=1}^{N} \frac{Y_i}{X_i} D \nabla X_i = \sum_{i=1}^{N} Y_i \sum_{j=1}^{N} X_j \mathbf{V}_j - \sum_{i=1}^{N} Y_i \mathbf{V}_i,$$

from which it follows that

$$\sum_{j=1}^{N} X_j \mathbf{V}_j = \sum_{i=1}^{N} Y_i D \nabla \ln X_i = \sum_{j=1}^{N} Y_j D \nabla \ln X_j, \tag{5.12.39}$$

where we have used the relations

$$\sum_{i=1}^{N} Y_i = 1, \quad \sum_{i=1}^{N} Y_i \mathbf{V}_i = 0.$$

When we substitute Eq. (5.12.39) into Eq. (5.12.38) and divide by X_i, we obtain

$$D \left(\nabla \ln X_i - \sum_{j=1}^{N} Y_j \nabla \ln X_j \right) = -V_i \quad (i = 1, 2, \ldots, N). \tag{5.12.40}$$

By making use of the relation

$$X_i = \frac{Y_i / W_i}{\sum_{j=1}^{N} (Y_j / W_j)}, \tag{5.12.41}$$

we obtain Fick's first law from Eq. (5.12.40):

$$\mathbf{V}_i = -D\nabla \ln Y_i \quad (i = 1, 2, \ldots, N), \tag{5.12.41a}$$

$$Y_i\mathbf{V}_i = -D\nabla Y_i \quad (i = 1, 2, \ldots, N), \tag{5.12.41b}$$

or

$$\rho_i V_i = -\rho D\nabla Y_i \quad (i = 1, 2, \ldots, N), \tag{5.12.42c}$$

which is in the same form as Eq. (5.12.7a). Thus Fick's first law is the simplified version of the rigorous form given in Eq. (5.12.35).

5.12.6 Shvab–Zel'dovich Approximations

In the derivation of Fick's first law, a number of simplifications were made. We neglected (1) body forces, (2) Soret and Dufour effects (terms involving α_i), (3) pressure-gradient diffusion, and (4) bulk viscosities. The so-called Shvab–Zel'dovich formulation requires additional simplifications: (1) steady flow, (2) viscous effects, and (3) constant pressure. Thus the energy equation, (5.12.27), reduces to

$$\rho(\mathbf{v}\cdot\nabla)H = \nabla\cdot(k\nabla T) - \nabla\cdot\left(\rho\sum_{i=1}^{N} H_i Y_i \mathbf{V}_i\right)$$

or

$$\nabla\cdot\left[\rho\sum_{i=1}^{N} H_i Y_i(\mathbf{v}+\mathbf{V}_i) - k\nabla T\right] = 0, \tag{5.12.43}$$

where we use the relations

$$H = \sum_{i=1}^{N} H_i Y_i, \quad \sum_{i=1}^{N} H_i Y_i \nabla\cdot(\rho\mathbf{v}) = 0.$$

The conservation-of-species equation, (5.12.17), for steady flow becomes

$$\nabla\cdot[\rho Y_i(\mathbf{v}+\mathbf{V}_i)] = \omega_i \quad (i = 1, 2, \ldots, N). \tag{5.12.44}$$

By substituting Eqs. (5.12.23), (5.12.42c), and (5.12.44) into Eq. (5.12.43), we obtain

$$\nabla\cdot\left[\rho\mathbf{v}\int_{T^0}^{T}\left(\sum_{i=1}^{N} Y_i c_{p_i}\right)dT - \rho D\sum_{i=1}^{N}\nabla Y_i\int_{T^0}^{T} c_{p_i}\,dT - \rho D\frac{k}{\rho c_p D}c_p\nabla T\right]$$
$$= -\sum_{i=1}^{N} H_i^0\omega_i. \tag{5.12.45}$$

Assuming that the Lewis number is unity, i.e., that

$$\mathrm{Le} = \frac{k}{\rho c_p D} = 1,$$

and by using the relation

$$\sum_{i=1}^{N} Y_i c_{p_i} = c_p$$

from Eq. (5.12.45), we derive

$$\nabla \cdot \left[\rho \mathbf{v} \int_{T^0}^{T} c_{p_i} dT - \rho D \left(\sum_{i=1}^{N} \nabla Y_i \int_{T^0}^{T} c_{p_i} dT + c_p \nabla T \right) \right] = -\sum_{i=1}^{N} H_i^0 \omega_i. \tag{5.12.46}$$

The second term on the left-hand side in Eq. (5.12.46) may be recast in the form

$$\begin{aligned} \sum_{i=1}^{N} (\nabla Y_i) \int_{T^0}^{T} c_{p_i} dT + c_p \nabla T &= \sum_{i=1}^{N} (\nabla Y_i) \int_{T^0}^{T} c_{p_i}(T) dT + \sum_{i=1}^{N} Y_i c_p(T) \nabla T \\ &= \sum_{i=1}^{N} (\nabla Y_i) \int_{T^0}^{T} c_{p_i}(T) dT + \sum_{i=1}^{N} Y_i \nabla \left(\int_{T^0}^{T} c_{p_i}(T) dT \right) \\ &= \nabla \left(\sum_{i=1}^{N} Y_i \right) \int_{T^0}^{T} c_{p_i} dT = \nabla \left(\int_{T^0}^{T} c_{p_i} dT \right). \end{aligned} \tag{5.12.47}$$

Substituting Eq. (5.12.47) into Eq. (5.12.46) yields

$$\nabla \cdot \left[\rho \mathbf{v} \int_{T^0}^{T} c_p dT - \rho D \mathbf{V} \left(\int_{T^0}^{T} c_{p_i} dT \right) \right] = -\sum_{i=1}^{N} H_i^0 \omega_i. \tag{5.12.48}$$

This is known as the Shvab–Zel'dovich energy equation. Likewise, substituting Eq. (5.12.42a) into Eq. (5.12.44) leads to the Shvab–Zel'dovich species equation:

$$\nabla \cdot (\rho \mathbf{v} Y_i - \rho D \nabla Y_i) = \omega_i. \tag{5.12.49}$$

Note that the specific heat or any transport coefficient of the mixture is not assumed to be constant and the specific heats of all the species are not necessarily considered equal.

If the specific heats are constant for a single species, then, in view of Eqs. (5.12.27) and (5.12.45), the energy equation may be written as

$$\begin{aligned} &\rho c_p \frac{\partial T}{\partial t} + \rho c_p (\mathbf{v} \cdot \nabla) T - \frac{\partial p}{\partial t} - (\mathbf{v} \cdot \nabla) p - p \nabla \cdot \mathbf{v} - \sigma_{ij} \frac{\partial v_j}{\partial x_i} \\ &- \rho c_p D (\nabla Y \cdot \nabla) T - \nabla \cdot (\rho c_p D \mathrm{Le} \nabla T) = -H^0 \omega. \end{aligned} \tag{5.12.50}$$

Similarly, the species equation becomes

$$\rho \frac{\partial Y}{\partial t} + \rho (\mathbf{v} \cdot \nabla) Y - \rho \nabla \cdot (D \nabla Y) = \omega. \tag{5.12.51}$$

For a single irreversible reaction step, we have

$$\sum_{i=1}^{N} \nu_i' M_i \to \sum_{i=1}^{N} \nu_i'' M_i, \tag{5.12.52}$$

$$\omega = \frac{\omega_i}{W_i(\upsilon_i'' - \upsilon_i')} \quad (i = 1, 2, \ldots, N). \tag{5.12.53}$$

If we denote

$$\xi_T = \frac{\int_{T^0}^{T} c_p dT}{\sum_{i=1}^{N} H_i^0 W_i(\nu_i' - \nu_i'')}, \tag{5.12.54}$$

$$\xi_i = \frac{Y_i}{W_i(\nu_i'' - \nu_i')} \quad (i = 1, 2, \ldots, N), \tag{5.12.55}$$

the energy equation, (5.12.48), becomes

$$\nabla \cdot (\rho \boldsymbol{v} \xi_T - \rho D \nabla \xi_T) = \omega \tag{5.12.56}$$

and the species equation, (5.12.49), is given by

$$\nabla \cdot (\rho \boldsymbol{v} \xi_i - \rho D \nabla \xi_i) = \omega. \tag{5.12.57}$$

It follows that the nonlinear term ω can be eliminated from N of the $N+1$ equations corresponding to

$$L(\xi) = \omega, \tag{5.12.58}$$

where L is the linear operator (if ρD is independent of ξ), such that

$$L(\xi) = \nabla \cdot (\rho \boldsymbol{v} \xi - \rho D \nabla \xi), \tag{5.12.59}$$

where ξ can be $\xi_T, \xi_1, \xi_2, \ldots, \xi_N$. For $\xi = \xi_1$, we have

$$L(\xi_1) = \omega, \tag{5.12.60}$$

from which other variables are determined by the linear function

$$L(\beta) = 0, \tag{5.12.61}$$

where β can be $\beta_T, \beta_2, \beta_3, \ldots, \beta_N$:

$$\begin{aligned} \beta_T &= \xi_T - \xi_1, \\ \beta_2 &= \xi_2 - \xi_1 \\ &\vdots \\ \beta_N &= \xi_N - \xi_1. \end{aligned}$$

It is obvious that the preceding approach (Shvab–Zel'dovich formulation) is efficient in a reacting combustion process for reactants that are unmixed initially. By solving the linear equations between flow variables, we can determine burning rates without solving the nonlinear equation.

Remarks

The governing equations for reacting fluids differ from those for nonreacting fluids mainly in the form of the respective continuity equations. There are $N-1$ species equations for the N species in addition to the continuity equation for the mixture. Therefore the variables to be solved are

$$\rho, Y_k \ (k = 1, 2, \ldots, N), \quad \mathbf{v} \ (i = 1, 2, \ldots, N), T, p.$$

The $N + 6$ equations consist of the following relations:

One overall mass continuity, Eq. (5.12.12);
$N - 1$ species equations, Eq. (5.12.49) or (5.12.51);
one equation relating all Y_i, $Y_1 + Y_2 + \cdots + Y_N = 1$;
three momentum equations, Eq. (5.12.22);
one energy equation, Eq. (5.12.48) or (5.12.50);
one equation of state, Eq. (5.12.2).

It is possible to solve the earlier energy and species equations, (5.12.18) and (5.12.27), with the diffusion velocities as unknown variables. In this case, however, there are other equations to be solved, including (5.12.35) or (5.12.37) and (5.12.41).

Many of the complicated problems of reacting flows remain unresolved because of the lack of understanding about such physical phenomena as a chemical kinetics, turbulence, and phase change. Furthermore, there are computational difficulties, such as nonlinearity caused by convection and "stiffness" of equations that is due to the widely disparate reaction rates that various species may exhibit. If reacting flows are involved in shock-wave discontinuities and turbulence, solutions of appropriate governing equations constitute perhaps the most challenging tasks in continuum mechanics. Some initial attempts are described in Chung (1993).

PROBLEMS

5.1 Use the first law of thermodynamics to derive FLT equation (5.3.12) and the most general form of the governing equations (continuity, momentum, and energy) for a nonisothermal compressible viscous flow in Cartesian coordinates.

5.2 Differentiate the CNS equation or the conservation form of the Navier–Stokes systems of equations, and show that the conservation of mass is prerequisite to the conservation of momentum and the conservation of both mass and momentum is prerequisite to the conservation of energy. Show also that the results of the first law of thermodynamics are recovered in this process.

5.3 Integrate the CNS equations, obtain the CVS equations, show all components of mass, momentum, and energy, and identify them on the three-dimensional free-body sketches, contributing to both the upstream and downstream sides of the square faces.

5.4 Use the results obtained in Problem 5.1 to show all governing equations for three-dimensional, compressible viscous flows completely expanded in terms of the primitive variables (u, v, w, p, ρ, T) and the independent variables (x, y, z).

5.5 Use the second law of thermodynamics to derive the energy equation with the thermal expansion coefficient.

5.6 Derive the Bernoulli equation, (5.5.23), describing all steps in detail.

5.7 Derive the Bernoulli equation equivalent to the energy equation, including the head losses that are due to internal energy, heat transfer, viscous dissipation (friction), and shaft work corresponding to the constant of integration. *Hint*: Use the third part of Eq. (5.3.27a) or (5.3.26), integrate over the domain to obtain the surface integral, and adjust the result for appropriate physical quantities.

5.8 Derive the vorticity transport equation for two-dimensional incompressible flows. Use the tensorial derivations to show that the steady-state momentum may be written in the form

$$\nabla^4\psi = \frac{1}{v}\left[\frac{\partial^3\psi}{\partial x\partial y^2}\frac{\partial\psi}{\partial y} - \frac{\partial^3\psi}{\partial x^2\partial y}\frac{\partial\psi}{\partial x} + \frac{\partial^3\psi}{\partial x^3}\frac{\partial\psi}{\partial y} - \frac{\partial^3\psi}{\partial y^3}\frac{\partial\psi}{\partial x}\right].$$

5.9 Derive the boundary conditions required in Problem 5.8. Provide comments on each boundary term.

5.10 Prove Eq. (5.6.9).

5.11 Derive the transport equations for turbulent kinetic energy and turbulent dissipation energy.

5.12 Carry out an order-of-magnitude analysis for the Navier–Stokes system and arrive at the laminar and turbulent boundary-layer equations, (5.8.12)–(5.8.17).

5.13 Derive the full potential equation for compressible inviscid flows.

5.14 Show mathematically that an incompressible flow can be defined by the flow speed with $M \ll 1$. *Hint*: Use the continuity equation, the Bernoulli equation, and the definition of the speed of sound.

5.15 Prove Eq. (5.10.24).

5.16 Derive the acoustic-wave equations in terms of pressure, velocity, and velocity potential. State the assumptions made in your derivation.

5.17 Derive all governing equations required for reacting fluids. State the assumptions made in your derivation.

PART II

Special Topics

6 Curvilinear Continuum

Equations of strains and stresses as well as all governing equations of solids and fluids may be written in curvilinear coordinates for curved geometries such as in cylinders, spheres, and large deformation problems. Mathematics involving curvilinear coordinates originated with Riemann in the middle of the 19th century and subsequently was used by Einstein in the early 20th century in describing curved spacetime geometries of the universe for the theory of relativity. The curvilinear coordinates were then utilized in engineering to describe the large deformations of solids (nonlinear elasticity), viscous fluids (non-Newtonian fluid dynamics), and differential geometries (shell structures).

The basic formulations and geometries of curvilinear continuum are employed in nonlinear continuum presented in Chap. 7 and differential geometry continuum discussed in Chap. 9, including shell theories and relativity theories.

Tensor analysis in curvilinear continuum, nonlinear continuum, and differential geometry continuum presented in Part II is more rigorous than in Part I. References for tensor analysis include Erickson (1960), Eringen (1962, 1967), and Gurtin (1981), among others.

6.1 Tensor Properties of Curvilinear Continuum

6.1.1 General

Recall that tangent vectors and metric tensors associated with deformed states were discussed in Section 2.2. In the analysis of cylinders or spheres, the undeformed state itself must be non-Cartesian. Figure 6.1.1 shows the use of a general Lagrangian non-Cartesian or curvilinear coordinate system for the undeformed state and its convected coordinates for the deformed state.

Tangent Vectors

The tangent vectors (base vectors) $\mathbf{g}_i$ are drawn tangent to the initial undeformed curvilinear coordinates ξ_i. The tangent vectors $\mathbf{G}_i$ on the deformed coordinates will not be used in this chapter except in Subsection 6.2.3. The tangent vectors $(\mathbf{g}_1, \mathbf{g}_2, \mathbf{g}_3)$ may be nonorthogonal to each other. The tangent vectors $\mathbf{g}_i$ on an

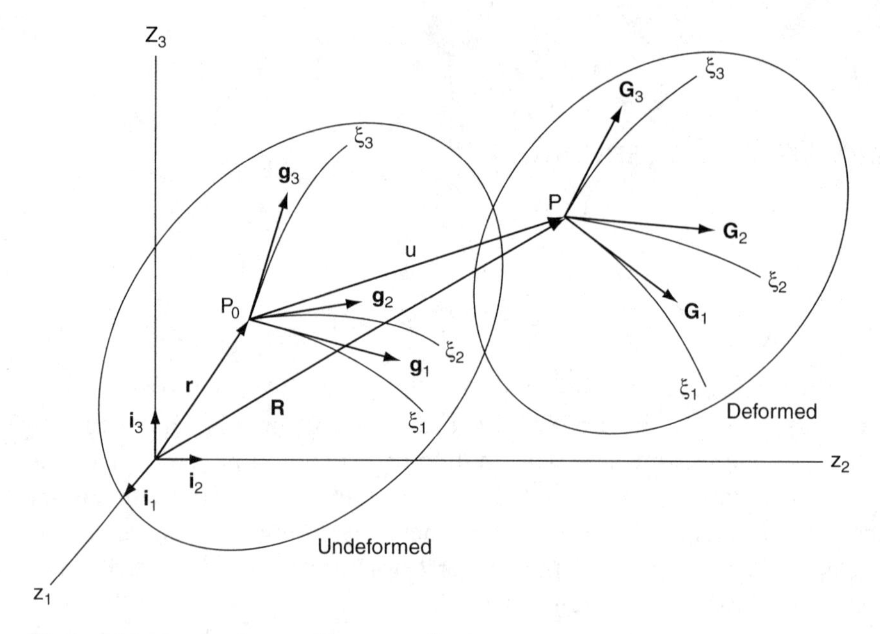

Figure 6.1.1. Lagrangian curvilinear coordinates.

undeformed state represent the rate of change of the position vector $\mathbf{r}$ with respect to the curvilinear coordinates ξ_i,

$$\mathbf{g}_i = \frac{\partial \mathbf{r}}{\partial \xi_i} = \frac{\partial z_m}{\partial \xi_i}\mathbf{i}_m, \tag{6.1.1}$$

whereas the *reciprocal tangent* vectors are defined as

$$\mathbf{g}^i = \frac{\partial \xi_i}{\partial z_m}\mathbf{i}_m. \tag{6.1.2}$$

These tangent vectors are related by

$$\mathbf{g}_i \cdot \mathbf{g}^j = \frac{\partial z_m}{\partial \xi_i}\mathbf{i}_m \cdot \frac{\partial \xi_j}{\partial z_n}\mathbf{i}_n = \frac{\partial z_m}{\partial \xi_i}\frac{\partial \xi_j}{\partial z_n}\delta_{mn} = \frac{\partial z_m}{\partial \xi_i}\frac{\partial \xi_j}{\partial z_m}, \tag{6.1.3a}$$

where $\mathbf{g}_i$ and $\mathbf{g}^j$ are referred to as the *covariant* and *contravariant* tangent (base) vectors, respectively. The contravariant (reciprocal) tangent vectors, $\mathbf{g}^1$, $\mathbf{g}^2$, $\mathbf{g}^3$, are orthogonal to the planes constructed by $\mathbf{g}_2$ and $\mathbf{g}_3$, $\mathbf{g}_3$ and $\mathbf{g}_1$, and $\mathbf{g}_1$ and $\mathbf{g}_2$, respectively (see Fig. 6.1.2). This definition requires that $\mathbf{g}_i \cdot \mathbf{g}^j = \delta_{ij}$ and

$$\frac{\partial z_m}{\partial \xi_i}\frac{\partial \xi_j}{\partial z_m} = \delta_i^j = \delta_{ij}, \tag{6.1.3b}$$

where the Kronecker delta δ_i^j performs the same way as δ_{ij}. We also define g_{ij} and g^{ij} as the covariant and contravariant metric tensors, respectively:

$$g_{ij} = \mathbf{g}_i \cdot \mathbf{g}_j = \frac{\partial z_m}{\partial \xi_i}\frac{\partial z_m}{\partial \xi_j}, \tag{6.1.4a}$$

$$g^{ij} = \mathbf{g}^i \cdot \mathbf{g}^j = \frac{\partial \xi_i}{\partial z_m}\frac{\partial \xi_j}{\partial z_m}. \tag{6.1.4b}$$

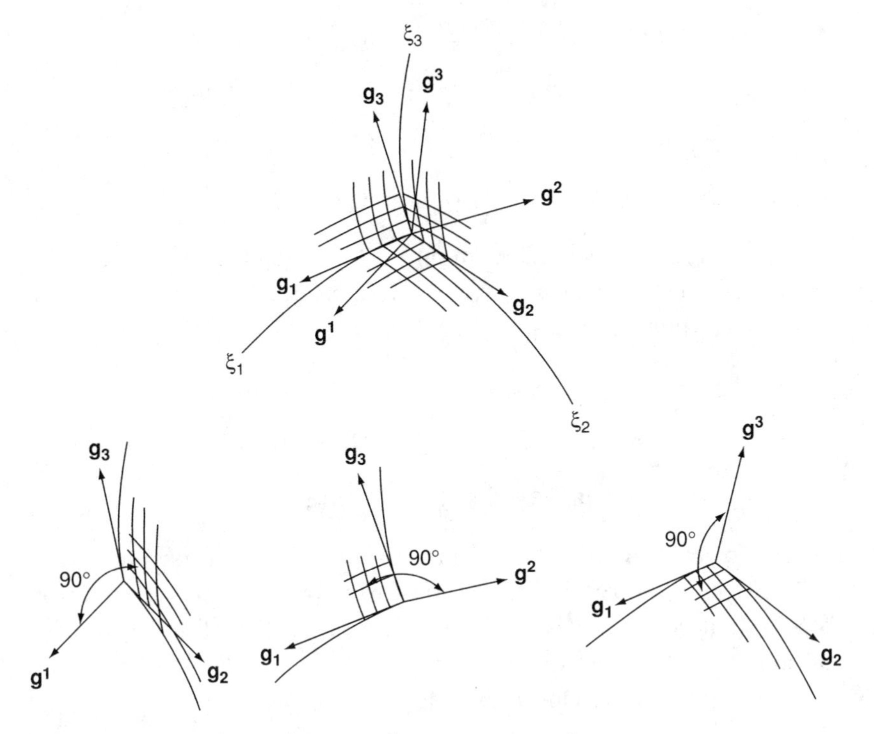

Figure 6.1.2. Covariant and contravariant components of tangent vectors.

Note also the following additional properties:

$$\mathbf{g}_i = g_{ij}\mathbf{g}^j, \quad \mathbf{g}^i = g^{ij}\mathbf{g}_j, \quad g_{ik}g^{ik} = \delta_i^j = \delta_{ij}, \tag{6.1.5a}$$

$$\begin{aligned}\mathbf{i}_p &= \frac{\partial \xi_i}{\partial z_p}\mathbf{g}_i = \frac{\partial \xi_i}{\partial z_p}g_{ik}\mathbf{g}^k = \frac{\partial \xi_i}{\partial z_p}\frac{\partial z_m}{\partial \xi_i}\frac{\partial z_m}{\partial \xi_k}\mathbf{g}^k \\ &= \delta_{mp}\frac{\partial z_m}{\partial \xi_k}\mathbf{g}^k = \frac{\partial z_p}{\partial \xi_k}\mathbf{g}^k.\end{aligned} \tag{6.1.5b}$$

The cross product of the covariant tangent vectors takes the form

$$\mathbf{g}_i \times \mathbf{g}_j = \sqrt{g}\epsilon_{ijk}\mathbf{g}^k, \tag{6.1.6}$$

where

$$g = |g_{ij}|,$$

$$\sqrt{g} = \frac{\partial z_m}{\partial \xi_1}\frac{\partial z_n}{\partial \xi_2}\frac{\partial z_p}{\partial \xi_3}\epsilon_{mnp} \equiv z_{m,1}z_{n,2}z_{p,3}\epsilon_{mnp}$$

[see the similar expression for $\sqrt{G}$ in Eq. (2.2.17)]. For curvilinear coordinates, all repeated indices are aligned diagonally, one is superscripted, and the other is subscripted, with the exception of indices involved in derivatives associated with coordinates (spatial coordinates z_i and curvilinear coordinates ξ_i). To prove the relation in Eq. (6.1.6), we may proceed by rewriting Eq. (6.1.6):

$$\begin{aligned}\mathbf{g}_i \times \mathbf{g}_j &= z_{m,i} z_{n,j} \epsilon_{mnp} \mathbf{i}_p = z_{m,i} z_{n,j} \epsilon_{mnp} z_{p,k} \mathbf{g}^k = z_{m,r} z_{n,s} z_{p,t} \delta_{ir} \delta_{js} \delta_{kt} \epsilon_{mnp} \mathbf{g}^k \\ &= [(z_{1,1} z_{2,1} z_{3,1} - z_{1,1} z_{3,1} z_{2,1} + \cdots +)\, \delta_{i1} \delta_{j1} \delta_{k1} + \cdots \\ &\quad + (\cdots + z_{3,3} z_{1,3} z_{2,3} - z_{3,3} z_{2,3} z_{1,3})\, \delta_{i3} \delta_{j3} \delta_{k3}]\, \mathbf{g}^k.\end{aligned}$$

The complete expansion results in 27 (sum on r, s, t) $\times 6$ (sum on m, n, p) $= 162$ terms. However, only those terms with $r, s, t = 1, 2, 3$; $1, 3, 2$; $2, 1, 3$; $2, 3, 1$; $3, 1, 2$; and $3, 2, 1$ survive. Thus, using Eq. (1.2.20), we obtain

$$\begin{aligned}\mathbf{g}_i \times \mathbf{g}_j &= z_{m,1} z_{n,2} z_{p,3} (\delta_{i1} \delta_{j2} \delta_{k3} - \delta_{i1} \delta_{j3} \delta_{k2} + \delta_{i2} \delta_{j3} \delta_{k1} \\ &\quad - \delta_{i2} \delta_{j1} \delta_{k3} + \delta_{i3} \delta_{j1} \delta_{k2} - \delta_{i3} \delta_{j2} \delta_{k1}) \epsilon_{mnp} \mathbf{g}^k \\ &= z_{m,1} z_{n,2} z_{p,3} \epsilon_{ijk} \epsilon_{mnp} \mathbf{g}^k = \sqrt{g}\, \epsilon_{ijk} \mathbf{g}^k.\end{aligned}$$

Again, it should be cautioned that the commas representing the partial derivative are used only when differentiations are performed with respect to independent variables. When confusion is likely to occur, as in Eq. (6.1.5b), the full derivative notation should be used.

Using a similar approach, we can show that

$$\mathbf{g}^i \times \mathbf{g}^j = \frac{1}{\sqrt{g}} \epsilon^{ijk} \mathbf{g}_k. \tag{6.1.7}$$

The permutation symbol ϵ^{ijk} performs identically as ϵ_{ijk} because their values are constants. The superscripts are used here merely for convenience so that the repeated indices can be placed diagonally, consistent with the basic rule for curvilinear coordinates.

A vector $\mathbf{V}$ can be written in terms of covariant components V_i or contravariant components V^i,

$$\mathbf{V} = V_i \mathbf{g}^i = V^i \mathbf{g}_i$$

or

$$V^i = \mathbf{V} \cdot \mathbf{g}^i, \quad V_i = \mathbf{V} \cdot \mathbf{g}_i,$$

in contrast with the Cartesian coordinates, in which we write

$$V_i = \mathbf{V} \cdot \mathbf{i}_i.$$

EXAMPLE 6.1.1. TANGENT VECTORS. Let us consider the functions z_i:

$$\begin{aligned} z_1 &= 3\xi_1 - \xi_2, \\ z_2 &= 2\xi_1 + 2\xi_2, \\ z_3 &= \xi_3. \end{aligned}$$

The covariant tangent vectors are

$$\mathbf{g}_i = \frac{\partial z_m}{\partial \xi_i}\mathbf{i}_m,$$

$$\mathbf{g}_1 = 3\mathbf{i}_1 + 2\mathbf{i}_2, \quad \mathbf{g}_2 = -\mathbf{i}_1 + 2\mathbf{i}_2, \quad \mathbf{g}_3 = \mathbf{i}_3.$$

The metric tensors are

$$g_{ij} = \mathbf{g}_i \cdot \mathbf{g}_j = \begin{bmatrix} 13 & 1 & 0 \\ 1 & 5 & 0 \\ 0 & 0 & 1 \end{bmatrix}.$$

Solving ξ_i from given z_i,

$$\xi_1 = \frac{1}{4}z_1 + \frac{1}{8}z_2,$$

$$\xi_2 = -\frac{1}{4}z_1 + \frac{3}{8}z_2,$$

$$\xi_3 = z_3,$$

we calculate the contravariant tangent vectors from $\mathbf{g}^i = \dfrac{\partial \xi_i}{\partial z_m}\mathbf{i}_m$:

$$\mathbf{g}^1 = \frac{1}{4}\mathbf{i}_1 + \frac{1}{8}\mathbf{i}_2, \qquad \mathbf{g}^2 = -\frac{1}{4}\mathbf{i}_1 + \frac{3}{8}\mathbf{i}_2, \qquad \mathbf{g}^3 = \mathbf{i}_3.$$

We now have the check

$$\mathbf{g}_i \cdot \mathbf{g}^j = \begin{bmatrix} 1 & 0 & 0 \\ 0 & 1 & 0 \\ 0 & 0 & 1 \end{bmatrix} = \delta_{ij}.$$

The reader is encouraged to verify Eqs. (6.1.3)–(6.1.7) by using the data provided in this example.

EXAMPLE 6.1.2. COVARIANT AND CONTRAVARIANT COMPONENTS. Given $\mathbf{g}_1 = 3\mathbf{i}_1$, $\mathbf{g}_2 = 6\mathbf{i}_1 + 8\mathbf{i}_2$, $\mathbf{g}_3 = \mathbf{i}_3$, and $\mathbf{V} = 4\mathbf{i}_1 + 3\mathbf{i}_2$ (see Fig. 6.1.3), calculate the metric tensors and covariant and contravariant components of $\mathbf{V}$.

First, we determine

$$g_{ij} = \begin{bmatrix} 9 & 18 & 0 \\ 18 & 100 & 0 \\ 0 & 0 & 1 \end{bmatrix}.$$

It follows from Eq. (6.1.5b) that

$$g^{ij} = [g_{ij}]^{-1} = \frac{1}{576}\begin{bmatrix} 100 & -18 & 0 \\ -18 & 9 & 0 \\ 0 & 0 & 576 \end{bmatrix}.$$

From

$$\mathbf{g}^i = g^{ij}\mathbf{g}_j$$

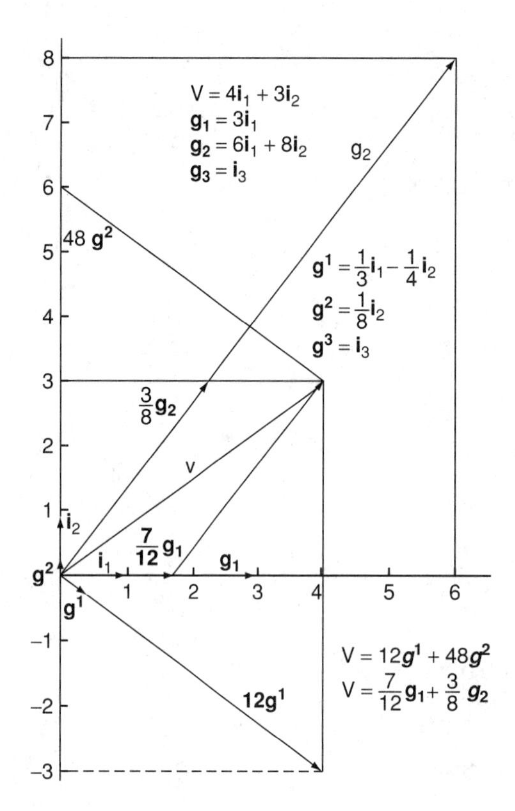

Figure 6.1.3. Relationships between covariant and contravariant components of a vector.

we obtain

$$\mathbf{g}^1 = \frac{1}{3}\mathbf{i}_1 - \frac{1}{4}\mathbf{i}_2, \quad \mathbf{g}^2 = \frac{1}{8}\mathbf{i}_2, \quad \mathbf{g}^3 = \mathbf{i}_3.$$

The covariant components of $\mathbf{V}$ take the form

$$\begin{aligned}
V_i &= \mathbf{V}\cdot\mathbf{g}_i,\\
V_1 &= (4\mathbf{i}_1 + 3\mathbf{i}_2)\cdot 3\mathbf{i}_1 = 12,\\
V_2 &= (4\mathbf{i}_1 + 3\mathbf{i}_2)\cdot(6\mathbf{i}_1 + 8\mathbf{i}_2) = 48,\\
V_3 &= 0.
\end{aligned}$$

Note that this gives the check

$$\mathbf{V} = V_i\mathbf{g}^i = 12\mathbf{g}^1 + 48\mathbf{g}^2 = 4\mathbf{i}_1 + 3\mathbf{i}_2.$$

Similarly, the contravariant components of $\mathbf{V}$ are

$$\begin{aligned}
V^i &= \mathbf{V}\cdot\mathbf{g}^i,\\
V^1 &= (4\mathbf{i}_1 + 3\mathbf{i}_2)\cdot\left(\frac{1}{3}\mathbf{i}_1 - \frac{1}{4}\mathbf{i}_2\right) = \frac{7}{12},\\
V^2 &= (4\mathbf{i}_1 + 3\mathbf{i}_2)\cdot\left(\frac{1}{8}\mathbf{i}_2\right) = \frac{3}{8},\\
V^3 &= 0.
\end{aligned}$$

Once again, we have a check:

$$\mathbf{V} = V^i \mathbf{g}_i = \frac{7}{12}\mathbf{g}_1 + \frac{3}{8}\mathbf{g}_2 = 4\mathbf{i}_1 + 3\mathbf{i}_2.$$

Derivatives of Tangent Vectors

Let us now consider the derivatives of the covariant tangent vectors:

$$\mathbf{g}_{i,j} = \frac{\partial \mathbf{g}_i}{\partial \xi_j} = \frac{\partial}{\partial \xi_j}\left(\frac{\partial z_m}{\partial \xi_i}\mathbf{i}_m\right) = \frac{\partial^2 z_m}{\partial \xi_j \partial \xi_i}\frac{\partial \xi_n}{\partial z_m}\mathbf{g}_n = \Gamma^n_{ij}\mathbf{g}_n, \tag{6.1.8}$$

where Γ^n_{ij} is called the Christoffel symbol of the second kind, defined as

$$\Gamma^n_{ij} = \frac{\partial^2 z_m}{\partial \xi_j \partial \xi_i}\frac{\partial \xi_n}{\partial z_m} = z_{m,ij}\frac{\partial \xi_n}{\partial z_m}. \tag{6.1.9}$$

Note also that

$$\frac{\partial}{\partial \xi_i}(\mathbf{g}^j \cdot \mathbf{g}_k) = \mathbf{g}_k \cdot \mathbf{g}^j_{,i} + \mathbf{g}^j \cdot \mathbf{g}_{k,i} = 0$$

or

$$\mathbf{g}_k \cdot \mathbf{g}^j_{,i} = -\mathbf{g}^j \cdot \mathbf{g}_{k,i} = -\mathbf{g}^j \cdot \mathbf{g}_n \Gamma^n_{ki} = -\Gamma^j_{ki}.$$

Thus the derivative of contravariant tangent vectors is of the form

$$\frac{\partial \mathbf{g}^j}{\partial \xi_i} = \mathbf{g}^j_{,i} = -\Gamma^j_{ik}\mathbf{g}^k. \tag{6.1.10}$$

The relation shown in Eq. (6.1.8) may be written in the alternative form

$$\frac{\partial \mathbf{g}_i}{\partial \xi_j} = \frac{\partial^2 z_m}{\partial \xi_i \partial \xi_j}\frac{\partial z_m}{\partial \xi_k}\mathbf{g}^k = \Gamma_{ijk}\mathbf{g}^k = \Gamma^k_{ij}\mathbf{g}_k, \tag{6.1.11}$$

where Γ_{ijk} is known as the Christoffel symbol of the first kind,

$$\Gamma_{ijk} = \frac{\partial^2 z_m}{\partial \xi_i \partial \xi_j}\frac{\partial z_m}{\partial \xi_k}, \tag{6.1.12}$$

which is related to the Christoffel symbol of the second kind as

$$\Gamma^k_{ij} = \Gamma_{ijm} g^{mk}. \tag{6.1.13}$$

Here the contravariant metric tensor transforms the Christoffel symbol of the first kind into that of the second kind.

With the definitions stated previously, the Christoffel symbols are related to the derivatives of metric tensors as

$$\Gamma_{ijk} = \frac{1}{2}(g_{ik,j} + g_{jk,i} - g_{ij,k}),$$

$$\Gamma^k_{ij} = \frac{1}{2}g^{km}(g_{mi,j} + g_{mj,i} - g_{ij,m}),$$

$$\Gamma^i_{ij} = \frac{1}{2}g^{im}g_{im,j}.$$

The usefulness of the Christoffel symbols (Christoffel, 1869a and 1869b) will be demonstrated in the examples of cylindrical and spherical coordinates given in Subsection 6.1.3.

6.1.2 Covariant Derivatives

We define the del operator in curvilinear coordinates as

$$\nabla = \mathbf{g}^i \frac{\partial}{\partial \xi_i}. \tag{6.1.14}$$

The divergence of a vector is given by

$$\begin{aligned}\nabla \cdot \mathbf{V} &= \mathbf{g}^i \frac{\partial}{\partial \xi_i} \cdot (V^j \mathbf{g}_j) = \mathbf{g}^i \cdot \mathbf{g}_j V^j_{,i} + \mathbf{g}^i \cdot \Gamma^k_{ji} \mathbf{g}_k V^j \\ &= V^i_{,i} + \Gamma^i_{ij} V^j = V^i_{|i},\end{aligned} \tag{6.1.15a}$$

where the subscripted vertical stroke stands for what is known as a *covariant derivative*, which indicates that the term contains the partial derivative plus the change of the tangent vector, as given by the Christoffel symbol.

Note that the contravariant metric tensor g^{ij} is the inverse of g_{ij} or the adjoint of g_{ij} divided by the determinant of g_{ij} (Cramer's rule):

$$g^{ij} = \frac{1}{g} \frac{\partial g}{\partial g_{ij}}.$$

Thus we may write Eqs. (6.1.5a) as

$$\nabla \cdot \mathbf{V} = V^i_{|i} = \frac{1}{\sqrt{g}} (V^i)_{,i}. \tag{6.1.15b}$$

The curl of a vector takes the form

$$\begin{aligned}\nabla \times \mathbf{V} &= \mathbf{g}^i \frac{\partial}{\partial \xi_i} \times V_j \mathbf{g}^j = \mathbf{g}^i \times \mathbf{g}^j V_{j,i} + \mathbf{g}^i \times \mathbf{g}^j_{,i} V_j \\ &= \mathbf{g}^i \times \mathbf{g}^j \left(V_{j,i} - \Gamma^r_{ij} V_r\right) = \frac{1}{\sqrt{g}} \epsilon^{ijk} \mathbf{g}_k V_{j|i},\end{aligned} \tag{6.1.16a}$$

with

$$V_{j|i} = V_{j,i} - \Gamma^r_{ij} V_r. \tag{6.1.16b}$$

We note that the vanishing of the divergence and the curl of a velocity vector $\mathbf{V}$ physically represents, respectively, the conditions of incompressibility and irrotationality in fluids.

The consequence of the covariant derivatives of a vector is that the changes of the components of any two constant parallel vectors at different points on a surface such as a sphere are nonzero, as calculated from the Christoffel symbols in Eqs. (6.1.5a) and (6.1.5b). These are known as *parallel transports*, prominent phenomena of motions in a curvature of shell structures or in the spacetime of general relativity to be discussed in Chap. 9.

The second derivative of a scalar or a vector can be carried out in a similar manner. For example, let us consider the Laplace equation

$$\nabla^2\psi = 0. \tag{6.1.17}$$

Because $\nabla^2\psi$ is the divergence of the gradient of ψ we obtain

$$\begin{aligned}
\nabla^2\psi &= (\nabla \cdot \nabla)\psi = \left(g^j \frac{\partial}{\partial \xi_j} \cdot g^i \frac{\partial}{\partial \xi_i}\right)\psi \\
&= g^j \cdot g^i_{,j}\psi_{,i} + g^j \cdot g^i \psi_{,ij} \\
&= g^j \cdot \left(g^{ik} g_k\right)_{,j} \psi_{,i} + g^{ij}\psi_{,ij} \\
&= g^{ij}_{,j}\psi_{,i} + g^{ij}\Gamma^k_{ki}\psi_{,j} + g^{ij}\psi_{,ij} \\
&= g^{ij}_{,i}\psi_{,j} + \psi_{|ji} g^{ij} = \frac{1}{\sqrt{g}}\left(\sqrt{g} g^{ij}\psi_{,j}\right)_{,i},
\end{aligned} \tag{6.1.18}$$

where $g^{ij}_{,i} = 0$ (see Problem 2.8) and $\psi_{|ji}$ is defined as the second-order covariant derivative of ψ:

$$\psi_{|ji} = \psi_{,ji} + \Gamma^k_{ki}\psi_{,j} = \frac{1}{\sqrt{g}}\left(\sqrt{g}\psi_{,j}\right)_{,i}. \tag{6.1.19}$$

If a tensor is of higher-order or mixed covariant contravariant form, then the covariant derivatives are more complicated. For example, consider a covariant derivative of A_{ij}. First, we write a dot product of two vectors,

$$\mathbf{V} \cdot \mathbf{W} = V_i \mathbf{g}^i \cdot W_j \mathbf{g}^j = A_{ij}\mathbf{g}^i \cdot \mathbf{g}^j,$$

and take the partial derivative of the preceding equation,

$$\begin{aligned}
\frac{\partial}{\partial \xi_s}(\mathbf{V} \cdot \mathbf{W}) &= A_{ij,s} g^{ij} - A_{ij}\Gamma^i_{sr} g^{rj} - A_{ij}\Gamma^j_{sr} g^{ri} \\
&= \left(A_{ij,s} - \Gamma^r_{is}A_{rj} - \Gamma^r_{js}A_{ir}\right) g^{ij} = A_{ij|s} g^{ij},
\end{aligned}$$

where the covariant derivative of the covariant tensor A_{ij} is of the form

$$A_{ij|s} = A_{ij,s} - \Gamma^r_{is}A_{rj} - \Gamma^r_{js}A_{ir}. \tag{6.1.20a}$$

Similarly, we set $\mathbf{V} \cdot \mathbf{W} = V^i \mathbf{g}_i \cdot W^j \mathbf{g}_j$ to obtain

$$\frac{\partial}{\partial \xi_s}(\mathbf{V} \cdot \mathbf{W}) = A^{ij}_{|s} g_{ij},$$

where the covariant derivative of the contravariant tensor A^{ij} is of the form

$$A^{ij}_{|s} = A^{ij}_{,s} + \Gamma^i_{rs}A^{rj} + \Gamma^j_{rs}A^{ir}. \tag{6.1.20b}$$

We may determine the covariant derivative of the mixed tensor by setting $\mathbf{V}\cdot\mathbf{W} = V^i\mathbf{g}_i \cdot W_j\mathbf{g}^j$. This gives

$$\frac{\partial}{\partial \xi_s}(\mathbf{V}\cdot\mathbf{W}) = A^i_{j|s}\delta^j_i,$$

where the covariant derivative of the mixed tensor A^i_j is of the form

$$A^i_{j|s} = A^i_{j,s} - \Gamma^r_{js}A^i_r + \Gamma^j_{rs}A^r_j. \tag{6.1.20c}$$

It follows from these observations that covariant derivatives of multiple-order mixed tensors can be written in the form

$$A^{j_1\ldots j_m}_{i_1\ldots i_n}|_s = A^{j_1\ldots j_m}_{i_1\ldots i_n,s} - \Gamma^r_{i_1 s}A^{j_1\ldots j_m}_{r i_2\ldots i_n} - \Gamma^r_{i_2 s}A^{j_1\ldots j_m}_{i_1 r i_3\ldots i_n} - \cdots - \Gamma^r_{i_n s}A^{j_1\ldots j_m}_{i_1\ldots r}$$
$$+\Gamma^{j_1}_{rs}A^{r j_2\ldots j_m}_{i_1\ldots i_n} + \Gamma^{j_2}_{rs}A^{j_1 r\ldots j_m}_{i_1\ldots i_n} + \cdots + \Gamma^{j_m}_{rs}A^{j_1\ldots r}_{i_1\ldots i_n}. \tag{6.1.21}$$

Covariant derivatives are useful in obtaining differential equations in curvilinear coordinates (cylindrical or spherical geometries) and also for automatic grid generation in finite-difference methods of computational mechanics. The most significant applications of covariant derivatives occur in the curved space-time geometry of general relativity. The covariant derivatives of various tensorial quantities resulting in Christoffel symbols are the process of parallel transports, as indicated earlier.

6.1.3 Physical Components of Tensors

It has been shown that tangent vectors are not unit vectors; therefore all tensor components must be reevaluated in terms of unit vectors to obtain physical components. To this end, we first consider the tangent vectors $\mathbf{g}_i$ or $\mathbf{g}^i$. Denoting $\bar{\mathbf{g}}_i$ and $\bar{\mathbf{g}}^i$ as unity, we write

$$\bar{\mathbf{g}}_i = \frac{\mathbf{g}_i}{|\mathbf{g}_i|} = \frac{\mathbf{g}_i}{\sqrt{\mathbf{g}_{(i)}\cdot\mathbf{g}_{(i)}}} = \frac{\mathbf{g}_i}{\sqrt{g_{(ii)}}},$$

$$\mathbf{g}_i = \sqrt{g_{(ii)}}\,\bar{\mathbf{g}}_i, \tag{6.1.22a}$$

$$\bar{\mathbf{g}}^i = \frac{\mathbf{g}^i}{|\mathbf{g}^i|} = \frac{\mathbf{g}^i}{\sqrt{\mathbf{g}^{(i)}\cdot\mathbf{g}^{(i)}}} = \frac{\mathbf{g}^i}{\sqrt{g^{(ii)}}},$$

$$\mathbf{g}^i = \sqrt{g^{(ii)}}\,\bar{\mathbf{g}}^i, \tag{6.1.22b}$$

where the index inside the parentheses does not imply a summing. Therefore any vector $\mathbf{V}$ can be written as

$$\mathbf{V} = V_i\mathbf{g}^i = \bar{V}_i\bar{\mathbf{g}}^i = \bar{V}_i\frac{\mathbf{g}^i}{\sqrt{g^{(ii)}}}.$$

This gives

$$V_i = \frac{\bar{V}_i}{\sqrt{g^{(ii)}}},$$

which implies that the physical components $\bar{V}_i$ of the tensor components V_i of the vector **V** are given by

$$\bar{V}_i = \sqrt{g^{(ii)}}\,V_i. \tag{6.1.23}$$

Similarly,

$$\bar{V}^i = \sqrt{g_{(ii)}}\,V^i. \tag{6.1.24}$$

It is easy to see that, for orthogonal coordinates in which all off-diagonal terms are zero, $g_{(ii)}$ can be written as

$$g_{(ii)} = \frac{1}{g^{(ii)}}. \tag{6.1.25}$$

Note that, with the conversion of tensor components into physical components, all units and dimensions are consistent throughout the differential equations.

Physical components of second-order tensors are discussed in the later chapters.

6.1.4 Cylindrical Coordinates

Consider the cylindrical coordinates $\xi_1 = r$, $\xi_2 = \theta$, and $\xi_3 = z$, $z_1 = r\cos\theta$, $z_2 = r\sin\theta$, and $z_3 = z$ [see Fig. 6.1.4(a)]. This gives

$$\begin{aligned}
\mathbf{r} &= z_i\mathbf{i}_i = r\,(\cos\theta)\,\mathbf{i}_1 + r\,(\sin\theta)\,\mathbf{i}_2 + z\mathbf{i}_3,\\
\mathbf{g}_1 &= \mathbf{r}_{,1} = (\cos\theta)\,\mathbf{i}_1 + (\sin\theta)\,\mathbf{i}_2,\\
\mathbf{g}_2 &= \mathbf{r}_{,2} = -r\,(\sin\theta)\,\mathbf{i}_1 + r\,(\cos\theta)\,\mathbf{i}_2,\\
\mathbf{g}_3 &= \mathbf{r}_{,3} = \mathbf{i}_3.
\end{aligned} \tag{6.1.26}$$

To calculate the Christoffel symbols of the second kind, we proceed as follows: The covariant metric tensors are computed by

$$\begin{aligned}
g_{11} &= \frac{\partial z_1}{\partial \xi_1}\frac{\partial z_1}{\partial \xi_1} + \frac{\partial z_2}{\partial \xi_1}\frac{\partial z_2}{\partial \xi_1} + \frac{\partial z_3}{\partial \xi_1}\frac{\partial z_3}{\partial \xi_1} = \cos^2\theta + \sin^2\theta = 1,\\
g_{22} &= r^2\sin^2\theta + r^2\cos^2\theta = r^2,\\
g_{33} &= 1,
\end{aligned} \tag{6.1.27}$$

with all other $g_{ij} = 0$.

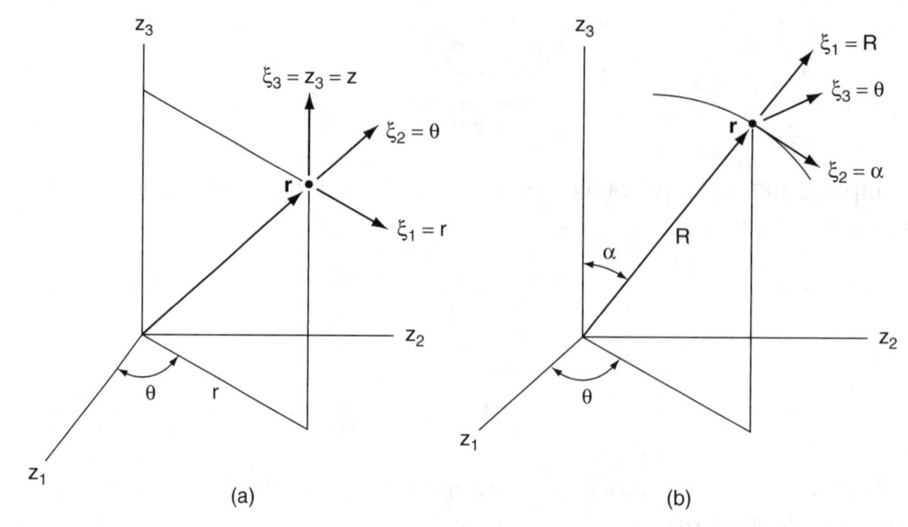

Figure 6.1.4. Curvilinear coordinates. (a) Cylindrical coordinates: r = radius of cylinder, θ = circumferential angle, z = axial coordinate. (b) Spherical coordinates: R = radius of sphere, α = meridian angle, θ = circumferential angle.

Thus

$$g_{ij} = \begin{bmatrix} 1 & 0 & 0 \\ 0 & r^2 & 0 \\ 0 & 0 & 1 \end{bmatrix}. \tag{6.1.28}$$

The contravariant metric tensors are given by the reciprocal of g_{ij} as

$$g^{ij} = \begin{bmatrix} 1 & 0 & 0 \\ 0 & \dfrac{1}{r^2} & 0 \\ 0 & 0 & 1 \end{bmatrix}. \tag{6.1.29}$$

Recall that the Christoffel symbols of the first and second kinds are

$$\Gamma_{ijs} = \frac{1}{2}\left(g_{si,j} + g_{sj,i} - g_{ij,s}\right), \tag{6.1.30}$$

$$\Gamma^r_{ij} = g^{rs}\Gamma_{ijs}. \tag{6.1.31}$$

The Christoffel symbols of the first kind are determined from derivatives of covariant metric tensors:

$$\Gamma_{122} = \frac{1}{2}(g_{21,2} + g_{22,1} - g_{12,2}) = \frac{1}{2}(0 + 2r - 0) = r.$$

In summary, the Christoffel symbols of the first kind are

$$\Gamma_{122} = \Gamma_{212} = r, \quad \Gamma_{221} = -r,$$

and all other $\Gamma_{ijk} = 0$. From these we calculate the Christoffel symbols of the second kind:

$$\Gamma^1_{22} = g^{11}\Gamma_{221} + g^{12}\Gamma_{222} = -r,$$
$$\Gamma^2_{21} = \Gamma^2_{12} = \frac{1}{r},$$

and all other $\Gamma^i_{jk} = 0$. The proof is left to the reader.

Note that, in Eq. (6.1.18),

$$g^{ij}_{,i}\psi_{,j} = 0, \tag{6.1.32}$$

so we can expand the Laplace equation in the form

$$\begin{aligned}\nabla^2\psi &= \psi_{|ji}g^{ij}\\ &= \psi_{|11} + \psi_{|22}\frac{1}{r^2} + \psi_{|33}\\ &= \psi_{,11} + \Gamma^1_{11}\psi_{,1} + \Gamma^2_{21}\psi_{,1} + \left(\psi_{,22} + \Gamma^1_{12}\psi_{,2} + \Gamma^2_{22}\psi_{,2}\right)\frac{1}{r^2} + \psi_{,33}\\ &\quad + \Gamma^1_{13}\psi_{,3} + \Gamma^2_{23}\psi_{,3} = 0\end{aligned} \tag{6.1.33a}$$

or

$$\nabla^2\psi = \frac{\partial^2\psi}{\partial r^2} + \frac{1}{r}\frac{\partial\psi}{\partial r} + \frac{1}{r^2}\frac{\partial^2\psi}{\partial\theta^2} + \frac{\partial^2\psi}{\partial z^2} = 0. \tag{6.1.33b}$$

6.1.5 Spherical Coordinates

Consider the spherical coordinates [see Fig. 6.1.4(b)] $\xi_1 = R, \xi_2 = \alpha, \xi_3 = \theta, z_1 = R\sin\alpha\cos\theta$, $z_2 = R\sin\alpha\sin\theta$, and $z_3 = R\cos\alpha$. The position vector takes the form

$$\mathbf{r} = R(\sin\alpha\cos\theta)\mathbf{i}_1 + R(\sin\alpha\sin\theta)\mathbf{i}_2 + R(\cos\alpha)\mathbf{i}_3. \tag{6.1.34}$$

Thus the tangent vectors are

$$\mathbf{g}_1 = \mathbf{r}_{,1} = (\sin\alpha\cos\theta)\mathbf{i}_1 + (\sin\alpha\sin\theta)\mathbf{i}_2 + (\cos\alpha)\mathbf{i}_3,$$
$$\mathbf{g}_2 = \mathbf{r}_{,2} = R(\cos\alpha\cos\theta)\mathbf{i}_1 + R(\cos\alpha\sin\theta)\mathbf{i}_2 - R(\sin\alpha)\mathbf{i}_3,$$
$$\mathbf{g}_3 = \mathbf{r}_{,3} = -R(\sin\alpha\sin\theta)\mathbf{i}_1 + R(\sin\alpha\cos\theta)\mathbf{i}_2.$$

From these the metric tensors are calculated as

$$g_{ij} = \begin{bmatrix}1 & 0 & 0\\ 0 & R^2 & 0\\ 0 & 0 & R^2\sin^2\alpha\end{bmatrix}. \tag{6.1.35}$$

Again, g^{ij} is the reciprocal of g_{ij}. The Christoffel symbols are

$$\Gamma^1_{22} = -R, \quad \Gamma^1_{33} = -R\sin^2\alpha,$$

$$\Gamma^2_{12} = \Gamma^2_{21} = \frac{1}{R}, \quad \Gamma^2_{33} = -\sin\alpha\cos\alpha,$$

$$\Gamma^3_{13} = \Gamma^3_{31} = \frac{1}{R}, \quad \Gamma^3_{32} = \Gamma^3_{23} = \cot\alpha,$$

with all other $\Gamma^i_{jk} = 0$. Therefore we obtain

$$\nabla^2\psi = \frac{\partial^2\psi}{\partial R^2} + \frac{2}{R}\frac{\partial\psi}{\partial R} + \frac{1}{R^2}\frac{\partial^2\psi}{\partial\alpha^2} + \frac{\cot\alpha}{R^2}\frac{\partial\psi}{\partial\alpha} + \frac{1}{R^2\sin^2\alpha}\frac{\partial^2\psi}{\partial\theta^2}. \tag{6.1.36}$$

6.1.6 Divergence and Curl of Vectors

Physical components of first-order tensors are computed as

$$u_i = \sqrt{g_{(ii)}}\bar{u}_i, \quad u^i = \sqrt{g^{(ii)}}\bar{u}^i.$$

These physical components are applied to cylindrical and spherical coordinates as follows:

Cylindrical coordinates

$$\begin{aligned}
&u_1 = \sqrt{g_{11}}\bar{u}_1 = \bar{u}_1 = u_r, \quad u_2 = \sqrt{g_{22}}\bar{u}_2 = r\bar{u}_2 = ru_\theta,\\
&u_3 = \sqrt{g_{33}}\bar{u}_3 = \bar{u}_3 = u_z,\\
&u^1 = \sqrt{g^{11}\bar{u}^1} = \bar{u}^1 = u_r, \quad u^2 = \sqrt{g^{22}}\bar{u}^2 = \frac{1}{r}\bar{u}^2 = \frac{1}{r}u_\theta,\\
&u^3 = \sqrt{g^{33}}\bar{u}^3 = \bar{u}^3 = u_z.
\end{aligned}$$

Spherical coordinates

$$\begin{aligned}
&u_1 = \bar{u}_1 = u_R, \quad u_2 = R\bar{u}_2 = Ru_\alpha, \quad u_3 = R(\sin\alpha)\bar{u}_3 = R(\sin\alpha)u_\theta,\\
&u^1 = \bar{u}^1 = u_R, \quad u^2 = \frac{\bar{u}^2}{R} = \frac{u_\alpha}{R}, \quad u^3 = \frac{\bar{u}^3}{R\sin\alpha} = \frac{u_\theta}{R\sin\alpha}.
\end{aligned}$$

Thus the divergence and the curl of a vector $\mathbf{u}$ in a cylindrical coordinates can be written as

$$\begin{aligned}
\nabla\cdot\mathbf{u} &= u^i_{|i}\\
&= u^i_{,i} + \Gamma^i_{ij}u^j\\
&= u^1_{,1} + u^2_{,2} + u^3_{,3} + \Gamma^1_{11}u^1 + \Gamma^1_{12}u^2 + \Gamma^1_{13}u^3 + \Gamma^2_{21}u^1 + \Gamma^2_{22}u^2\\
&\quad + \Gamma^2_{23}u^3 + \Gamma^3_{31}u^1 + \Gamma^3_{32}u^2 + \Gamma^3_{33}u^3\\
&= u^1_{,1} + u^2_{,2} + u^3_{,3} + \Gamma^2_{21}u^1
\end{aligned}$$

$$= (\sqrt{g^{11}}\bar{u}^1)_{,1} + (\sqrt{g^{22}}\bar{u}^2)_{,2} + (\sqrt{g^{33}}\bar{u}^3)_{,3} + \frac{1}{r}\sqrt{g^{11}}\bar{u}^1$$

$$= \frac{\partial u_r}{\partial r} + \frac{1}{r}\frac{\partial u_\theta}{\partial \theta} + \frac{\partial u_z}{\partial z} + \frac{u_r}{r}. \tag{6.1.37}$$

Similarly,

$$\nabla \times \mathbf{u} = \frac{1}{\sqrt{g}}\epsilon^{ijk}(u_{j,i} - \Gamma^r_{ji}u_r)\mathbf{g}_k$$

$$= \frac{1}{\sqrt{g}}(u_{2,1} - \Gamma^2_{21}u_2)\mathbf{g}_3 + \frac{1}{\sqrt{g}}(-u_{1,2} + \Gamma^2_{12}u_2)\mathbf{g}_3$$

$$+ \frac{1}{\sqrt{g}}(u_{1,3} - u_{3,1})\mathbf{g}_2 + \frac{1}{\sqrt{g}}(u_{3,2} - u_{2,31})\mathbf{g}_1$$

$$= \left(\frac{1}{r}\frac{\partial u_z}{\partial \theta} - \frac{\partial u_\theta}{\partial z}\right)\mathbf{e}_r + \left(\frac{\partial u_r}{\partial r} - \frac{\partial u_z}{\partial r}\right)\mathbf{e}_\theta$$

$$+ \left(\frac{\partial u_\theta}{\partial r} - \frac{1}{r}\frac{\partial u_r}{\partial \theta} + \frac{u_\theta}{r}\right)\mathbf{e}_z, \tag{6.1.38}$$

where

$$\sqrt{g} = \sqrt{|g_{ij}|} = r,$$

$$\mathbf{g}_1 = \sqrt{g_{11}}\bar{\mathbf{g}}_1 = \bar{\mathbf{g}}_1 = \mathbf{e}_r,$$

$$\mathbf{g}_2 = \sqrt{g_{22}}\bar{\mathbf{g}}_2 = r\bar{\mathbf{g}}_2 = r\mathbf{e}_\theta,$$

$$\mathbf{g}_3 = \sqrt{g_{33}}\bar{\mathbf{g}}_3 = \bar{\mathbf{g}}_3 = \mathbf{e}_z.$$

For spherical coordinates, we proceed in a similar manner and find

$$\nabla \cdot \mathbf{u} = \frac{\partial u_R}{\partial R} + \frac{1}{R}\frac{\partial u_\alpha}{\partial \alpha} + \frac{1}{R\sin\alpha}\frac{\partial u_\theta}{\partial \theta} + \frac{2u_R}{R} + \frac{u_\alpha \cot\alpha}{R}, \tag{6.1.39}$$

$$\nabla \times \mathbf{u} = \left[\frac{1}{R}\frac{\partial u_\theta}{\partial \alpha} - \frac{1}{R\sin\alpha}\frac{\partial u_\alpha}{\partial \theta} + \frac{(\cot\alpha)\,u_\theta}{R}\right]\mathbf{e}_R$$

$$+ \left(\frac{1}{R\sin\alpha}\frac{\partial u_R}{\partial \theta} - \frac{u_\theta}{R} - \frac{\partial u_\theta}{\partial R}\right)\mathbf{e}_\alpha$$

$$+ \left(\frac{\partial u_\alpha}{\partial R} - \frac{1}{R}\frac{\partial u_R}{\partial \alpha} + \frac{u_\alpha}{R}\right)\mathbf{e}_\theta, \tag{6.1.40}$$

where

$$\mathbf{g}_1 = \bar{\mathbf{g}}_1 = \mathbf{e}_R,$$
$$\mathbf{g}_2 = R\bar{\mathbf{g}}_2 = R\mathbf{e}_\alpha,$$
$$\mathbf{g}_3 = R\sin\alpha\bar{\mathbf{g}}_3 = R\sin\alpha\mathbf{e}_\theta.$$

Note that the material presented in this section is important information that is applicable to structured grid generations by means of the solution of elliptic partial differential equations in computational mechanics between the physical and the computational domains.

Discussions of the curvilinear coordinates that involve the deformed state are not rigorously pursued in this chapter, although strain-displacement relations associated with a deformed state (nonlinear strains) are discussed briefly in the next section. A complete treatment of this subject would require theories of curved surfaces or differential geometries (Chap. 9).

6.1.7 Strain Tensors

If an undeformed state is represented by curvilinear coordinates, then the position vectors are defined as

$$d\mathbf{r} = \frac{\partial \mathbf{r}}{\partial \xi_i} d\xi_i = \mathbf{r}_{,i} d\xi_i = \mathbf{g}_i d\xi_i, \tag{6.1.41}$$

$$d\mathbf{R} = \frac{\partial \mathbf{R}}{\partial \xi_i} d\xi_i = \frac{\partial}{\partial \xi_i} (\mathbf{r} + \mathbf{u})\, d\xi_i = (\mathbf{g}_i + \mathbf{u})\, d\xi_i, \tag{6.1.42}$$

where

$$\mathbf{u}_{,i} = (u_j \mathbf{g}^i)_{,i} = u_{j,i} \mathbf{g}^j - \Gamma^j_{ik} u_j \mathbf{g}^k \tag{6.1.43}$$

or

$$\mathbf{u}_{,i} = (u^j \mathbf{g}_j)_{,i} = u^j_{,i} \mathbf{g}_j + \Gamma^k_{ji} u^j \mathbf{g}_k. \tag{6.1.44}$$

Thus we obtain

$$\gamma_{ij} = \frac{1}{2}(G_{ij} - g_{ij}) = \frac{1}{2}\left(u_{i|j} + u_{j|i} + u^m_{|i} u_{m|j}\right), \tag{6.1.45}$$

with

$$u_{i|j} = \frac{\partial u_i}{\partial \xi_j} - \Gamma^m_{ij} u_m, \tag{6.1.46}$$

$$u^i_{|j} = \frac{\partial u^i}{\partial \xi_j} + \Gamma^i_{jm} u^m. \tag{6.1.47}$$

We discussed the transformation of the first-order tensor components into physical components in Subsection 6.1.2. Because the strain tensor is of the second order, we take a somewhat different approach. Note that the strain tensor is always a covariant form, although the coordinate lines ξ_i are expected to convect on the curvilinear coordinates and become contravariant. For this reason, it is appropriate to change ξ_i to ξ^i if confusion is likely to arise. Thus the analog of Eq. (2.3.3) for curvilinear coordinates is

$$ds^2 - ds_0^2 = 2\gamma_{ij} d\xi^i d\xi^j, \tag{6.1.48}$$

in which the repeated indices occur diagonally as they should for curvilinear coordinates and $d\xi^i$ is the contravariant component of a vector. The *physical component* $d\bar{\xi}^i$ is given by

$$d\bar{\xi}^i = \sqrt{g_{(ii)}} d\xi^i. \tag{6.1.49}$$

In view of Eqs. (6.1.48) and (6.1.49), it follows that

$$ds^2 - ds_0^2 = 2\bar{\gamma}_{ij} d\bar{\xi}^i d\bar{\xi}^j,$$

where

$$\bar{\gamma}_{ij} = \frac{1}{\sqrt{g_{(ii)} g_{(jj)}}} \gamma_{ij}, \tag{6.1.50a}$$

or, for curvilinear orthogonal coordinates,

$$\bar{\gamma}_{ij} = \sqrt{g^{(ii)} g^{(jj)}} \gamma_{ij}. \tag{6.1.50b}$$

An alternative approach is to use a second-order tensor as a product of two first-order tensors in the form

$$A_{ij} = V_i W_j,$$

with

$$V_i = \frac{1}{\sqrt{g^{(ii)}}} \bar{V}_i, \qquad W_i = \frac{1}{\sqrt{g^{(jj)}}} \bar{W}_i.$$

Thus,

$$A_{ij} = \frac{1}{\sqrt{g^{(ii)} g^{(jj)}}} \bar{V}_i \bar{W}_j = \sqrt{g_{(ii)} g_{(jj)}} \bar{A}_{ij}. \tag{6.1.50c}$$

Similarly, we can show the physical component of the strain tensor in the form of Eq. (6.1.50a) by setting $A_{ij} = \gamma_{ij}$, $\bar{A}_{ij} = \bar{\gamma}_{ij}$.

We can obtain explicit forms of the strain tensor by evaluating Eq. (6.1.45) for cylindrical and spherical coordinates, neglecting nonlinear terms.

Cylindrical coordinates

$$u_{i|j} = u_{i,j} - \Gamma_{ij}^m u_m,$$

$$\gamma_{ij} = \frac{1}{2}(u_{i,j} + u_{j,i}) - \Gamma_{ij}^m u_m.$$

The tensor components of the strain are

$$\gamma_{11} = u_{1,1} = \frac{\partial u_1}{\partial r},$$

$$\gamma_{22} = u_{2,2} - \Gamma^1_{22}u_1 - \Gamma^2_{22}u_2 = u_{2,2} + ru_1 = \frac{\partial u_2}{\partial \theta} + ru_1,$$

$$\gamma_{33} = u_{3,3} = \frac{\partial u_3}{\partial z},$$

$$\gamma_{12} = \frac{1}{2}(u_{1,2} + u_{2,1}) - \Gamma^1_{12}u_1 - \Gamma^2_{12}u_2$$

$$= \frac{1}{2}\left(\frac{\partial u_1}{\partial \theta} + \frac{\partial u_2}{\partial r}\right) - \frac{u_2}{r},$$

$$\gamma_{23} = \frac{1}{2}(u_{2,3} + u_{3,2}) = \frac{1}{2}\left(\frac{\partial u_2}{\partial z} + \frac{\partial u_3}{\partial \theta}\right),$$

$$\gamma_{31} = \frac{1}{2}(u_{3,1} + u_{1,3}) = \frac{1}{2}\left(\frac{\partial u_3}{\partial r} + \frac{\partial u_1}{\partial z}\right).$$

The physical components of the displacements are

$$u_1 = \bar{u}_1 = u_r, \quad u_2 = r\bar{u}_2 = ru_\theta, \quad u_3 = \bar{u}_3 = u_z.$$

Finally, the physical components of the strain tensor are as follows:

$$\bar{\gamma}_{11} = \gamma_{rr} = \sqrt{g^{(11)}g^{(11)}}\gamma_{11} = \frac{\partial u_r}{\partial r},$$

$$\bar{\gamma}_{22} = \gamma_{\theta\theta} = \sqrt{g^{(22)}g^{(22)}}\gamma_{22} = \frac{1}{r}\frac{\partial u_\theta}{\partial \theta} + \frac{u_r}{r},$$

$$\bar{\gamma}_{33} = \gamma_{zz} = \sqrt{g^{(33)}g^{(33)}}\gamma_{33} = \frac{\partial u_z}{\partial z},$$

$$\bar{\gamma}_{12} = \frac{1}{2}\gamma_{r\theta} = \sqrt{g^{(11)}g^{(22)}}\gamma_{12} = \frac{1}{2}\left(\frac{1}{r}\frac{\partial u_r}{\partial \theta} + \frac{\partial u_\theta}{\partial r}\right) - \frac{u_\theta}{r},$$

$$\bar{\gamma}_{23} = \frac{1}{2}\gamma_{\theta z} = \sqrt{g^{(22)}g^{(33)}}\gamma_{23} = \frac{1}{2}\left(\frac{\partial u_\theta}{\partial z} + \frac{1}{r}\frac{\partial u_z}{\partial \theta}\right),$$

$$\bar{\gamma}_{31} = \frac{1}{2}\gamma_{zr} = \sqrt{g^{(33)}g^{(11)}}\gamma_{31} = \frac{1}{2}\left(\frac{\partial u_z}{\partial r} + \frac{\partial u_r}{\partial z}\right).$$

Spherical coordinates

$$\gamma_{11} = \frac{\partial u_1}{\partial R} \qquad \gamma_{22} = \frac{\partial u_2}{\partial \alpha} + \frac{1}{R}u_1,$$

$$\gamma_{33} = \frac{\partial u_3}{\partial \theta} + R(\sin^2\alpha)u_1 + (\sin\alpha\cos\alpha)u_2,$$

$$\gamma_{12} = \frac{1}{2}\left(\frac{\partial u_1}{\partial \alpha} + \frac{\partial u_2}{\partial R}\right) - \frac{1}{R}u_2, \quad \gamma_{31} = \frac{1}{2}\left(\frac{\partial u_3}{\partial R} + \frac{\partial u_1}{\partial \theta}\right) - \frac{1}{R}u_3,$$

$$\gamma_{23} = \frac{1}{2}\left(\frac{\partial u_2}{\partial \theta} + \frac{\partial u_3}{\partial \alpha}\right) - (\cot\alpha)\, u_3,$$

$$u_1 = \bar{u}_1 = u_R, \quad u_2 = R\bar{u}_2 = Ru_\alpha, \quad u_3 = R(\sin\alpha)\bar{u}_3 = R(\sin\alpha)u_\theta,$$

$$\bar{\gamma}_{11} = \gamma_{RR} = \frac{\partial u_R}{\partial R}, \quad \bar{\gamma}_{22} = \gamma_{\alpha\alpha} = \frac{1}{R}\frac{\partial u_\alpha}{\partial \alpha} + \frac{u_R}{R},$$

$$\bar{\gamma}_{33} = \gamma_{\theta\theta} = \frac{1}{R\sin\alpha}\frac{\partial u_\theta}{\partial \theta} + \frac{u_R}{R} + \frac{\cot\alpha}{R}u_\theta,$$

$$\bar{\gamma}_{12} = \frac{1}{2}\gamma_{R_\alpha} = \frac{1}{2}\left(\frac{1}{R}\frac{\partial u_R}{\partial \alpha} + \frac{\partial u_\alpha}{\partial R} - \frac{u_\alpha}{R}\right),$$

$$\bar{\gamma}_{23} = \frac{1}{2}\gamma_{\alpha\theta} = \frac{1}{2}\left(\frac{1}{R\sin\alpha}\frac{\partial u_\alpha}{\partial \theta} + \frac{1}{R}\frac{\partial u_\theta}{\partial \alpha} - \frac{\cot\alpha}{r}u_\theta\right),$$

$$\bar{\gamma}_{31} = \frac{1}{2}\gamma_{\theta R} = \frac{1}{2}\left(\frac{1}{R\sin\alpha}\frac{\partial u_R}{\partial \theta} + \frac{\partial u_\theta}{\partial R} - \frac{u_\theta}{R}\right).$$

For nonlinear terms $u^m_{|i}u_{m|j}$ involved in the curvilinear coordinates, similar operations are repeated, and the reader is encouraged to verify the results given in Problem 6.11.

6.2 Linear Elasticity

6.2.1 Conservation of Linear and Angular Momenta

For curvilinear coordinates in the undeformed state, the unit vector is replaced with the tangent vector (Green and Zerna, 1954). Therefore the surface traction normal to the boundary surface must be written in terms of the contravariant stress tensor σ^{ij} so that

$$\sigma(n) = \sigma^{ij}\mathbf{g}_j n_i. \tag{6.2.1}$$

Here the stress tensor must be of contravariant form so that it can be based on the covariant curvilinear tangent vector $\mathbf{g}_j$ and the covariant normal vector component n_i. In view of Eq. (6.1.15a) and the Green–Gauss theorem, we write

$$\begin{aligned}\int_\Gamma \mathbf{V}\cdot\mathbf{n}\,d\Gamma &= \int_\Gamma V^i\mathbf{g}_i\cdot n_j\mathbf{g}^j\,d\Gamma \\ &= \int_\Gamma V^i n_j\delta^j_i\,d\Gamma = \int_\Gamma V^i n_i\,d\Gamma = \int_\Gamma V^i_{|i}\,d\Omega \\ &= \int_\Omega \nabla\cdot\mathbf{V}d\Omega.\end{aligned} \tag{6.2.2}$$

Similarly, by setting

$$\boldsymbol{\sigma}^i = \sigma^{ij}\mathbf{g}_j, \tag{6.2.3}$$

we obtain

$$\int_\Gamma \sigma^{ij}\mathbf{g}_j n_i \, d\Gamma = \int_\Gamma \boldsymbol{\sigma}^i n_i \, d\Gamma = \int_\Omega \boldsymbol{\sigma}^i_{|i} \, d\Omega, \tag{6.2.4}$$

where

$$\begin{aligned} \boldsymbol{\sigma}^i_{|i} &= \boldsymbol{\sigma}^i_{,i} + \Gamma^i_{ij}\boldsymbol{\sigma}^j = \left(\sigma^{ij}\mathbf{g}_j\right)_{,i} + \Gamma^i_{ij}\sigma^{jm}\mathbf{g}_m \\ &= \sigma^{ij}_{,i}\mathbf{g}_j + \sigma^{ij}\Gamma^m_{ij}\mathbf{g}_m + \Gamma^i_{ij}\sigma^{jm}\mathbf{g}_m \\ &= \sigma^{ij}_{|i}\mathbf{g}_j = \frac{1}{\sqrt{g}}\left(\sqrt{g}\sigma^{ij}\mathbf{g}_j\right)_{,i}, \end{aligned} \tag{6.2.5}$$

with

$$\sigma^{ij}_{|i} = \sigma^{ij}_{,i} + \Gamma^i_{im}\sigma^{mj} + \Gamma^j_{im}\sigma^{im}. \tag{6.2.6}$$

It is interesting to note that the covariant derivative of a second-order tensor expressed in Eq. (6.2.6) has a form identical to that of Eq. (6.1.20b), although it is derived in a different manner.

It follows from Eqs. (3.3.1) and (6.2.6) that the conservation of linear momentum becomes

$$\int_\Omega \left(\sigma^{ij}_{|i}\mathbf{g}_j + \rho\mathbf{F} - \rho\dot{\mathbf{v}}\right) d\Omega = 0.$$

Because the integrand must vanish, we obtain

$$\sigma^{ij}_{|i} + \rho F^j - \rho\dot{v}^j = 0. \tag{6.2.7}$$

This is Cauchy's first law of motion, or the equation of motion, for undeformed *curvilinear* coordinates.

From operations similar to those in Eq. (3.3.4), but using the curvilinear coordinates and noting that $\mathbf{r}_{|i} = \mathbf{r}_{,i} = \mathbf{g}_i$, we obtain the conservation of angular momentum (see Fig. 6.1.1):

$$\int_\Omega \mathbf{r} \times \left(\sigma^{ij}_{|i}\mathbf{g}_j + \rho\mathbf{F} - \rho\dot{\mathbf{v}}\right) d\Omega + \int_\Omega \mathbf{g}_i \times \sigma^{ij}\mathbf{g}_j \, d\Omega = 0. \tag{6.2.8}$$

Thus, once again the conservation of angular momentum requires that

$$\mathbf{g}_i \times \sigma^{ij}\mathbf{g}_j = 0, \tag{6.2.9}$$

and it follows from Eq. (6.1.6) that the previous expression takes the form

$$\sigma^{ij}\sqrt{g}\epsilon_{ijk}\mathbf{g}^k = 0. \tag{6.2.10}$$

With algebra similar to that in Eq. (3.3.7), we obtain

$$\sigma^{ij} = \sigma^{ji}, \tag{6.2.11}$$

which is Cauchy's second law of motion, indicating the symmetry of the stress tensor in the curvilinear coordinates.

The physical components $\bar{\sigma}^{ij}$ of the stress tensor may be derived in a manner similar to that of Eq. (6.1.50c). Thus, for cylindrical or spherical coordinates,

$$\sigma^{ij} = \sqrt{g^{(ii)}g^{(jj)}}\bar{\sigma}^{ij}. \tag{6.2.12}$$

Explicit Equations of Motion for Cylindrical and Spherical Coordinates

The equations of motion for small strain in cylindrical and spherical coordinates are derived in terms of physical components as

$$\sigma^{ij}_{,i} + \Gamma^{i}_{im}\sigma^{jm} + \Gamma^{j}_{mi}\sigma^{mi} + \rho F^{j} - \rho\dot{v}^{j} = 0. \tag{6.2.13}$$

For cylindrical coordinates

$$F^1 = F_r, \quad F^2 = \frac{F_\theta}{r}, \quad F^3 = F_z,$$

$$\dot{v}^1 = \ddot{u}_r, \quad \dot{v}^2 = \frac{\ddot{u}_\theta}{r}, \quad \dot{v}^3 = \ddot{u}_z,$$

$$\sigma^{11} = \sigma_{rr}, \quad \sigma^{22} = \frac{\sigma_{\theta\theta}}{r^2}, \quad \sigma^{33} = \sigma_{zz},$$

$$\sigma^{12} = \sigma^{21} = \frac{\sigma_{\theta r}}{r}, \quad \sigma^{13} = \sigma^{31} = \sigma^{zr}, \quad \sigma^{23} = \sigma^{32} = \frac{\sigma_{z\theta}}{r},$$

$$\frac{\partial\sigma_{rr}}{\partial r} + \frac{1}{r}\frac{\partial\sigma_{\theta r}}{\partial\theta} + \frac{\partial\sigma_{zr}}{\partial z} + \frac{\sigma_{rr} - \sigma_{\theta\theta}}{r} + \rho F_r - \rho\ddot{u}_r = 0,$$

$$\frac{\partial\sigma_{r\theta}}{\partial r} + \frac{1}{r}\frac{\partial\sigma_{\theta\theta}}{\partial\theta} + \frac{\partial\sigma_{z\theta}}{\partial z} + \frac{2}{r}\sigma_{r\theta} + \rho F_\theta - \rho\ddot{u}_\theta = 0,$$

$$\frac{\partial\sigma_{rz}}{\partial r} + \frac{1}{r}\frac{\partial\sigma_{\theta z}}{\partial\theta} + \frac{\partial\sigma_{zz}}{\partial z} + \frac{\sigma_{rz}}{r} + \rho F_z - \rho\ddot{u}_z = 0.$$

For spherical coordinates

$$F^1 = F_R, \quad F^2 = \frac{F_\alpha}{R}, \quad F^3 = \frac{F_\theta}{R\sin\alpha},$$

$$\dot{v}^1 = \ddot{u}_R, \quad \dot{v}^2 = \frac{\ddot{u}_\alpha}{R}, \quad \dot{v}^3 = \frac{\ddot{u}_\theta}{R\sin\alpha},$$

$$\sigma^{11} = \sigma_{RR}, \quad \sigma^{22} = \frac{\sigma_{\alpha\alpha}}{R^2}, \quad \sigma^{33} = \frac{\sigma_{\theta\theta}}{R^2\sin^2\alpha},$$

$$\sigma^{12} = \sigma^{21} = \frac{\sigma_{\alpha R}}{R}, \quad \sigma^{13} = \sigma^{31} = \frac{\sigma_{\theta R}}{R\sin\alpha}, \quad \sigma^{23} = \sigma^{32} = \frac{\sigma_{\theta\alpha}}{R^2\sin\alpha},$$

$$\frac{\partial \sigma_{RR}}{\partial R} + \frac{1}{R}\frac{\partial \sigma_{\alpha R}}{\partial \alpha} + \frac{1}{R \sin\alpha}\frac{\partial \sigma_{\theta R}}{\partial \theta} + \frac{2}{R}\sigma_{RR}$$
$$+ \frac{\cot\alpha}{R}\sigma_{\alpha R} - \frac{\sigma_{\alpha\alpha}}{R} - \frac{\sigma_{\theta\theta}}{R} + \rho F_R - \rho \ddot{u}_R = 0,$$
$$\frac{\partial \sigma_{R\alpha}}{\partial R} + \frac{1}{R}\frac{\partial \sigma_{\alpha\alpha}}{\partial \alpha} + \frac{1}{R \sin\alpha}\frac{\partial \sigma_{\theta\alpha}}{\partial \theta} + \frac{3}{R}\sigma_{R\alpha}$$
$$+ \frac{\cot\alpha}{R}\sigma_{\alpha\alpha} - \frac{(\cot\alpha)}{R}\sigma_{\theta\theta} + \rho F_\alpha - \rho \ddot{u}_\alpha = 0,$$
$$\frac{\partial \sigma_{R\theta}}{\partial R} + \frac{1}{R}\frac{\partial \sigma_{\alpha\theta}}{\partial \alpha} + \frac{1}{R \sin\alpha}\frac{\partial \sigma_{\theta\theta}}{\partial \theta} + \frac{3}{R}\sigma_{R\theta} + \frac{2}{R}(\cot\alpha)\sigma_{\alpha\theta}$$
$$+ \rho F_\theta - \rho \ddot{u}_\theta = 0.$$

6.2.2 Navier Equations

We may recast the Navier equations in curvilinear coordinates by utilizing the vector form of Eq. (4.2.2b). First, consider $\nabla^2 \mathbf{u}$ and $\nabla(\nabla \cdot \mathbf{u})$ in curvilinear coordinates,

$$\begin{aligned}
\nabla^2 \mathbf{u} = (\nabla \cdot \nabla)\mathbf{u} &= \left(\mathbf{g}^k \frac{\partial}{\partial \xi_k} \cdot \mathbf{g}^j \frac{\partial}{\partial \xi_j}\right)\mathbf{u} \\
&= \left(g^{kj}\frac{\partial^2}{\partial \xi_k \partial \xi_j} + \mathbf{g}^k \cdot \mathbf{g}^j_{,k}\frac{\partial}{\partial \xi_j}\right)\mathbf{u} \\
&= \left(g^{kj}\frac{\partial^2}{\partial \xi_k \partial \xi_j} - \mathbf{g}^k \cdot \Gamma^j_{km} g^{mn}\mathbf{g}_n \frac{\partial}{\partial \xi_j}\right)\mathbf{u} \\
&= g^{kj}\left[\left(u^i \mathbf{g}_i\right)_{,kj} - \Gamma^m_{kj}(u^i \mathbf{g}_i)_{,m}\right] = g^{kj}\mathbf{g}_i u^i_{|kj},
\end{aligned} \tag{6.2.14}$$

where

$$u^i_{|kj} = u^i_{,jk} + 2u^m_{,k}\Gamma^i_{mj} + u^m\left(\Gamma^i_{jm}\right)_{,k} + u^m\Gamma^n_{mj}\Gamma^i_{nk} - u^i_{,m}\Gamma^m_{kj} - u^n\Gamma^i_{nm}\Gamma^m_{kj},$$

$$\begin{aligned}
\nabla(\nabla \cdot \mathbf{u}) &= \mathbf{g}^k \frac{\partial}{\partial \xi_k}\left(\mathbf{g}^j \frac{\partial}{\partial \xi_j} \cdot u^i \mathbf{g}_i\right) = \mathbf{g}^k \frac{\partial}{\partial \xi_k}\left(u^i_{,i} + \mathbf{g}^j \cdot u^i \Gamma^m_{ij}\mathbf{g}_m\right) \\
&= \mathbf{g}^k \frac{\partial}{\partial \xi_k}\left(u^i_{,i} + u^j \Gamma^j_{ji}\right) = \mathbf{g}^k\left[u^i_{,ik} + u^i_{,k}\Gamma^j_{ji} + u^i\left(\Gamma^j_{ji}\right)_{,k}\right] = g^{ki}\mathbf{g}_i u^j_{|jk},
\end{aligned} \tag{6.2.15}$$

and where

$$u^j_{|jk} = u^j_{,jk} + u^j_{,k}\Gamma^m_{mj} + u^j\left(\Gamma^m_{mj}\right)_{,k}.$$

Using the physical components for cylindrical coordinates,

$$\Gamma_{22}^{1} = -r, \quad \Gamma_{12}^{2} = \Gamma_{21}^{2} = \frac{1}{r},$$

$$g^{11} = g^{33} = 1, \quad g^{22} = \frac{1}{r^2},$$

$$u^1 = u_r, \quad u^2 = \frac{u_\theta}{r}, \quad u^3 = u_z,$$

we obtain, for the radial direction ($i = 1$),

$$\mu\left(\frac{\partial^2 u_r}{\partial r^2} + \frac{1}{r}\frac{\partial u_r}{\partial r} + \frac{1}{r^2}\frac{\partial^2 u_r}{\partial \theta^2} - \frac{2}{r^2}\frac{\partial u_\theta}{\partial \theta} - \frac{u_r}{r^2} + \frac{\partial^2 u_r}{\partial z^2}\right)$$
$$+ (\lambda + \mu)\left(\frac{\partial^2 u_r}{\partial r^2} + \frac{1}{r}\frac{\partial^2 u_\theta}{\partial \theta \partial r} + \frac{\partial^2 u_z}{\partial r \partial z} - \frac{1}{r^2}\frac{\partial u_\theta}{\partial \theta} + \frac{1}{r}\frac{\partial u_r}{\partial r} - \frac{u_r}{r^2}\right)$$
$$+ \rho F_r - \rho \ddot{u}_r = 0. \tag{6.2.16a}$$

Similarly, for the tangential direction ($i = 2$), we obtain

$$\mu\left(\frac{\partial^2 u_\theta}{\partial r^2} + \frac{1}{r}\frac{\partial u_\theta}{\partial r} + \frac{1}{r^2}\frac{\partial^2 u_\theta}{\partial \theta^2} + \frac{2}{r^2}\frac{\partial u_r}{\partial \theta} + \frac{\partial^2 u_\theta}{\partial z^2} - \frac{u_\theta}{r^2}\right)$$
$$+ (\lambda + \mu)\left(\frac{1}{r}\frac{\partial^2 u_r}{\partial r \partial \theta} + \frac{1}{r^2}\frac{\partial^2 u_\theta}{\partial \theta^2} + \frac{1}{r}\frac{\partial^2 u_z}{\partial z \partial \theta} + \frac{1}{r^2}\frac{\partial u_r}{\partial \theta}\right)$$
$$+ \rho F_\theta - \rho \ddot{u}_\theta = 0, \tag{6.2.16b}$$

and, finally, for the axial direction ($i = 3$), we obtain

$$\mu\left(\frac{\partial^2 u_z}{\partial r^2} + \frac{1}{r}\frac{\partial u_z}{\partial r} + \frac{1}{r^2}\frac{\partial^2 u_z}{\partial \theta^2} + \frac{\partial^2 u_z}{\partial z^2}\right)$$
$$+ (\lambda + \mu)\left(\frac{\partial^2 u_r}{\partial r \partial \theta} + \frac{1}{r}\frac{\partial^2 u_\theta}{\partial \theta \partial z} + \frac{\partial^2 u_z}{\partial z^2} + \frac{1}{r}\frac{\partial u_r}{\partial z}\right) + \rho F_z - \rho \ddot{u}_z = 0. \tag{6.2.16c}$$

The Navier equations for spherical coordinates may be obtained in a similar manner. The results are as follows.

***R* direction (radial)**

$$\mu\left(\frac{\partial^2 u_R}{\partial R^2} + \frac{1}{R^2}\frac{\partial^2 u_r}{\partial \alpha^2} + \frac{1}{R^2 \sin^2 \alpha}\frac{\partial^2 u_R}{\partial \theta^2} - \frac{2}{R^2}\frac{\partial u_\alpha}{\partial \alpha} - \frac{2}{R^2 \sin \alpha}\frac{\partial u_\theta}{\partial \theta}\right.$$
$$\left. - \frac{2u_r}{R^2} - \frac{2u_\alpha \cot \alpha}{R^2}\frac{2}{R}\frac{\partial u_R}{\partial R} + \frac{\cot \alpha}{R^2}\frac{\partial u_R}{\partial \alpha}\right)$$
$$+ (\lambda + \mu)\left(\frac{\partial^2 u_R}{\partial R^2} - \frac{1}{R^2}\frac{\partial u_\alpha}{\partial \alpha} + \frac{1}{R}\frac{\partial^2 u_\alpha}{\partial \alpha \partial R} + \frac{1}{R^2 \sin \alpha}\frac{\partial u_\theta}{\partial \theta}\right.$$
$$\left. + \frac{1}{R \sin \alpha}\frac{\partial^2 u_\theta}{\partial R \partial \theta} - \frac{2u_R}{R^2} - \frac{\cot \alpha \, u_\alpha}{R^2} + \frac{2}{R}\frac{\partial u_R}{\partial R} + \frac{\cot \alpha}{R}\frac{\partial u_\alpha}{\partial R}\right)$$
$$+ \rho F_R - \rho \ddot{u}_R = 0. \tag{6.2.17a}$$

α direction (meridional)

$$\mu\left(\frac{2}{R}\frac{\partial u_\alpha}{\partial R}+\frac{\partial^2 u_\alpha}{\partial R^2}+\frac{1}{R^2}\frac{\partial^2 u_\alpha}{\partial \alpha^2}+\frac{1}{R^2\sin^2\alpha}\frac{\partial^2 u_\alpha}{\partial \theta^2}+\frac{2}{R^2}\frac{\partial u_R}{\partial \alpha}\right.$$
$$\left.-\frac{2\cos\alpha}{R^2\sin^2\alpha}\frac{\partial u_\theta}{\partial \theta}-\frac{u_\alpha}{R^2\sin^2\alpha}+\frac{\cot\alpha}{R^2}\frac{\partial u_\alpha}{\partial \alpha}\right)$$
$$+(\lambda+\mu)\left(\frac{1}{R}\frac{\partial^2 u_R}{\partial R\partial \alpha}+\frac{1}{R^2}\frac{\partial^2 u_\alpha}{\partial \alpha^2}-\frac{\cos\alpha}{R^2\sin^2\alpha}\frac{\partial u_\theta}{\partial \theta}+\frac{1}{R^2\sin\alpha}\frac{\partial^2 u_\theta}{\partial \alpha\partial \theta}\right.$$
$$\left.+\frac{2}{R^2}\frac{\partial u_R}{\partial \alpha}+\frac{\cot\alpha}{R^2}+\frac{\partial u_\alpha}{\partial \alpha}-\frac{u_\alpha}{R^2}-\frac{u_\alpha\cot^2\alpha}{R^2}\right)+\rho F_\alpha-\rho\ddot{u}_\alpha=0.$$

(6.2.17b)

θ direction (tangential)

$$\mu\left(\frac{2}{R}\frac{\partial u_\theta}{\partial R}+\frac{\partial^2 u_\theta}{\partial R^2}+\frac{\cot\alpha}{R^2}\frac{\partial u_\theta}{\partial \alpha}+\frac{1}{R^2}\frac{\partial^2 u_\theta}{\partial \alpha^2}+\frac{1}{R^2\sin^2\alpha}\frac{\partial^2 u_\theta}{\partial \theta^2}\right.$$
$$\left.-\frac{u_\theta}{R^2\sin^2\alpha}+\frac{2}{R^2\sin^2\alpha}\frac{\partial u_R}{\partial \theta}+\frac{2\cos\alpha}{R^2\sin^2\alpha}\frac{\partial u_\alpha}{\partial \theta}\right)$$
$$+(\lambda+\mu)\left[\frac{1}{R\sin\alpha}\left(\frac{\partial^2 u_R}{\partial R\partial \theta}+\frac{1}{R}\frac{\partial^2 u_\alpha}{\partial \alpha\partial \theta}+\frac{1}{R\sin\alpha}\frac{\partial^2 u_\theta}{\partial \theta^2}\right.\right.$$
$$\left.\left.+\frac{2}{R}\frac{\partial u_R}{\partial \theta}+\frac{\cot\alpha}{R}\frac{\partial u_\alpha}{\partial \theta}\right)\right]+\rho F_\theta-\rho\ddot{u}_\theta=0. \tag{6.2.17c}$$

6.2.3 Compatibility Equations

At this point, we return to our six strain-displacement equations, as defined in Eq. (2.3.5) and spelled out in Eq. (2.3.6), and consider them as partial differential equations to be solved. If three displacement components u_i are given, then it is possible to determine the six strain components. The six strain components, however, may sometimes not uniquely define the three displacement components. There are more equations available than there are unknowns. Thus the existence of unique displacements cannot be guaranteed unless certain compatibility conditions are satisfied. It will be shown that determination of such compatibility conditions requires curvilinear coordinates dealing with large deformations.

Consider displacements of two points A and B. Let A_1 and B_1 denote the rectangular coordinates at A and B, respectively. We see that the displacement at B may be related to the displacement at A through the expression

$$\mathbf{u}_B=\mathbf{u}_A+\int_A^B d\mathbf{u}=\mathbf{u}_A+\int_A^B \mathbf{u}_{,i}dx_i=\mathbf{u}_A+\int_A^B \mathbf{u}_{,i}d\left(x_i-B_i\right).$$

Integrating by parts, we get

$$\mathbf{u}_B = \mathbf{u}_A + (B_i - x_i)\,\mathbf{u}_{,i} + \int_A^B (B_i - x_i)\mathbf{u}_{,ij}dx_j$$
$$= \mathbf{u}_A + (B_i - x_i)\,\mathbf{u}_{,i} + \int_A^B (B_i - x_i)\,\mathbf{G}_{i,j}dx_j. \qquad (6.2.18)$$

Here, the definition given in Eq. (2.2.8) is used so that $\mathbf{u}_{,ij} = \mathbf{G}_{i,j}$. Let $\mathbf{W}_n = (B_i - x_i)\mathbf{G}_{i,n}$. Then, in a simply connected region, $\int_A^B \mathbf{W}_n dx_n$ is independent of the path if and only if

$$\frac{\partial \mathbf{W}_n}{\partial x_m} - \frac{\partial \mathbf{W}_m}{\partial x_n} = 0.$$

Therefore the existence of **u** is ensured only if

$$\frac{\partial}{\partial x_m}[(B_i - x_i)\mathbf{G}_{i,n}] - \frac{\partial}{\partial x_n}[(B_i - x_i)\mathbf{G}_{i,m}] = 0,$$
$$(\Gamma_{nmp} - \Gamma_{mnp})\mathbf{G}^p + (B_i - x_i)[(\Gamma_{inr}\mathbf{G}^r)_{,m} - (\Gamma_{imr}\mathbf{G}^r)_{,n}] = 0$$

or

$$\{(\Gamma_{nmp} - \Gamma_{mnp}) + (B_i - x_i)[\Gamma_{inp,m} - \Gamma_{imp,n}$$
$$+ \mathbf{G}^{qr}(-\Gamma_{inr}\Gamma_{mpq} + \Gamma_{imr}\Gamma_{npq})]\}\mathbf{G}^P = 0, \qquad (6.2.19)$$

where

$$\Gamma_{nmp} = \frac{\partial^2 z_r}{\partial x_n \partial x_m}\frac{\partial z_r}{\partial x_p} = \gamma_{np,m} + \gamma_{mp,n} - \gamma_{nm,p} = \Gamma_{mnp}.$$

For $B_i \neq x_i$, Eq. (6.2.19) is satisfied if and only if the bracketed terms vanish:

$$R_{ijkq} = \gamma_{jk,qi} + \gamma_{qi,jk} - \gamma_{jq,ki} - \gamma_{ki,jq} + G^{rs}(\Gamma_{jkr}\Gamma_{qis} - \Gamma_{jqr}\Gamma_{kis}) = 0, \quad (6.2.20)$$

where R_{ijkq} is called the *Riemann–Christoffel tensor*. Because of the symmetry of γ_{ij}, we have only six independent equations corresponding to

$$R_{2112}, \quad R_{3113}, \quad R_{2332}, \quad R_{2113}, \quad R_{2331}, \quad R_{1223} = 0.$$

These represent the integrability conditions (with regard to the existence of a solution) or compatibility conditions of the strain components. The component R_{2331} will be used in torsion in Subsection 6.2.4.

For two-dimensional space, if we neglect higher-order terms (small strains), then Eq. (6.2.20) assumes the form

$$R_{2112} = \gamma_{22,11} + \gamma_{11,22} - \gamma_{12,12} - \gamma_{21,21}$$

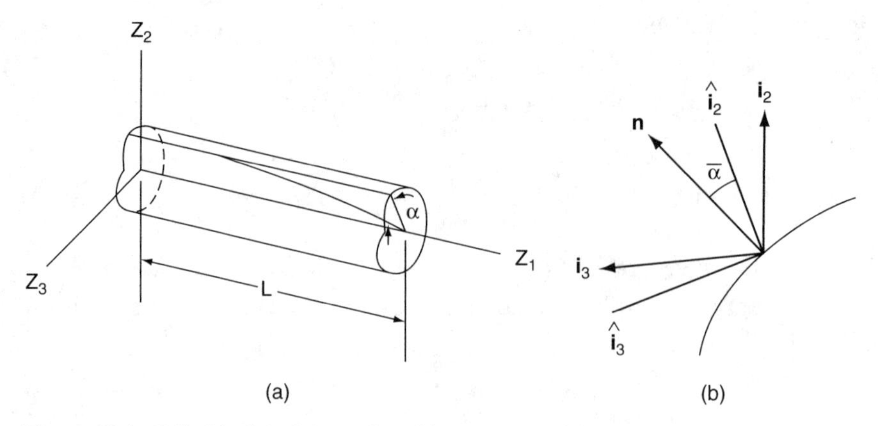

Figure 6.2.1. Cylindrical shaft in torsion: (a) twist angle, (b) boundaries.

or

$$R_{2112} = \frac{\partial^2 \gamma_{11}}{\partial x_2^2} + \frac{\partial^2 \gamma_{22}}{\partial x_1^2} - 2\frac{\partial^2 \gamma_{12}}{\partial x_1 \partial x_2} = 0. \tag{6.2.21}$$

This is called the *compatibility equation*. If the strain components vanish, then the integrand of Eq. (6.2.18) must vanish. This means that the displacement at B is the sum of the displacement at A and the relative displacement that is due to rigid-body motion.

The usefulness of the compatibility equation is demonstrated for obtaining the governing equation for torsion in the following subsection.

6.2.4 Torsion

Torsional deformations must be consistent with the compatibility equation based on the Riemann–Christoffel tensor discussed in Subsection 6.2.3. This requires an application of curvilinear coordinates as the curvatures of surface must be defined in order to analyze the distorted surfaces of a body.

If a system of forces acting on a small portion of the surface of an elastic body is replaced with another statically equivalent system of force acting on the same portion of the surface, this redistribution of loading produces substantial changes in stress in only the immediate neighborhood of the loading. At the same time, stresses are essentially the same in the parts of the body that are at large distances in comparison with the linear dimensions of the surface on which the forces are changed. This is known as Saint-Venant's principle.

Consider the cylindrical bar fixed at one end and free at the other shown in Fig. 6.2.1. If the free end is subjected to twisting action, the net twist angle at the free end is

$$\alpha = \theta z_1 = \theta x_1, \tag{6.2.22}$$

where θ is a twist angle per unit length. We then have

$$u_1 = \theta\psi(x_2, x_3), \quad u_2 = -\alpha x_3, \quad u_3 = \alpha x_2,$$

where ψ is the warping function. From $z_i = x_i + u_i$, we write

$$z_1 = x_1 + \theta\psi(x_2, x_3), \tag{6.2.23a}$$

$$z_2 = x_2\cos(\theta x_1) - x_3\sin(\theta x_1), \tag{6.2.23b}$$

$$z_3 = x_2\sin(\theta x_1) + x_3\cos(\theta x_1). \tag{6.2.23c}$$

Because

$$\mathbf{G}_i = z_{m,i}\mathbf{i}_m,$$

we have

$$\begin{aligned}
\mathbf{G}_1 &= \hat{\mathbf{i}}_1 - [\theta x_2\sin(\theta x_1) + \theta x_3\cos(\theta x_1)]\hat{\mathbf{i}}_2 \\
&\quad + [\theta x_2\cos(\theta x_1) - \theta x_3\sin(\theta x_1)]\hat{\mathbf{i}}_3 \\
&= \hat{\mathbf{i}}_1 + \theta(-x_3\hat{\mathbf{i}}_2 + x_2\hat{\mathbf{i}}_3), \\
\mathbf{G}_2 &= \theta\psi_{,2}\hat{\mathbf{i}}_1 + \hat{\mathbf{i}}_2, \\
\mathbf{G}_3 &= \theta\psi_{,3}\hat{\mathbf{i}}_1 + \hat{\mathbf{i}}_3,
\end{aligned}$$

where the $\hat{\mathbf{i}}_i$'s are the unit vectors on the deformed surface:

$$\begin{aligned}
\hat{\mathbf{i}}_1 &= \mathbf{i}_1, \\
\hat{\mathbf{i}}_2 &= \mathbf{i}_2\cos(\theta x_1) + \mathbf{i}_3\sin(\theta x_1), \\
\hat{\mathbf{i}}_3 &= -\mathbf{i}_2\sin(\theta x_1) + \mathbf{i}_3\cos(\theta x_1).
\end{aligned}$$

The components of the strain tensor are

$$\begin{aligned}
\gamma_{ij} &= \frac{1}{2}(G_{ij} - \delta_{ij}), \\
\gamma_{11} &= \frac{\theta^2}{2}\left(x_2^2 + x_3^2\right), \quad \gamma_{22} = \frac{\theta^2}{2}\psi_{,2}^2, \quad \gamma_{33} = \frac{\theta^2}{2}\psi_{,3}^2, \\
\gamma_{12} &= \frac{\theta}{2}(\psi_{,2} - x_3), \quad \gamma_{13} = \frac{\theta}{2}(\psi_{,3} - x_2), \quad \gamma_{23} = \frac{\theta^2}{2}\psi_{,2}\psi_{,3}.
\end{aligned}$$

For a small unit twist angle, we may neglect θ^2. Thus,

$$\gamma_{11} = \gamma_{22} = \gamma_{33} = \gamma_{23} = 0, \tag{6.2.24a}$$

$$\sigma_{11} = \sigma_{22} = \sigma_{33} = \sigma_{23} = 0, \tag{6.2.24b}$$

$$\gamma_{12} = \frac{\theta}{2}(\psi_{,2} - x_3), \quad \gamma_{13} = \frac{\theta}{2}(\psi_{,3} - x_2), \tag{6.2.24c}$$

$$\sigma_{12} = G\theta(\psi_{,2} - x_3), \quad \sigma_{13} = G\theta(\psi_{,3} - x_2), \tag{6.2.24d}$$

where $G = \mu$. With these approximations, we may conclude that

$$\mathbf{G}_i = \hat{\mathbf{i}}_i, \quad \boldsymbol{\sigma}_i = \sigma_{ij}\hat{\mathbf{i}}_j,$$
$$\frac{\partial \sigma_{ij}}{\partial x_i}\hat{\mathbf{i}}_j + \rho_0 F_i \hat{\mathbf{i}}_i = 0. \tag{6.2.25}$$

Let

$$\mathbf{n} = n_i \hat{\mathbf{i}}_i = (\cos\bar{\alpha})\,\hat{\mathbf{i}}_2 + (\sin\bar{\alpha})\,\hat{\mathbf{i}}_3; \tag{6.2.26}$$

then the surface traction becomes

$$\boldsymbol{\sigma} = \sigma_{ij} n_i \hat{\mathbf{i}}_j = (\sigma_{21}\cos\bar{\alpha} + \sigma_{31}\sin\bar{\alpha})\hat{\mathbf{i}}_1. \tag{6.2.27}$$

Because the bar is subjected to twisting at the ends ($x_1 = 0, L$) only, we have

$$\frac{\partial \sigma_{i1}}{\partial x_i} = 0, \tag{6.2.28}$$

and all the traction vanishes on the lateral surface:

$$\sigma_{21}\cos\bar{\alpha} + \sigma_{31}\sin\bar{\alpha} = 0. \tag{6.2.29}$$

In view of Eqs. (6.2.24d) and (6.2.28), it follows that

$$\frac{\partial \sigma_{21}}{\partial x_2} + \frac{\partial \sigma_{31}}{\partial x_3} = \psi_{,22} + \psi_{,33} = 0. \tag{6.2.30}$$

From Eqs. (6.2.29) and (6.2.27) on the boundary (Fig. 6.2.1), we have

$$(\psi_{,2} - x_3)\cos\bar{\alpha} + (\psi_{,3} + x_2)\sin\bar{\alpha} = 0. \tag{6.2.31}$$

It is obvious that the warping function ψ must satisfy the Laplace equation [Eq. (6.2.30)].

If $\bar{n}$ denotes distance along a normal to the boundary element $d\bar{n}$, we have, from Eq. (6.2.31),

$$\frac{\partial \psi}{\partial \bar{n}} = \frac{\partial \psi}{\partial x_2}\frac{dx_2}{d\bar{n}} + \frac{\partial \psi}{\partial x_3}\frac{dx_3}{d\bar{n}} = x_3\cos\bar{\alpha} - x_2\sin\bar{\alpha}.$$

In view of Eqs. (6.2.24a), (6.2.30), and (6.2.31), the resultant force on a cross section vanishes:

$$\int_\Omega \sigma_{21}\, d\Omega = \int_\Omega \sigma_{31}\, d\Omega = 0. \tag{6.2.32}$$

The resultant torsional moment T on the ends of the bar that is due to the assumed stress distribution is

$$\begin{aligned} T &= \int_\Omega (x_2\sigma_{13} - x_3\sigma_{12})\, d\Omega \\ &= G\theta \int_\Omega \left(x_2^2 + x_3^2 + x_2\psi_{,3} - x_3\psi_{,2}\right) d\Omega = G\theta J, \end{aligned} \tag{6.2.33}$$

in which $G = \mu$ and

$$J = \int_\Omega \left(x_2^2 + x_3^2 + x_2\psi_{,3} - x_3\psi_{,2}\right) d\Omega \tag{6.2.34}$$

is called a torsional constant. Let us assume that ψ may be given by a constant: $\psi = C$. The boundary condition [Eq. (6.2.31)] becomes

$$-x_3 \cos\alpha + x_2 \sin\bar{\alpha} = 0.$$

Integrating with respect to $\bar{\alpha}$, we have

$$x_3 \sin\bar{\alpha} + x_2 \cos\bar{\alpha} = 0,$$

$$x_3 \frac{dx_3}{d\bar{n}} + x_2 \frac{dx_2}{d\bar{n}} = 0,$$

$$\frac{d}{d\bar{n}}\left(\frac{x_2^2 + x_3^2}{2}\right) = 0,$$

or

$$x_2^2 + x_3^2 = \text{constant},$$

which is an expression for a circle. Thus the function of $\psi = C$ gives us the solution of a circular bar. Now, substituting this function into Eq. (6.2.34) and assuming $r = r_0$ gives

$$J = \int_\Omega \left(x_2^2 + x_3^2\right) d\Omega = \frac{1}{2}\pi r_0^4,$$

which corresponds to the polar moment of inertia. The shear stresses are

$$\sigma_{13} = G\theta x_2 = \frac{Tx_2}{J}, \quad \sigma_{12} = -G\theta x_3 = -\frac{Tx_3}{J}. \tag{6.2.35}$$

Prandtl proposed that Eq. (6.2.30) would be satisfied if we chose a stress function ϕ such that

$$\sigma_{21} = G\theta\phi_{,3}, \quad \sigma_{31} = -G\theta\phi_{,2}, \tag{6.2.36a}$$

$$\gamma_{21} = \frac{\theta}{2}\phi_{,3}, \quad \gamma_{31} = -\frac{\theta}{2}\phi_{,2}. \tag{6.2.36b}$$

Let us consider a section of a bar that is simply connected. At this time we invoke the compatibility equation developed in Subsection 6.2.3. If the compatibility conditions of Eq. (6.2.20) are satisfied, then

$$\begin{aligned} R_{2331} &= \gamma_{33,12} + \gamma_{12,33} - \gamma_{31,32} - \gamma_{32,31} \\ &= (\phi_{,22} + \phi_{,33})_{,3} = 0, \\ R_{1223} &= -(\phi_{,22} + \phi_{,33})_{,2} = 0. \end{aligned}$$

Integrating, we have

$$\phi_{,22} + \phi_{,33} = C. \tag{6.2.37}$$

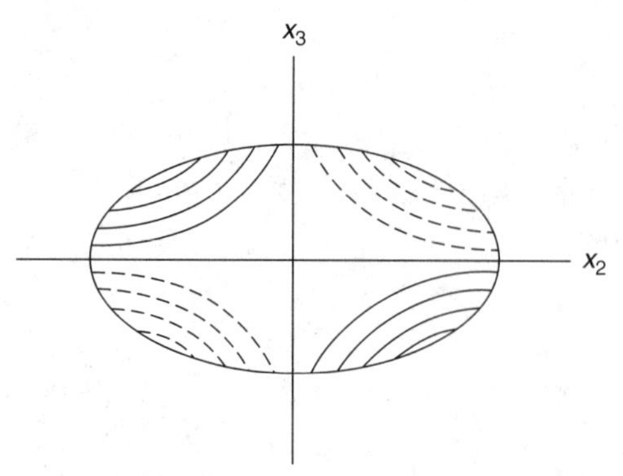

Figure 6.2.2. Lines of constant warping, application of Saint-Venant's principle.

We can determine the constant C by equating the right-hand sides of Eqs. (6.2.24c) and (6.2.36b),

$$\phi_{,3} = \psi_{,2} - x_3, \quad \phi_{,2} = -\psi_{,3} - x_2, \tag{6.2.38}$$

and by substituting these into Eq. (6.2.37), which gives $C = -2$. As a result, we have

$$\phi_{,22} + \phi_{,33} = -2, \tag{6.2.39}$$

subject to the boundary condition $\phi = 0$ on the surface. This is known as the strain compatibility equation for a solid undergoing torsional deformations. The solution of Eq. (6.2.37), together with Eqs. (6.2.36) and (6.2.33), will provide the state of stress and associated torsional properties.

It has been shown that, viewed from a position on the z_1 axis, a straight line of the cross section appears to remain straight but, in fact, the line is warped because particles of the line displace axially, which is an application of the Saint-Venant principle (Saint-Vénant, 1864).

As an example, let us consider a bar with elliptical cross section, Fig. 6.2.2. Determine the tangential stress and warping function. Let the boundary of the cross section be given by the following step.

Step 1.

$$\left(\frac{x_2}{a}\right)^2 - \left(\frac{x_3}{b}\right)^2 - 1 = 0.$$

The solution of Eq. (6.2.38) yields the next step.

Step 2.

$$\phi = k\left[\left(\frac{x_2}{a}\right)^2 + \left(\frac{x_3}{b}\right)^2 - 1\right],$$

which satisfies the boundary condition $\phi = 0$ on C such that we obtain Step 3.

Step 3.

$$k = -\frac{a^2 b^2}{a^2 + b^2}.$$

It follows from Eqs. (6.2.36a) that the tangential stress is of the form given in the next step.

Step 4.

$$\sigma_1 = \sqrt{(\sigma_{21})^2 + (\sigma_{31})^2} = \frac{2G\theta}{1 + \left(\frac{a}{b}\right)^2} \sqrt{(x_2)^2 + \left(\frac{a}{b}\right)^4 (x_3)^2}.$$

The warping function ψ is determined from Eq. (6.2.38) and Step (2).

Step 5.

$$\psi_{,2} = \frac{b^2 - a^2}{a^2 + b^2} x_3.$$

Step 6.

$$\psi_{,3} = \frac{b^2 - a^2}{a^2 + b^2} x_2.$$

Integrating Steps 5 and 6 yields

$$\psi = \frac{b^2 - a^2}{a^2 + b^2} x_2 x_3.$$

Contour lines of constant displacement along the x_3 axis are hyperbolas as shown in Fig. 6.2.2. The solid curves in the figure indicate where u_1 is positive, the dotted curves where u_1 is negative.

6.3 Newtonian Fluid Mechanics

6.3.1 Navier–Stokes System of Equations

Details of derivation for Newtonian fluids for Cartesian coordinates were presented in Chap. 5. The first law of thermodynamics for fluids given in (5.3.12) is written now for curvilinear coordinates as

$$\frac{\partial \rho}{\partial t} + (\rho V^i)_{|i} = 0, \tag{6.3.1}$$

$$\rho \frac{\partial V^j}{\partial t} + \rho V^i_{|i} V^i - \sigma^{ij}_{|i} - \rho F^j = 0, \tag{6.3.2}$$

$$\rho \frac{\partial \epsilon}{\partial t} + \rho \epsilon_{|i} V^i - q^i_{|i} - \rho r = 0, \tag{6.3.3}$$

where, for isotropic fluids,

$$\sigma_{ij} = -pg^{ij} + \tau^{ij}, \tag{6.3.4}$$

$$\begin{aligned}\tau^{ij} &= E^{ijkm} d_{km} = \lambda g^{ij} g^{km} d_{km} + \mu(g^{ik} g^{jm} + g^{im} g^{jk}) d_{km} \\ &= \lambda d_k^k g^{ij} + 2\mu d^{ij} = -\frac{2}{3}\mu d_k^k g^{ij} + 2\mu d^{ij} \\ &= \mu\left(g^{ik} V_{|k}^j + g^{ik} V_{|k}^i - \frac{2}{3} g^{ij} V_{|k}^k\right).\end{aligned} \tag{6.3.5}$$

Recall that Kronecker deltas were replaced with metric tensors. The rate-of-deformation tensor is of the form

$$d_{ij} = \frac{1}{2}(V_{i|j} + V_{j|i}). \tag{6.3.6}$$

Multiply Eq. (6.3.2) by $\mathbf{g}_j$ to obtain

$$\left(\rho \frac{\partial V^j}{\partial t} + \rho V_{|i}^j V^i - \rho F^j - \sigma_{|i}^{ij}\right) \mathbf{g}_j = 0 \tag{6.3.7}$$

or

$$\rho \frac{\partial \mathbf{V}}{\partial t} + \rho(\mathbf{V} \cdot \nabla)\mathbf{V} - \rho \mathbf{F} - \sigma_{|i}^{ij} \mathbf{g}_j = 0. \tag{6.3.8a}$$

In view of Eqs. (6.3.4)–(6.3.6), it follows that

$$\begin{aligned}\sigma_{|i}^{ij} \mathbf{g}_j &= \left[(-pg^{ij})_{|i} + \tau_{,i}^{ij} + \Gamma_{mi}^j \tau^{mi} + \Gamma_{mi}^i \tau^{jm}\right] \mathbf{g}_j \\ &= \left\{-p_{,i} g^{ij} + \mu\left[\left(V_{|m}^i g^{jm} + V_{|m}^j g^{im} - \frac{2}{3} V_{|k}^k g^{ij}\right)_{,i}\right.\right. \\ &\quad + \Gamma_{mi}^j \left(V_{|k}^m g^{ik} + V_{|k}^i g^{mk} - \frac{2}{3} V_{|k}^k g^{mi}\right) \\ &\quad \left.\left. + \Gamma_{mi}^i \left(V_{|k}^j g^{mk} + V_{|k}^m g^{ik} - \frac{2}{3} V_{|k}^k g^{im}\right)\right]\right\} \mathbf{g}_j.\end{aligned} \tag{6.3.8b}$$

Be cautioned that the tangent vector $\mathbf{g}_j$ must be included in the algebra so that, when all tensor quantities are converted into physical components, they will be based on the unit vectors in the final form.

Instead of using Eq. (6.3.8a), we may work with vector notation in the form (Aris, 1962)

$$\sigma_{|i}^{ij} \mathbf{g}_j = -\nabla p + \mu\left[\nabla^2 \mathbf{V} + \frac{1}{3}\nabla(\nabla \cdot \mathbf{V})\right], \tag{6.3.9}$$

where

$$\nabla p = \mathbf{g}^i \frac{\partial p}{\partial \xi_i} = p_{,i} \mathbf{g}^i = p_{,i} g^{ij} \mathbf{g}_j, \tag{6.3.10}$$

with

$$\nabla^2 \mathbf{V} = (\nabla \cdot \nabla)\mathbf{V} = \left(\mathbf{g}^k \frac{\partial}{\partial \xi_k} \cdot \mathbf{g}^j \frac{\partial}{\partial \xi_j}\right) V^i \mathbf{g}_i = g^{jk} V^i_{|jk} \mathbf{g}_i,$$

$$V^i_{|jk} = -\Gamma^n_{kj} V^i_{,n} - \Gamma^m_{kj}\Gamma^i_{nm} V^n + V^i_{,jk} + 2V^m_{,k}\Gamma^i_{mj} + V^m \left(\Gamma^i_{jm}\right)_{,k} + V^m \Gamma^n_{mj}\Gamma^i_{nk}, \tag{6.3.11}$$

$$\nabla(\nabla \cdot \mathbf{V}) = \mathbf{g}^k \frac{\partial}{\partial \xi_k}\left(\mathbf{g}^j \frac{\partial}{\partial \xi_j} \cdot V^i \mathbf{g}_i\right) = V^i_{|ik}\mathbf{g}^k = V^j_{|jk} g^{ki} \mathbf{g}_i, \tag{6.3.12}$$

with

$$\nabla(\nabla \cdot \mathbf{V}) = \mathbf{g}^k \frac{\partial}{\partial \xi_k}\left(\mathbf{g}^j \frac{\partial}{\partial \xi_j} \cdot V^i \mathbf{g}_i\right) = V^i_{|ik}\mathbf{g}^k = V^j_{|jk} g^{ki} \mathbf{g}_i,$$

$$V^j_{|jk} = V^j_{,jk} + V^j_{jk}\Gamma^m_{mj} + V^j \left(\Gamma^m_{mj}\right)_{,k}.$$

The reader may find the approach of Eq. (6.3.9) more straightforward than that of Eq. (6.3.8b).

The equations of continuity and energy in curvilinear coordinates can be obtained similarly. Thus the governing equations for compressible viscous fluids in cylindrical coordinates are written as follows.

Continuity

$$\frac{\partial \rho}{\partial t} + V_r \frac{\partial \rho}{\partial r} + \frac{V_\theta}{r}\frac{\partial \rho}{\partial \theta} + V_z \frac{\partial \rho}{\partial z} + \rho\left(\frac{\partial V_r}{\partial r} + \frac{1}{r}\frac{\partial V_\theta}{\partial \theta} + \frac{\partial V_z}{\partial z} + \frac{V_r}{r}\right) = 0. \tag{6.3.13}$$

Momentum in the *r* direction

$$\begin{aligned}
&\rho \frac{\partial V_r}{\partial t} + \rho\left(\frac{\partial V_r}{\partial r} V_r + \frac{1}{r}\frac{\partial V_r}{\partial \theta} V_\theta + \frac{\partial V_r}{\partial z} V_z - \frac{V_\theta^2}{r}\right) - \rho F_r + \frac{\partial p}{\partial r} \\
&- \mu\left[\frac{\partial^2 V_r}{\partial r^2} + \frac{1}{r}\frac{\partial V_r}{\partial r} + \frac{1}{r^2}\frac{\partial^2 V_r}{\partial \theta^2} - \frac{2}{r^2}\frac{\partial V_\theta}{\partial \theta} - \frac{V_r}{r^2} + \frac{\partial^2 V_r}{\partial z^2}\right. \\
&\left.+ \frac{1}{3}\left(\frac{\partial^2 V_r}{\partial r^2} + \frac{1}{r}\frac{\partial^2 V_\theta}{\partial \theta \partial r} + \frac{\partial^2 V_z}{\partial r \partial z} - \frac{1}{r^2}\frac{\partial V_\theta}{\partial \theta} + \frac{1}{r}\frac{\partial V_r}{\partial r} - \frac{V_r}{r^2}\right)\right] = 0.
\end{aligned} \tag{6.3.14a}$$

Momentum in the θ direction

$$\begin{aligned}
&\rho \frac{\partial V_\theta}{\partial t} + \rho\left(\frac{\partial V_\theta}{\partial r} V_r + \frac{1}{r}\frac{\partial V_\theta}{\partial \theta} V_\theta + \frac{\partial V_\theta}{\partial z} V_z + \frac{V_r V_\theta}{r}\right) - \rho F_\theta + \frac{1}{r}\frac{\partial p}{\partial \theta} \\
&- \mu\left[\frac{\partial^2 V_\theta}{\partial r^2} + \frac{1}{r}\frac{\partial V_\theta}{\partial r} + \frac{1}{r^2}\frac{\partial^2 V_\theta}{\partial \theta^2} - \frac{2}{r^2}\frac{\partial V_r}{\partial \theta} + \frac{\partial^2 V_\theta}{\partial z^2} - \frac{V_\theta}{r^2}\right. \\
&\left.+ \frac{1}{3}\left(\frac{1}{r}\frac{\partial^2 V_r}{\partial r \partial \theta} + \frac{1}{r^2}\frac{\partial^2 V_\theta}{\partial \theta^2} + \frac{1}{r}\frac{\partial^2 V_z}{\partial z \partial \theta} + \frac{1}{r^2}\frac{\partial V_r}{\partial \theta}\right)\right] = 0.
\end{aligned} \tag{6.3.14b}$$

Momentum in the z direction

$$\rho\frac{\partial V_z}{\partial t}+\rho\left(\frac{\partial V_z}{\partial r}V_r+\frac{1}{r}\frac{\partial V_z}{\partial \theta}V_\theta+\frac{\partial V_z}{\partial z}V_z\right)-\rho F_z+\frac{\partial p}{\partial z}$$
$$-\mu\left[\frac{\partial^2 V_z}{\partial r^2}+\frac{1}{r}\frac{\partial V_z}{\partial r}+\frac{1}{r^2}\frac{\partial^2 V_z}{\partial \theta^2}+\frac{\partial^2 V_z}{\partial z^2}\right.$$
$$\left.+\frac{1}{3}\left(\frac{\partial^2 V_r}{\partial r\partial z}+\frac{1}{r}\frac{\partial^2 V_\vartheta}{\partial \theta\partial z}+\frac{\partial^2 V_z}{\partial z^2}+\frac{1}{r}\frac{\partial V_r}{\partial z}\right)\right]=0. \tag{6.3.14c}$$

Energy equation

$$\rho c_p\frac{\partial T}{\partial t}+\rho c_p(\mathbf{V}\cdot\nabla)T-\frac{\partial p}{\partial t}-(\mathbf{V}\cdot\nabla)p-p(\nabla\cdot\mathbf{V})-\sigma^{ij}V_{j|i}-\nabla\cdot\mathbf{q}-\rho r=0, \tag{6.3.15a}$$

or

$$\rho c_p\frac{\partial T}{\partial t}+\rho c_p\left(\frac{\partial T}{\partial r}V_r+\frac{V_\theta}{r}\frac{\partial T}{\partial \theta}+\frac{\partial T}{\partial z}V_z\right)-\frac{\partial p}{\partial t}-\frac{\partial p}{\partial r}V_r-\frac{V_\theta}{r}\frac{\partial p}{\partial \theta}-\frac{\partial p}{\partial z}V_z$$
$$-\tau^{ij}V_{j|i}-k\left(\frac{\partial^2 T}{\partial r^2}+\frac{1}{r^2}\frac{\partial^2 T}{\partial \theta^2}+\frac{\partial^2 T}{\partial z^2}+\frac{1}{r}\frac{\partial T}{\partial r}\right)-\rho r=0, \tag{6.3.15b}$$

where

$$\tau^{ij}V_{j|i}=\mu\left(g^{ik}V^j_{|k}V_{j|i}+V^i_{|k}V^k_{|i}-\frac{2}{3}V^i_{|i}V^k_{|k}\right)$$
$$=\mu\left\{2\left(\frac{\partial V_r}{\partial r}\right)^2+\frac{2}{r^2}\left(\frac{\partial V_\theta}{\partial \theta}+V_r\right)^2+2\left(\frac{\partial V_z}{\partial z}\right)^2\right.$$
$$+\left(\frac{\partial V_\theta}{\partial r}\right)^2+\left(\frac{\partial V_z}{\partial r}\right)^2+\frac{1}{r^2}\left(\frac{\partial V_r}{\partial \theta}+V_\theta\right)^2+\frac{1}{r^2}\left(\frac{\partial V_z}{\partial \theta}\right)^2$$
$$+\left(\frac{\partial V_r}{\partial z}\right)^2+\left(\frac{\partial V_\theta}{\partial z}\right)^2+\frac{2}{r}\frac{\partial V_\theta}{\partial r}\left(\frac{\partial V_r}{\partial \theta}-V_\theta\right)+\frac{2}{r}\frac{\partial V_\theta}{\partial z}\frac{\partial V_z}{\partial \theta}$$
$$\left.+2\frac{\partial V_z}{\partial r}\frac{\partial V_r}{\partial z}-\frac{2}{3}\left[\frac{\partial V_r}{\partial r}+\frac{1}{r}\left(\frac{\partial V_\theta}{\partial \theta}+V_r\right)+\frac{\partial V_z}{\partial z}\right]^2\right\}. \tag{6.3.16}$$

Equations (6.3.13)–(6.3.15) and equations of state (5.1.2) are the governing equations for compressible viscous flow in cylindrical coordinates.

The governing equations in spherical coordinates are derived in a similar manner. The results are as follows.

Continuity

$$\frac{\partial p}{\partial t}+V_R\frac{\partial \rho}{\partial R}+\frac{V_\alpha}{R}\frac{\partial \rho}{\partial \alpha}+\frac{V_\theta}{R\sin\alpha}\frac{\partial \rho}{\partial \theta}$$
$$+\rho\left(\frac{\partial V_R}{\partial R}+\frac{1}{R}\frac{\partial V_\alpha}{\partial \alpha}+\frac{1}{R\sin\alpha}\frac{\partial V_\theta}{\partial \theta}+\frac{2V_R}{R}+\frac{V_\alpha}{R}\cot\alpha\right)=0. \tag{6.3.17}$$

***R* momentum**

$$\rho\frac{\partial V_R}{\partial t}+\rho\left(V_R\frac{\partial V_R}{\partial R}+\frac{V_\alpha}{R}\frac{\partial V_R}{\partial \alpha}-\frac{V_\alpha^2}{R}+\frac{V_\theta}{R\sin\alpha}\frac{\partial V_R}{\partial\theta}-\frac{V_\theta^2}{R}\right)$$

$$-\rho F_R+\frac{\partial p}{\partial R}-\mu\left[\frac{\partial^2 V_R}{\partial R^2}+\frac{1}{R^2}\frac{\partial^2 V_R}{\partial\alpha^2}+\frac{1}{R^2\sin^2\alpha}\frac{\partial^2 V_R}{\partial\theta^2}\right.$$

$$-\frac{2}{R^2}\frac{\partial V_\alpha}{\partial\alpha}-\frac{2}{R^2\sin\alpha}\frac{\partial V_\theta}{\partial\theta}-\frac{2V_R}{R^2}-\frac{2V_\alpha\cot\alpha}{R^2}+\frac{2}{R}\frac{\partial V_R}{\partial R}$$

$$+\frac{\cot\alpha}{R^2}\frac{\partial V_R}{\partial\alpha}+\frac{1}{3}\left(\frac{\partial^2 V_R}{\partial R^2}-\frac{1}{R^2}\frac{\partial V_R}{\partial\alpha}+\frac{1}{R}\frac{\partial^2 V_\alpha}{\partial\alpha\partial R}-\frac{1}{R^2\sin\alpha\partial\theta}\frac{\partial V_\theta}{\partial\theta}\right.$$

$$\left.\left.+\frac{1}{R\sin\alpha}\frac{\partial^2 V_\theta}{\partial\theta\partial R}-\frac{2V_R}{R^2}+\frac{2}{R}\frac{\partial V_R}{\partial R}-\frac{V_\alpha\cot\alpha}{R^2}+\frac{\cot\alpha}{R}\frac{\partial V_\alpha}{\partial R}\right)\right]=0 \quad (6.3.18a)$$

α momentum

$$\rho\frac{\partial V_\alpha}{\partial t}+\rho\left(V_R\frac{\partial V_R}{\partial R}+\frac{V_\alpha}{R}\frac{\partial V_\alpha}{\partial\alpha}+\frac{V_\alpha V_R}{R}+\frac{V_\theta}{R\sin\alpha}\frac{\partial V_\alpha}{\partial\theta}-\frac{V_\theta^2\cot\alpha}{R}\right)$$

$$-\rho F_\alpha+\frac{1}{R\sin^2\alpha}\frac{\partial p}{\partial\alpha}-\mu\left[\frac{2}{R}\frac{\partial V_\alpha}{\partial R}+\frac{1}{R^2}\frac{\partial^2 V_\alpha}{\partial R^2}+\frac{1}{R^2}\frac{\partial^2 V_\alpha}{\partial\alpha^2}\right.$$

$$+\frac{1}{R^2\sin^2\alpha}\frac{\partial^2 V_\alpha}{\partial\theta^2}+\frac{2}{R^2}\frac{\partial V_R}{\partial\alpha}-\frac{2\cos\alpha}{R^2\sin^2\alpha}-\frac{V_\alpha}{R^2\sin^2\alpha}$$

$$+\frac{\cot\alpha}{R^2}\frac{\partial V_\alpha}{\partial\alpha}+\frac{1}{3}\left(\frac{1}{R}\frac{\partial^2 V_R}{\partial R\partial\alpha}+\frac{1}{R^2}\frac{\partial^2 V_R}{\partial\alpha^2}-\frac{\cos\alpha}{R^2\sin^2\alpha}\frac{\partial V_\theta}{\partial\theta}\right.$$

$$\left.\left.+\frac{1}{R^2\sin\alpha}\frac{\partial^2 V_\theta}{\partial\alpha\partial\theta}+\frac{2}{R^2}\frac{\partial V_R}{\partial\alpha}+\frac{\cot\alpha}{R^2}\frac{\partial V_\alpha}{\partial\alpha}-\frac{V_\alpha}{R_2}-\frac{V_\alpha\cot^2\alpha}{R^2}\right)\right]=0. \quad (6.3.18b)$$

θ momentum

$$\rho\frac{\partial V_\theta}{\partial t}+\rho\left(V_R\frac{\partial V_\theta}{\partial R}+\frac{V_\alpha}{R}\frac{\partial V_\theta}{\partial\alpha}+\frac{V_\theta}{R\sin\alpha}\frac{\partial V_\theta}{\partial\theta}+\frac{V_\theta V_R}{R}+\frac{V_\alpha V_\theta}{R}\cot\alpha\right)$$

$$-\rho F_\theta+\frac{1}{R\sin\alpha}\frac{\partial p}{\partial\theta}-\mu\left[\frac{2}{R}\frac{\partial V_\theta}{\partial R}+\frac{\partial^2 V_\theta}{\partial R^2}+\frac{\cot\alpha}{R^2}\frac{\partial V_\theta}{\partial\alpha}+\frac{1}{R^2}\frac{\partial^2 V_\theta}{\partial\alpha^2}\right.$$

$$+\frac{1}{R^2\sin^2\alpha}\frac{\partial^2 V_\theta}{\partial\theta^2}-\frac{V_\theta}{R^2\sin^2\alpha}+\frac{2}{R^2\sin\alpha}\frac{\partial V_R}{\partial\theta}+\frac{2\cos\alpha}{R^2\sin^2\alpha}\frac{\partial V_\alpha}{\partial\theta}$$

$$\left.+\frac{1}{3R\sin\alpha}\left(\frac{\partial^2 V_R}{\partial R\partial\theta}+\frac{1}{R}\frac{\partial^2 V_\alpha}{\partial\alpha\partial\theta}+\frac{1}{R\sin\alpha}\frac{\partial^2 V_\theta}{\partial\theta^2}+\frac{2}{R}\frac{\partial V_R}{\partial\theta}+\frac{\cot\alpha}{R}\frac{\partial V_\alpha}{\partial\theta}\right)\right]=0.$$

(6.3.18c)

Energy

$$
\begin{aligned}
&\rho c_p \frac{\partial T}{\partial t} - \frac{\partial p}{\partial t} + \rho c_p \left(V_R \frac{\partial T}{\partial R} + \frac{V_\alpha}{R}\frac{\partial T}{\partial \alpha} + \frac{V_\theta}{R\sin\alpha}\frac{\partial T}{\partial \theta}\right) - V_R\frac{\partial p}{\partial R} - \frac{V_\alpha}{R}\frac{\partial p}{\partial \alpha} - \frac{V_\theta}{R\sin\alpha}\frac{\partial p}{\partial \theta}\\
&- \mu \left\{ \frac{4}{3}\left(\frac{\partial V_R}{\partial R}\right)^2 + \frac{2}{R}\frac{\partial V_\alpha}{\partial R}\frac{\partial V_R}{\partial \alpha} - \frac{2V_\alpha}{R}\frac{\partial V_\alpha}{\partial R} + \frac{1}{R\sin\alpha}\frac{\partial V_\theta}{\partial R}\left(\frac{\partial V_R}{\partial \theta} - V_\theta \sin\alpha\right)\right.\\
&+ \left(\frac{1}{R}\frac{\partial V_\alpha}{\partial \alpha} + \frac{V_R}{R}\right)^2 + \frac{2}{R\sin\alpha}\frac{\partial V_\theta}{\partial \alpha}\left(\frac{1}{R}\frac{\partial V_\alpha}{\partial \theta} - \frac{V_\theta\cos\alpha}{R}\right)\\
&+ \left(\frac{1}{R\sin\alpha}\frac{\partial V_\theta}{\partial \theta} + \frac{V_\alpha}{R}\cot\alpha + \frac{V_R}{R}\right)^2 + \left(\frac{\partial V_\alpha}{\partial R}\right)^2 + \left(\frac{\partial V_\theta}{\partial R}\right)^2\\
&+ \frac{1}{R^2}\left[\left(\frac{\partial V_R}{\partial \alpha} - V_\alpha\right)^2 + \left(\frac{\partial V_\alpha}{\partial \alpha} + V_R\right)^2 + \left(\frac{\partial V_\theta}{\partial \alpha}\right)^2\right] + \frac{1}{R^2\sin^2\alpha}\left[\left(\frac{\partial V_R}{\partial \theta} - V_\theta\sin\alpha\right)^2\right.\\
&+ \left(\frac{\partial V_\alpha}{\partial \theta} - V_\theta\cos\alpha\right)^2 + \left(\frac{1}{R\sin\alpha}\frac{\partial V_\theta}{\partial \theta} + \frac{V_\alpha}{R}\cot\alpha + \frac{V_R}{R}\right)\\
&\times \left.\left(R\sin\alpha\frac{\partial V_\theta}{\partial \theta} + RV_R\sin^2\alpha + RV_R\sin\alpha\cos\alpha\right)\right]\\
&- \frac{2}{3}\left[\frac{2}{R}\frac{\partial V_R}{\partial R}\left(\frac{\partial V_\alpha}{\partial \alpha} + V_R\right) + \frac{2}{R}\frac{\partial V_R}{\partial R}\left(\frac{1}{\sin\alpha}\frac{\partial V_\theta}{\partial \theta} + V_\alpha\cot\alpha + V_R\right)\right.\\
&+ \frac{1}{R^2}\left(\frac{\partial V_\alpha}{\partial \alpha} + V_R\right)^2 + \frac{2}{R^2}\left(\frac{\partial V_\alpha}{\partial \alpha} + V_R\right)\left(\frac{1}{\sin\alpha}\frac{\partial V_\theta}{\partial \theta} + V_\alpha\cot\alpha + V_R\right)\\
&\left.\left. + \frac{1}{R^2}\left(\frac{1}{\sin\alpha}\frac{\partial V_\theta}{\partial \theta} + V_\alpha\cot\alpha + V_R\right)^2\right]\right\}\\
&- k\left(\frac{\partial^2 T}{\partial R^2} + \frac{1}{R^2}\frac{\partial^2 T}{\partial \alpha^2} + \frac{1}{R^2\sin^2\alpha}\frac{\partial^2 T}{\partial \theta^2} + \frac{2}{R}\frac{\partial T}{\partial R} + \frac{\cot\alpha}{R^2}\frac{\partial T}{\partial \alpha}\right) - \rho r = 0. \qquad (6.3.19)
\end{aligned}
$$

6.3.2 Stream Functions for Cylindrical and Spherical Coordinates

The existence of stream functions for Cartesian coordinates was discussed in Section 5.5. This can be extended to curvilinear coordinates similarly. For axisymmetric cylindrical motion ($V_\theta = 0$) [see Fig. 6.1.4(a)], we have $x_1 = rx$, $x_2 = \theta$, $x_3 = z$, $V^1 = V_r$, $V^2 = V_\theta/r$, and $V^3 = V_z$. Thus it follows that (see Subsection 5.5.2)

$$
\frac{dz}{V_z(z,r)} = \frac{dr}{V_r(z,r)} = \frac{r d\theta}{0}. \qquad (6.3.20)
$$

Because we must have $d\theta = 0$, and thus $\theta =$ constant, we define

$$
\psi^{(1)} = \psi, \quad \psi^{(2)} = \theta.
$$

These conditions lead to

$$\begin{aligned}
\mathbf{V} = \nabla\psi^{(1)} \times \nabla\psi^{(2)} &= \frac{\partial \psi^{(1)}}{\partial \xi_i}\frac{\partial \psi^{(2)}}{\partial \xi_j}\mathbf{g}^i \times \mathbf{g}^j \\
&= \psi^{(1)}_{,i}\psi^{(2)}_{,j}\frac{1}{\sqrt{g}}\epsilon^{ijk}\mathbf{g}_k \\
&= \frac{1}{\sqrt{g}}\{[\psi^{(1)}_{,2}\psi^{(2)}_{,3} - \psi^{(1)}_{,3}\psi^{(2)}_{,2}]\,\mathbf{g}_1 + [\psi^{(1)}_{,3}\psi^{(2)}_{,1} - \psi^{(1)}_{,1}\psi^{(2)}_{,3}]\,\mathbf{g}_2 \\
&\quad + [\psi^{(1)}_{,1}\psi^{(2)}_{,2} - \psi^{(1)}_{,2}\psi^{(2)}_{,1}]\,\mathbf{g}_3\} \\
&= \frac{1}{r}[-\psi^{(1)}_{,3}\psi^{(2)}_{,2}\mathbf{g}_1 - \psi^{(1)}_{,1}\psi^{(2)}_{,2}\mathbf{g}_3] \\
&= -\frac{1}{r}\frac{\partial \psi}{\partial z}\mathbf{e}_r + \frac{1}{r}\frac{\partial \psi}{\partial r}\mathbf{e}_z. \qquad (6.3.21)
\end{aligned}$$

Because $\mathbf{V} = V_r\mathbf{e}_r + V_z\mathbf{e}_z$, the velocity–stream-function relations are expressed by

$$V_r = -\frac{1}{r}\frac{\partial \psi}{\partial z}, \qquad V_z = \frac{1}{r}\frac{\partial \psi}{\partial r}. \qquad (6.3.22)$$

Note that, once again, the condition for incompressibility, Eq. (5.5.1), is satisfied. Here streamlines are turned into stream tubes in the three-dimensional domain.

Similarly, for axisymmetric spherical motion [see Fig. 6.1.4(b)], we have $x_1 = R$, $x_2 = \alpha$, $x_3 = \theta$, $V^1 = V_R$, $V^2 = V_\alpha/R$, and $V^3 = v_\theta/(R\sin\alpha)$. Thus,

$$\frac{dR}{V_R\,(R,\alpha)} = \frac{R\,d\alpha}{V_\alpha\,(R,\alpha)} = \frac{R\sin\alpha\,d\theta}{0}, \qquad (6.3.23)$$

which requires

$$d\theta = 0, \quad \theta = \text{constant}, \quad \psi^{(1)} = \psi\,(R,\alpha), \quad \psi^{(2)} = \theta,$$

$$\begin{aligned}
\mathbf{V} = \nabla\psi^{(1)} \times \nabla\psi^{(2)} &= \frac{\partial \psi^{(1)}}{\partial \xi_i}\frac{\partial \psi^{(2)}}{\partial \xi_j}\mathbf{g}^i \times \mathbf{g}^j \\
&= \psi^{(1)}_{,i}\psi^{(2)}_{,j}\frac{1}{\sqrt{g}}\epsilon^{ijk}\mathbf{g}_k \\
&= \frac{1}{\sqrt{g}}\{[\psi^{(1)}_{,2}\psi^{(2)}_{,3} - \psi^{(1)}_{,3}\psi^{(2)}_{,2}]\,\mathbf{g}_1 + [\psi^{(1)}_{,3}\psi^{(2)}_{,1} - \psi^{(1)}_{,1}\psi^{(2)}_{,3}]\,\mathbf{g}_2 \\
&\quad + [\psi^{(1)}_{,1}\psi^{(2)}_{,2} - \psi^{(1)}_{,2}\psi^{(2)}_{,1}]\,\mathbf{g}_3\} \\
&= \frac{1}{R^2\sin\alpha}[\psi^{(1)}_{,2}\psi^{(2)}_{,3}\mathbf{g}_1 - \psi^{(1)}_{,1}\psi^{(2)}_{,3}\mathbf{g}_2] \\
&= \frac{1}{R^2\sin\alpha}\frac{\partial \psi}{\partial \alpha}\mathbf{e}_R - \frac{1}{R\sin\alpha}\frac{\partial \psi}{\partial R}\mathbf{e}_\alpha. \qquad (6.3.24)
\end{aligned}$$

This gives

$$V_R = \frac{1}{R^2 \sin\alpha}\frac{\partial\psi}{\partial\alpha}, \qquad V_\alpha = -\frac{1}{R\sin\alpha}\frac{\partial\psi}{\partial R}, \tag{6.3.25}$$

and the condition for incompressibility, Eq. (5.5.1), is also satisfied.

PROBLEMS

6.1 Prove Eqs. (6.1.6) and (6.1.7) with complete details. Show that the proof of these equations may be obtained by use of the determinants $|\partial z_i/\partial\xi_j|$for Eq. (6.1.6) and $|\partial\xi_i/\partial z_j|$ for Eq. (6.1.7).

6.2 Given

$$\begin{aligned} \mathbf{v} &= 3\mathbf{i}_1 + 6\mathbf{i}_2, \\ \mathbf{g}_1 &= 4\mathbf{i}_1 + 2\mathbf{i}_2, \\ \mathbf{g}_2 &= 3\mathbf{i}_1 + 2\mathbf{i}_2, \\ \mathbf{g}_3 &= \mathbf{i}_3, \end{aligned}$$

(a) calculate the metric tensors;

(b) calculate the covariant and contravariant components of **v**;

(c) show that the original vector **v** as given can be recovered by use of the results obtained in parts (a) and (b); and

(d) draw a sketch to show all your results.

6.3 Prove the following equations:

(a) $\mathbf{g}^i = g^{ij}\mathbf{g}_j$,

(b) $g_{ik}g^{kj} = \delta_{ij}$,

(c) $\Gamma_{ijk} = \frac{1}{2}(g_{ik,j} + g_{jk,i} - g_{ij,k})$,

(d) $\Gamma^k_{ij} = \frac{1}{2}g^{km}(g_{mi,j} + g_{mj,i} - g_{ij,m})$,

(e) $\Gamma^i_{ij} = \frac{1}{2}g^{ik}g_{ik,j}$,

(f) $g^{ij} = \frac{1}{g}\frac{\partial g}{\partial g_{ij}}$,

(g) $\Gamma^i_{ij} = \left(\ln\sqrt{g}\right)_{,j}$.

6.4 Prove that $g^{ij}_{,i} = 0$. *Hint*: Use the definition given in Eq. (6.1.4b)

6.5 Prove Eqs. (6.1.20a)–(6.1.20c).

6.6 Prove that the Laplace equation takes the following form in spherical coordinates:

$$\nabla^2\phi = \frac{\partial^2\phi}{\partial R^2} + \frac{2}{R}\frac{\partial\phi}{\partial R} + \frac{1}{R^2}\frac{\partial^2\phi}{\partial\alpha^2} + \frac{\cot\alpha}{R^2}\frac{\partial\phi}{\partial\alpha} + \frac{1}{R^2\sin^2\alpha}\frac{\partial^2\phi}{\partial\theta^2}.$$

6.7 Derive $\nabla \cdot \mathbf{u}$, the divergence of $\mathbf{u}$, in cylindrical coordinates in terms of physical components.

6.8 Repeat Problem 6.7 for spherical coordinates.

6.9 Derive $\nabla \times \mathbf{u}$, the curl of $\mathbf{u}$, in cylindrical coordinates in terms of physical components.

6.10 Repeat Problem 6.9 for spherical coordinates.

6.11 Show that the components of the Green–Saint-Venant strain tensor in cylindrical coordinates in terms of physical components, including the nonlinear terms, take the following forms:

$$\bar{\gamma}_{11} = \gamma_{rr} = \frac{\partial u_r}{\partial r} + \frac{1}{2}\left[\left(\frac{\partial u_r}{\partial r}\right)^2 + \left(\frac{\partial u_\theta}{\partial r}\right)^2 + \left(\frac{\partial u_z}{\partial r}\right)^2\right],$$

$$\bar{\gamma}_{22} = \gamma_{\theta\theta} = \frac{1}{r}\frac{\partial u_\theta}{\partial \theta} + \frac{1}{2}\left[\left(\frac{1}{r}\frac{\partial u_r}{\partial \theta}\right)^2 + \left(\frac{1}{r}\frac{\partial u_\theta}{\partial \theta}\right)^2 + \left(\frac{1}{r}\frac{\partial u_z}{\partial \theta}\right)^2 - \frac{2u_\theta}{r^2}\frac{\partial u_r}{\partial \theta} + \frac{2u_r}{r^2}\frac{\partial u_\theta}{\partial \theta} + \left(\frac{u_\theta}{r}\right)^2 + \frac{2u_r}{r} + \left(\frac{u_r}{r}\right)^2\right],$$

$$\bar{\gamma}_{33} = \gamma_{zz} = \frac{\partial u_z}{\partial z} + \frac{1}{2}\left[\left(\frac{\partial u_r}{\partial z}\right)^2 + \left(\frac{\partial u_\theta}{\partial z}\right)^2 + \left(\frac{\partial u_z}{\partial z}\right)^2\right],$$

$$\bar{\gamma}_{12} = \frac{1}{2}\gamma_{r\theta} = \frac{1}{2}\left(\frac{1}{r}\frac{\partial u_r}{\partial \theta} + \frac{\partial u_\theta}{\partial r} - \frac{u_\theta}{r} + \frac{1}{r}\frac{\partial u_r}{\partial r}\frac{\partial u_r}{\partial \theta} + \frac{1}{r}\frac{\partial u_\theta}{\partial r}\frac{\partial u_\theta}{\partial \theta} + \frac{1}{r}\frac{\partial u_z}{\partial r}\frac{\partial u_z}{\partial \theta} + \frac{u_r}{r}\frac{\partial u_\theta}{\partial r} - \frac{u_\theta}{r}\frac{\partial u_r}{\partial r}\right),$$

$$\bar{\gamma}_{23} = \frac{1}{2}\gamma_{\theta z} = \frac{1}{2}\left(\frac{\partial u_\theta}{\partial z} + \frac{1}{r}\frac{\partial u_z}{\partial \theta} + \frac{1}{r}\frac{\partial u_r}{\partial \theta}\frac{\partial u_r}{\partial z} + \frac{1}{r}\frac{\partial u_\theta}{\partial \theta}\frac{\partial u_\theta}{\partial z} + \frac{1}{r}\frac{\partial u_z}{\partial \theta}\frac{\partial u_z}{\partial z} - \frac{u_\theta}{r}\frac{\partial u_r}{\partial z} + \frac{u_r}{r}\frac{\partial u_\theta}{\partial z}\right),$$

$$\bar{\gamma}_{31} = \frac{1}{2}\gamma_{zr} = \frac{1}{2}\left(\frac{\partial u_z}{\partial r} + \frac{\partial u_r}{\partial z} + \frac{\partial u_r}{\partial z}\frac{\partial u_r}{\partial r} + \frac{\partial u_\theta}{\partial z}\frac{\partial u_\theta}{\partial r} + \frac{\partial u_z}{\partial z}\frac{\partial u_z}{\partial r}\right),$$

where $\gamma_{r\theta}$, $\gamma_{\theta z}$, and γ_{zr} are the total shear strains.

6.12 Derive Cauchy's first law of motion in curvilinear coordinates and verify that

$$\frac{1}{\sqrt{g}}\left(\sqrt{g}\sigma^{ij}\mathbf{g}_j\right)_{,i} = \sigma^{ij}_{|i}\,\mathbf{g}_j.$$

6.13 Derive Cauchy's second law of motion in curvilinear coordinates.

6.14 Repeat Problem (5.1) for curvilinear coordinates and obtain the governing equations for continuity, momentum, and energy for cylindrical coordinates in terms of physical components.

6.15 Repeat Problem 6.14 for spherical coordinates.

6.16 Derive the velocity and stream-function relations for cylindrical coordinates.

6.17 Repeat Problem 6.16 for axisymmetric spherical motions.

7 Nonlinear Continuum

Nonlinearity occurs in solid mechanics and fluid mechanics. Nonlinear elasticity describes an elastic material behavior with large deformations, known as geometric nonlinearity (Section 7.1). This is distinguished from material nonlinearity as occurs in plasticity (Section 7.4). For fluids, the nonlinear behavior refers to higher-order velocity gradients in the viscous stresses. This is the subject of non-Newtonian fluid mechanics (Section 7.2). Such nonlinearity is in addition to the nonlinearity arising from convection and thermoviscous dissipation discussed in Section 5.4. A similar behavior in solids leads to viscoelasticity with time dependency of the strain tensor (Section 7.3). Finally we consider the combined effects of thermoelasticity, viscoelasticity, and plasticity, known as thermoviscoelastoplasticity (Section 7.5). The curvilinear coordinates introduced in Chap. 6 will be utilized in order to deal with curvatures that may arise in large deformations.

7.1 Nonlinear Elasticity

7.1.1 Stresses with Large Strains

Definitions of stresses with large strains on deformed surfaces are intimately associated with the deformed coordinates chosen. This has been a controversial subject in literature. Some of the methods for defining stresses on deformed surfaces are uncompromising with computational aspects. It is the purpose of this section to review historical developments and demonstrate some practical approaches for handling stresses with large strains (Green and Shield, 1950; Green and Adkins, 1960; Green and Naghdi, 1965; Ogden, 1984).

Recall that the subject of undeformed and deformed states of the Lagrangian coordinate system was given in Subsection 2.2.1, with Fig. 2.2.1 showing that the unit vectors $\mathbf{i}_i$ on the undeformed state change into the tangent vectors $\mathbf{G}_i$ on the deformed state. Consider a deformed surface in comparison with an undeformed surface, as shown in Fig. 7.1.1. The undeformed infinitesimal area dA_{01} constructed by $dx_2\mathbf{i}_2$ and $dx_3\mathbf{i}_3$ is

$$dA_{01} = |dx_2\mathbf{i}_2 \times dx_3\mathbf{i}_3| = |dx_2 dx_3\mathbf{i}_1| = dx_2 dx_3\hat{n}_1,$$

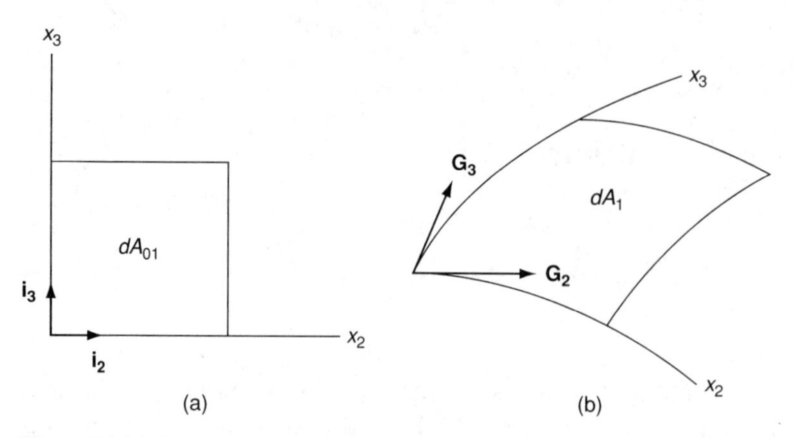

Figure 7.1.1. Undeformed and deformed areas: (a) undeformed area, (b) deformed area.

where $\hat{n}_1$ denotes a component of the unit vector normal to the surface dA_{01}. Similarly, the infinitesimal areas constructed by other coordinate lines are

$$dA_{02} = dx_3 dx_1 \hat{n}_2,$$
$$dA_{03} = dx_1 dx_2 \hat{n}_3.$$

For the deformed infinitesimal area dA_1 constructed by $dx_2\mathbf{G}_2$ and $dx_3\mathbf{G}_3$,

$$dA_1 = |dx_2\mathbf{G}_2 \times dx_3\mathbf{G}_3|,$$

with

$$|\mathbf{G}_2 \times \mathbf{G}_3| = |\sqrt{G}\mathbf{G}^1| = \sqrt{G}\sqrt{\mathbf{G}^1 \cdot \mathbf{G}^1} = \sqrt{GG^{11}},$$

we obtain the relationship between the deformed and undeformed areas:

$$dA_1 = \sqrt{GG^{11}} dx_2 dx_3 = \sqrt{GG^{11}} dA_{01}.$$

Thus, in general,

$$dA_i = \sqrt{GG^{(ii)}} dA_{0i}, \tag{7.1.1}$$

with

$$dA_{0i} = dA_0 \hat{n}_i, \tag{7.1.2}$$

where $\hat{n}_i$ is the component of a vector normal to the undeformed area. Note here that $\mathbf{G}^i$ is the contravariant tangent vector in the same sense as $\mathbf{g}^i$ for the curvilinear coordinates described earlier. On examining the inclined surface area dA of the deformed volume, as shown in Fig. 7.1.2, we find that

$$\begin{aligned} dA\mathbf{n} &= (\mathbf{G}_1 dx_1 - \mathbf{G}_3 dx_3) \times (\mathbf{G}_2 dx_2 - \mathbf{G}_3 dx_3) \\ &= \sqrt{G}\mathbf{G}^1 dx_2 dx_3 + \sqrt{G}\mathbf{G}^2 dx_1 dx_3 + \sqrt{G}\mathbf{G}^3 dx_1 dx_2 \\ &= \sqrt{G}\mathbf{G}^i dA_{0i}. \end{aligned}$$

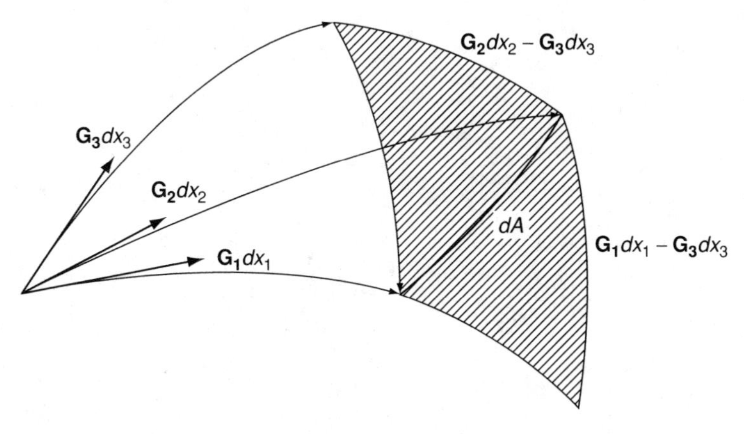

Figure 7.1.2. Deformed surface.

We use the relation given in Eq. (7.1.1) to obtain

$$dA\mathbf{n} = \frac{dA_i}{\sqrt{G^{(ii)}}}\mathbf{G}^i. \tag{7.1.3}$$

With $\mathbf{n} = n_i\mathbf{G}^i$, we have

$$dA_i = \sqrt{G^{(ii)}}dAn_i. \tag{7.1.4}$$

The stress vector acting on dA (Fig. 7.1.3) is of the form

$$\boldsymbol{\sigma}dA = \boldsymbol{\sigma}^i dA_i,$$

where $\boldsymbol{\sigma}^i$ is the contravariant stress vector and

$$\boldsymbol{\sigma} = \boldsymbol{\sigma}^i\frac{dA_i}{dA} = \boldsymbol{\sigma}^i\sqrt{G^{(ii)}}n_i. \tag{7.1.5}$$

By definition, as noted in Eq. (6.2.1), the stress vector normal to the surface is

$$\boldsymbol{\sigma} = \boldsymbol{\sigma}(n) = \sigma^{ij}n_i\mathbf{G}_j, \tag{7.1.6}$$

where σ^{ij} is the contravariant stress tensor acting on the deformed surface measured in coordinates applicable to the deformed state.

In general, the geometry of the undeformed configuration is known *a priori*. For this reason, it is convenient to use a stress tensor t^{ij} in terms of convected coordinates in the deformed state but measured per unit of undeformed area instead of deformed area (Rivlin, 1948; Green and Zerna, 1954; Rivlin, 1960). Thus we use Eqs. (7.1.1), (7.1.3), and (7.1.6) to obtain

$$\boldsymbol{\sigma}dA = \sigma^{ij}n_i\mathbf{G}_jdA = \sigma^{ij}\mathbf{G}_j\sqrt{G}n_idA_0, \tag{7.1.7a}$$

$$\mathbf{t}dA_0 = t^{ij}\hat{n}_i\mathbf{G}_jdA_0. \tag{7.1.7b}$$

Equating Eqs. (7.1.7a) and (7.1.7b) gives

$$t^{ij} = \sqrt{G}\sigma^{ij}, \tag{7.1.8}$$

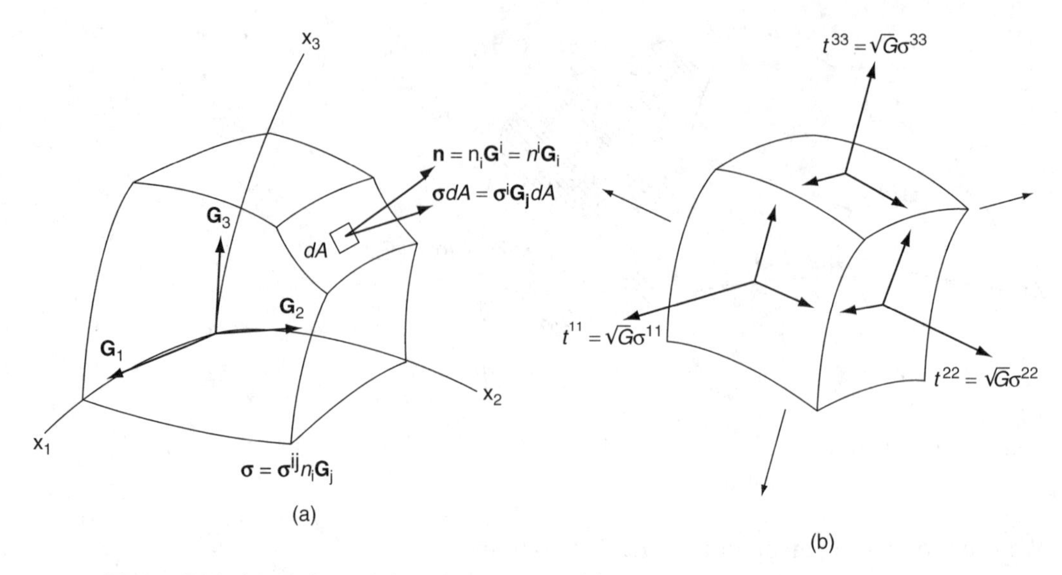

Figure 7.1.3. (a) Deformed boundary surface, (b) components of Kirchhoff stress tensor on deformed state, measured in terms of undeformed state.

where σ^{ij} and t^{ij} are referred to as the Cauchy stress tensor and the Kirchhoff stress tensor, respectively. If $\sqrt{G} = 1$, then the undeformed state of stress prevails. In this case, we may place the indices in the subscript position so that $t_{ij} = \sigma_{ij}$.

We may devise an alternative way of measuring stresses in deformed states in terms of an undeformed area by defining the relation between the undeformed and deformed areas by using a different approach from that of Eq. (7.1.1). The normal vector $\hat{\mathbf{n}}$ and the stress vector $\mathbf{T}_{(1)}$ on undeformed surface dA_0, as related to the normal vector $\mathbf{n}$ and the stress vector $\boldsymbol{\sigma}$ on deformed surface dA_0, are schematically shown in Fig. 7.1.4. We begin with the product of the undeformed area and the normal vector defined as

$$\hat{\mathbf{n}} dA_0 = dx_i dx_j \epsilon_{ijk} \mathbf{i}_k$$

or

$$n_k \mathbf{i}_k dA_0 = \frac{\partial x_i}{\partial z_m} \frac{\partial x_j}{\partial z_n} dz_m dz_n \epsilon_{ijk} \mathbf{i}_k.$$

We eliminate $\mathbf{i}_k$ and multiply by $\partial x_k / \partial z_p$ to get

$$\frac{\partial x_k}{\partial z_p} \hat{n}_k dA_0 = \frac{\partial x_i}{\partial z_m} \frac{\partial x_j}{\partial z_n} \frac{\partial x_k}{\partial z_n} dz_m dz_n \epsilon_{ijk} = \frac{1}{\sqrt{G}} \epsilon_{mnp} dz_m dz_n. \tag{7.1.9}$$

Similarly,

$$\mathbf{n} dA = dz_i dz_j \epsilon_{ijk} \mathbf{i}_k$$

or

$$n_k dA = dz_i dz_j \epsilon_{ijk}. \tag{7.1.10}$$

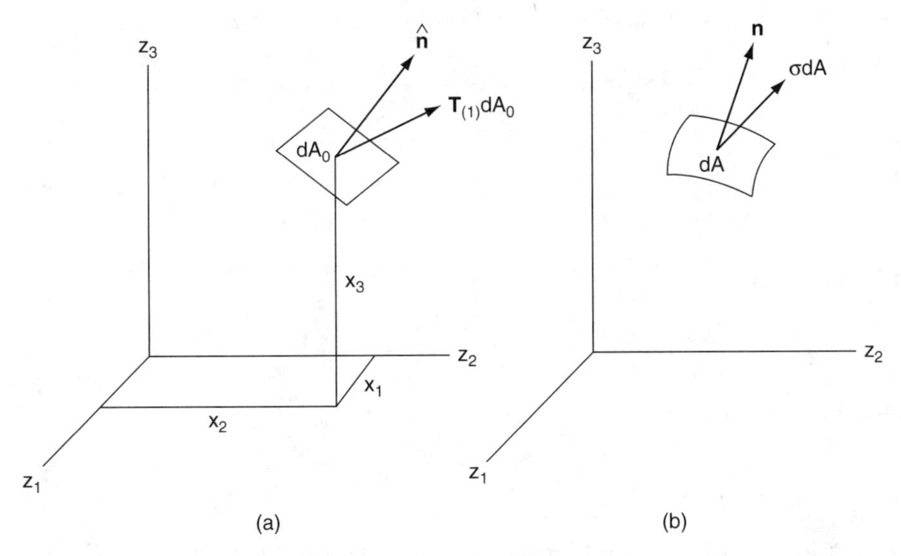

Figure 7.1.4. (a) First Piola–Kirchhoff stress vector measured in terms of the undeformed area, and (b) Cauchy stress vector measured in terms of the deformed area.

The substitution of Eq. (7.1.10) into Eq. (7.1.9) yields

$$\frac{\partial x_k}{\partial z_i}\hat{n}_k dA_0 = \frac{1}{\sqrt{G}} n_i dA,$$

which gives

$$n_i dA = \sqrt{G}\frac{\partial x_k}{\partial z_i}\hat{n}_k dA_0. \tag{7.1.11}$$

Let $\mathbf{T}_{(1)}$ and $\boldsymbol{\sigma}$ be defined as the stress vectors measured in terms of the undeformed and deformed areas, respectively. Then

$$\mathbf{T}_{(1)} dA_0 = \boldsymbol{\sigma} dA.$$

The foregoing relation can be written as

$$T_{(1)}^{ij}\hat{n}_i \mathbf{G}_j dA_0 = \sigma^{ij} n_i \mathbf{G}_j dA = \sigma^{ij}\mathbf{G}_j \sqrt{G}\frac{\partial x_k}{\partial z_i}\hat{n}_k dA_0,$$

from which we obtain

$$T_{(1)}^{ij} = \sqrt{G}\frac{\partial x_i}{\partial z_k}\sigma^{kj}. \tag{7.1.12}$$

Here σ^{kj} and $T_{(1)}^{ij}$ are identified as the Cauchy stress tensor and the first Piola–Kirchhoff stress tensor, respectively. Now, let Eq. (7.1.11) be multiplied by $(\partial x_k/\partial z_j)\sigma^{ji}$:

$$n_i dA \frac{\partial x_k}{\partial z_j}\sigma^{ij} = \sqrt{G}\frac{\partial x_n}{\partial z_i}\hat{n}_n dA_0 \frac{\partial x_k}{\partial z_j}\sigma^{ji}.$$

We write the left-hand side of this equation for the undeformed area as

$$\frac{\partial x_k}{\partial z_j} T^{ij}_{(1)} \hat{n}_i dA_0 = \sqrt{G} \frac{\partial x_n}{\partial z_i} \frac{\partial x_k}{\partial z_j} \sigma^{ji} \hat{n}_n dA_0.$$

If we define

$$T^{jk}_{(2)} = \frac{\partial x_k}{\partial z_i} T^{ji}_{(1)},$$

then we may write

$$T^{ik}_{(2)} \hat{n}_i dA_0 = \sqrt{G} \frac{\partial x_n}{\partial z_i} \frac{\partial x_k}{\partial z_j} \sigma^{ji} \hat{n}_n dA_0,$$

which leads to

$$T^{ij}_{(2)} = \sqrt{G} \frac{\partial x_i}{\partial z_k} \frac{\partial x_j}{\partial z_n} \sigma^{kn}, \tag{7.1.13}$$

where $T^{ij}_{(2)}$ is called the second Piola–Kirchhoff stress tensor. Thus the Cauchy stress tensor can be determined from Eq. (7.1.13):

$$\sigma^{ij} = \frac{1}{\sqrt{G}} \frac{\partial z_i}{\partial x_k} \frac{\partial z_j}{\partial x_n} T^{kn}_{(2)}. \tag{7.1.14}$$

Similarly,

$$T^{ij}_{(2)} = \sqrt{G} \left(\frac{\partial x_i}{\partial z_k} \frac{\partial x_j}{\partial z_n} \right)^{-1} \sigma^{kn}. \tag{7.1.15}$$

The substitution of Eq. (7.1.14) into Eq. (7.1.12) yields

$$T^{ij}_{(1)} = \frac{\partial z_j}{\partial x_k} T^{ik}_{(2)}. \tag{7.1.16}$$

Note that the relation given in Eq. (7.1.8) results from the definition in Eq. (7.1.2), whereas the first and second Piola–Kirchhoff stresses are based on

$$dA_{0i} = \frac{\partial x_k}{\partial z_i} \hat{n}_k dA_0. \tag{7.1.17}$$

The first Piola–Kirchhoff stress tensor is asymmetric and awkward for use in computations. The second Piola–Kirchhoff stress tensor, although symmetric, is still not practical for numerical analysis. The two are, however, useful in the development of constitutive theories. The Kirchhoff stress tensor, as defined by Eq. (7.1.8), is the most convenient in numerical analysis (Oden, 1972), because calculations for deformed states can be carried out in terms of undeformed areas only by means of $\sqrt{G}$.

7.1.2 Equations of Motion for Large Strains

To deal with large strains, we recast the balance laws discussed in Section 6.2 in Lagrangian convective coordinates with the tangent vectors $\mathbf{G}_i$. The algebra required toward this end is identical to that of Eqs. (6.2.6) and (6.2.7), with the substitution of $\mathbf{G}_i$ for $\mathbf{g}_i$. Thus Cauchy's first law of motion becomes

$$\frac{1}{\sqrt{G}}(\sqrt{G}\sigma^{ij}\mathbf{G}_j)_{,i} + \rho\mathbf{F} - \rho\dot{\mathbf{v}} = 0. \tag{7.1.18}$$

Note that this equation is not practical for use in computation with the Cauchy stress tensor. By virtue of Eq. (7.1.8) and

$$\mathbf{F} = F^m\mathbf{G}_m = F_m^0\mathbf{i}_m,$$
$$\dot{\mathbf{v}} = \dot{\mathbf{v}}^m\mathbf{G}_m = \ddot{u}_m\mathbf{i}_m,$$
$$\rho_0 = \sqrt{G}\rho,$$

it is possible to rewrite Eq. (7.1.18) in the form

$$(t^{ij}z_{m,j})_{,i} + \rho_0 F_m^0 - \rho_0\ddot{u}_m = 0 \tag{7.1.19}$$

or

$$[t^{ij}(\delta_{mj} + u_{m,j})]_{,i} + \rho_0 F_m^0 - \rho_0.\ddot{u}_m = 0.$$

Finally, we have the alternative Cauchy's first law of motion,

$$t^{im}_{,i} + t^{ij}_{,i}u_{m,j} + t^{ij}u_{m,ji} + \rho_0 F_m^0 - \rho_0\ddot{u}_m = 0, \tag{7.1.20}$$

which contains nonlinear terms in each direction. Note also that in terms of the Kirchhoff stress tensor t^{ij} we are able to describe stresses on the deformed state in terms of the undeformed area. This is the reason why the physical components of σ^{ij} in the deformed curvilinear coordinates need not be evaluated, with σ^{ij} being replaced with t^{ij}. The nonlinear behavior that is due to large deformations is characterized by those terms with the first and second derivatives of displacements. Using the Kirchhoff stress tensor t^{ij}, we discuss finite elasticity in the next subsection.

The body undergoing large deformations may be accompanied by large rotations and stretches defined by the polar decomposition theorem. This subject is included in Section 7.2, on non-Newtonian fluid mechanics.

7.1.3 Finite Elasticity

If a material ceases to be linearly elastic after reaching a certain magnitude of strain and at that point exhibits nonlinear stress–strain relations, the generalized form of Hooke's law no longer applies. Here, we require a nonlinear constitutive law. Materials such as rubber fall under the category of nonlinear elasticity, commonly known as *finite elasticity*, a term that indicates the material can undergo a finite deformation and remain elastic (Green and Adkins, 1960; Hart-Smith, 1966). For an isothermal condition, the stress tensor is given in terms of the derivative of the internal energy density ϵ with respect to the strain tensor:

$$\sigma^{ij} = \rho \frac{\partial \epsilon}{\partial \gamma_{ij}}.$$

Consider an elastic potential $W = \hat{W}(\gamma_{ij})$ in the form

$$W = \rho_0 \epsilon,$$

with ρ_0 given by Eq. (2.2.24b),

$$\rho_0 = \sqrt{G}\rho.$$

We then have the Cauchy stress tensor written in the form

$$\sigma^{ij} = \frac{1}{\sqrt{G}} \frac{\partial W}{\partial \gamma_{ij}}$$

or we can use the Kirchhoff stress tensor on the deformed surface in terms of the undeformed area to obtain

$$t^{ij} = \frac{\partial W}{\partial \gamma_{ij}}. \tag{7.1.21}$$

Using the approach introduced in Section 2.5, we can derive the principal strain invariants of G_{ij} as

$$I_1 = G_{ii} = \delta^{ij} G_{ij} = \delta^{ij}(2\gamma_{ji} + \delta_{ji}) = 2\gamma_{ii} + 3,$$

$$I_2 = \frac{1}{2}(\delta^{ir}\delta^{js} G_{ri} G_{sj} - \delta^{ir}\delta^{js} G_{rj} G_{si}) = 2(\gamma_{ii}\gamma_{jj} - \gamma_{ij}\gamma_{ji}) + 4\gamma_{ii} + 3,$$

$$I_3 = |G_{ij}| = \frac{4}{3}\epsilon_{ijk}\epsilon_{rst}\gamma_{ir}\gamma_{js}\gamma_{kt} + 2(\gamma_{ii}\gamma_{jj} - \gamma_{ij}\gamma_{ji}) + 2\gamma_{ii} + 1,$$

and because W is a function of γ_{ij} and also is a function of the principal strain invariants of G_{ij}, we obtain

$$\begin{aligned}
t^{ij} &= \frac{\partial W}{\partial I_1}\frac{\partial I_1}{\partial \gamma_{ij}} + \frac{\partial W}{\partial I_2}\frac{\partial I_2}{\partial \gamma_{ij}} + \frac{\partial W}{\partial I_3}\frac{\partial I_3}{\partial \gamma_{ij}} \\
&= 2\frac{\partial W}{\partial I_1}\delta^{ij} + 4\frac{\partial W}{\partial I_2}[\delta^{ij}(1+\gamma_{rr}) + \delta^{ir}\delta^{js}\gamma_{rs}] \\
&\quad + 2\frac{\partial W}{\partial I_3}[\delta^{ij}(1+2\gamma_{rr}) - \delta^{ir}\delta^{js}\gamma_{rs} + 2\,\epsilon^{imn}\epsilon^{jrs}\gamma_{mr}\gamma_{ns}].
\end{aligned} \tag{7.1.22}$$

Here W is given by [see Eqs. (4.1.2)]

$$W = \frac{1}{2}E^{ijkm}\gamma_{ij}\gamma_{km} + \frac{1}{6}E^{ijkmnp}\gamma_{ij}\gamma_{km}\gamma_{np} + \cdots + . \tag{7.1.23}$$

If we expand the invariants in powers, then W takes the form

$$W = \sum_{r=0}^{\infty}\sum_{s=0}^{\infty}\sum_{t=0}^{\infty} C_{rst}(I_1 - 3)^r (I_2 - 3)^s (I_3 - 1)^t. \tag{7.1.24}$$

A choice of $r = s = t = 1$ gives

$$W = C_{100}(I_1 - 3) + C_{010}(I_2 - 3) + C_{001}(I_3 - 1). \tag{7.1.25}$$

Alternatively, a polynomial approximation may be written in the form

$$\begin{aligned} W = {} & A_1 K_1^2 + A_2 K_2 + A_3 K_1^3 + A_4 K_1 K_2 + A_5 K_3 \\ & + A_6 K_1^4 + A_7 K_1^2 K_2 + A_8 K_1 K_3 + A_9 K_2^2, \end{aligned} \tag{7.1.26}$$

in which $A_1, \ldots, A_9$ are material constants and

$$K_1 = \gamma_{ii}, \quad K_2 = \gamma_{ij}\gamma_{ij}, \quad K_3 = \gamma_{ij}\gamma_{ir}\gamma_{jr}.$$

Many materials undergo finite deformations without an appreciable change in volume. These materials are said to be *incompressible*. The most significant characteristic of incompressibility is that the stress tensor is not completely determined by the deformation. The addition of hydrostatic pressure to an incompressible elastic solid affects the stress, but the strain is not altered. The incompressibility condition is given by

$$I_3 = |G_{ij}| = 1. \tag{7.1.27}$$

For hyperelastic materials we may write

$$W = W(\gamma_{ij}) + \lambda(I_3 - 1),$$

where λ is a Lagrange multiplier. Thus the Kirchhoff stress tensor takes the form

$$t^{ij} = \frac{\partial W}{\partial \gamma_{ij}} + \lambda \frac{\partial I_3}{\partial \gamma_{ij}} + \frac{\partial W}{\partial \gamma_{ij}} + hG^{ij}.$$

Here, $h = 2\lambda$ is the hydrostatic pressure. In view of Eq. (7.1.27), we see that W is independent of I_3, or that

$$W = W(I_1, I_2)$$

and

$$\begin{aligned} t^{ij} &= \frac{\partial W}{\partial I_1}\frac{\partial I_1}{\partial \gamma_{ij}} + \frac{\partial W}{\partial I_2}\frac{\partial I_2}{\partial \gamma_{ij}} + \frac{\partial W}{\partial I_3}\frac{\partial I_3}{\partial \gamma_{ij}} \\ &= \frac{\partial W}{\partial I_1}\frac{\partial I_1}{\partial \gamma_{ij}} + \frac{\partial W}{\partial I_2}\frac{\partial I_2}{\partial \gamma_{ij}} + h\frac{\partial I_3}{\partial \gamma_{ij}} \\ &= 2\frac{\partial W}{\partial I_1}\delta^{ij} + 4\frac{\partial W}{\partial I_2}[\delta^{ij}(1 + \gamma_{rr}) - \delta^{ir}\delta^{js}\gamma_{rs}] \\ &\quad + 2h[\delta^{ij}(1 + 2\gamma_{rr}) - \delta^{ir}\delta^{js}\gamma_{rs} + 2\epsilon^{imn}\epsilon^{jrs}\gamma_{mr}\gamma_{ns}]. \end{aligned} \tag{7.1.28}$$

Thus the polynomial approximations become

$$W = \sum_{r=0}^{\infty} \sum_{s=0}^{\infty} C_{rs}(I_1 - 3)^r (I_2 - 3)^s, \quad C_{00} = 0. \tag{7.1.29}$$

For example, the well-known Mooney material (or rubbery solids) (Mooney, 1940; Alexander, 1968) is characterized as

$$W = C_1(I_1 - 3) + C_2(I_2 - 3), \tag{7.1.30}$$

in which $C_1 = C_{10}$ and $C_2 = C_{01}$.

In summary, the subject of finite elasticity depends on experimental data (the evaluation of constants in the expression for elastic potential) that characterize material properties. Once the appropriate forms of elastic potential are available, we then return to the Kirchhoff stress tensor of the form expressed in Eq. (7.1.28) and proceed with the analysis indicated in Section 4.4. Furthermore, finite elasticity involves large rotations and stretches governed by the polar decomposition theorem. This subject is discussed in the next section, on non-Newtonian fluid mechanics.

7.2 Non-Newtonian Fluid Mechanics

7.2.1 General

This section is an exposition of non-Newtonian fluid mechanics, in which the stress tensor is regarded as consisting of higher-order functions of the velocity gradients. This is in contrast to the assumption for Newtonian fluids, in which the stress tensor is a linear function of the velocity gradients. Non-Newtonian fluids include polymer melts, polymer solutions, and suspensions. They are referred to as *rheological materials* (Engler, 1991).

Although the basic principles for constitutive equations are similar to those stated earlier for solid mechanics (Subsection 4.4.4), the rules specifically applied to the non-Newtonian fluids are summarized as follows:

(1) *Coordinate invariance.* Physical laws are invariant with changes of the coordinate system.
(2) *Determinism.* The stress is determined by the history of the material.
(3) *Material objectivity.* The response of a material to a given history of motion is independent of any motion of the observer and is said to be objective.
(4) *Fading memory.* The recent history of the motion should have a more predominant effect on a functional of the motion of the body than the distant history does.

(5) *Local action.* The stress at a given point in a material is uniquely determined by the history of deformation of an arbitrarily small neighborhood of that point.
(6) *Nonexistence of a natural state.* This implies the formalization of the intuitive but elusive concept of fluidity: a fluid material has no preferred configuration or "natural state." This principle requires a fluid to be isotropic.

Noll (1958) postulates that the most general constitutive equation for stress is of the form

$$\hat{\boldsymbol{\tau}}(\mathbf{x}, t) = \hat{F}(t, \mathbf{x})(\chi). \tag{7.2.1}$$

This implies that the stress $\hat{\boldsymbol{\tau}}(\mathbf{x}, t)$ at time t and material point $\mathbf{x}$ is determined by the history of the motion χ of the whole body. Here, $\hat{F}$ is known as a functional whose argument is the function χ. We may restrict Eq. (7.2.1) by postulating the stress at material point $\mathbf{x}$ (according to the principle of local action). Thus Eq. (7.2.1) becomes

$$\tau(\mathbf{x}, t) = \underset{s=0}{\overset{s=\infty}{\hat{F}}} \chi_{t-s}(\mathbf{x}). \tag{7.2.2}$$

Here s is the time variable, $0 \leq s \leq \infty$, and $\chi_{t-s}(x)$ is the motion in an arbitrarily small neighborhood around $\mathbf{x}$ and at time $t - s$. What this means is that the stress at particle $\mathbf{x}$ and time t is determined by the motion over all past times up to and including the present of an arbitrarily small region around $\mathbf{x}$.

Although Eq. (7.2.2) is more specific than Eq. (7.2.1), it is still unclear how we would proceed to obtain complete characterization of a material to provide computational utility. Oldroyd (1950) presented the invariant forms of rheological equations of state for a homogeneous continuum, suitable for application to all conditions of motion and stress. Green and Rivlin (1957) and Green, Rivlin, and Spencer (1959) showed how one can account for memory effects in materials by writing a constitutive functional as a series of integrals over past time. Noll (1958) studied a different mathematical formalism in order to effect a compromise between the generality and the utility. In what follows, we discuss these and other topics of interest with regard to non-Newtonian fluids.

7.2.2 Kinematics

7.2.2.1 Stretch and Rotation

Recall that the differential position vector $d\mathbf{R}$ for the deformed state can be written in the form

$$d\mathbf{R} = dz_i \mathbf{i}_i = \frac{\partial \mathbf{R}}{\partial x_j} dx_j = \frac{\partial z_i}{\partial x_j} dx_j \mathbf{i}_i \tag{7.2.3a}$$

or

$$dz_i = \frac{\partial z_i}{\partial x_j} dx_j. \qquad (7.2.3b)$$

Using the so-called *direct tensor* notation, we may write Eq. (7.2.3b) as

$$d\mathbf{z} = \mathbf{F}\, d\mathbf{x}. \qquad (7.2.3c)$$

Here $\mathbf{F}$ represents the deformation gradient $\partial z_i/\partial x_j$ that is a second-order tensor, whereas $d\mathbf{z}$ and $d\mathbf{x}$ denote the first-order tensors corresponding to dz_i and dx_i, respectively. The direct tensor notation as used in (7.2.3c) appears to have broken all the rules for vectors and tensors that we have closely followed (Chap. 1). Tensors of all orders without indices written in boldface (here they are *not* vectors) are confusing to the beginner. The direct tensor notation, however, has been used extensively in theoretical research because of its simplicity. Nevertheless, the reader may find direct tensors useful in the discussion of stretch, rotation, and non-Newtonian fluids that we study in this section.

Assuming that $\mathbf{F}^{-1}$ exists in Eq. (7.2.3c), which is manifest for continuity or mass conservation, we can determine $d\mathbf{x}$ as

$$d\mathbf{x} = \mathbf{F}^{-1}\, d\mathbf{z}.$$

At this time, a further discussion of the deformation gradient is in order. We may write $\mathbf{F}$ in the following form, known as the *polar decomposition theorem*:

$$\mathbf{F} = \mathbf{RU} = \mathbf{VR}, \qquad (7.2.4)$$

where $\mathbf{U}$ and $\mathbf{V}$ are called the *right stretch* and *left stretch*, respectively, and $\mathbf{R}$ is the proper orthogonal matrix such that

$$\mathbf{R}^{-1} = \mathbf{R}^T.$$

We establish the following relations:

$$\mathbf{U}^2 = \mathbf{F}^T\mathbf{F}, \quad \mathbf{V}^2 = \mathbf{F}\mathbf{F}^T,$$

$$\mathbf{U} = \mathbf{R}^T\mathbf{V}\mathbf{R}, \quad \mathbf{V} = \mathbf{R}\mathbf{U}\mathbf{R}^T.$$

Because $\mathbf{F}^T\mathbf{F}$ is symmetric, there exists a proper orthogonal matrix (say $\mathbf{A}$) that transforms $\mathbf{F}^T\mathbf{F}$ into a diagonal form:

$$\mathbf{U}^2 = \mathbf{F}^T\mathbf{F} = \mathbf{A}\ \mathrm{diag}\left[\lambda_{(1)}^2,\ \lambda_{(2)}^2,\ \lambda_{(3)}^2\right]\mathbf{A}^T, \qquad (7.2.5)$$

where $\lambda_{(i)}^2$ represents the eigenvalues of $\mathbf{U}^2$ and $\mathbf{A} = [\mathbf{n}^{(i)}]$, with each $\mathbf{n}^{(i)}$ being the eigenvector for one eigenvalue of $\mathbf{U}^2$. Thus

$$\mathbf{U} = \mathbf{A}\ \mathrm{diag}\left[\lambda_{(1)},\ \lambda_{(2)},\ \lambda_{(3)}\right]\mathbf{A}^T. \qquad (7.2.6)$$

Furthermore, from

$$\mathbf{R} = \mathbf{F}\mathbf{U}^{-1},$$

we have

$$\mathbf{R}^T\mathbf{R} = \mathbf{U}^{-1}\mathbf{F}^T\mathbf{F}\mathbf{U}^{-1} = \mathbf{U}^{-1}\mathbf{U}^2\mathbf{U}^{-1} = \mathbf{I}.$$

Thus

$$\mathbf{F} = \mathbf{R}\mathbf{U} = \mathbf{R}\mathbf{U}\mathbf{R}^T\mathbf{R} = (\mathbf{R}\mathbf{U}\mathbf{R}^T)\mathbf{R},$$

which gives

$$\mathbf{V} = \mathbf{R}\mathbf{U}\mathbf{R}^T, \tag{7.2.7a}$$

$$\mathbf{U} = \mathbf{R}^T\mathbf{V}\mathbf{R}. \tag{7.2.7b}$$

This indicates that the eigenvalues of **U** and **V** are identical. See Truesdell and Noll (1965) for further details on the polar decomposition theorem.

Let us now examine some of the properties of right and left stretches in strain-displacement relations. We write

$$d\mathbf{z} = \mathbf{R}\mathbf{U}\,d\mathbf{x} = \mathbf{V}\mathbf{R}\,d\mathbf{x},$$

$$\mathbf{C} = \mathbf{U}^2 = \mathbf{F}^T\mathbf{F} = \mathbf{U}\mathbf{R}^T\mathbf{R}\mathbf{U}, \tag{7.2.8a}$$

$$\mathbf{B} = \mathbf{V}^2 = \mathbf{F}\mathbf{F}^T = \mathbf{R}\mathbf{U}\mathbf{U}\mathbf{R}^T, \tag{7.2.8b}$$

where **C** and **B** are called the right Cauchy–Green matrix and the left Cauchy–Green matrix, respectively. Thus,

$$ds^2 - ds_0^2 = d\mathbf{x}^T(\mathbf{C} - \mathbf{I})\,d\mathbf{x} = 2d\mathbf{x}^T\boldsymbol{\gamma}\,d\mathbf{x},$$

with

$$\boldsymbol{\gamma} = \frac{1}{2}(\mathbf{C} - \mathbf{I}), \tag{7.2.9}$$

which is equivalent to Eq. (2.3.4). Returning to the tensorial index notation, we may write the deformation gradient in the form

$$\begin{aligned} F_{ij} &= \frac{\partial z_i}{\partial x_j} = \delta_{ij} + \frac{\partial u_i}{\partial x_j} \\ &= \delta_{ij} + \frac{1}{2}\left(\frac{\partial u_i}{\partial x_j} + \frac{\partial u_j}{\partial x_i}\right) + \frac{1}{2}\left(\frac{\partial u_i}{\partial x_j} - \frac{\partial u_j}{\partial x_i}\right), \end{aligned}$$

$$F_{ij} = \delta_{ij} + E_{ij} + \Omega_{ij}, \tag{7.2.10}$$

where E_{ij} and Ω_{ij} are called the small strain tensor and the spin tensor, respectively:

$$E_{ij} = \frac{1}{2}\left(\frac{\partial u_i}{\partial x_j} + \frac{\partial u_j}{\partial x_i}\right), \quad \text{symmetric;} \tag{7.2.11a}$$

$$\Omega_{ij} = \frac{1}{2}\left(\frac{\partial u_i}{\partial x_j} + \frac{\partial u_j}{\partial x_i}\right), \quad \text{antisymmetric.} \tag{7.2.11b}$$

Alternatively,

$$\mathbf{F} = \mathbf{I} + \mathbf{E} + \boldsymbol{\Omega} \approx (\mathbf{I} + \boldsymbol{\Omega})(\mathbf{I} + \mathbf{E}), \tag{7.2.12}$$

where the product $\boldsymbol{\Omega}\mathbf{E}$ is considered negligibly small in comparison with the nonlinear terms dropped from Eq. (2.3.5). Comparing Eqs. (7.2.12) and (7.2.4) we note that

$$\mathbf{I} + \boldsymbol{\Omega} \approx \mathbf{R},$$
$$\mathbf{I} + \mathbf{E} \approx \mathbf{U} \approx \mathbf{V}.$$

These results are also evident from

$$\mathbf{U}^2 = \mathbf{C} = \mathbf{I} + 2\mathbf{E},$$
$$\mathbf{U} = \sqrt{\mathbf{C}} \approx \mathbf{I} + \mathbf{E}.$$

We also note that

$$\mathbf{B} \approx \mathbf{C} = \mathbf{I} + 2\mathbf{E},$$
$$\mathbf{B}^{-1} \approx \mathbf{I} - 2\mathbf{E}.$$

With these observations, we conclude that, for small strain approximations, we have

$$\mathbf{U} = \mathbf{V}, \qquad \mathbf{C} = \mathbf{B}.$$

Calculation of Right and Left Stretch. As an example, let us consider

$$F = \begin{bmatrix} \frac{2}{5} & -1 \\ \frac{11}{5} & 2 \end{bmatrix} \tag{7.2.13}$$

and calculate right and left stretches

$$\mathbf{U}^2 = \mathbf{F}^T\mathbf{F} = \begin{bmatrix} 5 & 4 \\ 4 & 5 \end{bmatrix}.$$

The eigenvalues are $\lambda_{(1)}^2 = 1$, $\lambda_{(2)}^2 = 9$. The corresponding eigenvectors are

$$\mathbf{n}^{(1)} = \begin{bmatrix} \frac{1}{\sqrt{2}} \\ -\frac{1}{\sqrt{2}} \end{bmatrix}, \quad \mathbf{n}^{(2)} = \begin{bmatrix} \frac{1}{\sqrt{2}} \\ \frac{1}{\sqrt{2}} \end{bmatrix},$$

$$\mathbf{A} = \left[\mathbf{n}^{(1)}\ \mathbf{n}^{(2)}\right] = \begin{bmatrix} \frac{1}{\sqrt{2}} \\ -\frac{1}{\sqrt{2}} \end{bmatrix},$$

$$\mathbf{U}^2 = \begin{bmatrix} \frac{1}{\sqrt{2}} & \frac{1}{\sqrt{2}} \\ -\frac{1}{\sqrt{2}} & \frac{1}{\sqrt{2}} \end{bmatrix} \begin{bmatrix} 1 & 0 \\ 0 & 9 \end{bmatrix} \begin{bmatrix} \frac{1}{\sqrt{2}} & -\frac{1}{\sqrt{2}} \\ \frac{1}{\sqrt{2}} & \frac{1}{\sqrt{2}} \end{bmatrix},$$

$$\mathbf{U} = \begin{bmatrix} \frac{1}{\sqrt{2}} & \frac{1}{\sqrt{2}} \\ -\frac{1}{\sqrt{2}} & \frac{1}{\sqrt{2}} \end{bmatrix} \begin{bmatrix} 1 & 0 \\ 0 & 3 \end{bmatrix} \begin{bmatrix} \frac{1}{\sqrt{2}} & -\frac{1}{\sqrt{2}} \\ \frac{1}{\sqrt{2}} & \frac{1}{\sqrt{2}} \end{bmatrix} = \begin{bmatrix} 2 & 1 \\ 1 & 2 \end{bmatrix},$$

$$\mathbf{U}^{-1} = \begin{bmatrix} \frac{2}{3} & -\frac{1}{3} \\ -\frac{1}{3} & \frac{2}{3} \end{bmatrix}, \quad \mathbf{R} = \mathbf{F}\mathbf{U}^{-1} = \begin{bmatrix} \frac{3}{5} & -\frac{4}{5} \\ \frac{4}{5} & \frac{3}{5} \end{bmatrix},$$

$$\mathbf{V} = \mathbf{F}\mathbf{R}^T = \begin{bmatrix} \frac{26}{25} & -\frac{7}{25} \\ -\frac{7}{25} & \frac{74}{25} \end{bmatrix}.$$

Deformations induce not only stretches but also rotations. The spin tensor appearing in Eq. (7.2.10) can be redefined as a vector:

$$\boldsymbol{\Omega} = \frac{1}{2}\nabla \times \mathbf{u} = \frac{1}{2}\epsilon_{ijk}u_{k,j}\mathbf{i}_i \tag{7.2.14a}$$

or

$$\boldsymbol{\Omega} = \frac{1}{2}\epsilon_{ijk}\Omega_{kj}\mathbf{i}_i = \Omega_i\mathbf{i}_i, \tag{7.2.14b}$$

where

$$\Omega_i = \frac{1}{2}\epsilon_{ijk}\Omega_{kj},$$

$$\Omega_{ij} = \Omega_k\epsilon_{kji} = \begin{bmatrix} 0 & -\Omega_3 & \Omega_2 \\ \Omega_3 & 0 & -\Omega_1 \\ -\Omega_2 & \Omega_1 & 0 \end{bmatrix}, \tag{7.2.15}$$

which is antisymmetric.

The deformation, as shown in Fig. 7.2.1, represents both stretch and rotation:

$$d\mathbf{R} = d\mathbf{r} + d\mathbf{u} = d\mathbf{r} + \frac{\partial u_i}{\partial x_j}dx_j\mathbf{i}_i = d\mathbf{r} + (E_{ij} + \Omega_{ij})\, dx_j\mathbf{i}_i.$$

If $E_{ij} \ll \Omega_{ij}$, then

$$d\mathbf{R} = d\mathbf{r} + \Omega_{ij}\, dx_j\mathbf{i}_i \tag{7.2.16a}$$

or

$$d\mathbf{R} = d\mathbf{r} + \boldsymbol{\Omega} \times d\mathbf{r}, \tag{7.2.16b}$$

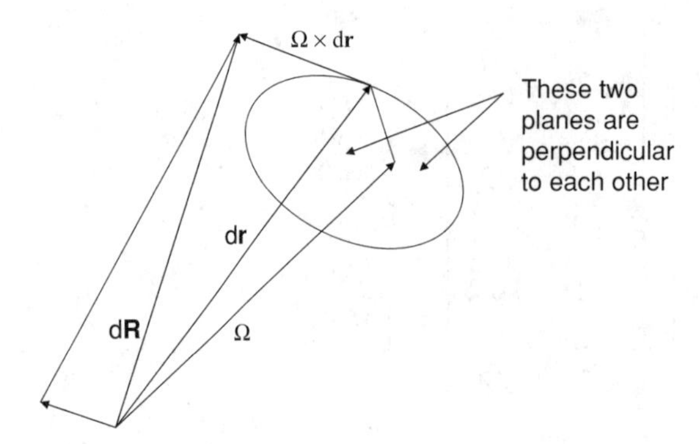

Figure 7.2.1. Deformed state $d\mathbf{R}$ results from the sum of translational and rotational contributions; $\boldsymbol{\Omega} \times d\mathbf{r}$ lies in the plane of circle that is normal to the plane constructed by $\boldsymbol{\Omega}$ and $d\mathbf{r}$.

which indicates that $d\mathbf{R}$ consists of translation and rotation. It is seen from Eq. (2.3.5) that

$$\gamma_{ij} = E_{ij} + \frac{1}{2}(E_{mi} + \Omega_{mi})(E_{mj} + \Omega_{mj}).$$

If $E_{ij} \ll \Omega_{ij}$, as assumed in Eq. (7.2.16a), then

$$\gamma_{ij} = E_{ij} + \frac{1}{2}\Omega_{mi}\Omega_{mj}, \tag{7.2.17}$$

in which the so-called *moderate rotation* is represented by

$$\frac{1}{2}\Omega_{mi}\Omega_{mj}.$$

With an assumption of small rotation, however, we have once again

$$\gamma_{ij} = E_{ij} = \frac{1}{2}\left(\frac{\partial u_i}{\partial x_j} + \frac{\partial u_j}{\partial x_i}\right), \tag{7.2.18}$$

which is the small strain tensor.

7.2.2.2 Rates of Strain: Stretching and Spin

We recall the deformation gradient $\mathbf{F}$ as given by Eq. (7.2.3c):

$$d\mathbf{z} = \mathbf{F}\, d\mathbf{x}. \tag{7.2.19}$$

We note also that the polar decomposition theorem [Eq. (7.2.4)] is of the form

$$\mathbf{F} = \mathbf{RU}, \quad \mathbf{F} = \mathbf{VR}, \tag{7.2.20}$$

$$\mathbf{C} = \mathbf{F}^T\mathbf{F}. \tag{7.2.21}$$

At time $\tau = t$, we have $\mathbf{F} = \mathbf{I}$ and thus

$$\mathbf{U} = \mathbf{R} = \mathbf{I}.$$

Differentiating Eqs. (7.2.20) with respect to time and evaluating at $\tau = t$, we obtain the velocity gradient $\mathbf{L}$:

$$\mathbf{L} = \dot{\mathbf{F}} = \dot{\mathbf{R}}\mathbf{U} + \mathbf{R}\dot{\mathbf{U}} = \dot{\mathbf{R}} + \dot{\mathbf{U}}. \tag{7.2.22}$$

Here the velocity gradient $\mathbf{L}$ can be considered as comprising a rate-of-spin tensor and a rate-of-stretching tensor:

$$\mathbf{L} = \frac{1}{2}(\mathbf{L} - \mathbf{L}^T) + \frac{1}{2}(\mathbf{L} + \mathbf{L}^T). \tag{7.2.23}$$

Furthermore, we note that

$$\begin{aligned} \mathbf{R}\mathbf{R}^T &= I, \\ \dot{\mathbf{R}} + \dot{\mathbf{R}}^T &= 0, \end{aligned} \tag{7.2.24}$$

which represents the definition of an antisymmetric tensor. Thus, in view of Eqs. (7.2.22)–(7.2.24), we can write the rate-of-spin tensor $\mathbf{W}$ and the rate-of-stretching tensor $\mathbf{D}$ as follows:

$$\mathbf{W} = \dot{\mathbf{R}} = \frac{1}{2}(\mathbf{L} - \mathbf{L}^T), \tag{7.2.25a}$$

$$\mathbf{D} = \dot{\mathbf{U}} + \frac{1}{2}(\mathbf{L} + \mathbf{L}^T). \tag{7.2.25b}$$

The higher material time derivatives may be taken indefinitely:

$$\mathbf{D}_1 = \mathbf{D}, \quad \mathbf{D}_2 = \dot{\mathbf{D}} = \ddot{\mathbf{U}}, \quad \text{etc.,}$$

$$\mathbf{D}_n = \overset{(n)}{\mathbf{U}}, \tag{7.2.26a}$$

$$\mathbf{W}_n = \overset{(n)}{\mathbf{R}}. \tag{7.2.26b}$$

For example, the nth Rivlin–Ericksen tensor is given by

$$\mathbf{A}_n = \overset{(n)}{\mathbf{C}}. \tag{7.2.27}$$

The material time derivative equation, (7.2.8a), assumes the form

$$\dot{\mathbf{C}} = \mathbf{F}^T\mathbf{F} + \mathbf{F}^T\mathbf{F}. \tag{7.2.28}$$

If Eq. (7.2.28) is evaluated at $t = t_0$, we obtain

$$A_1 = \dot{\mathbf{C}} = \mathbf{L} + \mathbf{L} = 2D, \tag{7.2.29a}$$

$$A_2 = \ddot{\mathbf{C}} = 2\mathbf{L}^T\mathbf{L} + \mathbf{L}_2 + \mathbf{L}_2^T. \tag{7.2.29b}$$

In general, it can be shown that

$$\mathbf{A}_n = \mathbf{L}_n + \mathbf{L}_n^T + \sum_{i=1}^{n-1} \frac{n!}{(n-i)!i!} \mathbf{L}_i^T \mathbf{L}_{(n-i)}, \tag{7.2.30}$$

$$\mathbf{D}_n = \frac{1}{2}\left[\mathbf{A}_n - \sum_{i=1}^{n-1} \frac{n!}{(n-i)!i!} \mathbf{D}_i \mathbf{D}_{(n-i)}\right]. \tag{7.2.31}$$

With these preliminaries, we are now prepared to develop kinematics of non-Newtonian fluids. Historically, there are two schools of thought: the convective coordinate approach advanced by Oldroyd (1950) and the vector space approach pioneered by Coleman (1965) and Coleman, Markovit, and Noll (1966). Index notation is more convenient for the convective coordinate approach whereas direct tensors are preferred in the vector space approach, as will be seen in the discussions presented in the following subsections.

7.2.2.3 Convective Coordinate Approach

We have dealt with direct methods of vector spaces; we will continue to utilize them for the rest of this chapter. However, the convective coordinate approach developed by Oldroyd (1950), which has enjoyed widespread acceptance in the literature, is worthy of mention in some detail. We begin with a new definition of coordinates $dx(t)$ and $dz(t)$ in the form

$$d\mathbf{x} = dx^i \hat{\mathbf{g}}_i, \tag{7.2.32a}$$

$$d\mathbf{z} = dz^i \mathbf{g}_i, \tag{7.2.32b}$$

where the $\hat{\mathbf{g}}_i$'s are the base vectors for curvilinear coordinate lines dx and the $\mathbf{g}_i$'s are the base vectors for coordinate lines dz^i. Substituting Eqs. (7.2.32a) and (7.2.32b) into Eq. (7.2.19) leads to

$$dz^i \mathbf{g}_i = \mathbf{F}\, dx^i \hat{\mathbf{g}}_i. \tag{7.2.33}$$

Multiplying both sides of Eq. (7.2.33) by $\mathbf{g}^n$, we obtain

$$\mathbf{g}^n \cdot dz^i \mathbf{g}_i = \delta_i^n dz^i = \mathbf{g}^n \cdot \mathbf{F}\, dx^j \hat{\mathbf{g}}_j,$$

$$dz^n = \mathbf{g}^n \cdot \mathbf{F}\, dx^i \hat{\mathbf{g}}_i. \tag{7.2.34}$$

Note that

$$dz^n = \frac{\partial z^n}{\partial x^j} dx^j. \tag{7.2.35}$$

Comparing Eqs. (7.2.34) and (7.2.35) gives

$$\frac{dz^n}{dx^j} = \mathbf{g}^n \cdot \mathbf{F}\hat{\mathbf{g}}_j. \tag{7.2.36a}$$

$$\frac{dz^m}{dx^i} = \hat{\mathbf{g}}_i \cdot \mathbf{F}^T \mathbf{g}_m. \tag{7.2.36b}$$

By definition,

$$C_i = \mathbf{C}\hat{\mathbf{g}}_i,$$

$$C_{ij} = \mathbf{C}\hat{\mathbf{g}}_i \cdot \hat{\mathbf{g}}_j = \mathbf{F}^T\mathbf{F}\hat{\mathbf{g}}_i \cdot \hat{\mathbf{g}}_j. \tag{7.2.37}$$

Alternatively,

$$C_{ij} = \mathbf{F}^T\mathbf{F}(\hat{\mathbf{g}}_i \cdot \hat{\mathbf{g}}_j)(\mathbf{g}^m \cdot \mathbf{g}_m)(\mathbf{g}^n \cdot \mathbf{g}_n),$$

$$C_{ij} = (\hat{\mathbf{g}}_i \cdot \mathbf{F}^T \hat{\mathbf{g}}_m)(\mathbf{g}_m \cdot \mathbf{g}_n)(\mathbf{g}^n \cdot \mathbf{F}\hat{\mathbf{g}}_j),$$

or

$$C_{ij} = \frac{\partial z^m}{\partial x^i}\frac{\partial z^n}{\partial x^j} g_{mn}, \tag{7.2.38}$$

where g_{mn} is the metric tensor at $d\mathbf{z}(t)$ and C_{ij} represents the components of the Cauchy tensor at $d\mathbf{x}(t_0)$. An inverse of the Cauchy tensor is given by

$$(C_{ij})^{-1} = \frac{\partial x^i}{\partial z^m}\frac{\partial x^j}{\partial z^n} g^{mn}. \tag{7.2.39}$$

This is known as the *Finger tensor*.

Let us now consider a fluid particle in motion. We introduce a coordinate system ξ^i defined as follows:

$$\xi^i = z^i(\tau)\,|_{\tau=t} = x^i(t).$$

Here, we consider the ξ^i's to be coordinates embedded in the material or convected coordinates that relate the metric tensor G_{ij} to the square of a line element:

$$ds^2 = G_{ij}dx^i dx^j = G_{ij}d\xi^i d\xi^j, \tag{7.2.40}$$

where

$$G_{ij} = C_{ij} = \frac{\partial z^m}{\partial \xi^i}\frac{\partial z^n}{\partial \xi^j} g_{mn}. \tag{7.2.41}$$

Here it is important to note that the natural basis, $\mathbf{g}_i$, $\hat{\mathbf{g}}_i$, may change along the trajectory of a material point in dealing with the substantial time derivative. This is because the coordinate system changes with time.

The area of a surface element is related by

$$dA^2 = G^{ij} dA_i dA_j, \tag{7.2.42}$$

where dA_i stands for the convected covariant components of the area vector that do not change with deformation. It follows from Eq. (7.2.42) that

$$G^{ij} = (C^{ij})^{-1}. \tag{7.2.43}$$

The $N + n$ time derivatives of Eqs. (7.2.40) and (7.2.42) lead, respectively, to

$$\frac{d^n}{dt^n}(ds^2) = \frac{d^n}{dt^n}(G_{ij}) dx^i dx^j \tag{7.2.44a}$$

or

$$\frac{d^n}{dt^n}(ds^2) = (\mathbf{A}_n)_{ij} dx^i dx^j \tag{7.2.44b}$$

and

$$\frac{d^n}{dt^n}(da^2) = \frac{d^n}{dt^n}(G^{ij}) dA_i dA_j \tag{7.2.45a}$$

or

$$\frac{d^n}{dt^n}(dA^2) = (-\mathbf{E}_n)^{ij} dA_i dA_j. \tag{7.2.45b}$$

Here, $\mathbf{A}_n$ is the nth Rivlin–Ericksen tensor. Note also that $(d^n/dt^n)(G_{ij})$ is the nth derivative of G_{ij} and that we cannot obtain $(d^n/dt^n)(G^{ij})$ by raising the indices (changing from subscript to superscript) on $(d^n/dt^n)(G_{ij})$ because time differentiations and raising of indices do not commute in this case.

The choice between the convected coordinates and the vector space approach is a matter of personal preference. Although no special emphasis is placed on either one of the methods in this book, the majority of the discussions in what follows are based on the vector space approach.

7.2.2.4 Constant-Stretch History

We consider here flows for which the history of stretch does not depend on the instant of observation t, but only on the *time lag* $s = t - \tau$. In a memory fluid, the stress at the instant of observation is influenced by the entire deformation history of the neighborhood of the material point considered. In constant-stretch flows, however, the stress is independent of the instant of observation.

We consider an orthogonal transformation $\mathbf{Q}(t)$ and flows governed by

$$\mathbf{C}_{(t)}(t - s) \equiv \mathbf{C}_t(s) = \mathbf{Q}^{(t)} \mathbf{C}_{(0)}(0 - s) \mathbf{Q}^T(t), \tag{7.2.46}$$

where $\mathbf{Q}(0) = \mathbf{I}$. Note here that $\mathbf{C}_{(t)}(t - s)$ depends on the time interval s from the present into the past, but not on t.

Following Noll (1958), we begin with Eq. (7.2.21):

$$\mathbf{C}(\tau) = \mathbf{F}^T(t)\mathbf{C}_{(t)}(\tau)\mathbf{F}(t). \tag{7.2.47}$$

Taking a special configuration, say, $t = t'$, or more specifically, $t' = 0$, and substituting Eq. (7.2.46) into Eq. (7.2.47), we have

$$\mathbf{C}_{(0)}(t-s) = \mathbf{F}^T_{(0)}\mathbf{Q}(t)\mathbf{C}_{(0)}(0-s)\mathbf{Q}^T(t)\mathbf{F}_{(0)}(t), \tag{7.2.48}$$

$$\mathbf{C}_{(0)}(t-s) = \mathbf{M}^T(t)\mathbf{C}_{(0)}(0-s)\mathbf{M}^T, \tag{7.2.49}$$

with $\mathbf{M}(t) = \mathbf{Q}^T(t)\mathbf{F}_{(0)}(t)$. Taking a time derivative of Eq. (7.2.49) leads to

$$\dot{\mathbf{C}}_{(0)}(t-s) = \dot{\mathbf{M}}^T(t)\mathbf{C}_{(0)}(0-s)\mathbf{M}(t) + \mathbf{M}^T(t)\mathbf{C}_{(0)}(0-s)\dot{\mathbf{M}}(t). \tag{7.2.50}$$

If we set $t = 0$ and $[d\mathbf{M}(t)/dt]_{t=0} = k\mathbf{N}_0$, then the expression in Eq. (7.2.50) takes the form

$$-\frac{d}{ds}\left[\mathbf{C}_{(0)}(0-s)\right] = \mathbf{k}\mathbf{N}_0^T\mathbf{C}_{(0)}(0-s) + \mathbf{C}_{(0)}(0-s)\mathbf{k}\mathbf{N}_0. \tag{7.2.51}$$

Note here that $\mathbf{M}(0) = I$ because $\mathbf{Q}(0) = \mathbf{F}_{(0)}(0) = I$. The solution of a differential equation of the type shown in Eq. (7.2.51) is of the form

$$\mathbf{C}_{(0)}(0-s) = \exp\left(-s\mathbf{k}\mathbf{N}_0^T\right)\exp\left(-s\mathbf{k}\mathbf{N}_0\right). \tag{7.2.52}$$

In view of Eqs. (7.2.46) and (7.2.52), we have

$$\mathbf{C}_{(t)}(t-s) = \mathbf{Q}(t)\exp\left(-s\mathbf{k}\mathbf{N}_0^T\right)\exp(-s\mathbf{k}\mathbf{N}_0)\,\mathbf{Q}^T(t). \tag{7.2.53}$$

Note that tensor exponents can be written as

$$\exp(\mathbf{A}) = \mathbf{I} + \sum_{N=1}^{\infty}\frac{1}{N!}\mathbf{A}^N \tag{7.2.54}$$

and also as

$$[\mathbf{Q}\ \ \mathbf{N}_0\ \ \mathbf{Q}^T]^n = \mathbf{Q}\ \ \mathbf{N}^n\ \ \mathbf{Q}^T.$$

From these we finally obtain

$$\mathbf{C}_{(t)}(t-s) = \exp(-s\mathbf{k}\mathbf{N}^T)\exp(-s\mathbf{k}\mathbf{N}), \tag{7.2.55}$$

where

$$\mathbf{N} = \mathbf{Q}(t)\mathbf{N}_0\mathbf{Q}^T(t). \tag{7.2.56}$$

Equation (7.2.55) represents the constant-stretch history. To elucidate our discussion, some examples are subsequently presented. Consider

$$\mathbf{N} \neq 0 \quad \text{but} \quad \mathbf{N}^2 = 0.$$

This implies that $\mathbf{N}^n = 0$ for $n \geq 3$. Thus Eq. (7.2.55) becomes

$$\mathbf{C}_{(t)}(t-s) = \mathbf{I} - s\mathbf{k}(\mathbf{N} + \mathbf{N}^T) + (s\mathbf{k})^2\mathbf{N}^T\mathbf{N}. \tag{7.2.57}$$

Differentiating Eq. (7.2.57) with respect to s and setting $s = 0$, we have

$$\mathbf{A}_n(t) = \mathbf{C}_{(t)}(t-s)\,|_{s=0}, \tag{7.2.58}$$

$$\mathbf{A}_1 = 2\mathbf{D} = \mathbf{k}(\mathbf{N} + \mathbf{N}^T), \tag{7.2.59a}$$

$$\mathbf{A}_2 = 2\mathbf{k}^2\mathbf{N}^T\mathbf{N}, \tag{7.2.59b}$$

$$\mathbf{A}_3 = 0. \tag{7.2.59c}$$

To evaluate the components of $\mathbf{N}$, we consider the symmetric tensor $\mathbf{M} = \mathbf{N} + \mathbf{N}^T$, where $|\mathbf{N}| = 1$ and $\mathbf{N}^2 = 0$. It follows that

$$\mathbf{M}^2 = \mathbf{N}\mathbf{N}^T + \mathbf{N}^T\mathbf{N}.$$

Here, $\mathbf{M}^2$ is symmetric, tr $\mathbf{M}^2 = 2$, and

$$\det\mathbf{N} = (\det\mathbf{N}\det\mathbf{N}^T)^{1/2} = (\det\mathbf{N}^2)^{1/2} = 0,$$
$$\det\mathbf{M} = 0. \tag{7.2.60}$$

We may find an orthogonal basis for $\mathbf{M}$ such that

$$m_{ij} = \begin{bmatrix} a & 0 & 0 \\ 0 & b & 0 \\ 0 & 0 & 0 \end{bmatrix}, \tag{7.2.61}$$
$$\det\mathbf{M}^2 = a^2 + b^2 = 2.$$

From the Cayley–Hamilton theorem applied to $\mathbf{N}$ and by using $\mathbf{N}^2 = 0$, we have

$$I_2\mathbf{N} - I_3 I = 0, \tag{7.2.62}$$

where

$$I_2 = \frac{1}{2}[(\text{tr}\,\mathbf{N})^2 - \text{tr}\,\mathbf{N}^2],$$
$$I_3 = \det\mathbf{N}.$$

In view of Eqs. (7.2.60) and (7.2.62), we have

$$\frac{1}{2}(\text{tr}\,\mathbf{N})^2\mathbf{N} = 0,$$

which gives tr $\mathbf{N}$ = tr $\mathbf{M} = a + b = 0$. Thus,

$$a = \pm 1,$$
$$b = \pm 1.$$

This implies that a 45° rotation of the base vectors given in Eq. (7.2.61) leads to

$$\mathbf{M} = m_{ij} = \begin{bmatrix} 0 & 1 & 0 \\ 1 & 0 & 0 \\ 0 & 0 & 0 \end{bmatrix} \tag{7.2.63}$$

or

$$\mathbf{N} = n_{ij} = \begin{bmatrix} 0 & 1 & 0 \\ 0 & 0 & 0 \\ 0 & 0 & 0 \end{bmatrix}. \tag{7.2.64}$$

Note that these matrices satisfy

$$\mathbf{M} = \mathbf{N} + \mathbf{N}^T$$

and $|\mathbf{N}| = 1$ and $\mathbf{N}^2 = 0$. It is also seen that $\mathbf{A}_1$ and $\mathbf{A}_2$ in Eq. (7.2.59) assume the respective forms

$$\mathbf{A}_1 = \begin{bmatrix} 0 & \gamma & 0 \\ k & 0 & 0 \\ 0 & 0 & 0 \end{bmatrix}, \tag{7.2.65}$$

$$\mathbf{A}_2 = \begin{bmatrix} 0 & 0 & 0 \\ 0 & 2\gamma^2 & 0 \\ 0 & 0 & 0 \end{bmatrix}. \tag{7.2.66}$$

Flows that meet this criterion are known as *viscometric* flows.

Fourth-Order Flow. Various superposing viscometric flows can be generated from (Huilgol, 1968)

$$\mathbf{N}^3 = 0, \quad \mathbf{N}^2 \neq 0, \tag{7.2.67}$$

where

$$\mathbf{N} = \begin{bmatrix} 0 & \alpha_1 & \alpha_2 \\ 0 & 0 & \alpha_3 \\ 0 & 0 & 0 \end{bmatrix}, \tag{7.2.68}$$

with $\alpha_i \alpha_i = 1$, α_1 and α_3 being different from zero.

Extensional Flow. A steady extensional flow is represented by $\mathbf{N}^m \neq 0$ for any value of m. For illustration, we consider a simple shear flow (linear Couette flow),

$$v^1 = \gamma z^2,$$
$$v^2 = v^3 = 0.$$

Similarly,

$$\frac{dz^1}{d\tau} = z^2,$$

$$\frac{dz^2}{d\tau} = \frac{dz^3}{d\tau} = 0.$$

Noting that $z^i = x^i$ for $\tau = t$, we find that these relations lead to

$$z^1 = x^1 - \gamma x^2(t - \tau),$$

$$z^2 = x^2,$$

$$z^3 = x^3,$$

$$\frac{\partial z^i}{\partial x^j} = \begin{bmatrix} 1 & -\gamma(t-\tau) & 0 \\ 0 & 1 & 0 \\ 0 & 0 & 1 \end{bmatrix},$$

$$\mathbf{C} = \mathbf{F}^T\mathbf{F} = \begin{bmatrix} 1 & -\gamma(t-\tau) & 0 \\ -\gamma(t-\tau) & 1+\gamma^2(t-\tau^2) & 0 \\ 0 & 0 & 1 \end{bmatrix}. \tag{7.2.69}$$

The Rivlin–Ericksen tensors can be calculated from

$$\mathbf{A}_N(t) = \frac{d^N\mathbf{C}(t)}{dt^N} = \overset{(N)}{\mathbf{C}}(t), \tag{7.2.70}$$

$$\mathbf{A}_1 = \overset{(1)}{\mathbf{C}}(\tau) = \begin{bmatrix} 0 & \gamma & 0 \\ \gamma & -2\gamma^2(t-\tau) & 0 \\ 0 & 0 & 0 \end{bmatrix}, \tag{7.2.71a}$$

$$\mathbf{A}_2 = \overset{(2)}{\mathbf{C}}(\tau) = \begin{bmatrix} 0 & 0 & 0 \\ 0 & 2\gamma^2 & 0 \\ 0 & 0 & 0 \end{bmatrix}, \tag{7.2.71b}$$

$$\mathbf{A}_3 = \mathbf{A}_4 = \cdots = 0. \tag{7.2.71c}$$

For the kinematics of special symmetric flow toward a point sink, where $v^r = -ar^2$, $v^\theta = v^\phi = 0$, the Cauchy–Green matrix can be shown to be of the form

$$\mathbf{C} = \begin{bmatrix} \left[\dfrac{1+3a(t-\tau)}{r_t^3}\right]^{-4/3} & 0 & 0 \\ 0 & \left[r_t^3 + 3a(t-\tau)\right]^{2/3} & 0 \\ 0 & 0 & \left[r_t^3 + 3a(t-\tau)\right]^{2/3}\sin^2\theta \end{bmatrix}, \tag{7.2.72}$$

and the Rivlin–Ericksen tensors become

$$\mathbf{A}_1 = \overset{(1)}{\mathbf{C}}(\tau)$$

$$= \begin{bmatrix} \frac{4a}{r_t^3}\left[1+\frac{3a(t-\tau)}{r_2^3}\right]^{-1/3} & 0 & 0 \\ 0 & \frac{-2a}{r_t^3}\left[1+\frac{3a(t-\tau)}{r_2^3}\right]^{-1/3} & 0 \\ 0 & 0 & \frac{-2a}{r_t^3}\left[1+\frac{3a(t-\tau)}{r_t^3}\right]^{1/3} \end{bmatrix}. \tag{7.2.73}$$

We can obtain subsequent Rivlin–Ericksen tensors by differentiating Eq. (7.2.73), and we note that high-order tensors will not vanish, in contrast to the case of a simple shear.

7.2.3 Simple Fluids

7.2.3.1 General

Recall that in Chap. 5 a Newtonian fluid was defined by Eq. (5.2.39), which is a special case of the Reiner–Rivlin equation [Eq. (5.2.7)]. Any flow that fails to obey Eq. (5.2.39) may be defined as non-Newtonian. It is well known that memory fluids do not obey Eq. (5.2.39), because the stress is influenced by deformation before the instant of observation. The mechanical behavior of real materials can be described by the axiomatic approach, which provides rigor and generality but often lacks utility for solution. On the other hand, a phenomenological approach can be used to facilitate engineering solutions, although lacking the advantage of mathematical rigor.

Elastic solids possess a perfect memory and thus remember a preferred shape forever, whereas viscous fluids (or Reiner–Rivlin fluids) have no memory. It is logical to assume that certain materials somewhere between solids and fluids may have some memory of past deformation. This is a way of stating the concept of fading memory. Such materials, known as memory fluids, remember all shapes assumed in the past, but not forever.

Let us now consider a material for which the stress at a material point is determined by the history of the deformation gradient, evaluated at the material point in question over all past times up to and including the present, as characterized by Eq. (7.2.2). Materials of this kind are called *simple fluids*. In the discussion of a simple-fluid theory, we have to draw on axiomatic principles such as determinism of stress, local action, and fading memory. In addition, we draw on the principle of the nonexistence of a natural state, which defines fluidity as having no preferred configuration or natural state. This implies that all possible configurations are essentially equivalent, so that any difference in stress is due only to a difference in the history of deformation. It follows that every simple fluid is isotropic. Finally, the principle of fading memory is needed. This principle states that the influence of past deformations on the present stress is weaker for the distant past than for the recent past. This provides the concept of a *natural time* for any given material. For example, the Reiner–Rivlin theory, which is based on purely viscous fluids, has zero natural time.

7.2.3.2 Fading Memory

Let us now consider a material for which the stress at a material point is determined by the history of the deformation gradient over all past times up to and including the present, characterized by Eq. (7.2.2). A material of this kind is known as a simple material. For a constant-density simple fluid, we write

$$\sigma(t) = \overset{\infty}{\underset{s=0}{\hat{F}}}[\mathbf{F}(t-s)], \tag{7.2.74}$$

$$|\hat{F}(t-s)| = 1, \tag{7.2.75}$$

where $\mathbf{F}$ is a variable over time.

It follows from the principle of material objectivity that

$$\mathbf{Q}(t)\boldsymbol{\sigma}(t)\mathbf{Q}^T(t) = \overset{\infty}{\underset{s=0}{\hat{F}}}[\mathbf{Q}(t-s)\mathbf{F}_t(s)\mathbf{F}(t)], \tag{7.2.76}$$

where $\mathbf{F}_t(s) \equiv \mathbf{F}_t(t-s)$ and $\mathbf{Q}(t-s)$ is an arbitrary orthogonal matrix. Let us assume that

$$\mathbf{Q}(t-s) = \mathbf{R}_t^T(s). \tag{7.2.77}$$

From the polar decomposition theorem,

$$\mathbf{F}_t(s) = \mathbf{R}_t(s)\mathbf{U}_t(s), \tag{7.2.78}$$

and for

$$\mathbf{U}_t(t) = \mathbf{R}_t(t) = \mathbf{I} \tag{7.2.79}$$

we obtain

$$\boldsymbol{\sigma}(t) = \overset{\infty}{\underset{s=0}{\hat{F}}}[\mathbf{U}_t(s)\mathbf{R}(t)\mathbf{U}(t)] \tag{7.2.80}$$

or

$$\boldsymbol{\sigma}(t) = \overset{\infty}{\underset{s=0}{\hat{G}}}[\mathbf{C}_t(s),\ \mathbf{U}(t)]. \tag{7.2.81}$$

We note that the response of a fluid is sensitive to the history of its deformation but that the present state of deformation is important only to the extent that it affects the density. This will permit us to replace $\mathbf{U}(t)$ with $\rho(t)$, and it is convenient to isolate a scalar-valued function of the density as an isotropic portion of the response functional:

$$\mathbf{T}(t) = -p(\rho)\mathbf{I} + \overset{\infty}{\underset{s=0}{\hat{H}}}[\mathbf{C}(s),\ \rho]. \tag{7.2.82}$$

For an incompressible flow in which density is constant, we have

$$\mathbf{T}(t) = -p\mathbf{I} + \overset{\infty}{\underset{s=0}{\hat{H}}}[\mathbf{C}_t(s)]. \tag{7.2.83}$$

Furthermore, we assert that the stress in a simple fluid that has always been at rest is isotropic. Conversely, a simple fluid cannot sustain indefinitely a nonisotropic stress without eventually starting to flow (Coleman, Markovit, and Noll, 1966). Thus we have

$$\boldsymbol{\sigma}(t) = \overset{\infty}{\underset{s=0}{\hat{H}}}[\mathbf{0}(s)] = p\mathbf{I}, \tag{7.2.84}$$

where $\mathbf{0}(s)$ is a tensor-valued function whose value is the zero tensor for any value of the argument s.

With these preliminaries in mind, in the next subsection we examine constitutive equations, including the various aspects of a simple fluid.

7.2.4 Constitutive Equations

7.2.4.1 Integral Equations

As an alternative to the functional approach of Noll (1958), Green and Rivlin (1957) utilize a series of integrals over past time in order to account for memory effects. The multiple-integral approach is based on the approach of Frechet (Volterra, 1959); every function $\hat{H}_{t=a}^{t=b}[y(t)]$ takes the form

$$G_0 + \int_a^b \hat{G}_1(s)z(x)\,dx + \int_a^b \int_a^b \hat{G}_2(x_1, x_2)z(x_1)z(x_2)\,dx_1dx_2$$
$$+ \cdots + \int_a^b \cdots \int_a^b \hat{G}_n(x_1, \ldots, x_n)z(x_1)\cdots z(x_n)\,dx_1 \ldots dx_2 \ldots. \tag{7.2.85}$$

Extending this form to tensor-valued functionals for a simple fluid, we write

$$\boldsymbol{\sigma}(t) = \overset{\infty}{\underset{s=0}{\hat{H}}}[\mathbf{C}_t(s)] = \overset{\infty}{\underset{s=0}{\hat{G}}}[\mathbf{G}_t(s)], \tag{7.2.86}$$

where $\mathbf{G}_t(s) = \mathbf{C}_t(s) - \mathbf{I}$. Expanding Eq. (7.2.86) yields

$$\boldsymbol{\sigma}(t) = \sum_{n=1}^{\infty} \int_0^{\infty} \cdots \int_0^{\infty} \mathbf{M}(s_1 \ldots s_n)\mathbf{G}_t(s_1) \ldots \mathbf{G}_t(s_n) ds_1 \ldots ds_n. \tag{7.2.87}$$

In view of the invariance requirements of the principle of material objectivity (Truesdell and Noll, 1965), we write

$$\boldsymbol{\sigma}(t) = \int_0^\infty M_1(s)\mathbf{G}_t(s)\,ds + \int_0^\infty \int_0^\infty \mathbf{M}_2(s_1, s_2)\mathbf{G}_t(s_2)\,ds_1 dx_2$$
$$+ \int_0^\infty \int_0^\infty \int_0^\infty \{M_1(s_1, s_2, s_3)\mathbf{G}_t(s_1)\,\mathrm{tr}\,[\mathbf{G}_t(s_2)\mathbf{G}_t(s_3)]$$
$$+ M_4(s_1, s_2, s_3)\mathbf{G}_t(s_1)\mathbf{G}_t(s_3)\}ds_1\,ds_2\,ds_3 + \cdots +, \qquad (7.2.88)$$

where M_i must possess certain symmetry properties: specifically, $M_2(s_1, s_2) = M_2(s_2, s_1)$, and so forth.

Coleman and Noll (1959) have shown how the theory of simple fluids can be truncated systematically to the finite linear and infinitesimal theories of viscoelasticity. Toward this end, we maintain $\mathbf{G}_t(s)$ as the measure of strain and require the allowable motions to be such that all integrals in Eq. (7.2.88) beyond the first are negligible because of a sufficiently small norm for $\mathbf{G}_t(s)$. Thus we have

$$\boldsymbol{\sigma}(t) - \int_0^\infty M_1(s)\mathbf{G}_t(s)\,ds, \qquad (7.2.89)$$

which represents the finite linear viscoelasticity.

7.2.4.2 Differential Equations

It has been shown (Rivlin and Ericksen, 1955) that the stress tensor can be written as

$$\boldsymbol{\sigma}(t) = \mathbf{M}(\mathbf{A}_1, \mathbf{A}_2, \ldots, \mathbf{A}_n). \qquad (7.2.90)$$

In accordance with material objectivity, we must satisfy

$$\mathbf{Q} \cdot \mathbf{M}(\mathbf{A}_1, \ldots, \mathbf{A}_n) \cdot \mathbf{Q}^T = \mathbf{M}(\mathbf{Q} \cdot \mathbf{A}_1 \mathbf{Q}^T n \mathbf{Q}\mathbf{A}_n \mathbf{Q}^T). \qquad (7.2.91)$$

Here, $\mathbf{M}$ is isotropic and tensors $\mathbf{A}_1, \ldots, \mathbf{A}_n$ are symmetric. Thus, if $n = 2$, we obtain

$$\boldsymbol{\sigma}(t) = \alpha_0\mathbf{I} + \alpha_1\mathbf{A}_1 + \alpha_2\mathbf{A}_2 + \alpha_3\mathbf{A}_1^2 + \alpha_4\mathbf{A}_2^2 + \alpha_5\left(\mathbf{A}_1 \cdot \mathbf{A}_2\right.$$
$$\left.+ \mathbf{A}_2 \cdot \mathbf{A}_1\right) + \alpha_6\left(\mathbf{A}_1^2 \cdot \mathbf{A}_2 + \mathbf{A}_2 \cdot \mathbf{A}_1^2\right) + \alpha_7\left(\mathbf{A}_1 \cdot \mathbf{A}_2^2\right.$$
$$\left.+ \mathbf{A}_2^2 \cdot \mathbf{A}_2^2 \cdot \mathbf{A}_1\right) + \alpha_8\left(\mathbf{A}_1^2 \cdot \mathbf{A}_2^2 + \mathbf{A}_2^2 \cdot \mathbf{A}_1^2\right). \qquad (7.2.92)$$

Note that α_0 is an arbitrary scalar and that other scalar coefficients are invariant functions of $\mathbf{A}_1$ and $\mathbf{A}_2$. It is readily seen that, for $n \geq 2$, if the motion is brought to a halt, all tensors become zero, incapable of exhibiting stress relaxation. This is contrary to the case for most polymeric liquids or simple fluids.

Let us now examine a motion that has a history of the Cauchy tensor, $\mathbf{C}(\tau)$, continuous with all its derivatives at all times, and a second motion $\mathbf{C}'(\tau)$ having the same values of $\mathbf{C}(\tau)$ only in the interval $t' \leq \tau \leq t$. At $\tau = t'$, $\mathbf{C}'(\tau)$ or one of its derivatives will be discontinuous. On the other hand, the same stress is predicted at $\tau = t$, because the $\mathbf{A}_n$ is the same for both motions. This is not so if the general constitutive equation of the simple fluids [Eq. (7.2.81)] is used. However, one of the two motions, the one that has a history that is continuous with respect to all its derivatives, and if not that, the stress calculated from Eq. (7.2.81), must coincide with that from Eq. (7.2.90) with $n \to \infty$. Also, for $t' \ll t$ the past history is considered negligible; it is possible that the stress from Eqs. (7.2.81) and (7.2.90) is thus the nth-order approximation of Eq. (7.2.81):

$$\mathbf{T} = -p\mathbf{I}, \text{ zeroth order,} \tag{7.2.93a}$$

$$\mathbf{T} = -p\mathbf{I} + 2\mu\mathbf{D}, \text{ first order,} \tag{7.2.93b}$$

$$\mathbf{T} = -p\mathbf{I} + 2\mu_0\mathbf{D} + \beta_1\mathbf{D}^2 + \beta_2\mathbf{A}_2, \text{ second order,} \tag{7.2.93c}$$

and so forth. Given any history that is continuous at the instant of observation, the retarded history obtained from it by the factor α is a history that with decreasing α, becomes continuous with all its derivatives on a wider and wider interval of time previous to the instant of observation. This implies that a singular point in the past history tends to disappear with decreasing α, and consequently Eq. (7.2.90) becomes increasingly more adequate to predict the correct response.

7.2.4.3 Convective Coordinate Approach

If constitutive equations involve the history of deformation of material elements, then the equations should be formulated with respect to these elements. This process makes the physics of the problem more evident and thus removes the necessity of an explicit material objectivity requirement. On the basis of these arguments, Oldroyd (1950) and, subsequently, Lodge (1964) proposed a *convective subjectivity* that is automatically satisfied.

We consider the time derivative of vector $\mathbf{u}(x^i, \tau)$ with respect to τ as viewed by an observer in the frame with body-fixed basis $\mathbf{e}_i(\tau)$, which may be connected to the space-fixed frame with basis $\mathbf{e}_i(t)$ at $\tau = t$ and coordinates z^i:

$$\mathbf{u}(\xi^i, \tau) = u^j(\xi^i, \tau)\mathbf{e}_j(\tau) = u^j(z^i, \tau, t)\mathbf{e}_j(t), \tag{7.2.94}$$

where $\tau = t$ and $x^i = z^i$. Let us now assume that an observer in the body-fixed frame with basis $\mathbf{e}_j(\tau)$, which appears constant to an observer in that frame, records a time rate of change $[\partial \mathbf{u}(x^i, \tau)/\partial \tau]_{\tau=t}$of $\mathbf{u}$. We then wish to express components of the derivative in terms of quantities defined in the space-fixed frame basis $\mathbf{e}_i(t)$. Toward this end, we begin with the following preliminaries:

$$\mathbf{e}_i(\xi^j, \tau) = \frac{\partial \mathbf{r}}{\partial \xi^i}, \tag{7.2.95}$$

where $\mathbf{r}$ is the position vector to a material point identified by ξ^i, and

$$e_{ij}(\xi^k, \tau) = \mathbf{e}_i(\tau) \cdot \mathbf{e}_j(\tau), \tag{7.2.96}$$

$$\left(\frac{\partial \mathbf{r}}{\partial \tau}\right)_{\tau=t} = \mathbf{v}(\mathbf{z}, t), \tag{7.2.97a}$$

$$\mathbf{z} = \mathbf{z}^i \mathbf{e}_i(t). \tag{7.2.97b}$$

Then

$$\left(\frac{\partial \mathbf{e}_i}{\partial \tau}\right)_{\tau=t} = \left(\frac{\partial \mathbf{v}}{\partial \xi^i}\right)_{\tau=t} = \left(\frac{\partial \mathbf{v}}{\partial z^i}\right) = v^j_{,i}\mathbf{e}_j(\mathbf{z}, t). \tag{7.2.98}$$

In view of Eqs. (7.2.98) and (7.2.94), we obtain for $\tau = t$

$$\left\{\left[\frac{\partial u^j(\xi^i, \tau)}{\partial \tau}\right]_{\tau=t} + v^j_{,m}u^m\right\}\mathbf{e}_j(t) = \left[\frac{\partial u^j(z^i, \tau, t)}{\partial \tau}\right]\mathbf{e}_j(t), \tag{7.2.99}$$

with ξ^i = constant, $\tau = t$. The time derivative on the right-hand side is simply the usual material derivative, whereas the time derivative on the left-hand side is known as the convected derivative, denoted by $\hat{D}$ as follows:

$$\left[\frac{\partial u^j(\xi^i, \tau)}{\partial \tau}\right]_{\tau=t} = \frac{\hat{D}u^j(z^i, t)}{\hat{D}t} = \frac{\partial u_j}{\partial t} + v^i u^j_{,i} - v^j_{,i}u^i. \tag{7.2.100}$$

We note that this is not, in general, the same as a time derivative seen by an observer in a body-fixed frame with basis $\mathbf{e}^i(\tau)$. Expressing components of that time derivative in a space-fixed frame with basis $\mathbf{e}^i(t)$, we obtain

$$\frac{\hat{D}u^j(z^i, t)}{\hat{D}t} = \frac{\partial u_j}{\partial t} + v^i u_{j,i} + v^i_{,j}u_i. \tag{7.2.101}$$

For second-order tensors, we denote $\hat{D}a_{ij}/\hat{D}t$ as the time derivative of some tensor $\mathbf{A}(\xi^i, \tau)$ as seen by someone in the body-fixed frame with basis $\mathbf{e}^i(\tau)$, but expressed at time t in terms of components in a space-fixed frame with basis $\mathbf{e}^i(t)$ so that $\mathbf{A}(\xi^i, \tau) = a_{ij}(z^i, \tau, t)\mathbf{e}^i \times \mathbf{e}^j$. This leads to

$$\frac{\hat{D}a_{ij}}{\hat{D}t} = \frac{\partial a_{ij}}{\partial t} + v^k a_{ij,k} + v^k_{,i}a_{kj} + v^k_{,j}a_{ik}. \tag{7.2.102}$$

Similarly, for a time derivative in a frame with basis $\mathbf{e}_i(\tau)$, components expressed in terms of the space-fixed frame with basis $\mathbf{e}_i(t)$ are

$$\frac{\hat{D}a^{ij}}{\hat{D}t} = \frac{\partial a^{ij}}{\partial t} + v^k a^{ij}_{,k} - v^i_{,k}a^{kj} - v^j_{,k}a^{ik}. \tag{7.2.103}$$

In general, it can be shown that [see Eq. (6.2.21)]

$$\begin{aligned}\frac{\hat{D}a^{i_1 i_2 \ldots i_m}_{j_1 j_2 \ldots j_n}}{\hat{D}t} &= \frac{\partial a^{i_1 i_2 \ldots i_m}_{j_1 j_2 \ldots j_n}}{\partial t} + v^k\left(a^{i_1 i_2 \ldots i_m}_{j_1 j_2 \ldots j_n}\right)_{,k} \\ &+ v^k_{,j_1}a^{i_1 i_2 \ldots i_m}_{k j_2 \ldots j_n} + \cdots + v^k_{,j_n}a^{i_1 i_2 \ldots i_m}_{j_1 j_2 \ldots k} - v^{i_1}_{,k}a^{k i_2 \ldots i_m}_{j_1 j_2 \ldots j_n} - \ldots - v^{i_m}_{,k}a^{i_1 i_2 \ldots k}_{j_1 j_2 \ldots j_n}.\end{aligned} \tag{7.2.104}$$

Furthermore, at $\tau = t$, $e_{ij} = \mathbf{e}_i(t) \cdot \mathbf{e}_j(t)$, and $e_{ij,k} = 0$, we find that

$$\frac{\hat{D}a_{ij}}{\hat{D}t} = v_{i,j} + v_{j,i} = 2d_{ij}. \tag{7.2.105}$$

This indicates that the time rate of change of the body-fixed metric tensor is indeed just twice the local value of the rate of stretch. In fact, we have

$$\frac{\hat{D}^n e_{ij}}{\hat{D}t^n} = A_{ij}^{(n)}, \tag{7.2.106}$$

where $A_{ij}^{(n)}$ is the covariant component of the nth Rivlin–Ericksen tensor. Note also that

$$\frac{\hat{D}e^{ij}}{\hat{D}t} = -2d^{ij}. \tag{7.2.107}$$

It is concluded that the operation $\hat{D}/\hat{D}t$ does not commute with the interchange of superscripts for subscripts, i.e.,

$$\frac{\hat{D}a_{km}}{\hat{D}t} e^{ik} e^{jm} \neq \frac{\hat{D}a^{ij}}{\hat{D}t}. \tag{7.2.108}$$

We further demonstrate that, in general, $\hat{D}G_{ij}/\hat{D}t$ is not equal to $\hat{D}G^{ij}/\hat{D}t$. Toward this end, we begin with two adjacent surfaces in a material defined as

$$S(\xi) = k,$$

where k characterizes each surface. Then the distance between the two surfaces is expressed by

$$dh = \mathbf{n} \cdot d\boldsymbol{\xi}, \tag{7.2.109}$$

where $\mathbf{n}$ is the normal vector between the two surfaces,

$$\mathbf{n} = \frac{\nabla S}{(\nabla S \cdot \nabla S)^{1/2}},$$

and furthermore, we have

$$\nabla S \cdot d\boldsymbol{\xi} = dk,$$

$$dh = \frac{dk}{(\nabla S \cdot \nabla S)^{1/2}}, \tag{7.2.110}$$

where

$$\nabla S \cdot \nabla S = G^{ij} \frac{\partial S}{\partial \xi^i} \frac{\partial S}{\partial \xi^j}$$

and thus

$$\left(\frac{dk}{dh}\right)^2 = G^{ij}(\xi, \tau) \frac{\partial S}{\partial \xi^i} \frac{\partial S}{\partial \xi^j}. \tag{7.2.111}$$

Here, $G^{ij}(\xi, \tau)$ differs from the contravariant metric tensor G^{ij} in such a way that the time rate of change of the distance between material surfaces will not be the same as the corresponding change for material points.

Let us consider a basis $\mathbf{e}_i(\tau) + g_{ij}\mathbf{e}^j(\tau)$ where $g_{ij} = \mathbf{e}_i(t) \cdot \mathbf{e}_j(t)$ such that its time derivative becomes

$$\frac{d}{d\tau}(\mathbf{e}_i + g_{ij}\mathbf{e}^j) = 2W\mathbf{e}_i$$

where W is the antisymmetric rotational tensor. We now consider another basis, $\mathbf{e}_i(t) = \frac{1}{2}[\mathbf{e}_i(\tau) + g_{ij}\mathbf{e}^j(\tau)]$, which shares in the rotation of the material but not in its deformation, and its time derivatives in this corotational frame, which are expressed in a space-fixed frame with basis $\mathbf{e}_i(t) = \frac{1}{2}[\mathbf{e}_i(t) + g_{ij}\mathbf{e}^j(t)]$. Note that

$$\mathbf{A} = a^{ij}(\xi^k, \tau)\hat{\mathbf{e}}_i(\tau) \times \hat{\mathbf{e}}_j(\tau) = a^{ij}(z^k, t, t)\hat{\mathbf{e}}_i(t) \times \hat{\mathbf{e}}_i(t).$$

Also, note that the corotational time derivative is expressed in a space-fixed frame by $\hat{D}a^{ij}/\hat{D}t$:

$$\frac{\hat{D}a^{ij}}{\hat{D}t} = \frac{\partial a^{ij}}{\partial t} + v^k a^{ij}_{,k} - w_k a^{ik}_{,j} - w^j_{,k} a^{ik}, \tag{7.2.112}$$

where $\hat{D}/\hat{D}t$ is known as the Jaumann derivative. For $\tau = t$ the reciprocal basis $\mathbf{e}^i$ is

$$\mathbf{e}^i(\tau) = \frac{1}{2}[\mathbf{e}^i(\tau) + g^{ij}\mathbf{e}_j(\tau)],$$

which leads to

$$\frac{\hat{D}a_{ij}}{\hat{D}t} = \frac{\partial a_{ij}}{\partial t} + v^k a_{ij,k} - w^k_i a_{kj} w^k_j a_{ik}, \tag{7.2.113}$$

$$\frac{\hat{D}a_{ij}}{\hat{D}t} - g_{ik}g_{jn}\frac{\hat{D}a^{kn}}{\hat{D}t} = 0. \tag{7.2.114}$$

It was further shown by Oldroyd (1950) that integral expression for memory effects may be extended to finite deformations. Toward this end, we assume that it is the history of a material point, not a spatial point, that is recorded. We consider the stress tensor $\boldsymbol{\sigma}$, where

$$\boldsymbol{\sigma} = \sigma_{ij}\mathbf{e}^i \times \mathbf{e}^j,$$

$$\sigma_{ij}(\xi^k, t) = \int_{-\infty}^{t} \psi(t-\tau)\frac{\partial G_{ij}}{\partial \tau} d\tau. \tag{7.2.115}$$

We further consider a tensor $\mathbf{D}$, defined by

$$\mathbf{D} = \frac{1}{2}\frac{\partial G_{ij}}{\partial \tau}\mathbf{e}^i \times \mathbf{e}^j = d_{ij}(\hat{z}^k, \tau, t)\mathbf{e}^i(\hat{z}^k, t) \times \mathbf{e}^j(\hat{z}^k, t),$$

where we require that $\tau = t, \xi^k = z^k$. The $\hat{z}^k$ are coordinates associated with the space-fixed basis $\mathbf{e}_i(\hat{z}^k, t)$. Thus we have

$$\hat{\sigma}_{ij}(z^m, t) = 2\int_{-\infty}^{t} \psi(t-\tau)\frac{\partial \hat{z}^k}{\partial z^i}\frac{\partial \hat{z}^n}{\partial z^j} d_{kn}(\hat{z}^m, \tau)\, d\tau. \tag{7.2.116}$$

Note that $\sigma_{ij} = \hat{\sigma}_{ij}$ at $\tau = t$. It can further be shown that

$$\hat{\sigma}^{ij}(z^m, t) = 2\int_{-\infty}^{t} \psi(t-\tau)\frac{\partial z^i}{\partial \hat{z}^k}\frac{\partial z^j}{\partial \hat{z}^n} d^{kn}\, d\tau. \tag{7.2.117}$$

To illustrate the utilization of the foregoing approach, a simple example is examined in the following subsection.

7.2.4.4 Illustrations

In applications of the theory of non-Newtonian fluids in industrial problems, we may cite polymer processing that involves irregular geometries (Walters and Schowalter, 1972). Toward this end, we examine a rheological model based on convective coordinates.

Consider the constitutive relations given by

$$\sigma_{ij} = -p\delta_{ij} + \hat{\sigma}_{ij}, \tag{7.2.118}$$

$$\hat{\sigma}_{ij} + \lambda_1 \frac{\hat{D}\hat{\sigma}^{ij}}{\hat{D}t} + \mu_0 \hat{\sigma}^k_k d^{ij} = 2\eta_0\left(1 + \lambda_2 \frac{\hat{D}}{\hat{D}t}\right) d^{ij}, \tag{7.2.119}$$

where λ_1, λ_2, and μ_0 are material constants, each with the dimensions of time (with $\lambda_1 \geq \lambda_2 \geq 0,\ \mu_0 \geq 0$), and η_0 is the limiting viscosity at small rates of shear.

Following Walters (1975) and Perera and Walters (1980), we proceed with

$$\sigma_{12} = \gamma\eta(\gamma), \tag{7.2.120}$$

$$\eta(\gamma) = \eta_0 \frac{1 + \lambda_2\mu_0\gamma^2}{1 + \lambda_1\mu_0\gamma^2}, \tag{7.2.121}$$

$$\sigma_{11} - \sigma_{12} = \frac{2\eta_0\gamma^2(\lambda_1 - \lambda_2)}{1 + \mu_0\lambda_1\gamma^2}, \tag{7.2.122}$$

$$\sigma_{22} - \sigma_{33} = 0, \tag{7.2.123}$$

where γ is a constant shear rate associated with a steady simple shear flow with velocity components v_i:

$$v_1 = \gamma X_2, \quad v_2 = v_3 = 0,$$

We note that high-elasticity and long-range memory effects are associated with relatively high values of the relaxation time λ_1. The comparison between the behavior of an elastic liquid and that of an inelastic liquid having the same $\eta(\lambda)$ function may provide a meaningful assessment of relaxation effects arising from fluid elasticity.

First, we take $\lambda_1 = O(\lambda)$, $\lambda_2 = O(\lambda)$, and $\mu_0 = O(1/\lambda)$, where λ is a characteristic time, and operate on both sides of Eq. (7.2.119) with $1 - \lambda_1(\hat{D}/\hat{D}t)$:

$$\hat{\sigma}^{ij} = \left(2\eta_0 - \mu_0\hat{\sigma}_k^k\right) d^{ij} - 2\eta_0(\lambda_1 - \lambda_2)\frac{\hat{D}d^{ij}}{\hat{D}t} + \lambda_1\mu_0\frac{\hat{D}}{\hat{D}t}\left(\hat{\sigma}_k^k d^{ij}\right). \qquad (7.2.124)$$

In view of Eq. (7.2.114) and the continuity equation, we have

$$\hat{\sigma}_k^k = -2\eta_0\frac{(\lambda_1 - \lambda_2)\left(\hat{D}d_k^k/\hat{D}t\right)}{1 - \lambda_1\mu_0\left(\hat{D}d_k^k/\hat{D}t\right)}. \qquad (7.2.125)$$

This gives

$$\sigma^{ij} = 2\eta_0 d^{ij} + 2\eta_0\mu_0\frac{(\lambda_1 - \lambda_2)\left(\hat{D}d_k^k/\hat{D}t\right) d^{ij}}{(1 - \lambda_1\mu_0)\left(\hat{D}d_k^k/\hat{D}t\right)} - 2\eta_0(\lambda_1 - \lambda_2)\left(\hat{D}d^{ij}/\hat{D}t\right). \qquad (7.2.126)$$

For $\lambda \to 0$ and isochoric flow, where

$$\hat{D}d_k^k/\hat{D}t = -2d_m^k d_k^m, \qquad (7.2.127)$$

it follows that

$$\hat{\sigma}^{ij} = 2\eta_0\frac{\left(1 + 2\lambda_2\mu_0 d_m^k d_k^m\right) d^{ij}}{1 + 2\lambda_1\mu_0 d_m^k d_k^m}. \qquad (7.2.128)$$

This is equivalent to an explicit expression,

$$\hat{\sigma}^{ij} = 2\eta(I_2)d_{ij}, \qquad (7.2.129)$$

where I_2 is the second invariant of d_{ij}:

$$I_2 = 2d_m^k d_k^m, \qquad (7.2.130)$$

and η is as given by Eq. (7.2.121).

Explicit forms of the governing equations for a two-dimensional elasticoviscous problem are listed as follows:

Continuity equation

$$\frac{\partial u}{\partial x} + \frac{\partial v}{\partial y} = 0. \qquad (7.2.131)$$

Stress equation of motion

$$\rho\left(u\frac{\partial u}{\partial x} + v\frac{\partial u}{\partial x}\right) = -\frac{\partial p}{\partial x} + \frac{\partial\hat{\sigma}_x}{\partial x} + \frac{\partial\hat{\sigma}_{xy}}{\partial y}, \qquad (7.2.132a)$$

$$\rho\left(u\frac{\partial v}{\partial x} + v\frac{\partial v}{\partial y}\right) = -\frac{\partial p}{\partial y} + \frac{\partial\hat{\sigma}_{xy}}{\partial x} + \frac{\partial\hat{\sigma}_y}{\partial y}. \qquad (7.2.132b)$$

Rheological equations of state

$$\hat{\sigma}_x\left[1+(\mu_0-2\lambda_1)\frac{\partial u}{\partial x}\right]+\lambda_1\left(u\frac{\partial\hat{\sigma}_x}{\partial x}+v\frac{\partial\hat{\sigma}_x}{\partial y}\right)-2\lambda_1\frac{\partial u}{\partial y}\hat{\sigma}_{xy}+\mu_0\frac{\partial u}{\partial x}\hat{\sigma}_y$$
$$=2\eta_0\frac{\partial u}{\partial x}+2\lambda_2\eta_0\left[u\frac{\partial^2 u}{\partial x^2}+v\frac{\partial^2 u}{\partial x\partial y}-2\left(\frac{\partial u}{\partial x}\right)^2-\frac{\partial u}{\partial y}\left(\frac{\partial u}{\partial y}+\frac{\partial v}{\partial x}\right)\right], \tag{7.2.133a}$$

$$\mu_0\frac{\partial v}{\partial x}\hat{\sigma}_x-2\lambda_1\frac{\partial v}{\partial x}\hat{\sigma}_{xy}+\hat{\sigma}_y\left[1+(\mu_0-2\lambda_1)\frac{\partial v}{\partial y}\right]+\lambda_1\left(1+\frac{\partial y}{\partial x}+v\frac{\partial\hat{\sigma}_{xy}}{\partial y}\right)$$
$$=2\eta_0\frac{\partial v}{\partial y}+2\lambda_2\eta_0\left[u\frac{\partial^2 v}{\partial x\partial y}+v\frac{\partial^2 v}{\partial y^2}-2\left(\frac{\partial v}{\partial y}\right)^2-\frac{\partial u}{\partial x}\left(\frac{\partial v}{\partial x}+\frac{\partial u}{\partial y}\right)\right], \tag{7.2.133b}$$

$$\hat{\sigma}_x\left[\mu_0\left(\frac{\partial u}{\partial y}+\frac{\partial v}{\partial x}\right)-\lambda_1\frac{\partial u}{\partial x}\right]+\hat{\sigma}_y\left[\frac{\mu_0}{2}\left(\frac{\partial u}{\partial y}+v\frac{\partial v}{\partial x}\right)-\lambda_1\frac{\partial u}{\partial y}\right]$$
$$+\hat{\sigma}_{xy}+\lambda_1\left(u\frac{\partial\hat{\sigma}_{xy}}{\partial x}+v\frac{\partial\hat{\sigma}_{xy}}{\partial y}\right)$$
$$=\eta_0\left(\frac{\partial u}{\partial y}+\frac{\partial v}{\partial x}\right)+\lambda_2\eta_0\left[\left(u\frac{\partial}{\partial x}+v\frac{\partial}{\partial y}\right)\left(\frac{\partial u}{\partial y}+\frac{\partial u}{\partial x}\right)\right.$$
$$\left.-2\frac{\partial v}{\partial x}\frac{\partial u}{\partial x}-2\frac{\partial u}{\partial y}+\frac{\partial v}{\partial y}\right]. \tag{7.2.133c}$$

Viscoinelastic equations

$$\hat{\sigma}_x=2\eta(I_2)\frac{\partial u}{\partial x}, \tag{7.2.134a}$$

$$\hat{\sigma}_y=2\eta(I_2)\frac{\partial v}{\partial y}, \tag{7.2.134b}$$

$$\hat{\sigma}_{xy}=\eta(I_2)\left(\frac{\partial u}{\partial y}+\frac{\partial v}{\partial x}\right), \tag{7.2.134c}$$

$$I_2=\left(\frac{\partial v}{\partial x}+\frac{\partial u}{\partial y}\right)^2-4\frac{\partial u}{\partial x}\frac{\partial v}{\partial y}. \tag{7.2.134d}$$

Note that we may also derive Eq. (7.2.134) from Eq. (7.2.133) by letting $\lambda_1\to 0$, $\lambda_2\to 0$, where $\lambda_1\eta_0$ and $\lambda_2\eta_0$ are fixed constants.

We may rewrite Eq. (7.2.122) in the form

$$\rho\left(u\frac{\partial^2 u}{\partial x\partial y}+v\frac{\partial^2 u}{\partial y^2}-u\frac{\partial^2 v}{\partial x^2}-v\frac{\partial^2 v}{\partial x\partial y}\right)$$
$$=\frac{\partial^2\hat{\sigma}_x}{\partial x\partial y}+\frac{\partial^2\hat{\sigma}_{xy}}{\partial y^2}-\frac{\partial^2\hat{\sigma}_{xy}}{\partial x^2}-\frac{\partial^2\hat{\sigma}_y}{\partial x\partial y}. \quad (7.2.135)$$

The solution of these equations (Perera and Walters, 1980) indicates, in general, that elasticity works against inertia, reducing the pressure drop caused by the abrupt change in geometry and reducing the area of influence of the L-shaped bend for finite Reynolds numbers. Stress fields are critical in the corner region, but there is also a region of normal stresses that develops as a result of stretching rather than shearing.

7.3 Viscoelasticity

The mathematical theory of viscoelasticity, though old in origin, has rapidly become the center of widespread attention in recent years because of the advent of polymers. In general, most engineering materials, including metals, are known to creep under sustained constant loads and relax while being subjected to a constant strain. In so doing they exhibit a time-dependent inelastic behavior (Green and Rivlin, 1957; Bland, 1960; Lee, 1960; Rivlin, 1964; Rivlin, 1966; Christensen and Naghdi, 1967; Flügge, 1967; Christensen, 1971).

In contrast to the theory of elasticity, which deals with materials that have a capacity to store mechanical energy, the theory of viscoelasticity is associated with those materials capable of dissipating mechanical energy as well as storing it. If a load is suddenly applied – but not impulsively so as to excite a dynamic response – an elastic material will respond instantaneously and reach a final state of deformation. A Newtonian viscous fluid responds to a suddenly applied loading state of uniform shear stress by a steady flow process. There are, however, materials for which a suddenly applied and maintained state of uniform stress induces an instantaneous deformation followed by a flow process that may or may not be limited in magnitude as time elapses. This is the behavior that is referred to as viscoelastic.

The governing differential equations in viscoelasticity are linear if infinitesimal strains are considered, whereas they become nonlinear if nonlinear strains are involved. Thus we have both a linear and a nonlinear theory of viscoelasticity. The modern theory of nonlinear viscoelasticity, termed a functional theory, has been developed through the works of Truesdell and Noll (1965), Coleman (1964), and Noll (1958), among others. At present, however, the applications of their work in engineering are rather limited because of their mathematical complexities. For more practical applications one may resort to a phenomenological theory in which suitable mechanical models are used to describe the viscoelastic behavior.

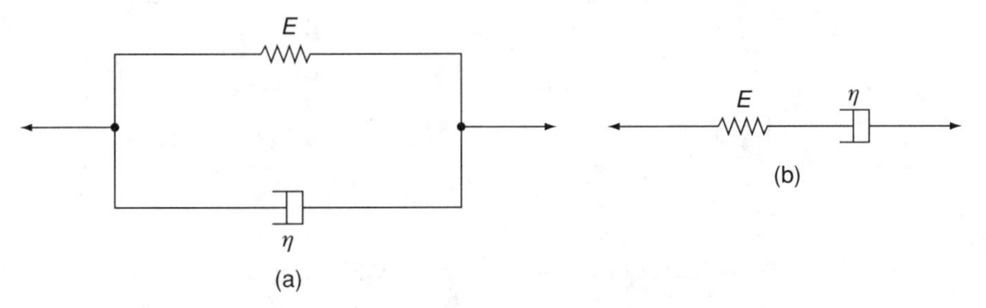

Figure 7.3.1. Viscoelastic models. (a) Kelvin solid, (b) Maxwell fluid.

7.3.1 Spring–Dashpot Models

Two simple models of viscoelasticity in materials are the *Kelvin solid* and the *Maxwell fluid*. The Kelvin solid model represents the material through a schematic diagram consisting of a spring and a dashpot connected in parallel [Fig. 7.3.1(a)], whereas the Maxwell fluid model uses a schematic showing a spring and a dashpot connected in series [Fig. 7.3.1(b)]. We first examine the differential equations for these models. In what follows we use, for convenience, the notation ϵ for the one-dimensional strain (this is simply due to tradition in viscoelasticity) until we deal with the multidimensional strain tensor γ_{ij} in Subsection 7.3.6.

Kelvin Model. Because the strain ϵ in a spring and a dashpot connected in parallel must equal the total stress σ, the Kelvin model provides the sum of the stresses in the spring and dashpot:

$$\sigma = \sigma_s + \sigma_d = E\epsilon + \eta\dot{\epsilon}, \tag{7.3.1}$$

where E and η are the spring constant and viscosity (damping) constant, respectively. Consider a constant stress applied to this system, $\sigma = \sigma_0$. The solution of Eq. (7.3.1) is then

$$\epsilon = Ae^{-\lambda t} + \frac{\sigma_0}{E}, \tag{7.3.2}$$

where $\lambda = 1/T$, T being the *relaxation time*, $T = \eta/E$. At $t = 0$, the stress σ jumps from 0 to σ_0 (Fig. 7.3.2). This requires that

$$\epsilon = 0 = A + \frac{\sigma_0}{E}$$

or

$$A = -\frac{\sigma_0}{E}. \tag{7.3.3}$$

Substituting Eq. (7.3.3) into Eq. (7.3.2) gives

$$\epsilon = \frac{\sigma_0}{E}(1 - e^{-\lambda t}). \tag{7.3.4}$$

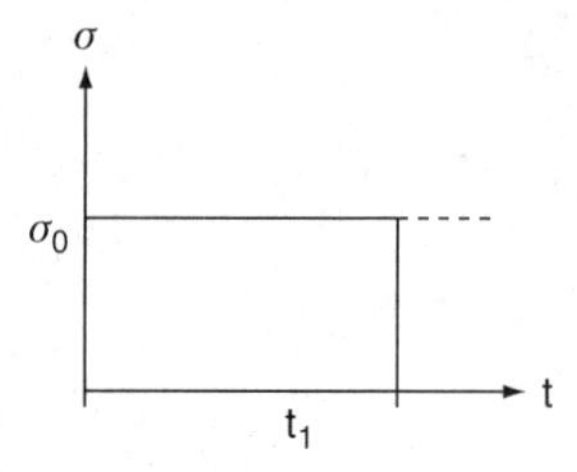

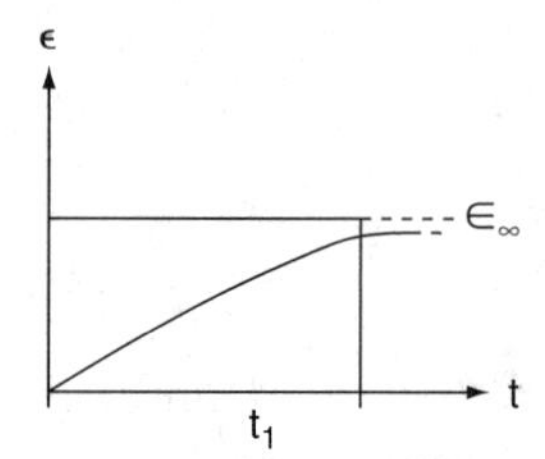

Figure 7.3.2. Constant-stress and strain time relation.

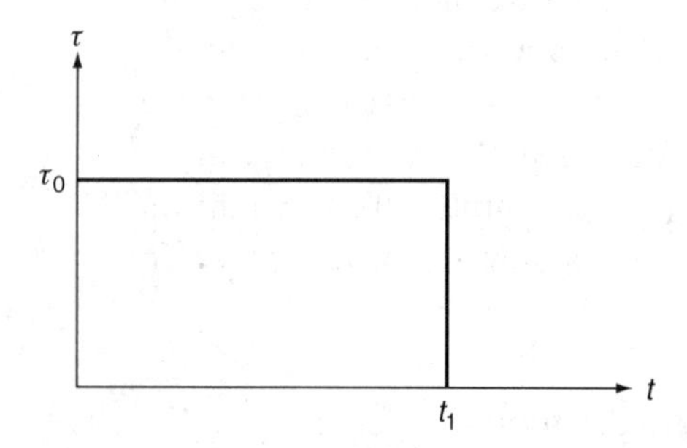

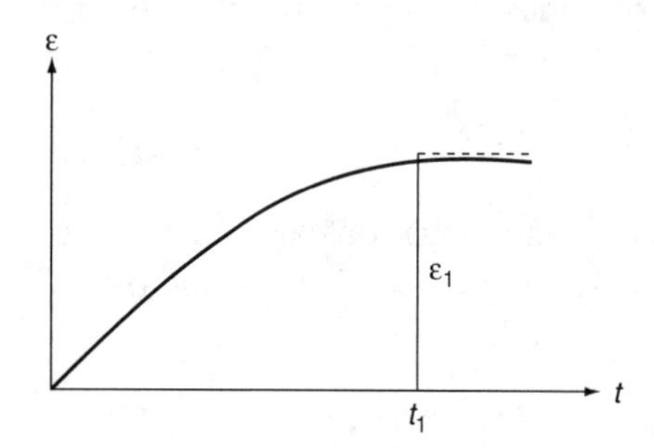

Figure 7.3.3. Removal of stress at t_1.

For $t = \infty$, we have

$$\epsilon_\infty = \frac{\sigma_0}{E_\infty},$$

where E_∞ is called the *asymptotic elastic modulus*. For $t > t_1$, we obtain

$$\sigma = E\epsilon_1 = \sigma_0(1 - e^{-\lambda t_1}). \tag{7.3.5}$$

On the other hand, if the stress σ_0 is removed completely at $t = t_1$ (Fig. 7.3.3), we have from Eq. (7.3.2)

$$\epsilon = Ae^{-\lambda t}, \quad \epsilon_1 = Ae^{-\lambda t_1}.$$

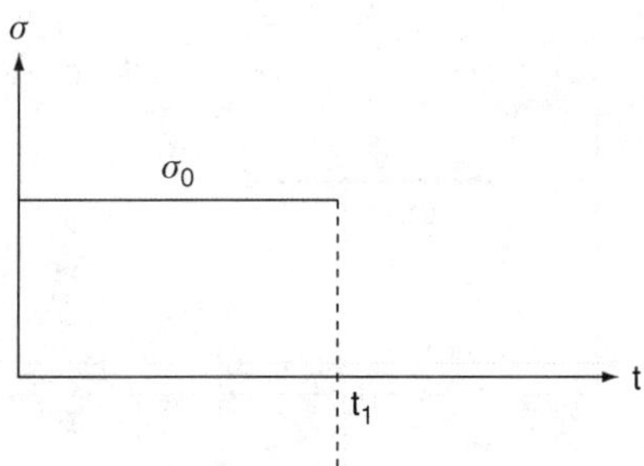

Figure 7.3.4. Creep and recovery.

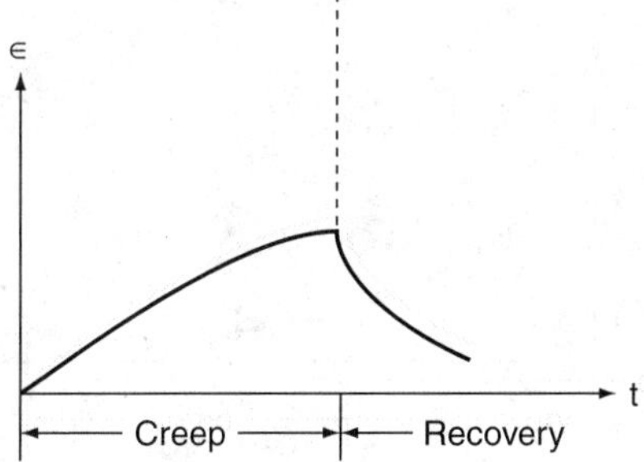

This gives $A = e^{\lambda t_1}\epsilon_1$, and

$$\epsilon = e^{\lambda t_1}\epsilon_1 e^{-\lambda t} = \frac{\sigma_0}{E}(1 - e^{-\lambda t_1})\, e^{-\lambda(t-t_1)}, \tag{7.3.6}$$

which represents the *creep* ($t < t_1$) and *recovery* ($t > t_1$), with the separation point at $t = t_1$ (Fig. 7.3.4).

Maxwell Model. The total strain in the Maxwell model is the sum of the *strains* in the spring and dashpot. Thus the strain rate becomes

$$\dot{\epsilon} = \frac{\dot{\sigma}}{E} + \frac{\sigma}{\eta}. \tag{7.3.7}$$

If we apply the load so that the total strain ϵ is maintained at $\epsilon = \epsilon_1$ (Fig. 7.3.5), then

$$\dot{\sigma} + \lambda\sigma = 0, \tag{7.3.8}$$

which gives the solution

$$\sigma = Ae^{-\lambda t}. \tag{7.3.9}$$

Because we have $\sigma = \sigma_0$ at $t = 0$, we also have $A = \sigma_0$, and it follows that

$$\sigma = \sigma_0 e^{-\lambda t}. \tag{7.3.10}$$

If we maintain $\sigma = \sigma_0$ for $0 \le t < t_1$, then $\dot{\sigma} = 0$ in Eq. (7.3.7) and

$$\dot{\epsilon} = \frac{\sigma_0}{\eta}.$$

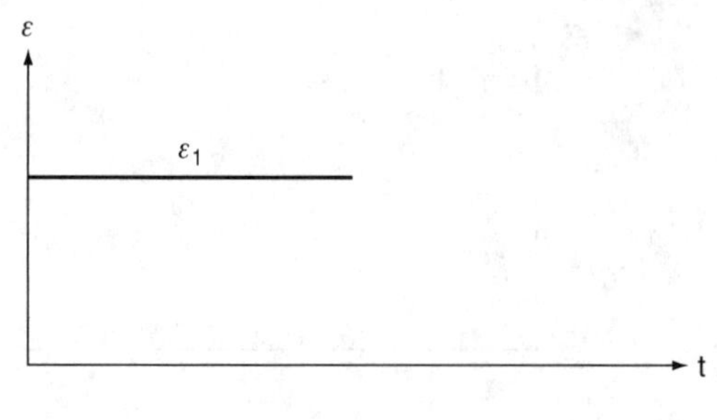

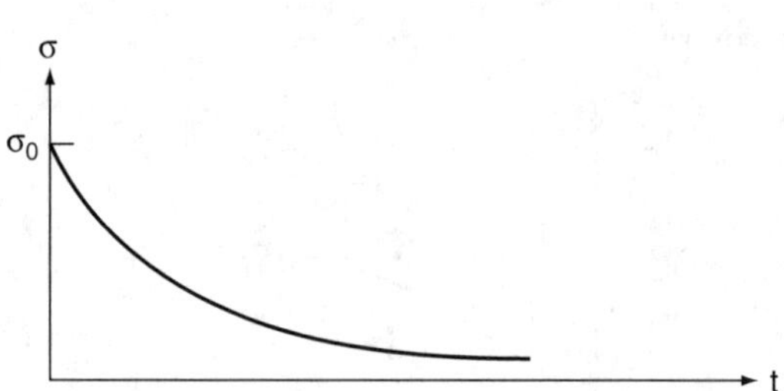

Figure 7.3.5. Constant-strain and stress time relations.

Solving this last differential equation, we obtain

$$\epsilon = \frac{\sigma_0}{\eta} t + C_1. \tag{7.3.11}$$

For $t = 0$, we have $\sigma = \sigma_0$, which gives $C_1 = \epsilon_0 = \sigma_0/E$. This result can be obtained alternatively as follows. The sudden application of stress σ_0 at $t = 0$ gives rise to a singularity at this point. Let us first integrate Eq. (7.3.7):

$$\int_{-\tau}^{+\tau} \sigma \, dt + \frac{1}{\lambda} [\sigma(+\tau) - \sigma(-\tau)] = \eta[\epsilon(+\tau) - \epsilon(-\tau)].$$

If $\tau \to 0$, we have

$$\frac{1}{\lambda}\sigma_0 = \eta\epsilon_0$$

and

$$\epsilon_0 = \frac{\sigma_0}{\lambda\eta} = \frac{\sigma_0}{E},$$

or

$$C_1 = \epsilon_0 = \frac{\sigma_0}{E}. \tag{7.3.12}$$

From Eqs. (7.3.12) and (7.3.11), we get

$$\epsilon = \frac{\sigma_0}{E}(1 + \lambda t) = \frac{\sigma_0}{\eta}(T + t). \tag{7.3.13}$$

If we now maintain $\epsilon = \epsilon_1$ at t_1, then the stress relaxes (Fig. 7.3.6) according to the relation

$$\sigma = \sigma_0 e^{-\lambda(t-t_1)}. \tag{7.3.14}$$

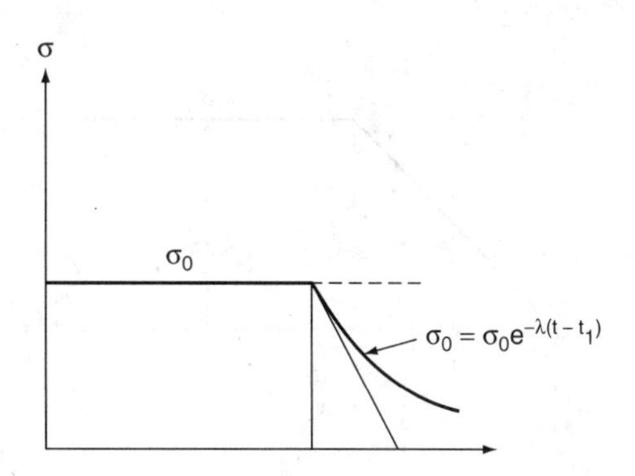

Figure 7.3.6. Stress relaxation.

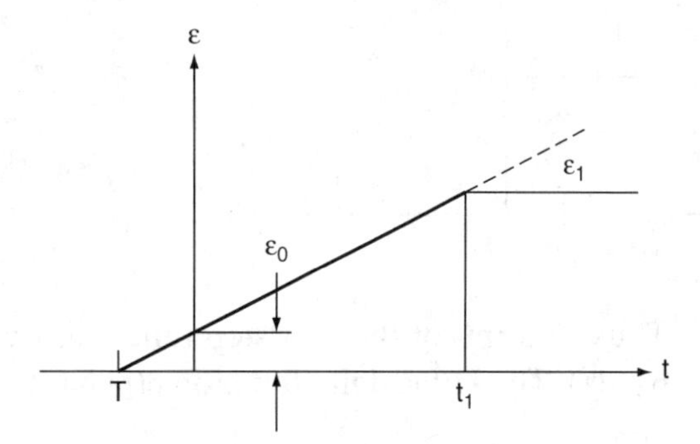

With these preliminaries, we shall discuss practical applications in the following subsections.

7.3.2 Multiple-Parameter Spring–Dashpot Models

In general, the simple models discussed in the preceding subsection are not capable of representing the true behavior of real materials. Various combinations of Kelvin and Maxwell models must be used to adequately represent the behavior of real materials as recorded in the laboratory experiments. Differential equations of such models, then, are complicated, and special techniques for solving these equations are needed. The Laplace transform is one of the most convenient methods for this purpose. Toward this end, we need to review the concept of the unit step function and the Dirac delta function in conjunction with introducing the Laplace transform solution in viscoelasticity.

Unit Step and Dirac Delta Functions. The sudden application of a stress σ can be represented by means of the unit step function and the Dirac delta function (Fig. 7.3.7). We define the unit step function $\Delta(t)$ as follows:

$$\Delta(t) = 0 \quad \text{for} \quad t < 0,$$
$$\Delta(t) = 1 \quad \text{for} \quad t > 0. \tag{7.3.15}$$

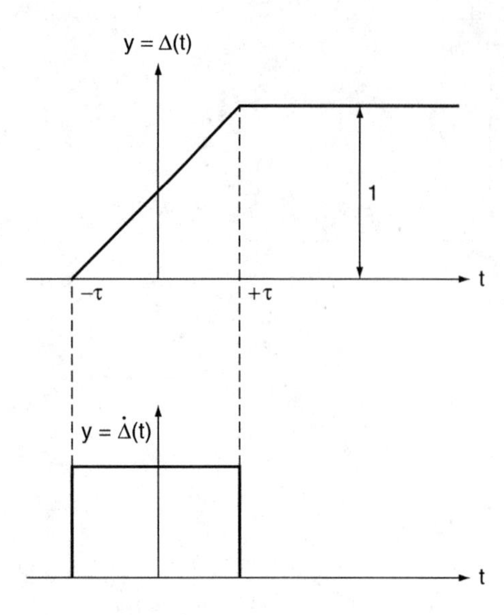

Figure 7.3.7. Unit step function.

Thus, in terms of the unit step function, we may define the creep phase as $\sigma = \sigma_0 \Delta(t)$. The Dirac delta function $\delta(t)$ has the properties

$$\delta(t) = \dot{\Delta}(t), \tag{7.3.16}$$

$$\delta(t) = 0 \text{ for } t \neq 0,$$

$$\delta(t) = \infty \text{ for } t = 0,$$

$$\int_{-\infty}^{\infty} \delta(t)\, dt = \int_{0^-}^{0^+} \delta(t)\, dt = 1. \tag{7.3.17}$$

Let us now examine the solution given by Eq. (7.3.4):

$$\epsilon = \frac{\sigma_0}{E}(1 - e^{\lambda t}), \tag{7.3.18}$$

where $\lambda = 1/T$. At $t = t_1$, let $\epsilon = \epsilon_1 =$ constant; then

$$\epsilon_1 = \frac{\sigma_0}{E}(1 - e^{\lambda t_1}) = \frac{\sigma_0}{E}\left[1 - 1 + \lambda t_1 - \frac{1}{2}(\lambda t_1)^2 + \cdots +\right] \approx \frac{\sigma_0}{\eta} t_1.$$

In the limit, we have $\sigma_0 t_1 = \epsilon_1 \eta$ and σ_0 must tend to infinity as t_1 tends to zero. This leads to

$$\sigma = \eta \epsilon_1 \delta(t) + E\epsilon_1 \Delta(t) \quad \text{for} \quad \epsilon = \epsilon_1 \Delta(t). \tag{7.3.19}$$

Laplace Transform. Consider any function $f(t)$ and define the Laplace transform of $f(t)$ as

$$\bar{f}(s) = \int_0^{\infty} f(t)\, e^{-st}\, dt. \tag{7.3.20}$$

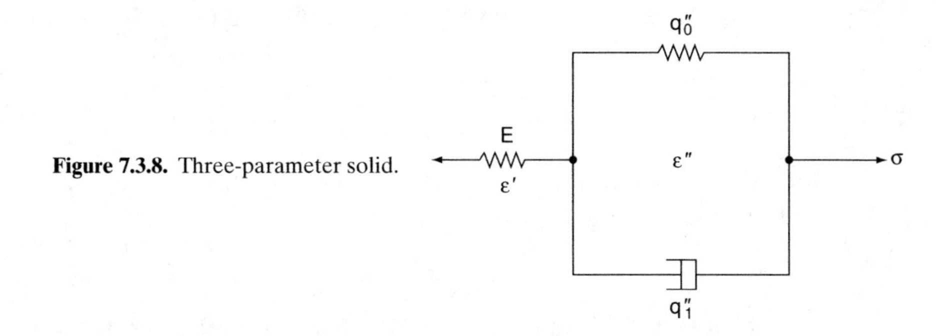

Figure 7.3.8. Three-parameter solid.

Apply Eq. (7.3.20) to $f(t)$ and integrate by parts:

$$\int_0^\infty f(t)\, e^{-st}\, dt = [f(t)\, e^{-st}]_0^\infty - \int_0^\infty f(t)(-s)\, e^{-st}\, dt$$
$$= -f(0) + s\bar{f}(s). \tag{7.3.21a}$$

Similarly,

$$\int_0^\infty \ddot{f}(t)e^{-st}\, dt = -\dot{f}(0) - s\bar{f}(0) + s^2\bar{f}(s), \tag{7.3.21a}$$

$$\int_0^\infty \dddot{f}\, e^{-st} dt = -\ddot{f}(0) - s\dot{f}(0) - s^2 f(0) + s^3\bar{f}(s). \tag{7.3.21b}$$

For functions that vanish for all $t < 0$, the nth derivative of $f(t)$ has as its Laplace transform $s^n\bar{f}(s)$. Consider the function $f(t) = e^{-\alpha t}$. Then

$$\bar{f}(s) = \int_0^\infty e^{-\alpha t} e^{-st}\, dt = -\frac{1}{\alpha + s} e^{-t(\alpha+s)}\Big|_0^\infty = \frac{1}{\alpha + s}. \tag{7.3.22a}$$

Likewise, we obtain

$$\bar{f}(s) = \int_0^\infty \Delta(t)e^{-st}\, dt = \frac{1}{s}, \tag{7.3.22b}$$

$$\bar{f}(s) = \int_0^\infty \delta(t)e^{-st}\, dt = 1, \tag{7.3.22c}$$

$$\bar{f}(s) = \int_0^\infty \frac{1}{\alpha}(1 - e^{\alpha t})e^{-st}\, dt = \frac{1}{s(\alpha + s)}. \tag{7.3.22d}$$

Three-Parameter Solid. Consider a three-parameter solid as shown in Fig. 7.3.8. We note that the stresses in the spring on the left and in the Kelvin model schematic on the right are, respectively,

$$\sigma = E\epsilon', \tag{7.3.23a}$$

$$\sigma = q_0''\epsilon'' + q_1''\dot{\epsilon}''. \tag{7.3.23b}$$

Taking the Laplace transforms of Eqs. (7.3.23a) and (7.3.23b), we obtain

$$\bar{\sigma} = E\bar{\epsilon}', \tag{7.3.24a}$$

$$\bar{\sigma} = (q_0'' + q_1''s)\bar{\epsilon}''. \tag{7.3.24b}$$

Here, the total strain ϵ is equal to the sum of the strains on the spring, left, and the Kelvin model, right:

$$\epsilon = \epsilon' + \epsilon''. \tag{7.3.25}$$

Solving for $\bar{\epsilon}'$ and $\bar{\epsilon}''$from Eqs. (7.3.24a) and (7.3.24b) and substituting into the Laplace transform of Eq. (7.3.25) yields

$$\bar{\epsilon} = \bar{\epsilon}' + \bar{\epsilon}'' = \left(\frac{1}{E} + \frac{1}{q_0'' + q_1''s}\right)\bar{\sigma} = \frac{(q_0'' + q_1''s + E)\bar{\sigma}}{E(q_0'' + q_1''s)}.$$

Rearranging, we obtain

$$\begin{aligned} \bar{\sigma}(E + q_0'' + q_1''s) &= E(q_0'' + q_1''s)(\bar{\epsilon}' + \bar{\epsilon}'') \\ &= E(q_0'' + q_1''s)\bar{\epsilon}. \end{aligned} \tag{7.3.26}$$

Taking an inverse Laplace transform, we get

$$\sigma + p_1\dot{\sigma} = q_0\epsilon + q_1\dot{\epsilon}, \tag{7.3.27}$$

where

$$p_1 = \frac{q_1''}{E + q_0''}, \quad q_0 = \frac{Eq_0''}{E + q_0''}, \quad q_1 = \frac{Eq_1''}{E + q_0''}. \tag{7.3.28}$$

From Eq. (7.3.28) we obtain the relationship

$$\frac{q_1}{p_1} - q_0 = \frac{E^2}{E + q_0''}. \tag{7.3.29}$$

It is seen that, as a consequence of Eq. (7.3.29), we require

$$\frac{q_1}{p_1} - q_0 > 0$$

or

$$q_1 > p_1 q_0 \tag{7.3.30}$$

Creep Phase. The creep phase [Fig. 7.3.9(a)] is governed by

$$\sigma = \sigma_0 \Delta(t), \quad \bar{\sigma} = \frac{\sigma_0}{s}.$$

From the Laplace transform

$$\bar{\sigma} + p_1 s\bar{\sigma} = q_0\bar{\epsilon} + q_1 s\bar{\epsilon}, \tag{7.3.31}$$

$$\sigma_0\left(\frac{1}{s} + p_1\right) = (q_0 + q_1 s)\bar{\epsilon}, \tag{7.3.32}$$

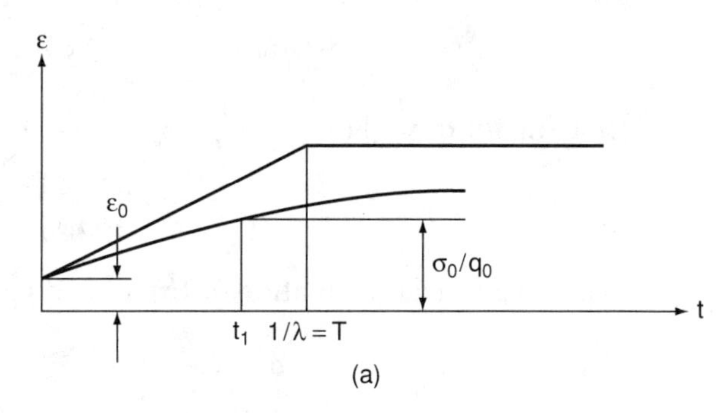

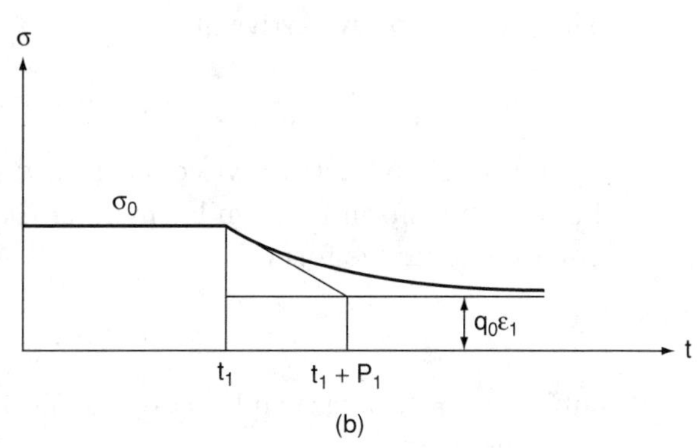

Figure 7.3.9. (a) Creep phase and (b) relaxation phase.

it follows that

$$\bar{\epsilon} = \frac{\sigma_0(1 + p_1 s)}{s(q_0 + q_1 s)} = \frac{\sigma_0}{q_1}\left(\frac{1}{s(s+\lambda)} + \frac{p_1}{s+\lambda}\right), \tag{7.3.33}$$

with $\lambda = q_0/q_1$. The inverse transform gives

$$\epsilon = \frac{\sigma_0}{q_1}\left[\frac{1}{\lambda}(1 - e^{-\lambda t}) + p_1 e^{-\lambda t}\right]$$

or

$$\epsilon = \frac{\sigma_0}{q_0}\left[1 - \left(1 - \frac{p_1 q_0}{q_1}\right)\exp\left(\frac{q_0 t}{q_1}\right)\right]. \tag{7.3.34}$$

Note that, for $t = 0$ and $t = \infty$, we have $\epsilon_0 = \sigma_0 p_1/q_1 = \sigma_0/E_0$ and $\epsilon_\infty = \sigma_0/q_0 = \sigma_0/E_\infty$, respectively. Here, E_0 is the instant elastic modulus and E_∞ is the asymptotic elastic modulus. If $\epsilon_\infty < \epsilon_0$, then it is seen that

$$\frac{\sigma_0}{q_0} < \sigma_0\frac{p_1}{q_1},$$

which is in violation of the usual requirement for a bar in tension, a phenomenon characteristic of a bar under sustained load.

Relaxation Phase. As depicted in Fig. 7.3.9(b), the constant σ_0 is removed at $\tau = t - t_1$. For $\tau \geq 0$, we have $\epsilon = \epsilon_1\Delta(\tau)$, $\bar{\epsilon} = \epsilon_1/s$. The Laplace transform for a three-parameter solid leads to

$$\bar{\sigma} + p_1(s\bar{\sigma} - \sigma_0) = q_0\bar{\epsilon} + q_1(s\bar{\epsilon} - \epsilon_1) = \frac{q_0\epsilon_1}{s}. \tag{7.3.35}$$

Solving for $\bar{\sigma}$, we have

$$\bar{\sigma} = \frac{q_0\epsilon_1}{s(1 + p_1 s)} + \frac{p_1\sigma_0}{1 + p_1 s}. \tag{7.3.36}$$

The inverse transformation of Eq. (7.3.36) gives

$$\sigma = q_0\epsilon_1(1 - e^{-\tau/p_1}) + \sigma_0 e^{-\tau/p_1}.$$

Thus, for $\tau = \infty$, we arrive at

$$\sigma_\infty = q_0\epsilon_1 = E_\infty\epsilon_1.$$

The simple models for viscoelasticity discussed earlier may be solved by the Laplace transform in a similar manner. For the Kelvin model, the differential equation takes the form

$$\sigma = q_0\epsilon + q_1\dot{\epsilon},$$

subject to $\sigma = \sigma_0\Delta(t)$ and $\bar{\sigma} = \sigma_0/s$. The Laplace transform of this differential equation yields

$$\begin{aligned}
\bar{\sigma} &= q_0\bar{\epsilon} + q_1 s\bar{\epsilon},\\
\frac{\sigma_0}{s} &= (q_0 + q_1 s)\bar{\epsilon},\\
\bar{\epsilon} &= \frac{\sigma_0}{s(q_0 + q_1 s)} = \frac{\sigma_0}{q_1}\frac{1}{(\lambda + s)s}.
\end{aligned}$$

Therefore it follows that

$$\epsilon = \frac{\sigma_0}{q_1}\left[\frac{1}{\lambda}(1 - e^{-\lambda t})\right] = \frac{\sigma_0}{q_0}(1 - e^{-\lambda t}).$$

For the Maxwell model, we proceed similarly:

$$\sigma + p_1\dot{\sigma} = q_1\dot{\epsilon},$$

with $\epsilon = \epsilon_0\Delta(t)$, $\bar{\epsilon} = \epsilon_1/s$, $\tau = t - t_1$. The Laplace transform is

$$\bar{\sigma} + p_1(s\bar{\sigma} - \sigma_0) = q_1(s\bar{\epsilon} - \epsilon_1) = q_1(\epsilon_1 - \epsilon_1) = 0.$$

This leads to the solution

$$\bar{\sigma} = \frac{p_1\sigma_0}{1 + p_1 s} = \frac{1}{\lambda + s}\sigma_0 = \sigma_0 e^{-\lambda t}$$

or

$$\sigma = \sigma_0 e^{-\lambda(t - t_1)}.$$

The simple results obtained here represent highly idealized cases. In the next subsection, we shall use a more realistic approach to viscoelastic behavior as it usually occurs through so-called *generalized models*.

7.3.3 General Form of the Differential Equations

It follows from the foregoing discussion that the general form of the differential equation may be written as

$$p_0\sigma + p_1\dot{\sigma} + p_2\ddot{\sigma} + \cdots + = q_0\epsilon + q_1\dot{\epsilon} + q_2\ddot{\epsilon} + \cdots +$$

or

$$\sum_0^m p_k \frac{d^k\sigma}{dt^k} = \sum_0^n q_k \frac{d^k\epsilon}{dt^k}, \tag{7.3.37}$$

with $p_0 = 1$. We rewrite Eq. (7.3.37) to read

$$\mathbf{P}\sigma = \mathbf{Q}\epsilon, \tag{7.3.38}$$

where **P** and **Q** are differential operators:

$$\mathbf{P} = \sum_0^m p_k \frac{d^k}{dt^k}, \quad \mathbf{Q} = \sum_0^n q_k \frac{d^k}{dt^k}. \tag{7.3.39}$$

The Laplace transforms $\bar{\sigma}(s)$ and $\bar{\epsilon}(s)$ are given by

$$\sum_0^m p_k s^k \bar{\sigma} = \sum_0^n q_k s^k \bar{\epsilon}$$

or

$$\mathbf{P}(s)\bar{\sigma} = \mathbf{Q}(s)\bar{\epsilon}. \tag{7.3.40}$$

Here, $\mathbf{P}(s)$ and $\mathbf{Q}(s)$ are polynomials in s:

$$\mathbf{P}(s) = \sum_0^m p_k s^k, \quad \mathbf{Q}(s)\sum_0^n q_k s^k. \tag{7.3.41}$$

A model that contains a dashpot connected in series as an independent strain unit is called a fluid model; otherwise, it is referred to as a solid model. It is not difficult to determine the form of the differential equations for multiparameter models. The following rules may be used to determine the nonzero **P** and **Q** terms:

For **P** terms:

(1) The first term, the one with $p_0 = 1$, is present for all models.
(2) The number of p_i terms, including the p_0 term, is equal to the total number of independent strain units.

Table 7.3.1. *Various viscoelastic models*

Model	Nonzero coefficient **P**	Nonzero coefficient **Q**	Differential equation	Creep compliance $J(t)$	Relation modulus $E(t)$
	p_0	q_0	$\sigma = q_0\epsilon$	$\frac{1}{q_0}$	q_0
	p_0	q_1	$\sigma = q_1\dot{\epsilon}_1$	$\frac{t}{q_1}$	$q_1\delta(t)$
	$p_0 p_1$	q_1	$\sigma + p_1\dot{\sigma} = q_1\epsilon + q_2\ddot{\epsilon}$	$\frac{(p_1 + t)}{q_1}$	$\frac{q_1}{p_1}e^{-t/p_1}$
	p_0	$q_0 q_1$	$\sigma = q_0\epsilon + q_1\dot{\epsilon}$	$\frac{1}{q_0}(1 - e^{\alpha t}),\ \alpha = \frac{q_0}{q_1}$	$q_0 + q_1\delta(t)$
	$p_0 p_1$	$q_0 q_1$	$\sigma + p_1\dot{\sigma} = q_0\epsilon + q_1\dot{\epsilon}$	$\frac{p_1}{q_1}e^{-\alpha t} + \frac{1}{q_0}(1 - e^{-\lambda t})$	$\frac{q_1}{p_1}e^{-t/p_1} + q_0(1 - e^{-t/p_1})$
	$p_0 p_1$	$q_1 q_2$	$\sigma + p_1\dot{\sigma} = q_1\dot{\epsilon}$	$\frac{t}{q_1} + \frac{p_1 q_1 - q_2}{q_1^2}(1 - e^{-\lambda t})$	$\frac{q_2}{p_1}\delta(t) + \frac{1}{p_1}\left(q_1 - \frac{q_2}{p_1}\right)e^{-t/p_1}$

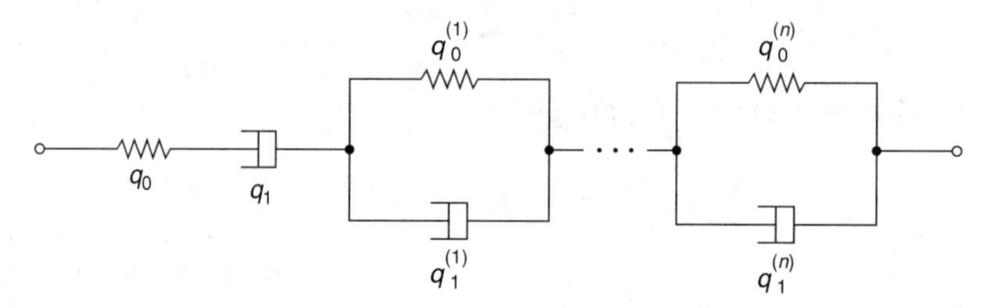

Figure 7.3.10. Generalized Kelvin model.

For **Q** terms:

(1) The q_0 term is present for all solid models but absent for all fluid models.
(2) The number of q_i terms is equal to the total number of dashpots.

The various models and nonzero p_i, q_i terms are shown in Table 7.3.1.

7.3.4 Constitutive Equations for Generalized Models

Generalized Kelvin Model. The generalized Kelvin model (Fig. 7.3.10) consists of a spring and dashpot in series with a number of Kelvin models connected in series. The expression for the strain is

$$\epsilon(t) = \left(\frac{1}{q_0} + \frac{1}{q_1}t\right)\sigma_0\Delta(t) + \sum_{r=1}^{n}\frac{1}{q_0^{(r)}}(1 - e^{-\lambda(r)t})\sigma_0\Delta(t). \tag{7.3.42}$$

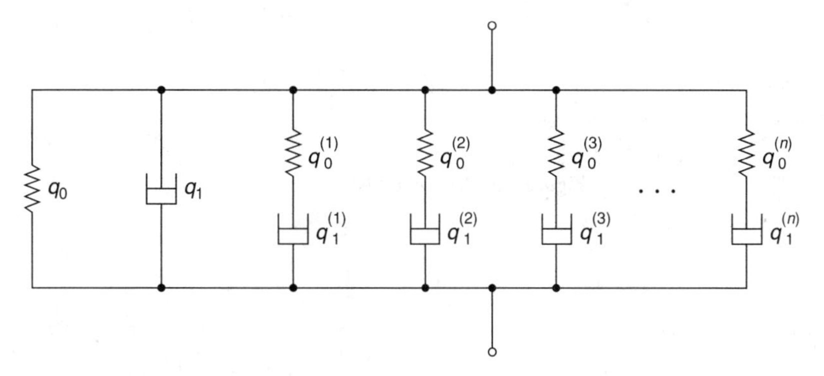

Figure 7.3.11. Generalized Maxwell model.

Generalized Maxwell Model. The generalized Maxwell model (Fig. 7.3.11) consists of a spring and dashpot in parallel with a number of Maxwell models connected in parallel. The expression for the stress is

$$\sigma(t) = q_0\epsilon_1\Delta(t) + q_1\epsilon_1\delta(t) + \sum_{r=1}^{n} q_0^{(r)} e^{-\lambda(r)t}\epsilon_1\Delta(t). \tag{7.3.43}$$

To characterize a viscoelastic material, either relaxation or creep tests are performed. In the relaxation test (Fig. 7.3.12) a constant strain $\epsilon = \epsilon_0$ is maintained and we measure the stress $\sigma(t)$. From this we determine the relaxation modulus $E(t) = \sigma(t)/\epsilon_0$. If $E(t)$ is found to be constant under different ϵ_0, then the material is said to be linearly viscoelastic.

In the case of a creep test (Fig. 7.3.13), on the other hand, $\sigma = \sigma_0$ is maintained and we measure $\epsilon(t)$. From this we obtain the creep compliance $J(t) = \epsilon(t)/\sigma_0$. If $J(t)$ is found to be constant under different σ_0, then the material is again categorized under linear viscoelasticity.

If the basic material constants (spring constants and viscosity constants) are available, then the creep compliance or relaxation modulus can be calculated from the formulas

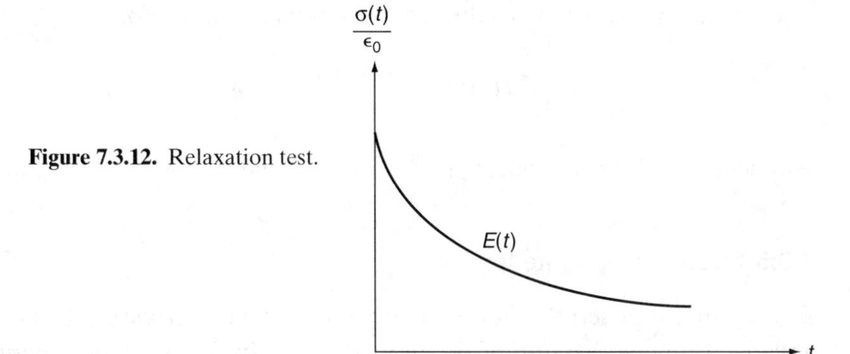

Figure 7.3.12. Relaxation test.

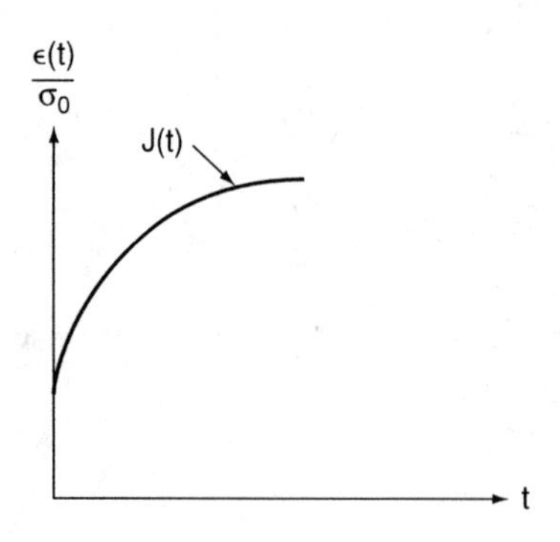

Figure 7.3.13. Creep test.

$$\bar{J}(s) = \frac{\mathbf{P}(s)}{s\mathbf{Q}(s)}, \quad \bar{E}(s) = \frac{\mathbf{Q}(s)}{s\mathbf{P}(s)}.$$

Combining these relations, we obtain

$$\bar{J}(s)\bar{E}(s) = \frac{1}{s^2}.$$

For example, consider three Kelvin units connected in series. The differential equation for this case reads

$$\sigma + p_1 \dot{\sigma} + p_2 \ddot{\sigma} = q_0 \epsilon + q_1 \dot{\epsilon} + q_2 \ddot{\epsilon} + q_3 \dddot{\epsilon}.$$

The corresponding creep compliance takes the form

$$J(s) = \frac{1 + p_1 s + p_2 s^2}{s\,(q_0 + q_1 s + q_2 s^2 + q_3 s^3)}$$

or

$$\bar{J}(s) = \frac{1}{q_3}\frac{1}{s}\left(\frac{a_1}{s-\lambda_1} + \frac{a_2}{s-\lambda_2} + \frac{a_3}{s-\lambda_3}\right),$$

where

$$a_1 = \frac{1 + p_1\lambda + p_2\lambda_1^2}{(\lambda_1 - \lambda_2)(\lambda_1 - \lambda_3)}, \text{ etc.}$$

The inverse transform gives the creep compliance in the form

$$J(t) = \frac{1}{q_3}\sum_{n=1}^{3} \frac{a_n}{-\lambda_n}(1 - e^{\lambda_n t}).$$

Applications of these models are discussed in the following subsections.

7.3.5 Superposition Integral

Boltzmann proposed that for a linear viscoelastic material the strain or stress at any given time is the sum of the individual strain or stress increments through

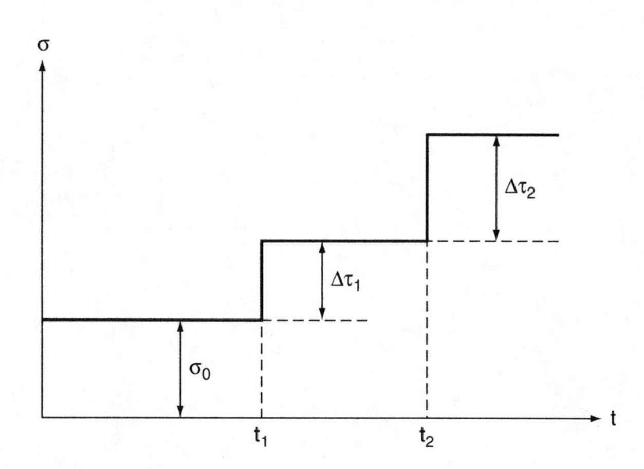

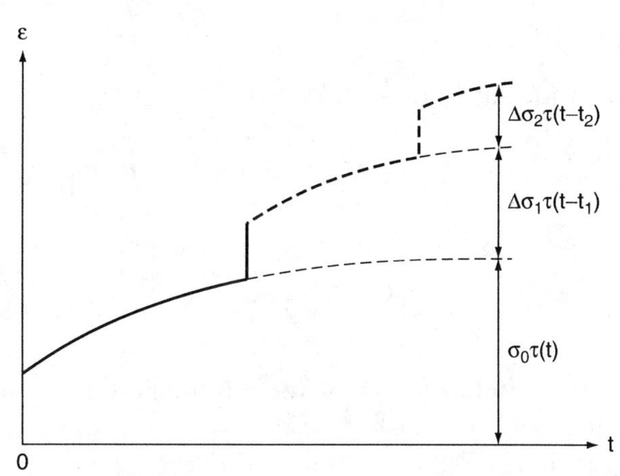

Figure 7.3.14. Superposition integral – Maxwell model.

the respective time intervals during which they have been applied. Suppose that stresses are applied in the following sequences:

$$\sigma = \sigma_0, \quad 0 \leq t \leq t_1,$$
$$\sigma = \sigma_1, \quad t_1 \leq t < t_2.$$

The total strain at t (Fig. 7.3.14) is the sum given by

$$\epsilon(t) = \sigma_0 J(t) + \Delta\sigma_1 J(t - t_1) + \Delta\sigma_2 J(t - t_2) + \cdots + .$$

For continuously varying stresses, it is possible to write

$$\epsilon(t) = \sigma_0 J(t) + \int_0^t J(t - t')\frac{d\sigma(t')}{dt'}dt'. \tag{7.3.44a}$$

Thus the strain at any given time is seen to be dependent on all past histories of stress. An integral of this kind is called a superposition (hereditary, convolution) integral. We may obtain another form of this integral by integrating by parts:

$$\epsilon(t) = \sigma_0 J(t) + [J(t - t')\sigma(t')]_0^t - \int_0^t \frac{dJ(t - t')}{dt'}\sigma(t')\,dt'$$

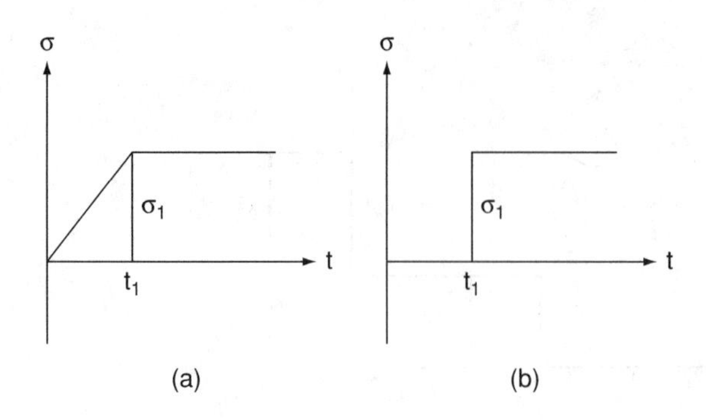

Figure 7.3.15. Example for superposition integral.

or

$$\epsilon(t) = \sigma(t')J(0) - \int_0^t \frac{dJ(t-t')}{dt'}\sigma(t')\, dt'.$$

Introducing $dJ(t-t')/dt' = -dJ(t-t')/d(t-t')$, we obtain still another form:

$$\epsilon(t) = \sigma(t)J(0) + \int_0^t \frac{dJ(t-t')}{d(t-t')}\sigma(t')\, dt'. \tag{7.3.44b}$$

It is also possible to write Eq. (7.3.44a) in the form

$$\epsilon(t) = \int_{-\infty}^{\infty} J(t-t')\, d\sigma(t'), \tag{7.3.44c}$$

which is referred to as a Stieltjes integral. In conjunction with this integral, consider the stress histories shown in Fig. 7.3.15. For a Maxwell material, the integral in Eq. (7.3.44b) subjected to the loading history in Fig. 7.3.15(a) for $t > t_1$ becomes

$$\epsilon(t) = \sigma_1\frac{p_1}{q_1} + \frac{\sigma_1}{t_1}\int_0^{t_1} t'\frac{1}{q_1}dt' + \sigma_1\int_{t_1}^{t}\frac{1}{q_1}dt' = \frac{\sigma_1}{q_1}\left(p_1 - \frac{t_1}{2} + t\right),$$

whereas the loading in Figure 7.3.15(b) leads to

$$\epsilon(t) = \sigma_1 J(t-t_1) = \frac{\sigma_1}{q_1}(p_1 + t - t_1).$$

In the case of a Kelvin material, however, the response for the loading history in Fig. 7.3.15(a) for $t > t_1$ becomes

$$\begin{aligned}\epsilon(t) &= \frac{\sigma_1}{t_1 q_1}\int_0^t t' e^{-\lambda(t-t')}dt' + \frac{\sigma_1}{q_1}\int_{t_1}^t e^{-\lambda(t-t')}dt' \\ &= \frac{\sigma_1}{q_0}\left[1 + \frac{q_1}{q_0 t_1}(1 - e^{\lambda t})\, e^{-\lambda t}\right].\end{aligned}$$

It is interesting to note that, for the Kelvin material at $t = \infty$, all the histories are wiped out, whereas for the Maxwell material, the initial irregularities in loading histories will remain undiminished no matter how much time elapses.

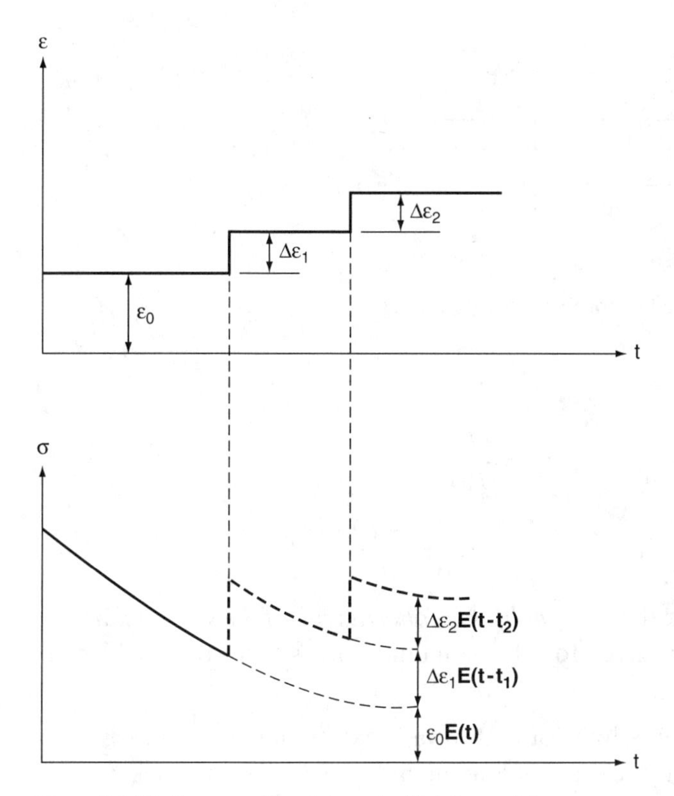

Figure 7.3.16. Superposition integral – Kelvin model.

Similarly, if the process is reversed in which the strains are increased in incremental steps, then the resulting stresses may be superposed as in Fig. 7.3.16. The hereditary integral becomes

$$\sigma(t) = \epsilon_0 E(t) + \int_0^t E(t - t')\frac{d\epsilon(t')}{dt'}dt', \tag{7.3.45a}$$

$$\sigma(t) = \epsilon(t)E(0) + \int_0^t \frac{dE(t - t')}{d(t - t')}\epsilon(t')\, dt', \tag{7.3.45b}$$

or

$$\sigma(t) = \int_{-\infty}^{\infty} E(t - t')\, d\epsilon(t'). \tag{7.3.45c}$$

Instead of differential equation (7.3.37), these integrals, Eqs. (7.3.44) and (7.3.45), may be used in solving viscoelastic problems. To this end, we may rewrite Eq. (7.3.44b) as

$$f(t) = \sigma(t) + \int_0^t K(t - t')\sigma(t')\, dt', \tag{7.3.46}$$

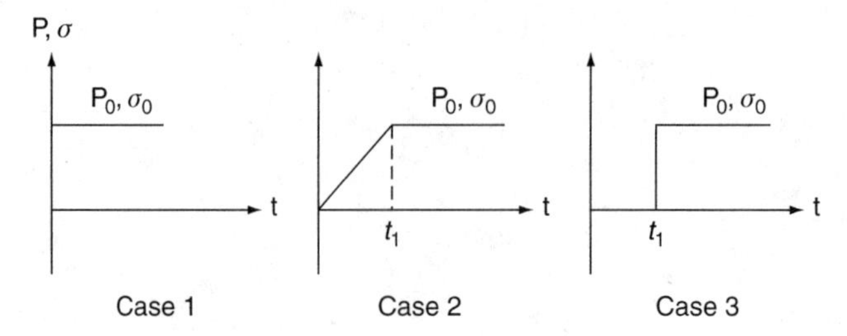

Figure 7.3.17. Three cases of loading for Examples 7.3.1–7.3.4.

with

$$f(t) = \frac{\epsilon(t)}{J(0)},$$

$$K(t-t') = \frac{1}{J(0)}\frac{dJ(t-t')}{d(t-t')}.$$

Equation (7.3.46) is called the *integral of the Volterra type of the second kind*. In what follows, examples are given for Maxwell fluids and Kelvin solids subjected to various loadings.

EXAMPLE 7.3.1. Derive expressions for $\epsilon(t)$ for a Maxwell material and a Kelvin material subjected to the three cases of loading in Fig. 7.3.17, using the superposition integral of the form

$$\epsilon(t) = \sigma(t)J(0) + \int_0^t \sigma(t')\frac{dJ(t-t')}{d(t-t')}\,dt'.$$

For the Maxwell model

$$J(t) = \frac{1}{q_t}(p_1 + t).$$

Case 1

$$\sigma(t) = \sigma_0\Delta(t),$$

$$\epsilon(t) = \sigma_0\frac{p_1}{q_1} + \int_0^t \sigma_0\frac{1}{q_1}dt' = \frac{\sigma_0}{q_1}(p_1 + t).$$

Case 2

$$\sigma(t) = \sigma_0\frac{t}{t_1} \text{ for } 0 \le t \le t_1,$$

$$\sigma(t) = \sigma_0 \text{ for } t > t_1.$$

For $0 \le t \le t_1$,

$$\epsilon(t) = \sigma_0\frac{t}{t_1}\frac{p_1}{q_1} + \int_0^t \sigma_0\frac{t'}{t_1}\frac{1}{q_1}dt' = \sigma_0\frac{p_1}{q_1}\frac{t}{t_1} + \sigma_0\frac{t^2}{2t_1q_1}$$

$$= \frac{\sigma_0 t}{t_1q_1}\left(p_1 + \frac{t}{2}\right).$$

For $t > t_1$,

$$\begin{aligned}\epsilon(t) &= \sigma_0 \frac{p_1}{q_1} + \int_0^1 \sigma_0 \frac{t'}{t_1} \frac{1}{q_1} dt' + \int_{t_1}^t \sigma_0 \frac{1}{q_1} dt' \\ &= \sigma_0 \frac{p_1}{q_1} + \frac{\sigma_0 t_1^2}{2 t_1 q_1} + \frac{\sigma_0}{q_1}(t - t_1) = \frac{\sigma_0}{q_1}\left(p_1 + \frac{t_1}{2} + t - t_1\right) \\ &= \frac{\sigma_0}{q_1}\left(p_1 + t - \frac{t_1}{2}\right).\end{aligned}$$

Case 3. Proceed the same way as in Case 1, but with t replaced with $t - t_1$ for $t \geq t_1$:

$$\epsilon(t) = \frac{\sigma_0}{q_1}(p_1 + t - t_1).$$

For the Kelvin model

$$J(t) = \frac{1}{q_0}(1 - e^{-\lambda t}), \quad \lambda = \frac{q_0}{q_1}.$$

Case 1

$$\sigma(t) = \sigma_0 \Delta(t),$$

$$\epsilon(t) = \sigma_0(0) - \int_0^t \sigma_0 \frac{\lambda e^{-\lambda(t-t')}}{q_0} dt' = \frac{\sigma_0}{q_0}(-e^{-\lambda t} + 1).$$

Case 2

$$\sigma(t) = \sigma_0 \frac{t}{t_1} \quad \text{for} \quad 0 \leq t \leq t_1,$$

$$\sigma(t) = \sigma_0 \quad \text{for} \quad t > t_1.$$

For $0 \leq t \leq t_1$,

$$\begin{aligned}\epsilon(t) &= \int_0^t \sigma_0 \frac{t'}{t_1} \frac{\lambda}{q_1} e^{-\lambda(t-t')} dt' \\ &= \frac{\sigma_0}{t_1 q_0} \lambda \frac{e^{-\lambda(t-t')}}{\lambda^2} (\lambda t' - 1)\Big|_0^t \\ &= \frac{\sigma_0}{t_1 q_0}\left[\frac{e^{-\lambda(0)}}{\lambda}(\lambda t - 1) - \frac{e^{-\lambda t}}{\lambda}(-1)\right] \\ &= \frac{\sigma_0}{t_1 q_0}\left(t - \frac{1}{\lambda} + \frac{e^{-\lambda t}}{\lambda}\right).\end{aligned}$$

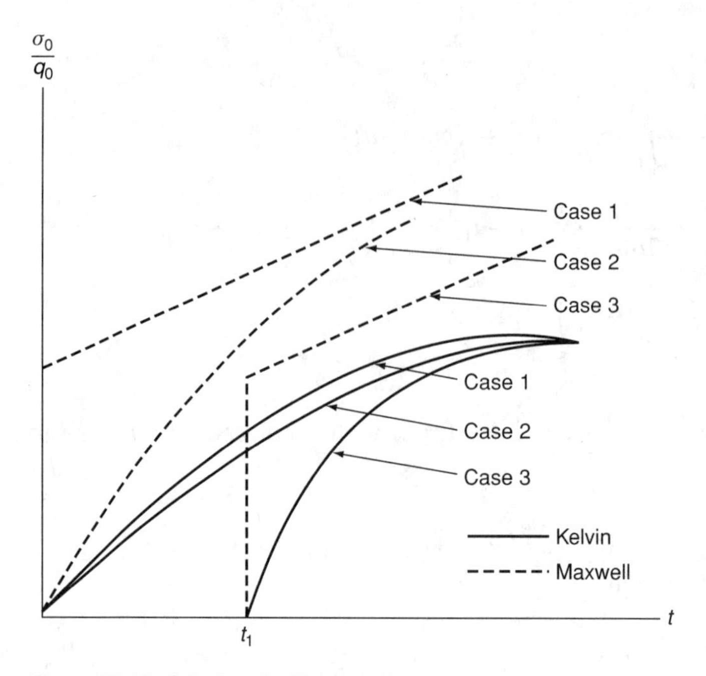

Figure 7.3.18. Solutions for Example 7.3.1.

For $t > t_1$,

$$\begin{aligned}
\epsilon(t) &= \int_0^t \sigma_0 \frac{t'}{t_1}\frac{\lambda}{q_1} e^{-\lambda(t-t')} dt' + \int_{t_1}^t \sigma_0 \frac{\lambda e^{-\lambda(t-t')}}{q_0} dt' \\
&= \frac{\sigma_0}{t_1 q_0}\lambda \frac{e^{-\lambda(t-t')}}{\lambda} (\lambda t' - 1)\Big|_0^{t_1} + \frac{\sigma_0}{q_0} e^{-\lambda(t-t')}\Big|_{t_1}^{t} \\
&= \frac{\sigma_0}{t_1 q_0}\frac{e^{-\lambda(t-t_1)}}{\lambda}(\lambda t - 1) + \frac{\sigma_0}{t_1 q_0}\frac{e^{-\lambda t}}{\lambda} + \frac{\sigma_0}{q_0}\left[1 - e^{-\lambda(t-t_1)}\right] \\
&= \frac{\sigma_0}{q_0}\left[e^{-\lambda(t-t_1)}\left(1 - \frac{1}{\lambda t_1} - 1\right) + 1 + e^{-\lambda t}\frac{1}{\lambda t_1}\right] \\
&= \frac{\sigma_0}{q_0}\left[1 + \frac{e^{-\lambda t}}{\lambda t_1}(1 - e^{\lambda t_1})\right].
\end{aligned}$$

Case 3

Again, this is the same as Case 1, but with t replaced with $t - t_1$:

$$\epsilon(t) = \frac{\sigma_0}{q_0}\left[-e^{-\lambda(t-t_1)} + 1\right].$$

Responses for all three cases for the Kelvin (solid) model are asymptotic to the line $\epsilon = \sigma_0/q_0$, which indicates that what has happened in the distant past is immaterial; whereas in the case of the Maxwell (fluid) model, the responses are divergent as t tends to infinity, as shown in Fig. 7.3.18.

EXAMPLE 7.3.2. A viscoelastic cantilever beam is subjected to a time-dependent load as shown in Fig. 7.3.17. Find the general solution for the displacement using the Kelvin model for Cases 1 and 2 of Fig. 7.3.17.

Solution. The elastic solution is of the form

$$w(x) = \frac{p_0}{24EI}(x^4 - 4l^3x + 3l^4) I = \text{moment of inertia.}$$

The creep compliance for the Kelvin model is

$$J(t) = \frac{1}{q_0}(1 - e^{-\lambda t}).$$

Case 1

The differential equation of an elastic beam is (with $E = 1$)

$$Iw''''(x) = p(x).$$

The Laplace transform of this fourth-order differential equation reads

$$I\mathbf{Q}(s)\bar{w}''''(x, s) = \mathbf{P}(s)\bar{p}(x, s)$$

or

$$I\frac{\mathbf{Q}(s)}{\mathbf{P}(s)}\bar{w}''''(x, s) = \bar{p}(x, s).$$

The solution of this last equation becomes

$$w(x, s) = \frac{\bar{p}}{24I}\frac{\mathbf{P}(s)}{\mathbf{Q}(s)}(x^4 - 4l^3x + 3l^4),$$

where

$$\bar{p} = \frac{p_0}{s}, \quad \mathbf{P}(s) = 1, \quad \mathbf{Q}(s) = q_0 + q_1 s.$$

Thus

$$\bar{w}(x, s) = \frac{p_0}{24I}\frac{1}{s(q_0 + q_1 s)}(x^4 - 4l^3x + 3l^4).$$

Taking an inverse Laplace transform gives

$$w(x, t) = \frac{p_0}{24q_0 I}(1 - e^{-\lambda t})(x^4 - 4l^3x + 3l^4).$$

Case 2

Let $R = (p_0/24I)(x^4 - 4l^3x + 3l^4)$. For $0 < t \leq t_1$, we have

$$p = p_0\left(\frac{t}{t_1}\right), \quad \bar{p} = \frac{p_0}{t_1 s^2},$$

$$w = RL^{-1}\left[\frac{1}{t_1 s^2}\frac{1}{(q_0 + q_1 s)}\right] = R\frac{1}{q_1 t_1}\left[\frac{t}{\lambda} - \frac{1}{\lambda^2}(1 - e^{-\lambda t})\right],$$

$$= R\frac{1}{q_0 t_1}\left[t - \frac{1}{\lambda}(1 - e^{-\lambda t})\right], \quad L^{-1} = \text{inverse Laplace transform.}$$

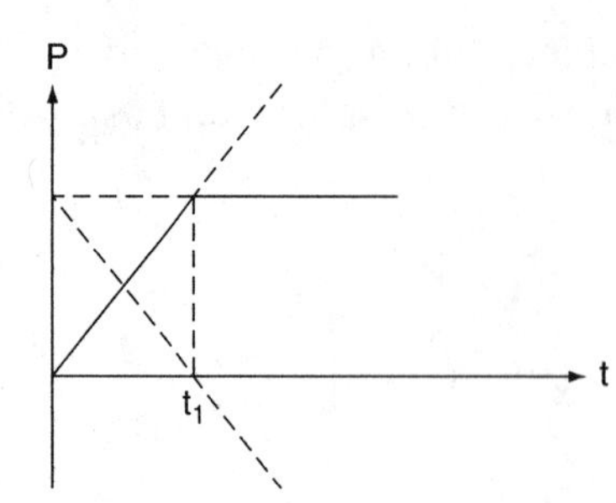

Figure 7.3.19. Two ramp functions (dashed lines) used in Case 2, Example 7.3.2.

For $t_1 < t < \infty$, the response can be considered the sum of the two ramp functions $p_0(t/t_1)$, the second of which is negative and delayed by time t_1, as shown in Fig. 7.3.19. Let $\tau = t = t_1$ for $t_1 < t < \infty$. Then the solution is obtained as follows:

$$p = p_0\left(\frac{\tau}{t_1}\right), \quad \bar{p} = p_0\left(-\frac{1}{t_1 s^2}\right)$$

$$\begin{aligned} w &= R\left\{\frac{1}{q_0 t_1}\left[t - \frac{1}{\lambda}(1 - e^{-\lambda t})\right] + L^{-1}\left[\frac{1}{t_1 s^2(q_0 + q_1 s)}\right]\right\} \\ &= R\left\{\frac{1}{q_0 t_1}\left[t - \frac{1}{\lambda}(1 - e^{-\lambda t})\right] + \frac{1}{q_0 t_1}\left[\tau - \frac{1}{\lambda}(1 - e^{-\lambda t})\right]\right\} \\ &= R\frac{1}{q_0 t_1}\left[1 + \frac{1}{t_1 \lambda}\left(e^{-\lambda t} - e^{-\lambda(t - t_1)}\right)\right]. \end{aligned}$$

7.3.6 Correspondence Principle

A viscoelastic problem may be solved by use of the so-called correspondence principle. In this approach, the problem is first converted to one of linear elasticity and a static solution is obtained. The solution is then combined with the integral transform that results from treating the problem as a time-dependent one. The reasoning will be explained after a brief description of the procedure.

For a linearly elastic beam subjected to distributed loads, p_i, $i = 1, 2, \ldots$, we have

$$\epsilon(x, y, z) = \frac{\sigma(x, y, z)}{E},$$

satisfying all kinematic conditions. Now consider a beam of the same shape but made of a viscoelastic material subjected to

$$p_i(t) = p_i \Delta(t).$$

Assume that the tentative solution is

$$\sigma(x, y, z; t) = \sigma(x, y, z)\Delta(t),$$

where $\sigma(x, y, z)$ is the stress in the hypothetical elastic beam. In the actual viscoelastic beam, the strains are

$$\epsilon(x, y, z; t) = \sigma(x, y, z)J(t),$$

which, at any time t, would be distributed like the strains in an elastic beam of modulus $E = 1/J(t)$ and would satisfy all kinematic conditions.

This process is valid because, if a viscoelastic beam is subjected to loads that are all applied simultaneously at $t = 0$ and then held constant, its stresses will be the same as those in an elastic beam under the same load, whereas its strains and displacements will be dependent on time and derivable from those of the hypothetical elastic problem by replacing E with $1/J(t)$. The process is thus said to follow the correspondence principle.

Similarly, if a viscoelastic beam is subjected to forced displacements of certain points that are all imposed at $t = 0$ and then held constant, all the point displacements and all the strains will be the same as in the corresponding elastic beam, whereas the stresses will derive from those of the elastic problem by multiplying them by $E(t)$:

$$\sigma(x, y, z; t) = \epsilon(x, y, z)E(t).$$

To illustrate, consider a simple beam with an elastic modulus $E = 1$ under a load $p(x)$; then

$$w(x, t) = w_0(x)J(t).$$

If the load applied is made time dependent, then

$$p(x, t) = p(x)f(t),$$

$$w(x, t) = w_0(x)\left[f(0^+)J(t) + \int_{0^+}^{t} J(t - t')\frac{df(t')}{dt'}dt'\right],$$

or

$$w(x, t) = w_0(x)\left[f(t)J(0^+) + \int_{0^+}^{t} \frac{dJ(t - t')}{d(t - t')} f(t')dt'\right].$$

EXAMPLE 7.3.3. Consider a Kelvin material with uniform load p applied as in Case 2, Fig. 7.3.17. The viscoelastic solution is easily obtained, as follows.

Solution. For $0 < t < t_1$,

$$w\left(\frac{l}{2}, t\right) = w_0\left(\frac{l}{2}\right)\frac{1}{q_0 t_1}\left[t_1 - \frac{1}{\lambda}(1 - e^{-\lambda t})\right].$$

For $t > t_1$

$$w\left(\frac{l}{2}, t\right) = w_0\left(\frac{l}{2}\right)\frac{1}{q_0 t_1}\left[t - \frac{1}{\lambda}(e^{\lambda t_1} - 1)e^{-\lambda t}\right],$$

where

$$w_0\left(\frac{l}{2}\right) = \frac{5pl^4}{384I}.$$

Note that the time $1/\lambda = q_1/q_0$ is characteristic of the creep of the material. It is the time at which the initial tangent of the creep curve reaches the asymptote $\epsilon = \epsilon_\infty$.

EXAMPLE 7.3.4. Repeat Example 7.3.2 for Case 1 using the correspondence principle.

Solution. The solution from the correspondence principle is of the form

$$w(x,t) = w_0(x)J(t),$$

where $w_0(x) = (p_0/24I)(x^4 - 4l^3x + 3l^4)$. The time-dependent load is

$$p(t) = p_0 f(t) = p_0\Delta(t).$$

Thus

$$J(t) = f(t)J(0^+) + \int_{0^+}^{t} \frac{dJ(t-t')}{d(t-t')} f(t')\,dt'$$
$$= \int_0^t \left(-\frac{1}{q_0}\lambda e^{-\lambda(t-t')}\right) dt' = \frac{1}{q_0}(1 - e^{-\lambda t}).$$

It is seen that the creep compliance $J(t)$ remains the same throughout the time domain. Thus

$$w(x,t) = w_0(x)\frac{1}{q_0}(1 - e^{-\lambda t}),$$

which is identical to the solution obtained by the Laplace transform in Example 7.3.2.

Remarks

Applications of the correspondence principle are limited to linear viscoelasticity. If the strain is large and/or the material kernel $E(t)$ is nonlinear, we must resort to the theory of nonlinear viscoelasticity. The correspondence principle will not be valid in this case. Examples of nonlinear viscoelasticity include materials with memory or thermorheologically simple materials. To study this subject, we must invoke the concept of irreversible thermodynamics.

7.3.7 Viscoelasticity Based on Irreversible Thermodynamics

Various components in machines, such as gears, bellows, mechanical couplings, bearing pads, and vibration pads, undergo cyclic loading. In recent years, along with the increasing use of polymers – for example, in solid rocket propellant grains, plastics, airplane components, and tires – advanced research on nonlinear viscoelastic behavior has been carried out. Polymers, when they are subjected

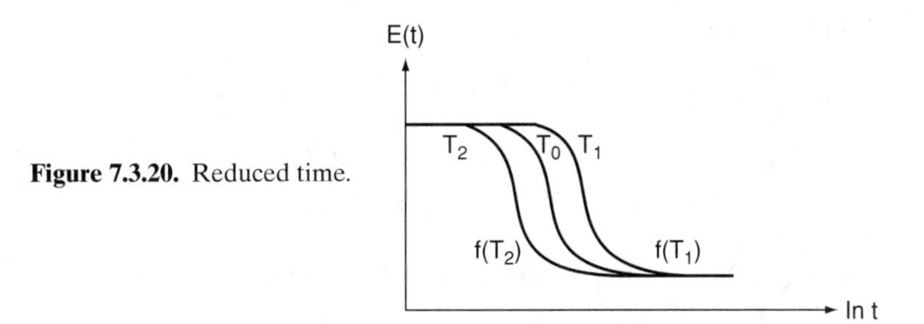

Figure 7.3.20. Reduced time.

to cyclic loading, exhibit significant viscoelastic behavior, in which the energy-dissipating capacity as well as the strain energy must be considered. Associated with cyclic loads and energy dissipation is the concept of irreversible thermodynamics. Time-dependent viscoelastic response is intimately related to time-dependent generation and dissipation of heat energy. The process is an irreversible one. The mechanical properties of viscoelastic materials are strongly dependent on temperature, and when the dissipated mechanical energy increases the temperature, the strength and rigidity of the material change markedly.

There are three basically different approaches to the thermomechanical theory of viscoelasticity: (1) Biot's (1956) approach, following Onsager's pioneering study on irreversible thermodynamics, is based on the assumption that entropy and internal energy in the equilibrium thermodynamic state can be used to explain the nonequilibrium state. (2) Coleman and Noll (1961) refer specifically to a nonequilibrium state in describing entropy and internal energy. The state of the material at time t is described by specification of how this state has been reached. This concept, based on the past history of the material, applies to materials with memory, and to materials with "fading memory" as well, which tend to "forget" what has happened in the distant past and "memorize" only the recent past. (3) The phenomenological theory of Staverman and Schwarzl (1952) is based on relaxation moduli or creep compliances. These three approaches lead to similar results for the stress–strain relation and dissipation energy for the so-called thermorheologically simple material in which temperature-dependent material properties are characterized by a *shift function* (Christensen and Naghdi, 1967).

Consider the case of isothermal deformations of an isotropic linearly viscoelastic solid:

$$\sigma_{ij} = \int_0^t E_1(t-s)\frac{\partial \gamma'_{ij}}{\partial s}ds + \frac{1}{3}\delta_{ij}\int_0^t E_2(t-s)\frac{\partial \gamma_{kk}}{\partial s}ds. \tag{7.3.47}$$

It is assumed that the moduli E_1 and E_2 are influenced by temperature and ln t (Fig. 7.3.20) in such a way that

$$E(t) = \hat{E}\left[\ln t + f(T)\right], \tag{7.3.48}$$

where $f(T)$ is a shift function relative to T_0. If we introduce a shift factor

$$b(t) = \exp\left[f(T)\right], \tag{7.3.49}$$

then

$$E(t) = \hat{E}\,[\ln t + \ln b(T)]$$
$$= \hat{E}\,[\ln t\, b(T)]$$
$$= \hat{E}(\xi), \tag{7.3.50}$$

where ξ is called the *reduced time*, which is defined as

$$\xi = tb(T). \tag{7.3.51}$$

Inverting Eq. (7.3.51) gives the real time:

$$t = g(\xi). \tag{7.3.52}$$

It is now possible to write Eq. (7.3.47) in terms of the reduced time:

$$\hat{\sigma}_{ij} = \int_{\sigma}^{\xi} E_1(\xi - \hat{s})\frac{\partial \hat{\gamma}_{ij}}{\partial \hat{s}} d\hat{s} + \frac{1}{3}\delta_{ij}\int_{\sigma}^{\xi} E_2(\xi - \hat{s})\frac{\partial \hat{\gamma}_{kk}}{\partial \hat{s}} d\hat{s}. \tag{7.3.53}$$

We may also write Eq. (7.3.51) as

$$\Delta\xi = b[T(X, t)]\,\Delta t,$$

so that

$$g(X, t) = \int_0^t b[T(X, s)]\,ds. \tag{7.3.54}$$

All quantities transform between the real and reduced time as follows:

$$\hat{\sigma}_{ij}(\mathbf{X}, \xi) = \sigma_{ij}(\mathbf{X}, t), \quad \hat{\theta}(\mathbf{X}, \xi) = \theta(\mathbf{X}, t),$$
$$\hat{\gamma}_{ij}(\mathbf{X}, \xi) = \gamma_{ij}(\mathbf{X}, t), \quad \hat{\eta}(\mathbf{X}, \xi) = \eta(\mathbf{X}, t).$$

For example, consider

$$\hat{f}(\mathbf{X}, \xi) = f(\mathbf{X}, t),$$
$$\dot{f} = \frac{d}{dt} f(\mathbf{X}, t) = \frac{d}{dt}\hat{f}(\mathbf{X}, \xi)$$
$$= \frac{\partial \hat{f}}{\partial \mathbf{X}}\frac{d\mathbf{X}}{dt} + \frac{\partial \hat{f}}{\partial \xi}\frac{d\xi}{dt} = \frac{\partial \hat{f}}{\partial \xi} b[T(\mathbf{X}, t)]$$
$$= \dot{\hat{f}}(\mathbf{X}, \xi)\hat{b}[\hat{T}(\mathbf{X}, \xi)].$$

Also, rates of displacement, velocity, and temperature are transformed similarly:

$$\dot{u} = b[\hat{\theta}(\mathbf{X}, \xi)]\dot{\hat{u}},$$
$$\ddot{u} = \hat{b}^2[\hat{\theta}(\mathbf{X}, \xi)]\dot{\hat{u}} + \hat{b}[\hat{\theta}(\mathbf{X}, \xi)]\dot{\hat{b}}[\hat{\theta}(\dot{\mathbf{X}}, \xi)]\hat{u},$$
$$\dot{\theta} = \hat{b}[\hat{\theta}(\mathbf{X}, \xi)]\dot{\hat{\theta}},$$
$$\dot{\eta} = \hat{b}[\hat{\theta}(\mathbf{X}, \xi)]\dot{\hat{\eta}},$$
$$\dot{q}^i = \hat{b}[\hat{\theta}(\mathbf{X}, \xi)]\dot{\hat{q}}^i,$$
$$\dot{D} = \hat{b}[\hat{\theta}(\mathbf{X}, \xi)]\dot{\hat{D}}.$$

Experiments by Williams, Landel, and Ferry (1955) and other experimental data (Eringen, 1967) may be used to evaluate the shift factor in terms of real and reduced times. The reader should consult these references for further details.

To establish the forms of the stress tensor and dissipation in reduced time for a thermorheologically simple material from a conventional viscoelastic model, we proceed with the phenomenological approach by using a generalized Maxwell model of the form

$$R(\xi) = \sum_{r=1}^{n} q_0^{(r)} e^{-\lambda_{(r)}\xi}, \quad \text{with} \quad \lambda_{(r)} = \frac{q_0^{(r)}}{q_1^{(r)}}, \tag{7.3.55}$$

where ξ is the reduced time. The superposition integral for the stress tensor is given by

$$\hat{\sigma}_{ij}(\xi) = \int_0^{\xi} R(\xi - \xi') \frac{\partial \hat{\gamma}_{ij}}{\partial \xi'} d\xi'. \tag{7.3.56}$$

It is possible to write the free energy per unit volume in the form

$$\begin{aligned}
\hat{\phi}(\xi) &= \frac{1}{2}\hat{\sigma}_{ij}\hat{\gamma}_{ij} = \frac{1}{2}\sum_{r=1}^{n} \sigma_{ij} \frac{1}{q_0^{(r)}} \sigma_{ij} \\
&= \frac{1}{2}\sum_{r=1}^{n} \frac{1}{q_0^{(r)}} \int_0^{\xi} R(\xi - \xi') \frac{\partial \hat{\gamma}_{ij}}{\partial \xi'} d\xi' \int_0^{\xi} R(\xi - \xi'') \frac{\partial \hat{\gamma}_{ij}}{\partial \xi''} d\xi'' \\
&= \frac{1}{2}\sum_{r=1}^{n} \frac{1}{q_0^{(r)}} \int_0^{\xi}\int_0^{\xi} \left(q_0^{(r)}\right)^2 e^{-\lambda_{(r)}(2\xi - \xi' - \xi'')} \frac{\partial \hat{\gamma}_{ij}}{\partial \xi'} \frac{\partial \hat{\gamma}_{ij}}{\partial \xi''} d\xi' d\xi'' \\
&= \frac{1}{2}\int_0^{\xi}\int_0^{\xi} R(2\xi - \xi' - \xi'') \frac{\partial \hat{\gamma}_{ij}}{\partial \xi'} \frac{\partial \hat{\gamma}_{ij}}{\partial \xi''} d\xi' d\xi''.
\end{aligned} \tag{7.3.57}$$

The dissipation is defined as

$$\begin{aligned}
\hat{D}(\xi) &= \hat{\sigma}_{ij}\dot{\hat{\gamma}}_{ij} = \sum_{r=1}^{n} \sigma_{ij} \frac{1}{q_0^{(r)}} \dot{\sigma}_{ij} \\
&= \sum_{r=1}^{n} \int_0^{\xi}\int_0^{\xi} \frac{1}{q_0^{(r)}} \left(q_0^{(r)}\right)^2 e^{-\lambda_{(r)}(2\xi - \xi' - \xi'')} \frac{\partial \hat{\gamma}_{ij}}{\partial \xi'} \frac{\partial \hat{\gamma}_{ij}}{\partial \xi'} d\xi' d\xi'' \\
&\quad + \frac{1}{2}\int_0^{\xi}\int_0^{\xi} \frac{\partial^2 R(2\xi - \xi' - \xi'')}{\partial \xi' \partial \xi''} \hat{\gamma}_{ij}(\xi')\hat{\gamma}_{ij}(\xi'') d\xi' d\xi'' \\
&= \frac{1}{2}\int_0^{\xi} R(\xi - \xi')\hat{\gamma}_{ij}(\xi) \frac{\partial \hat{\gamma}_{ij}}{\partial \xi'} d\xi' - \frac{1}{2} R(0)\hat{\gamma}_{ij}(\xi)\hat{\gamma}_{ij}(\xi) \\
&\quad + \frac{1}{2}\int_0^{\xi} R(\xi - \xi')\hat{\gamma}_{ij}(\xi) \frac{\partial \hat{\gamma}_{ij}(\xi'')}{\partial \xi''} d\xi'' \\
&\quad + \frac{1}{2}\int_0^{\xi}\int_0^{\xi} \frac{\partial^2 R(2\xi - \xi' - \xi'')}{\partial \xi' \partial \xi''} \hat{\gamma}_{ij}(\xi')\hat{\gamma}_{ij}(\xi'') d\xi' d\xi''.
\end{aligned} \tag{7.3.58}$$

Therefore,

$$\hat{\sigma}_{ij}(\xi) = \frac{\partial \hat{\phi}(\xi)}{\partial \hat{\gamma}_{ij}(\xi)}$$

$$= \int_0^{\xi} R(\xi - \xi') \frac{\partial \hat{\gamma}_{ij}}{\partial \xi'} d\xi' - R(0)\hat{\gamma}_{ij}(\xi), \tag{7.3.59}$$

where $R_0(0)\hat{\gamma}_{ij}(\xi) = \sigma_{ij}(0) = 0$ is the initial stress. Finally, we obtain

$$\hat{\sigma}_{ij}(\xi) = \int_0^{\xi} R(\xi - \xi') \frac{\partial \hat{\gamma}_{ij}}{\partial \xi'} d\xi',$$

which is identical to Eq. (7.3.56). This demonstrates that Eq. (7.3.57) is an acceptable form of free energy.

Expanding Eq. (7.3.56) in deviatoric and dilatational parts, we obtain

$$\hat{\sigma}_{ij}(\xi) = \int_0^{\xi} 2G(\xi - \xi') \frac{\partial \hat{\gamma}'_{ij}}{\partial \xi'} d\xi' + \frac{1}{3}\delta_{ij} \int_0^{\xi} 3K(\xi - \xi') \frac{\partial \hat{\gamma}'_{kk}}{\partial \xi'} d\xi',$$

where $\hat{\gamma}'_{ij}$ is the deviatoric stress tensor:

$$\hat{\gamma}'_{ij} = \gamma'_{ij} - \frac{1}{3}\delta_{ij}\hat{\gamma}_{kk}.$$

Substituting, we get

$$\hat{\sigma}_{ij}(\xi) = \int_0^{\xi} 2G(\xi - \xi') \frac{\partial \hat{\gamma}_{ij}}{\partial \xi'} d\xi' + \frac{1}{3}\delta_{ij} \int_0^{\xi} [3K(\xi - \xi')$$

$$-2G\,(\xi - \xi')]\,\frac{\partial \hat{\gamma}_{kk}}{\partial \xi'} d\xi'$$

$$= -\frac{1}{2} \int_0^{\xi} \int_0^{\xi} \left(q_0^{(r)} e^{-\lambda_{(r)}(2\xi - \xi' - \xi'')}\right) \frac{\partial \hat{\gamma}_{ij}}{\partial \xi'} \frac{\partial \hat{\gamma}_{ij}}{\partial \xi''} d\xi' d\xi''$$

$$= -\frac{1}{2} \int_0^{\xi} \int_0^{\xi} \frac{\partial}{\partial \xi} R(2\xi - \xi' - \xi'') \frac{\partial \hat{\gamma}_{ij}}{\partial \xi'} \frac{\partial \hat{\gamma}_{ij}}{\partial \xi''} d\xi' \xi''. \tag{7.3.60}$$

To see that Eq. (7.3.57) satisfies the requirement

$$\hat{\sigma}_{ij} = \frac{\partial \hat{\phi}}{\partial \hat{\gamma}_{ij}},$$

we first integrate $\hat{\phi}$ by parts:

$$\hat{\phi} = \frac{1}{2} \int_0^{\xi} R\,(2\xi - \xi' - \xi'') \frac{\partial \hat{\gamma}_{ij}\,(\xi')}{\partial \xi'} d\xi' \hat{\gamma}_{ij}\,(\xi'')\Big|_{\xi''=0}^{\xi'=\xi}$$

$$= -\frac{1}{2} \int_0^{\xi} \int_0^{\xi} \frac{\partial}{\partial \xi''} R\,(2\xi - \xi' - \xi'') \frac{\partial \hat{\gamma}_{ij}\,(\xi')}{\partial \xi'} d\xi' \hat{\gamma}_{ij}\,(\xi'')\, d\xi''$$

$$= \frac{1}{2} \int_0^{\xi} R\,(\xi - \xi') \frac{\partial \hat{\gamma}_{ij}\,(\xi')}{\partial \xi'} d\xi' \hat{\gamma}_{ij}\,(\xi)$$

$$
-\frac{1}{2}\int_0^{\xi}\frac{\partial}{\partial\xi''}R\left(2\xi-\xi'-\xi''\right)\hat{\gamma}_{ij}\left(\xi'\right)\hat{\gamma}_{ij}\left(\xi''\right)d\xi''\Bigg|_{\xi'=0}^{\xi'=\xi}
$$

$$
+\frac{1}{2}\int_0^{\xi}\int_0^{\xi}\frac{\partial^2 R\left(2\xi-\xi'-\xi''\right)}{\partial\xi'\partial\xi''}\hat{\gamma}_{ij}\left(\xi'\right)\hat{\gamma}_{ij}\left(\xi''\right)d\xi' d\xi''
$$

$$
=\frac{1}{2}\int_0^{\xi}R\left(\xi-\xi'\right)\hat{\gamma}_{ij}\left(\xi\right)\frac{\partial\hat{\gamma}_{ij}\left(\xi'\right)}{\partial\xi'}d\xi'-\frac{1}{2}\int_0^{\xi}\frac{\partial}{\partial\xi''}R\left(2\xi-\xi''\right)\hat{\gamma}_{ij}\left(\xi\right)\hat{\gamma}_{ij}\left(\xi''\right)d\xi''
$$

$$
+\frac{1}{2}\int_0^{\xi}\int_0^{\xi}\frac{R\left(2\xi-\xi'-\xi''\right)}{\partial\xi'\partial\xi''}\hat{\gamma}_{ij}\left(\xi'\right)\hat{\gamma}_{ij}\left(\xi''\right)d\xi' d\xi''
$$

$$
=\frac{1}{2}\int_0^{\xi}R\left(\xi-\xi'\right)\hat{\gamma}_{ij}\left(\xi\right)\frac{\partial\hat{\gamma}_{ij}\left(\xi'\right)}{\partial\xi'}d\xi'-\frac{1}{2}R\left(\xi-\xi''\right)\hat{\gamma}_{ij}\left(\xi\right)\hat{\gamma}_{ij}\left(\xi''\right)\Bigg|_{\xi''=0}^{\xi''=\xi}
$$

$$
+\frac{1}{2}\int_0^{\xi}R\left(\xi-\xi'\right)\hat{\gamma}_{ij}\left(\xi\right)\frac{\partial\hat{\gamma}_{ij}\left(\xi''\right)}{\partial\xi''}d\xi''.
$$

Adding the thermal stress, we obtain

$$
\hat{\sigma}_{ij}(\xi)=\int_0^{\xi}2G(\xi-\xi')\frac{\partial\hat{\gamma}_{ij}}{\partial\xi'}d\xi'+\frac{1}{3}\delta_{ij}\int_0^{\xi}[3K(\xi-\xi')
$$

$$
-2G(\xi-\xi')]\frac{\partial\hat{\gamma}_{kk}}{\partial\xi'}d\xi'-\delta_{ij}\int_0^{\xi}3\alpha K(\xi-\xi')\frac{\partial\hat{\theta}}{\partial\xi'}d\xi'. \quad (7.3.61)
$$

The form of free energy that may lead to this result is found to be

$$
\hat{\phi}(\xi)=-\frac{1}{2}\int_0^{\xi}\int_0^{\xi}2G(2\xi-\xi'-\xi'')\frac{\partial\hat{\gamma}'_{ij}}{\partial\xi'}\frac{\partial\hat{\gamma}'_{ij}}{\partial\xi''}d\xi' d\xi''
$$

$$
+\frac{1}{6}\int_0^{\xi}\int_0^{\xi}3K(2\xi-\xi'-\xi'')\frac{\partial(\hat{\gamma}'_{kk}-3\alpha\hat{\theta})}{\partial\xi'}\frac{\partial(\hat{\gamma}'_{kk}-3\alpha\hat{\theta})}{\partial\xi''}d\xi' d\xi''
$$

$$
=\frac{1}{4}(\delta_{ij}\delta_{jk}+\delta_{ik}\delta_{jl})\int_0^{\xi}\int_0^{\xi}2G(2\xi-\xi'-\xi'')\frac{\partial\hat{\gamma}'_{ij}}{\partial\xi'}\frac{\partial\hat{\gamma}'_{kl}}{\partial\xi''}d\xi' d\xi''
$$

$$
+\frac{1}{6}\delta_{ij}\delta_{kl}\int_0^{\xi}\int_0^{\xi}[3K\left(2\xi-\xi'-\xi''\right)
$$

$$
-2G\left(2\xi-\xi'-\xi''\right)]\frac{\partial\hat{\gamma}'_{ij}}{\partial\xi'}\frac{\partial\hat{\gamma}'_{kl}}{\partial\xi''}d\xi' d\xi''
$$

$$
-\delta_{ij}\int_0^{\xi}\int_0^{\xi}3\alpha K(2\xi-\xi'-\xi'')\frac{\partial\hat{\gamma}'_{ij}}{\partial\xi'}\frac{\partial\hat{\theta}}{\partial\xi''}d\xi' d\xi''
$$

$$
+\frac{1}{6}\int_0^{\xi}\int_0^{\xi}27\alpha^2 K(2\xi-\xi'-\xi'')\frac{\partial\hat{\theta}}{\partial\xi'}\frac{\partial\hat{\theta}}{\partial\xi''}d\xi' d\xi''. \quad (7.3.62)
$$

Likewise, the dissipation is of the form

$$\hat{D}(\xi) = -\frac{1}{2}\int_0^{\xi}\int_0^{\xi}\frac{\partial}{\partial\xi}[2G(2\xi-\xi'-\xi'')]\frac{\partial\hat{\gamma}_{ij}}{\partial\xi'}\frac{\partial\hat{\gamma}_{ij}}{\partial\xi''}d\xi'd\xi''$$
$$-\frac{1}{6}\int_0^{\xi}\int_0^{\xi}\frac{\partial}{\partial\xi}[3K(2\xi-\xi'-\xi'')-2G(2\xi-\xi'-\xi'')]\frac{\partial\hat{\gamma}_{ii}}{\partial\xi'}\frac{\partial\hat{\gamma}_{jj}}{\partial\xi''}d\xi'd\xi''$$
$$+\int_0^{\xi}\int_0^{\xi}\frac{\partial}{\partial\xi}[3\alpha K(2\xi-\xi'-\xi'')]\frac{\partial\hat{\gamma}_{kk}}{\partial\xi'}\frac{\partial\hat{\gamma}_{kk}}{\partial\xi''}d\xi'd\xi''$$
$$-\frac{1}{2}\int_0^{\xi}\int_0^{\xi}\frac{\partial}{\partial\xi}[9\alpha^2K(2\xi-\xi'-\xi'')]\frac{\partial\hat{\theta}}{\partial\xi'}\frac{\partial\hat{\theta}}{\partial\xi''}d\xi'd\xi''. \quad (7.3.63)$$

Using a similar approach, we can obtain the following relations for the generalized Kelvin model:

$$\phi(t) = -\frac{1}{2}\int_0^{t}\int_0^{t}J(2t-\xi-\zeta)\frac{\partial\sigma_{ij}(\xi)}{\partial\xi}\frac{\partial\sigma_{ij}(\zeta)}{\partial\zeta}d\xi d\zeta$$
$$+\sigma_{ij}\int_0^{t}J(t-\xi)\frac{\partial\sigma_{ij}}{\partial\xi}d\xi, \quad (7.3.64)$$

$$D(t) = \frac{1}{2}\int_0^{t}\int_0^{t}\frac{d}{dt}[J(2t-\xi-\zeta)]\frac{\partial\sigma_{ij}(\xi)}{\partial\xi}\frac{\partial\sigma_{ij}(\zeta)}{\partial\zeta}d\xi d\zeta. \quad (7.3.65)$$

In summary, we have derived the appropriate forms of free energy and the corresponding stress–strain relations and dissipation energy for a thermorheologically simple material from the viewpoint of irreversible thermodynamics. These constitutive equations, because of their complexity, are difficult to apply in analytical solutions. Appropriate numerical methods may be introduced to solve the governing equilibrium equations together with these constitutive relations.

7.4 Plasticity

7.4.1 General

The mechanics of plastic behavior is overwhelmingly complex. To use it in engineering requires a great number of simplifying assumptions. There are two approaches to the study of plasticity: one is to look at the material on a microscopic level, and the other, on a macroscopic level. Consider a material as consisting of many individual particles oriented in any arbitrary manner. When each particle is excited by external loads, it moves in translation, rotation, or both. On release of the loads, the particle may or may not return to its original position. The study of these actions at the microscopic level is categorized as the theory of *dislocation*, whereas the macroscopic view of such phenomena leads to the plasticity theory. The dislocation theory has been studied predominantly by physicists

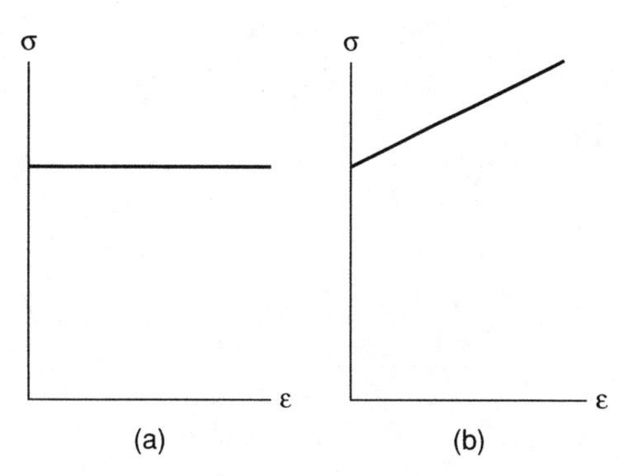

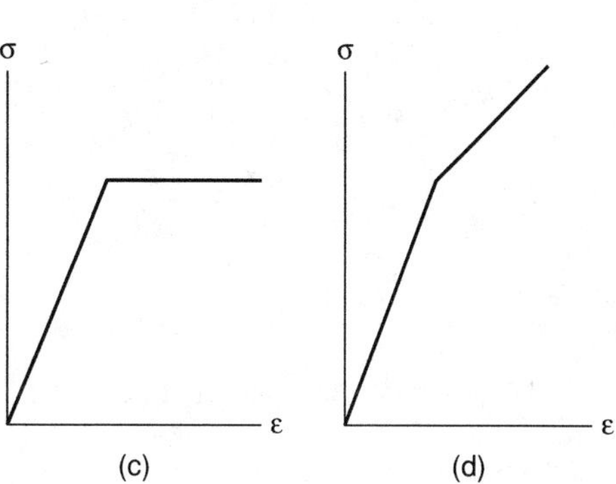

Figure 7.4.1. Various plastic behavior: (a) rigid perfectly plastic, (b) rigid linear strain hardening, (c) elastic perfectly plastic, (d) elastic linear strain hardening.

in connection with crystallography. The microscopic view of material behavior is considered either unnecessary or impractical to engineers who are primarily concerned with the analysis and design of load-carrying structural members.

The subject of plasticity theory has been studied by scientists and engineers for the past two centuries. It began with the works of Rankine (1851), Tresca (1864), von Mises (1913), Prandtl (1924), Reuss (1930), and others. More recent contributions to the literature are those of Prager (1945), Hill (1950), Drucker (1950), Koiter (1953), Ziegler (1959), Naghdi (1960), Ilyushin (1963), Mroz (1967), and Valanis (1980), among others.

7.4.2 Plasticity Theories

Idealization of Stress–Strain Relations. To obtain a solution to a deformation problem, it is necessary to idealize the stress–strain relation. There are several criteria which may be used in idealizing the stress–strain relation. The four parts of Fig. 7.4.1 show the strain–stress relations for rigid perfectly plastic material, elastic perfectly plastic material, rigid strain-hardening material, and elastic strain-hardening material (Hill, 1950; Phillips, 1956; Johnson and Mellor, 1962). An analysis based on these idealizations must be justified for each type of material before adoption.

For some materials the yield points in tension and compression are the same. The strain hardening occurring under this situation is called *isotropic hardening*.

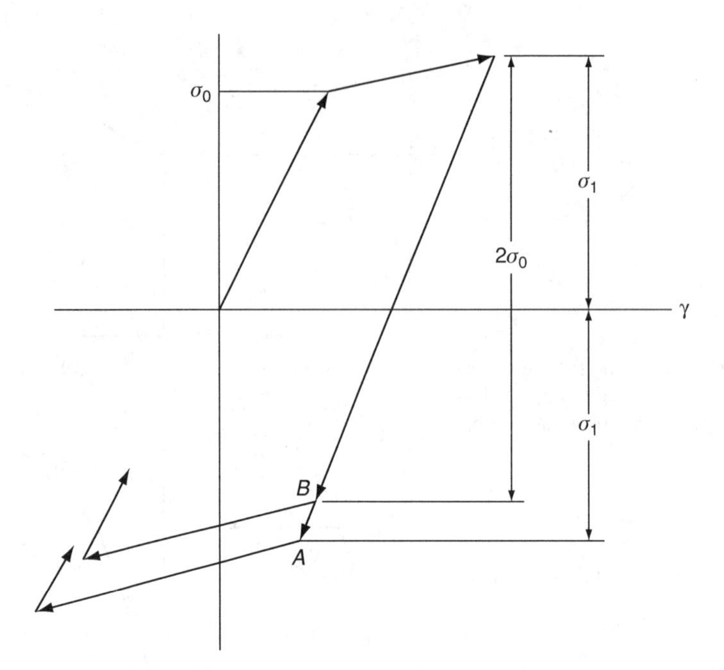

Figure 7.4.2. Bauschinger effect.

If metal is first deformed in tension and then reloaded in compression, the yield in compression may be less than that in tension. This is due to residual stress left in tension or to anisotropy of the dislocation field in reversal of the load. This phenomenon is called the *Bauschinger effect*; it causes the yield point in compression to occur somewhere between A and B in Fig. 7.4.2. The strain hardening associated with the Bauschinger effect is called *kinematic hardening*.

Failure Theories. In the design of a structure, we are concerned with the load-carrying capacity; more basically, the point of yield or failure. In earlier days, yielding of a material was defined in various ways, based mainly on experimental results. In what follows, we discuss various failure theories that have evolved over the past two centuries.

Maximum Stress Theory. Rankine (1851) proposed that yielding occurs when one of the principal stresses becomes equal to the yield stress σ_0 in simple tension or σ_{0c} in compression. This theory is shown schematically in Fig. 7.4.3. The rectangular block called the *yield surface* becomes a square if $\sigma_0 = \sigma_{0c}$. It is seen that any state of stress on or outside the yield surface represents yielding of the material.

The Rankine theory, commonly known as the *maximum stress theory*, is not considered dependable for many of the engineering materials in use today.

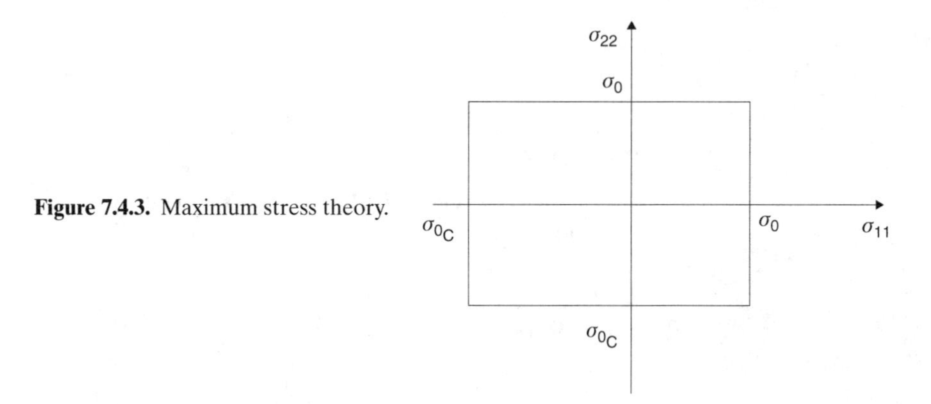

Figure 7.4.3. Maximum stress theory.

Maximum Strain Theory. This theory was first proposed by Saint-Venant in 1870. The maximum strain theory states that yielding occurs when the maximum value of the principal strain equals the value of the yield strain in simple tension or compression.

Let γ_0 be the yield strain and σ_0 the corresponding yield stress. Then, from Hooke's law, we have

$$\gamma_0 = \frac{\sigma_0}{E}, \tag{7.4.1}$$

where E is Young's modulus. For a biaxial state of stress, the principal strains are

$$\gamma_{11} = \frac{\sigma_{11}}{E} - v\frac{\sigma_{22}}{E}, \quad \gamma_{22} = \frac{\sigma_{22}}{E} - v\frac{\sigma_{11}}{E},$$

$$\sigma_{11} = E\gamma_{11} + v\sigma_{22}, \tag{7.4.2a}$$

$$\sigma_{22} = E\gamma_{22} + v\sigma_{11}, \tag{7.4.2b}$$

where v is the Poisson ratio. If the principal strains become equal to the yield strains, i.e., if $\gamma_{11} = \gamma_0 = \sigma_0/E$ and $\gamma_{22} = \gamma_0 = \sigma_0/E$, then Eqs. (7.4.2a) and (7.4.2b) assume the form

$$\sigma_{11} = \sigma_0 + v\sigma_{22}, \tag{7.4.3a}$$

$$\sigma_{22} = \sigma_0 + v\sigma_{11}. \tag{7.4.3b}$$

Plotting Eqs. (7.4.3a) and (7.4.3b) in the principal stress space, we obtain the diamond-shaped yield surface shown in Fig. 7.4.4.

The maximum strain theory was extensively used for gun-barrel (thick-wall cylinder) design during the 19th century.

Maximum Shear Theory (Tresca Theory). The maximum shear theory, known as the Tresca theory (1864), was first conceived in 1773 by Coulomb, who applied it to the design of soil foundations. The theory states that yielding occurs when the shear stress reaches the value of the maximum yield shear stress in a tension

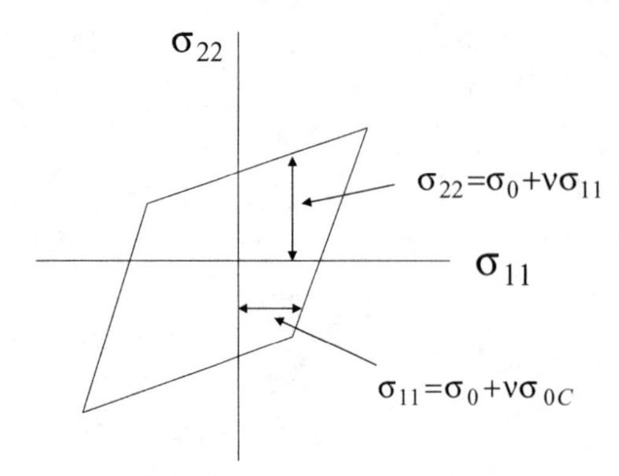

Figure 7.4.4. Maximum strain theory.

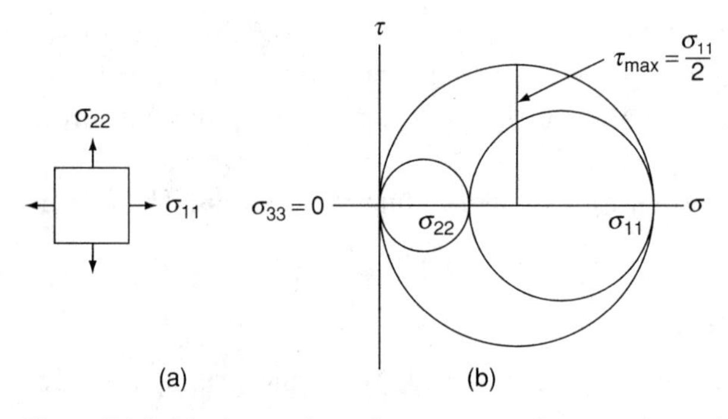

Figure 7.4.5. Maximum shear theory – normal stresses in tension.

test. In a triaxial state of stress, this criterion can be written as follows:

$$\frac{\sigma_{11} - \sigma_{22}}{2} = \frac{\sigma_0}{2} = k, \tag{7.4.4a}$$

$$\frac{\sigma_{22} - \sigma_{33}}{2} = \frac{\sigma_0}{2} = k, \tag{7.4.4b}$$

$$\frac{\sigma_{33} - \sigma_{11}}{2} = \frac{\sigma_0}{2} = k, \tag{7.4.4c}$$

where k is the yield stress in pure shear. For simplicity, let us consider a biaxial state of stress. First, assume that both σ_{11} and σ_{22} are in tension, as shown in Fig. 7.4.5(a). The Mohr circle representation is shown in Fig. 7.4.5(b). The maximum shear is found to be

$$\tau_{max} = \frac{\sigma_{11}}{2}. \tag{7.4.5}$$

Here, on yielding, we have

$$\left|\frac{\sigma_{11}}{2}\right| = \frac{\sigma_0}{2}. \tag{7.4.6}$$

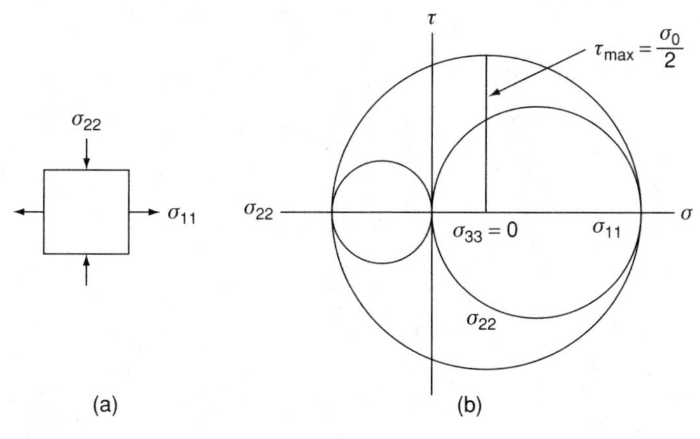

Figure 7.4.6. Maximum shear theory – σ_{22} in compression.

Similarly, if both σ_{11} and σ_{22} are in compression, the Mohr circle is plotted to the left of the τ axis in Fig. 7.4.5(b) and we get, on yielding,

$$\left|\frac{\sigma_{22}}{2}\right| = \frac{\sigma_0}{2}. \tag{7.4.7}$$

Let us now consider the case in which σ_{11} is in tension while σ_{22} is in compression, as shown in Fig. 7.4.6(a). From the Mohr circle for this case, we note that

$$\tau_{max} = \frac{|\sigma_{22}| + |\sigma_{11}|}{2}, \tag{7.4.8}$$

$$\frac{|\sigma_{11}| + |\sigma_{22}|}{2} = \frac{\sigma_0}{2}. \tag{7.4.9}$$

It is seen that Eq. (7.4.9) may be written in the form

$$\frac{\sigma_{11}}{\sigma_0} - \frac{\sigma_{22}}{\sigma_0} = \pm 1. \tag{7.4.10}$$

Plots of Eqs. (7.4.6) (7.4.7) and (7.4.10) result in a hexagon, as shown in Fig. 7.4.7. Note that Eqs. (7.4.6) and (7.4.7) represent the first and third quadrants, respectively, whereas Eq. (7.4.10) represents the second quadrant for $\sigma_{11}^{(-)}, \sigma_{22}^{(+)}$ and the fourth quadrant for $\sigma_{11}^{(+)}, \sigma_{22}^{(-)}$.

The Tresca theory has been popular because of its superiority over the maximum stress or maximum strain theory in predicting the failure of ductile materials.

Maximum Distortion-Energy Theory (von Mises Theory). The maximum distortion-energy theory, also known as the von Mises theory, perhaps the most widely used method of defining failure in ductile materials, was contributed by Huber (1904), von Mises (1913), Levy (1921), and Hencky (1924). In this theory,

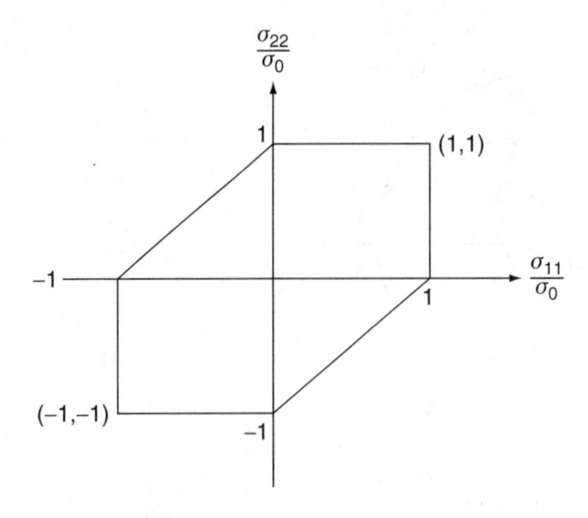

Figure 7.4.7. Maximum shear hexagon.

we assume that yielding occurs when the distortion energy equals the maximum yield distortion energy in simple tension.

Consider a state of triaxial stresses. The total strain energy is given by

$$U_T = \frac{1}{2}\sigma_{ij}\gamma_{ij}. \tag{7.4.11}$$

Expanding Eq. (7.4.11), we obtain

$$\begin{aligned} U_T &= \frac{1}{2}(\sigma_{11}\gamma_{11} + \sigma_{22}\gamma_{22} + \sigma_{33}\gamma_{33} + 2\sigma_{12}\gamma_{12} + 2\sigma_{23}\gamma_{23} + 2\sigma_{31}\gamma_{31}) \\ &= \frac{1}{2E}\left(\sigma_{11}^2 + \sigma_{22}^2 + \sigma_{33}^2\right) - \frac{\nu}{E}(\sigma_{11}\sigma_{22} + \sigma_{22}\sigma_{33} + \sigma_{33}\sigma_{11}) \\ &\quad + \frac{1}{2G}\left(\sigma_{12}^2 + \sigma_{23}^2 + \sigma_{31}^2\right), \end{aligned} \tag{7.4.12}$$

where $G = E/2(1 + \upsilon)$ is the shear modulus. If shear stresses are zero, i.e., if

$$\sigma_{12} = \sigma_{23} = \sigma_{31} = 0$$

and if we let $\sigma_{11} = \sigma_{22} = \sigma_{33} = p$, then we obtain the dilatation part of the total strain energy in the form

$$U_{\text{dilatation}} = \frac{3(1 - 2\nu)p^2}{2E} = \frac{1 - 2\nu}{6E}(\sigma_{11} + \sigma_{22} + \sigma_{33})^2. \tag{7.4.13}$$

Subtracting Eq. (7.4.13) from Eq. (7.4.12) yields the distortion part of the total strain energy in the form

$$U_{\text{distortion}} = \frac{1}{12G}[(\sigma_{11} - \sigma_{22})^2 + (\sigma_{22} - \sigma_{33})^2 + (\sigma_{33} - \sigma_{11})^2]. \tag{7.4.14}$$

If one of the principal stresses reaches the yield stress while all other principal stresses are zero, as in a simple tension test, then Eq. (7.4.14) becomes

$$U_{\text{distortion}} = \frac{\sigma_0^2}{6G}. \tag{7.4.15}$$

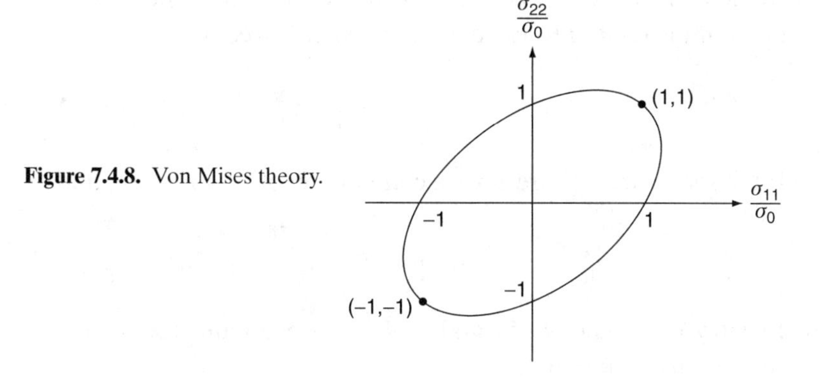

Figure 7.4.8. Von Mises theory.

Combining Eqs. (7.4.14) and (7.4.15) yields

$$\frac{1}{2}[(\sigma_{11}-\sigma_{22})^2+(\sigma_{22}-\sigma_{33})^2+(\sigma_{33}-\sigma_{11})^2]=\sigma_0^2. \tag{7.4.16}$$

Equation (7.4.16) is the von Mises yield criterion in a three-dimensional state of stress.

For a plane stress, where $\sigma_{33}=0$, Eq. (7.4.16) assumes the form

$$\frac{\sigma_{11}^2}{\sigma_0}-\frac{\sigma_{11}\sigma_{22}}{\sigma_0^2}+\frac{\sigma_{22}^2}{\sigma_0}=1. \tag{7.4.17}$$

The yield surface of Eq. (7.4.17) is plotted as an ellipse in Fig. 7.4.8. Note that the von Mises theory depends on the squares of stress, a form of energy, and it can be related to such useful quantities in continuum mechanics as the second deviatoric stress invariant or the square of the octahedral shear stress. To examine these relationships, consider the deviatoric stress defined as

$$\hat{\sigma}_{ij}=\sigma_{ij}-\frac{1}{3}\sigma_{kk}\delta_{ij}.$$

The first, second, and third deviatoric stress invariants are

$$J_1=\hat{\sigma}_{ii}=0, \tag{7.4.18a}$$

$$J_2=\frac{1}{2}\hat{\sigma}_{ij}\hat{\sigma}_{ij}, \tag{7.4.18b}$$

$$J_3=\frac{1}{3}\hat{\sigma}_{ij}\hat{\sigma}_{jk}\hat{\sigma}_{ki}. \tag{7.4.18c}$$

The octahedral shear stress is defined as

$$\tau_{\text{oct}}=\frac{1}{3}\sqrt{(\sigma_{11}-\sigma_{22})^2+(\sigma_{22}-\sigma_{33})^2+(\sigma_{33}-\sigma_{11})^2+6\,[(\sigma_{12})^2+(\sigma_{23})^2+(\sigma_{31})^2]}. \tag{7.4.19}$$

Expanding Eq. (7.4.18b), we get

$$J_2=\frac{1}{6}[(\sigma_{11}-\sigma_{22})^2+(\sigma_{22}-\sigma_{33})^2+(\sigma_{33}-\sigma_{11})^2]+\sigma_{12}^2+\sigma_{23}^2+\sigma_{31}^2. \tag{7.4.20}$$

In view of Eqs. (7.4.19) and (7.4.20), we note that the second deviatoric stress invariant is related to the octahedral shear stress by

$$J_2 = \frac{3}{2}\tau_{\text{oct}}^2. \tag{7.4.21}$$

If the shear stresses are zero, a comparison of Eqs. (7.4.14) and (7.4.21) leads to

$$U_{\text{distortion}} = \frac{1}{2G}J_2 = \frac{3}{4G}\tau_{\text{oct}}^2. \tag{7.4.22}$$

In view of Eqs. (7.4.15) and (7.4.22), we obtain, for a pure shear condition ($\sigma_{11} = -\sigma_{22} = \kappa = \sigma_0$),

$$J_2 = \frac{\sigma_{11}^2}{3}\frac{\sigma_0^2}{3} = \kappa^2.$$

Thus the yield stress in pure shear is $1/\sqrt{3}$ times the yield stress in simple tension. Also, from the Tresca yield criterion [Eq. (7.4.4)], we observe that the von Mises theory gives the yield stress in pure shear as 15% higher than does the Tresca theory.

It should be noted that, in the von Mises theory, yielding is independent of the first and third deviatoric stress invariants. This implies that hydrostatic pressure does not contribute to yielding. Furthermore, because the third deviatoric stress invariant is an odd function, it would add unnecessary algebraic difficulties were it present in the yield criterion.

Among the main advantages of the von Mises theory is the possibility of functional representation of yield phenomena not only at yield but also beyond the yield point. Therefore this theory permits a convenient mathematical modeling of strain-hardening phenomena leading to the so-called elastoplastic structural analysis.

7.4.3 Stress-Space Representation

The two-dimensional stress space is shown in Figs. 7.4.3, 7.4.4, 7.4.7, and 7.4.8 for the various failure theories. For comparison, they are plotted together in Fig. 7.4.9.

For a three-dimensional state of stress, we can obtain the representation of the stress space by plotting Eqs. (7.4.4b) and (7.4.4c) for the Tresca theory and Eq. (7.4.16) for the von Mises theory, as shown in Fig. 7.4.10. Note that the Tresca yield surface is the regular hexagonal prism whose six plane faces are defined by Eqs. (7.4.4), with κ replaced with $\sigma_0/2$, whereas the von Mises yield surface $J_2 = \sigma_0^2/3$ is the right circular cylinder defined by Eq. (7.4.16). It is seen that the projections of the three axes, σ_{11}, σ_{22}, and σ_{33}, on the cross section of the cylinder called the *deviatoric plane* ($\sigma_{11} + \sigma_{22} + \sigma_{33} = 0$) are 120° apart, making the angle

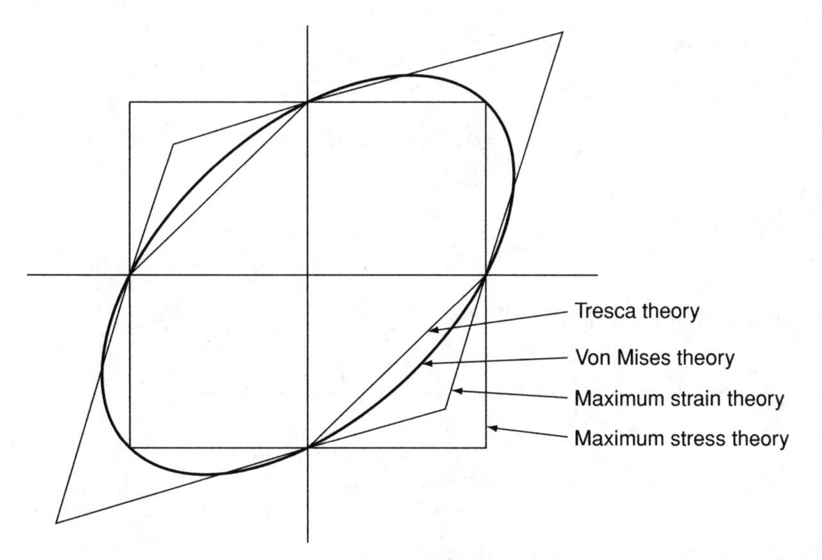

Figure 7.4.9. Stress-space representation.

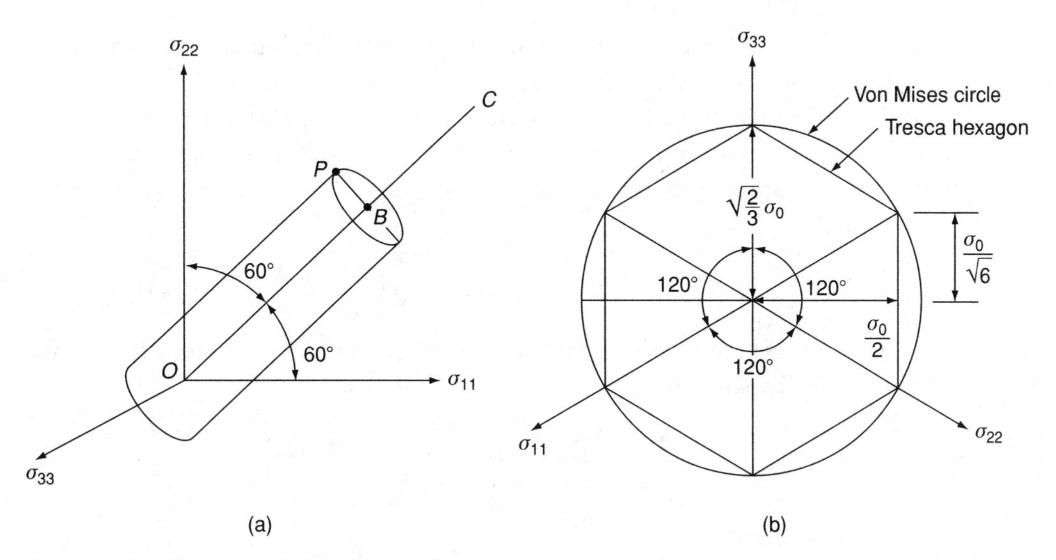

Figure 7.4.10. Von Mises circle and Tresca hexagon.

between the hydrostatic line $\overline{OC}$ and any of the stress axes 60°. The lengths $\overline{OP}$ and $\overline{OB}$ are given as

$$\overline{OP} = \left(\sigma_{11}^2 + \sigma_{22}^2 + \sigma_{33}^2\right)^{1/2},$$

$$\overline{OB} = \frac{1}{\sqrt{3}}(\sigma_{11} + \sigma_{22} + \sigma_{33}).$$

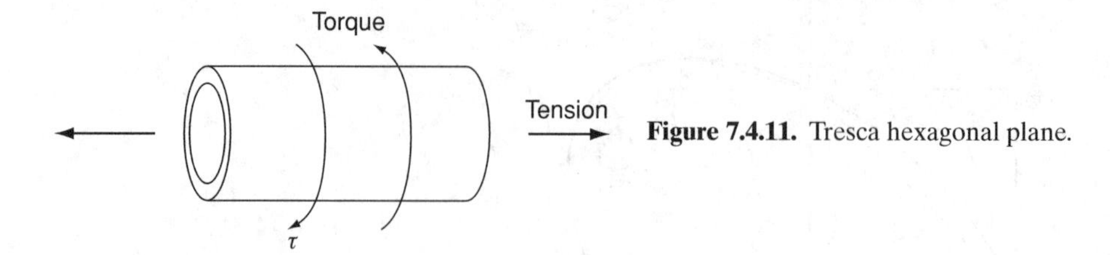

Figure 7.4.11. Tresca hexagonal plane.

Therefore the radius of the circle is

$$\overline{BP} = (\overline{OP}^2 - \overline{OB}^2)^{1/2} = \left[\sigma_{11}^2 + \sigma_{22}^2 + \sigma_{33}^2 - \frac{(\sigma_{11} + \sigma_{22} + \sigma_{33})^2}{3}\right]^{1/2}$$

$$= \left\{\frac{1}{3}.[(\sigma_{11} - \sigma_{22})^2 + (\sigma_{22} - \sigma_{33})^2 + (\sigma_{33} - \sigma_{11})^2]\right\}^{1/2}.$$

Comparing this relation with Eq. (7.4.16), we obtain

$$\overline{BP} = (2J_2)^{1/2} = \left(\frac{2}{3}\sigma_0^2\right)^{1/2} = \sqrt{\frac{2}{3}}\sigma_0.$$

The geometry of the Tresca hexagonal plane may be similarly evaluated; its dimensions are as shown in Fig. 7.4.10(b).

Instead of representing the yield surface in the principal stress space, we may also show the yield surface in the normal stress–shear stress ($\sigma\tau$) space. Consider a thin-walled tube subjected to tension and torque as depicted in Fig. 7.4.11. Note that

$$\sigma_{11} = \frac{\sigma}{2} + \sqrt{\left(\frac{\sigma}{2}\right)^2 + \tau^2}, \quad \sigma_{22} = \frac{\sigma}{2} - \sqrt{\left(\frac{\sigma}{2}\right)^2 + \tau^2}, \quad \sigma_{33} = 0.$$

Substituting these equalities into the Tresca theory and the von Mises theory, we obtain for the Tresca theory

$$\frac{\sigma_{11} - \sigma_{22}}{2} = \sqrt{\left(\frac{\sigma}{2}\right)^2 + \tau^2} = \frac{\sigma_0}{2}$$

or

$$\left(\frac{\sigma}{\sigma_0}\right)^2 + \left(\frac{2\tau}{\sigma_0}\right)^2 = 1, \tag{7.4.23}$$

and for the von Mises theory,

$$\frac{1}{2}[(\sigma_{11} - \sigma_{22})^2 + (\sigma_{22} - \sigma_{33})^2 + (\sigma_{33} - \sigma_{11})^2] = (\sigma^2 + 3\tau^2) = \sigma_0^2$$

or

$$\left(\frac{\sigma}{\sigma_0}\right)^2 + 3\left(\frac{\tau}{\sigma_0}\right)^2 = 1. \tag{7.4.24}$$

The plots of Eqs. (7.4.23) and (7.4.24) are shown in Fig. 7.4.12.

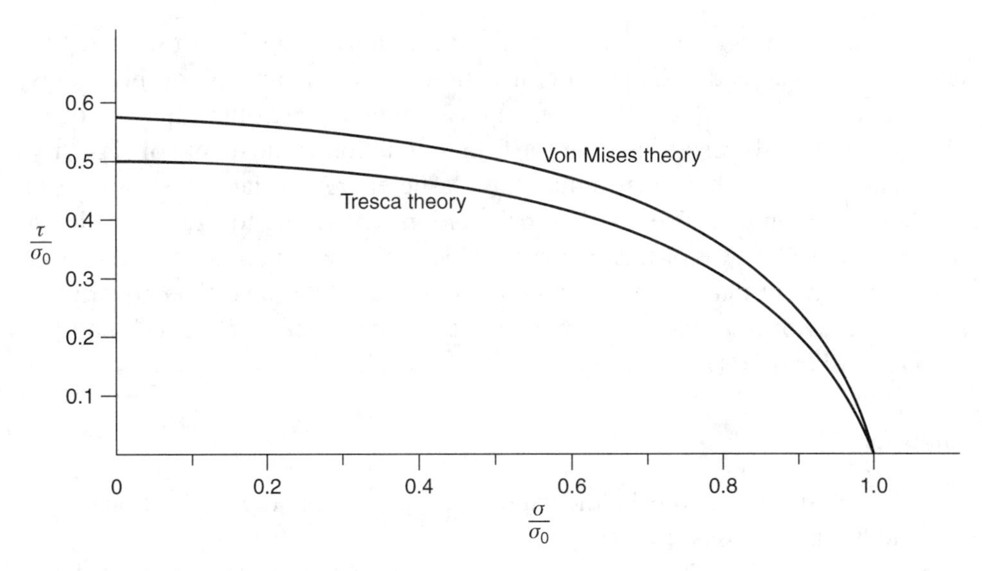

Figure 7.4.12. Tresca and von Mises theories.

It should be noted that the yield-surface representations discussed here are limited to isotropic materials that have equal yield stresses in tension and compression. For anisotropic materials having various directional properties coupled with different tensile and compressive yield stresses, we expect to have accordingly different yield-surface representations. The discussion of such materials is beyond the scope of this text.

In the following discussion, we are confined to the incremental theory of plasticity in which the so-called flow rule governing the incremental plastic strains is used. It can be adapted well to the numerical analysis. The deformation theory of plasticity advanced by Budiansky (1959), on the other hand, uses the total strain in defining the plastic deformations. Its application to the numerical analysis is seriously limited.

7.4.4 Plastic Potential Function

In the study of elasticity theory, one deals with an elastic strain-energy function $W = \hat{W}(\gamma_{ij})$ (see Subsection 4.2.1). The use of a Taylor series expansion of W with the second term retained, together with the virtual work relation, leads to the constitutive equation in the form of the generalized Hooke's law

$$\sigma_{ij} = \frac{\partial W}{\partial \gamma_{ij}} = E_{ijkl}\gamma_{kl}, \tag{7.4.25}$$

where E_{ijkl} is the tensor of elastic moduli.

In a similar manner, if the material reaches the elastic limit and further deformations take place, we may still be concerned with the energy stored in the plastically deformed body. This energy is referred to as the plastic energy. For the case of elastic energy, commonly known as the strain energy, full conservation or storage is guaranteed during the deformation: The entire amount of this energy is available to work back to the original position on release of the applied loads. This fact is directly responsible for the firm foundation of the theory of elasticity. The plastic energy, however, unlike the elastic energy, is liable to dissipation, perhaps through heat or particle reorientation that is due to excessive sliding and rotation. Because of this uncontrollable factor, it is necessary to establish rules that permit mathematical manipulation and at the same time govern the plastic behavior in a systematic manner, and so we come to *Drucker's stability and normality rules* (Drucker, 1959).

Stability

- The plastic work done by the external agency during the application of the additional stresses is *positive*.
- The net total work performed by the external agency during the cycle of adding and removing stresses is *nonnegative*.

If these criteria are not met, the material is said to be *unstable*.

Normality

- The incremental plastic strain vector is *normal* to the yield surface.

If this rule is violated, the material is not considered stable

Let us now examine these rules in the following examples. A stress–strain diagram is shown in Fig. 7.4.13(a). The plastic strain ranges are shown for stable and unstable materials in Figs. 7.4.13(b) and 7.4.13(c), respectively.

The elastic limit is reached at σ_0 in Fig. 7.4.13(a). Consider a point at σ_1 and a small increment $d\sigma$. The unloading from these points to σ_0^* and $\overset{*}{\sigma}_1$ and the associated energy for this process, shown as the shaded areas in Figs. 7.4.13(a) and 7.4.13(b), may also be depicted on the yield surfaces shown in Fig. 7.4.14. The total energy expended in this action is given by

$$\begin{aligned} dW_T &= \int_0^{t_1} \sigma_{ij} d\gamma_{ij}^e \, dt + \int_{t_1}^{t_1+\delta t} \sigma_{ij}\left(d\gamma_{ij}^e + d\gamma_{ij}^p\right) dt + \int_{t_1+\delta t}^{\overset{*}{t}} \sigma_{ij} d\gamma_{ij}^e \, dt \\ &= \oint \sigma_{ij} d\gamma_{ij}^e \, dt + \int_{t_1}^{t_1+\delta t} \sigma_{ij} d\gamma_{ij}^p \, dt = \int_{t_1}^{t_1+\delta t} \sigma_{ij} d\gamma_{ij}^p \, dt. \end{aligned} \tag{7.4.26}$$

Here, t_1, δt, and $\overset{*}{t}$ are the times corresponding to σ_1, $d\sigma$, and $\overset{*}{\sigma}$, respectively. Note that the net elastic work done on loading and unloading is zero. Because the internal energy is

$$dW_0 = \int_t^{t_1+\delta t} \overset{*}{\sigma}_{ij} d\gamma_{ij}^p \, dt, \tag{7.4.27}$$

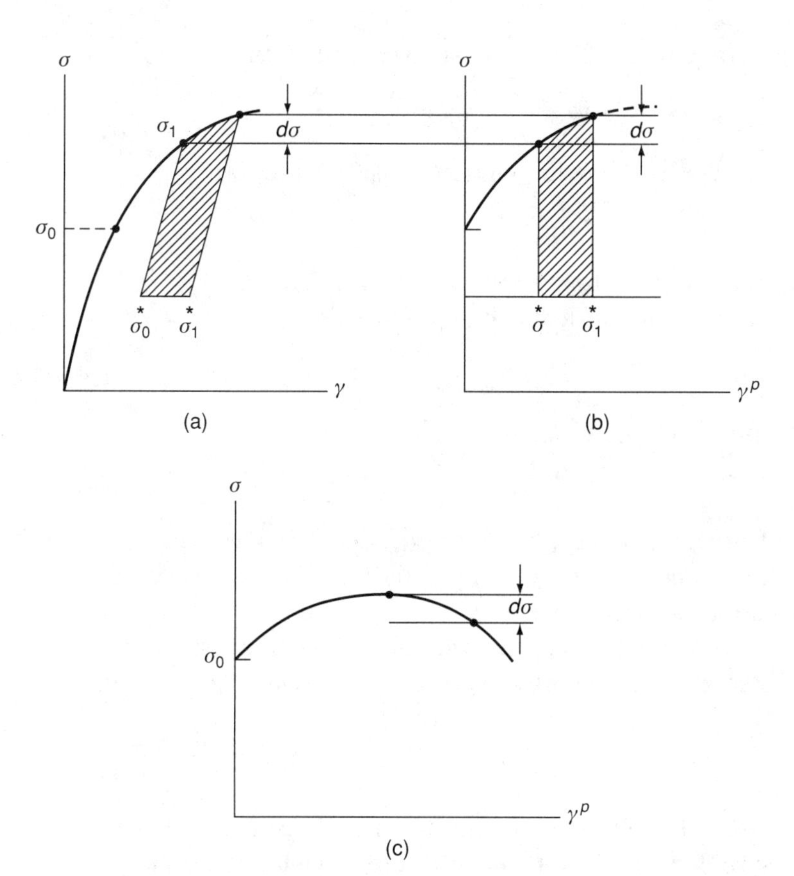

Figure 7.4.13. Stress–strain relationships.

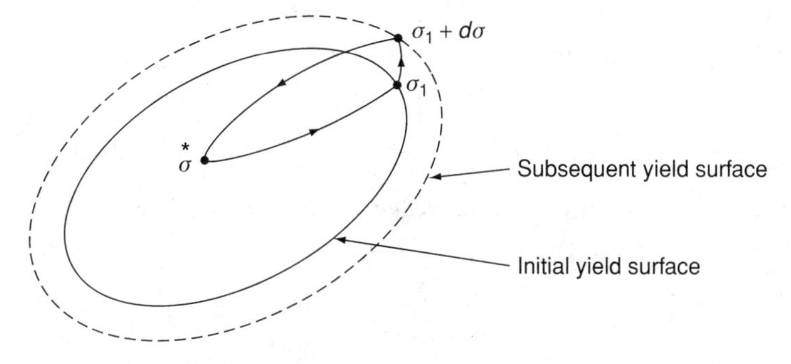

Figure 7.4.14. Yield surface.

the external energy assumes the form

$$dW_{\text{ext}} = dW_T - dW_0 = \int_{t}^{t_1+\delta t} (\sigma_{ij} - \overset{*}{\sigma}_{ij})\, d\gamma_{ij}^{p}\, dt. \tag{7.4.28}$$

According to Drucker's stability rule, we must have

$$dW_{\text{ext}} \geq 0. \tag{7.4.29}$$

Therefore the integrand of Eq. (7.4.28) must assume the form

$$(\sigma_{ij} - \overset{*}{\sigma}_{ij})d\gamma_{ij}^{p} \geq 0. \tag{7.4.30}$$

Noting that $\sigma_{ij} - \overset{*}{\sigma}_{ij}$ is $d\sigma_{ij}$, we write relation (7.4.30) in the form

$$d\sigma_{ij} d\gamma_{ij}^{p} \geq 0. \tag{7.4.31}$$

This is known as the condition for a *plastically stable material.*

Mathematically, for strain-hardening material, we define

$$d\sigma_{ij} d\gamma_{ij}^{p} > 0 \tag{7.4.32}$$

and for perfectly plastic material,

$$d\sigma_{ij} d\gamma_{ij}^{p} = 0. \tag{7.4.33}$$

If the condition of either inequality (7.4.32) or Eq. (7.4.33) is not satisfied, then the material is considered unstable and, consequently, the mathematical description of plastic deformation is not possible.

Let us now take a material that satisfies relation (7.4.31) and assume that there exists a scalar function f, called a *plastic potential function* or a yield function, such that

$$f = \hat{f}(\sigma_{ij}), \tag{7.4.34}$$

which is in contrast to the elastic potential function [Eq. (4.1.1)]. Because the hydrostatic pressure is assumed not to contribute to yielding, we may rewrite Eq. (7.4.34) in terms of the deviatoric stress as

$$f = \hat{f}(\hat{\sigma}_{ij}) \tag{7.4.35}$$

or, with deviatoric stress invariants [see Subsection 3.1.5 and Eqs. (3.5.3b) and (3.5.3c)],

$$f = f(\hat{J}_2, \hat{J}_3). \tag{7.4.36}$$

Here $\hat{J}_3$ is an odd function and thus precludes isotropic expansion of a yield function, and so we exclude it from Eq. (7.4.36). Thus

$$f = \hat{f}(\hat{J}_2). \tag{7.4.37}$$

Here, if the material is *elastic*,

$$f < 0 \quad \text{and} \quad df = \frac{\partial f}{\partial \sigma_{ij}} d\sigma_{ij} < 0, \tag{7.4.38}$$

and if the material is *plastic*,

$$f = 0 \quad \text{and} \quad df = \frac{\partial f}{\partial \sigma_{ij}} d\sigma_{ij} \geq 0. \tag{7.4.39}$$

Here $f = 0$ represents the yield surface and $df \geq 0$ refers to the plastic range. These conditions are depicted in Fig. 7.4.15.

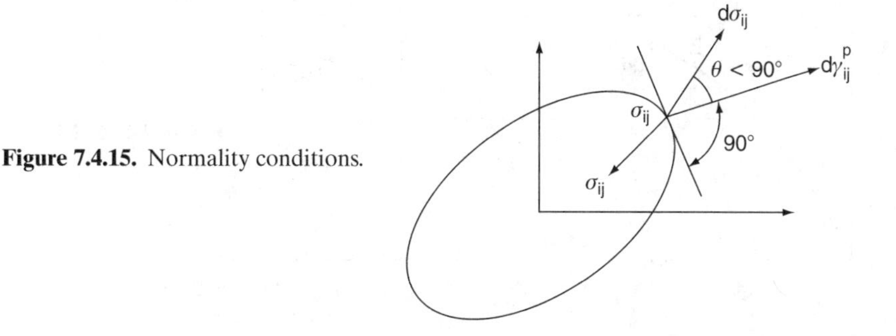

Figure 7.4.15. Normality conditions.

From Eq. (7.4.31), the plastic energy on the verge of yield is

$$d\sigma_{ij} d\gamma_{ij}^{p} = 0 \tag{7.4.40}$$

with the constraint condition

$$df = \frac{\partial f}{\partial \sigma_{ij}} d\sigma_{ij} = 0 \tag{7.4.41}$$

on the yield surface. Now we introduce a Lagrange multiplier $d\lambda$ such that the condition of yield can be represented by

$$d\sigma_{ij} d\gamma_{ij}^{p} - d\lambda \frac{\partial f}{\partial \sigma_{ij}} d\sigma_{ij} = 0$$

or

$$\left(d\gamma_{ij}^{p} - \frac{\partial f}{\partial \sigma_{ij}} d\lambda \right) d\sigma_{ij} = 0.$$

Because this relation must hold for all arbitrary values of $d\sigma_{ij}$, we require

$$d\gamma_{ij}^{p} = \frac{\partial f}{\partial \sigma_{ij}} d\lambda, \tag{7.4.42}$$

where $d\lambda$ may now be considered a positive parameter or proportionality constant. Equation (7.4.42) is known as the *Prandtl–Reuss flow rule*. Drucker interpreted this rule as a normality condition such that the plastic strain vector must be normal to the yield surface if the material is stable and capable of plastic deformations as governed by Eq. (7.4.42). Along with the requirement of normality of the plastic strain vector, the yield surface must be *convex*. The relation between the normality of the plastic strain vector and the convex yield surface is shown in Fig. 7.4.15. It is seen that through every point σon the yield surface there is a plane such that all points interior to the surface lie on the same side of the plane; that is, the yield surface is convex. Furthermore, on the yield surface, relation (7.4.31), $d\sigma_{ij} d\gamma_{ij}^{p} > 0$, requires that the angle θ must be obtuse, which is a further requirement of convexity of the yield surface and normality of the plastic strain increment to the yield surface.

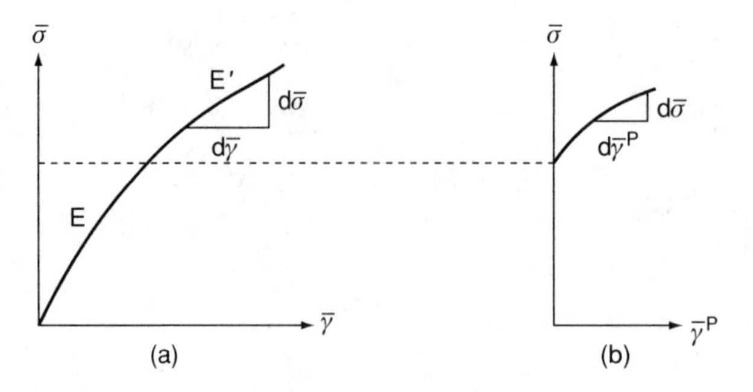

Figure 7.4.16. Elastoplastic behavior.

7.4.5 Isotropic Hardening

As suggested in Eq. (7.4.25), if there exists an elastic strain-energy function, we obtain the total of the stresses within the elastic range by differentiating this function with respect to the strains. We may attempt to follow a similar procedure to obtain the constitutive equation applicable to a plastic range. Obviously, because of the predicted nonlinearity, our constitutive equation must be of an incremental form. Therefore an incremental stress rather than total stress should be defined in conjunction with the already established incremental plastic strain [Eq. (7.4.42)]. Toward this end, we first make use of the result of Eq. (7.4.25) in the incremental form,

$$d\sigma_{ij} = E_{ijkl} d\gamma_{kl}^e = E_{ijkl} \left(d\gamma_{kl} - d\gamma_{kl}^p \right). \tag{7.4.43}$$

The total incremental strain $d\gamma_{kl}$ is assumed to be the sum of the elastic and the plastic strain components:

$$d\gamma_{kl} = d\gamma_{kl}^e + d\gamma_{kl}^p. \tag{7.4.44}$$

Consider the laboratory data for a stress–strain curve shown in Fig. 7.4.16(a). Let E and E' be Young's moduli for the elastic range and the plastic range, respectively. Here $\bar{\sigma}$ and $\bar{\gamma}$ are the stress and strain in a uniaxial test specimen. We define E' as follows:

$$E' = \frac{d\bar{\sigma}}{d\bar{\gamma}} = \frac{d\bar{\sigma}}{d\bar{\gamma}^e + d\bar{\gamma}^p} = \frac{d\bar{\sigma}}{d\sigma/E + d\bar{\gamma}^p}. \tag{7.4.45}$$

Here the direction of stress and strain is arbitrary; we may refer to a chosen direction as a reference direction. $\bar{\sigma}$ and $\bar{\gamma}$ are commonly known as *equivalent stress* and *equivalent strain*. If $\bar{\sigma}$ and $\bar{\gamma}$ are defined in the plastic range, we define $d\bar{\sigma}$ and $d\bar{\gamma}^p$ as the incremental equivalent stress and incremental equivalent plastic strain, respectively [Fig. 7.4.16(b)]. We define the slope of the curve in the plastic strain range as E_p, which we call the *plastic* modulus:

$$E_p = \frac{d\bar{\sigma}}{d\bar{\gamma}^p}. \tag{7.4.46}$$

Combining Eqs. (7.4.45) and (7.4.46) yields

$$E_p = \frac{EE'}{E - E'}. \tag{7.4.47}$$

The plastic potential function may be related to the von Mises yield condition as follows:

$$f = \bar{\sigma}^2 = 3J_2, \tag{7.4.48}$$

where the yield stress σ_0 is equated to the equivalent stress $\bar{\sigma}$. The incremental plastic work is defined as

$$dW^p = \sigma_{ij} d\gamma_{ij}^p \tag{7.4.49}$$

or

$$dW^p = \bar{\sigma} d\bar{\gamma}^p. \tag{7.4.50}$$

Inserting the flow rule of Eq. (7.4.42) into Eq. (7.4.49) yields

$$dW^p = \sigma_{ij} \frac{\partial f}{\partial \sigma_{ij}} d\lambda = 2\bar{\sigma}^2 d\lambda. \tag{7.4.51}$$

Combining Eqs. (7.4.50) and (7.4.51), we find that $d\lambda$ takes the form

$$d\lambda = \frac{d\bar{\gamma}^p}{2\bar{\sigma}}. \tag{7.4.52}$$

In view of Eqs. (7.4.43), (7.4.44), (7.4.42), and (7.4.52), the incremental stress is determined to be

$$d\sigma_{ij} = E_{ijkl}(d\gamma_{kl} - Z_{kl} d\bar{\gamma}^p), \tag{7.4.53}$$

where

$$Z_{kl} = \frac{1}{2\bar{\sigma}} \frac{\partial f}{\partial \sigma_{ij}} = \frac{3}{2\bar{\sigma}} \frac{\partial J_2}{\partial \sigma_{ij}}. \tag{7.4.54}$$

Differentiating Eq. (7.4.51), we determine the incremental equivalent stress:

$$d\bar{\sigma} = \frac{1}{2\bar{\sigma}} \frac{\partial f}{\partial \sigma_{ij}} d\sigma_{ij} = Z_{ij} d\sigma_{ij}. \tag{7.4.55}$$

Using the relations in Eqs. (7.4.46), (7.4.55), and (7.4.53), we may solve for $d\bar{\gamma}^p$ in the form

$$d\bar{\gamma}^p = \frac{E_{ijkl} Z_{ij} d\gamma_{kl}}{E_p + Z_{rs} Z_{tu} E_{rstu}}. \tag{7.4.56}$$

Substituting Eq. (7.4.56) into Eq. (7.4.53) leads to

$$d\sigma_{ij} = (E_{ijkl} + E^*_{ijkl}) d\gamma_{kl}, \tag{7.4.57}$$

with

$$E^*_{ijkl} = -\frac{E_{ijmn} Z_{mn} Z_{pq} E_{pqkl}}{E_p + Z_{rs} Z_{tu} E_{rstu}}. \tag{7.4.58}$$

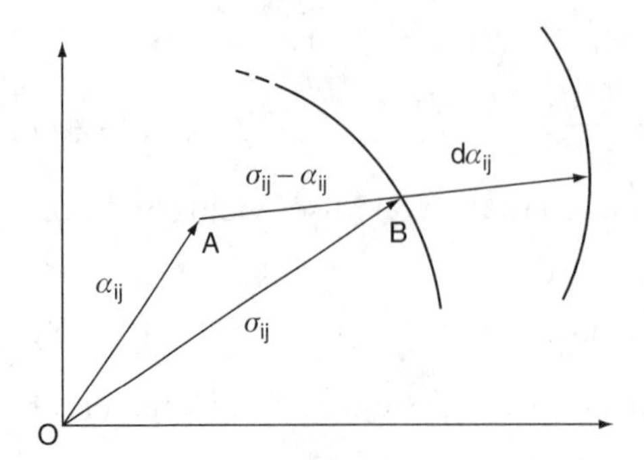

Figure 7.4.17. Kinematic hardening.

Here, E^*_{ijkl} is called the tensor of plastic moduli. The relation in Eq. (7.4.57) is the constitutive equation in the plastic range relating the incremental stresses and strains.

7.4.6 Kinematic Hardening

In our previous discussions, we dealt with isotropic solids and assumed that the material remains isotropic throughout the plastic deformations. The Bauschinger effect, the process referred to as anisotropic hardening, was not taken into account during the strain hardening. Prager (1956) first noted that a material may turn anisotropic because of excessive straining and the subsequent yield surfaces would translate without expansion. The translation of yield surfaces is known as *kinematic hardening*.

The mechanics of kinematic hardening is shown schematically in Fig. 7.4.17. The line $\overline{OB}$ represents a component of stress tensor σ_{ij} connecting the center of the initial yield surface and a point on the current yield surface. Suppose that the center of the initial yield surface moves to a new position A, which becomes the center of the subsequent yield surface. The translation $\overline{OA}$ is the result of what may be called the translation stress tensor α_{ij}. The difference $\sigma_{ij} - \alpha_{ij}$, representing a vector $\overline{AB}$, locates a subsequent yield surface at C by an extension $d\alpha_{ij}$, referred to as the incremental translation stress tensor, such that

$$d\alpha_{ij} = E_p d\gamma^p_{ij}, \tag{7.4.59}$$

in which the incremental plastic strain $d\gamma^p_{ij}$ still follows the normality or flow rule as in isotropic hardening:

$$d\gamma^p_{ij} = \frac{\partial f}{\partial \sigma_{ij}} d\lambda. \tag{7.4.60}$$

Ziegler (1959) modified Prager's rule [relation (7.4.39)] by suggesting that the incremental translation stress tensor may also be written as

$$d\alpha_{ij} = (\sigma_{ij} - \alpha_{ij}) d\mu, \tag{7.4.61}$$

where $d\mu$ is a positive constant known as the *kinematic hardening parameter*.

From the definition of kinematic hardening, the subsequent yield surface according to the von Mises yield criterion is given by

$$f = f(\sigma_{ij} - \alpha_{ij}) = \bar{\sigma}^2 = \frac{1}{2}\{[(\sigma_{11} - \alpha_{11}) - (\sigma_{22} - \alpha_{22})]^2$$
$$+ [(\sigma_{22} - \alpha_{22}) - (\sigma_{33} - \alpha_{33})^2]^2$$
$$+ [(\sigma_{33} - \alpha_{33}) - (\sigma_{11} - \alpha_{11})]^2 + 6[(\sigma_{12} - \alpha_{12})^2$$
$$+ (\sigma_{23} - \alpha_{23})^2 + (\sigma_{31} - \alpha_{31})^2]\}. \tag{7.4.62}$$

Because the subsequent yield surface does not expand, we require

$$df = 0 = \frac{\partial f}{\partial \sigma_{ij}} d\sigma_{ij} + \frac{\partial f}{\partial \alpha_{ij}} d\alpha_{ij}. \tag{7.4.63}$$

Differentiating Eq. (7.4.62) with respect to σ_{ij} and α_{ij}, we find that

$$\frac{\partial f}{\partial \sigma_{ij}} = -\frac{\partial f}{\partial \alpha_{ij}}. \tag{7.4.64}$$

Solving for $d\alpha_{ij}$ from Eq. (7.4.64) and using the relations given in Eqs. (7.4.59), (7.4.60), and (7.4.64) yields

$$d\lambda = \frac{1}{E_p} \frac{(\partial f/\partial \sigma_{ij}) d\sigma_{ij}}{(\partial f/\partial \sigma_{kl})(\partial f/\partial \sigma_{kl})}. \tag{7.4.65}$$

Combining Eqs. (7.4.63) and (7.4.64) yields

$$\frac{\partial f}{\partial \sigma_{ij}}(d\sigma_{ij} - d\alpha_{ij}) = 0. \tag{7.4.66}$$

In view of Eqs. (7.4.61) and (7.4.66), we obtain

$$d\mu = \frac{(\partial f/\partial \sigma_{ij}) d\sigma_{ij}}{(\sigma_{kl} - \alpha_{kl})(\partial f/\partial \sigma_{kl})}. \tag{7.4.67}$$

Equations (7.4.61) and (7.4.67) represent the kinematic hardening rule.

Substituting Eq. (7.4.59) into (7.4.66) and letting

$$Z_{ij} = \frac{\partial f}{\partial \sigma_{ij}}, \tag{7.4.68}$$

we get

$$Z_{ij}\left(d\sigma_{ij} - E_p d\gamma_{ij}^p\right) = 0. \tag{7.4.69}$$

Recall that

$$d\sigma_{ij} = E_{ijkl} d\gamma_{kl}^e = E_{ijkl}\left(d\gamma_{kl} - d\gamma_{kl}^p\right). \tag{7.4.70}$$

Multiplying Eq. (7.4.70) by Z_{ij} and comparing the latter with Eq. (7.4.69), we obtain

$$Z_{ij} E_p d\gamma_{ij}^p = Z_{ij} E_{ijkl}\left(d\gamma_{kl} - d\gamma_{kl}^p\right). \tag{7.4.71}$$

Note that the plastic work is given by

$$\sigma_{ij} d\gamma_{ij}^p = \bar{\sigma} d\bar{\gamma}^p. \tag{7.4.72}$$

Thus substituting Eq. (7.4.60) into Eq. (7.4.71) leads to

$$d\lambda = \frac{\bar{\sigma} d\bar{\gamma}^p}{\sigma_{ij} Z_{ij}}. \tag{7.4.73}$$

Once again, from Eq. (7.4.60) and using Eq. (7.4.73), we write

$$d\gamma_{ij}^p = \frac{Z_{ij} \bar{\sigma} d\bar{\gamma}^p}{\sigma_{kl} Z_{kl}}. \tag{7.4.74}$$

It follows from Eqs. (7.4.74) and (7.4.71) that

$$d\bar{\gamma}^p = \frac{\sigma_{mn} Z_{mn}}{\bar{\sigma}} \frac{Z_{ij} E_{ijkl} d\gamma_{kl}}{Z_{rs} Z_{rs} E_p + Z_{rs} Z_{tu} E_{rstu}}. \tag{7.4.75}$$

In view of Eqs. (7.4.70) and (7.4.74), the incremental stress takes the form

$$d\sigma_{ij} = E_{ijkl} \left(d\gamma_{kl} - \frac{Z_{kl} \bar{\sigma} d\bar{\gamma}^p}{\sigma_{rs} Z_{rs}} \right). \tag{7.4.76}$$

Substituting Eq. (7.4.75) into Eq. (7.4.76) finally leads to the incremental stress–strain law,

$$d\sigma_{ij} = \left(E_{ijkl} + E_{ijkl}^*\right) d\gamma_{kl}, \tag{7.4.77}$$

where E_{ijkl}^* is the tensor of plastic moduli of the form

$$E_{ijkl}^* = \frac{E_{ijmn} Z_{mn} Z_{pq} E_{pqkl}}{Z_{rs} Z_{rs} E_p + Z_{rs} Z_{tu} E^{rstu}}. \tag{7.4.78}$$

In the numerical analysis for the kinematic hardening process we set $\alpha_{ij} = 0$ initially, calculate $d\mu$ from Eq. (7.4.67), and substitute the latter into Eq. (7.4.61) to determine the incremental translation stress tensor. These magnitudes determine the translation of yield surfaces. In the subsequent calculations, we note that α_{ij} at the nth step is

$$\alpha_{ij}^{(n)} = \alpha_{ij}^{(n-1)} + d\alpha_{ij}^{(n)}. \tag{7.4.79}$$

It was noted earlier that the possible induced anisotropy during plastic deformation is responsible for kinematic hardening. For this reason, initially anisotropic materials, such as fiber-reinforced composites, are more susceptible to kinematic hardening, as the fiber orientations are likely to alter considerably on deformation.

7.4.7 Multisurface Models of Flow Plasticity

The multisurface model was proposed by Mroz (1967) based on the kinematic hardening rule. Mroz's hardening rule is depicted in Fig. 7.4.18 as

$$f_m\left(\sigma_{ij} - \alpha_{ij}^{(m)}\right) = Y_m^2, \tag{7.4.80}$$

with $m = 0, 1, 2, \ldots$, and

$$d\alpha_m = (\sigma_{m+1} - \sigma_m)\, d\mu, \tag{7.4.81}$$

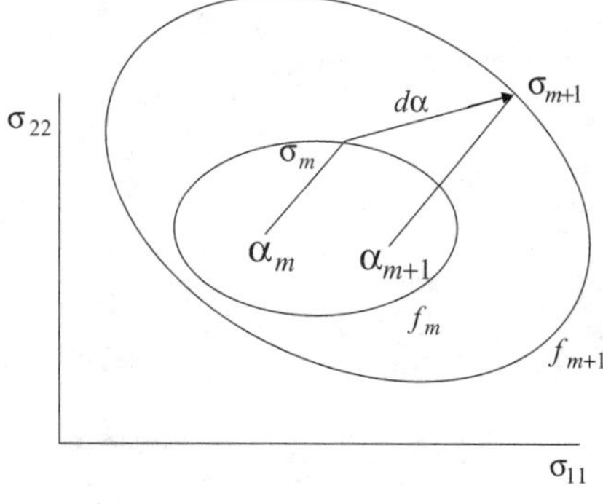

Figure 7.4.18. Mroz's hardening rule.

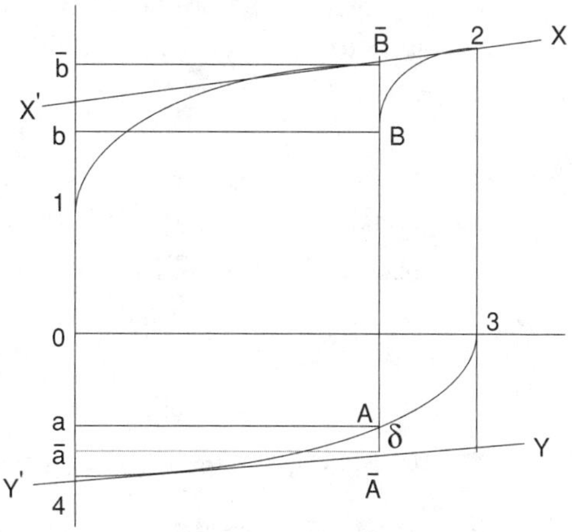

Figure 7.4.19. Bounds and parameters of two-surface model.

where $d\mu$ is a scalar to be determined by the consistence condition; $d\alpha_m$ refers to the translation of the active surface f_m.

Dafalias and Popov (1976) proposed a two-surface model and a continuous variation of the plastic modulus $d\sigma/d\gamma^P = E^P$ defined between these two surfaces (Fig. 7.4.19) with

$$E^P = E^P(\delta, \delta_{in}), \tag{7.4.82}$$

so that

$$E^P \rightarrow \infty \quad \text{for} \quad \delta \rightarrow \delta_{in}, \tag{7.4.83}$$

$$\bar{E}^P \rightarrow E^P \quad \text{for} \quad \delta \rightarrow 0, \tag{7.4.84}$$

where $\bar{E}^P$ is the slope of the bound line and E^P monotonically increases as δ decreases.

The distance δ is from point a on the yield surface $f = 0$ to a similar point $\bar{a}$ on the bounding surface $\bar{f} = 0$ (Fig. 7.4.20). At similar points, the normals to the respective surfaces are equal. Let the stress at point a be σ_{ij} and the stress at point $\bar{a}$ be $\bar{\sigma}_{ij}$; then σ is defined as

$$\sigma = [(\bar{\sigma}_{ij} - \sigma_{ij})(\bar{\sigma}_{ij} - \sigma_{ij})]^{1/2}. \tag{7.4.85}$$

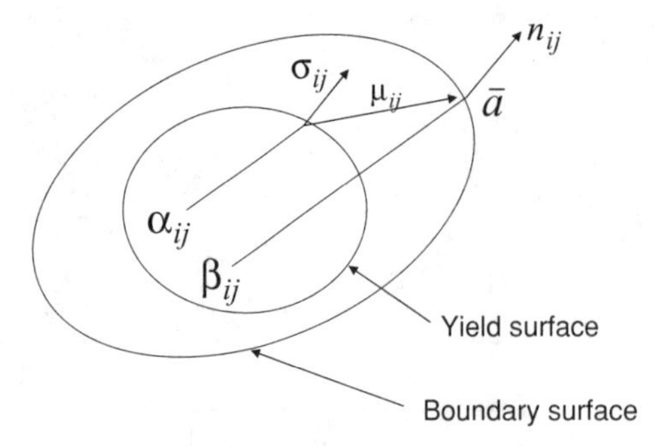

Figure 7.4.20. Two-surface model in two-dimensional stress space (Dafalias and Popov, 1976).

Assume that the bounding surface is an isotropic expansion of $f = 0$, but both surfaces are allowed to move. The centers for the two surfaces are α_{ij} and β_{ij}, respectively. Thus

$$\bar{\sigma}_{ij} - \beta_{ij} = m(\sigma_{ij} - \alpha_{ij}), \tag{7.4.86}$$

where m is a proportional factor as a function of internal variables.

It is seen that the multisurface models of Mroz and Dafalias–Popov represent an alternative of the single-surface theory given by (7.4.61). A similar procedure may be followed to determine the plastic moduli similar to (7.4.78).

7.4.8 The Strain Trajectory Models

The total strain approach was proposed by Ilyushin (1963) using stress and deviatoric strain tensors in a five-dimensional vector space to investigate the characteristics of the stress response to a predetermined strain trajectory. In contrast, Valanis (1971, 1980) developed an endochronic theory of plasticity using the plastic strain approach and irreversible thermodynamics of internal variables.

The total strain approach of Ilyushin begins with the deviator of the total strain tensor (not plastic strain) with the idea that the plastic strain is expected to be included in the total strain when large deformations occur:

$$\sum_{i,j=1,2,3} e_{ij}e_{ij} = \sum_{i=1}^{5} \gamma_i\gamma_i, \tag{7.4.87}$$

with

$$\gamma_3 = \sqrt{2}e_{12}, \quad \gamma_4 = \sqrt{2}\gamma_{23}, \quad \gamma_5 = \sqrt{2}\gamma_{31}. \tag{7.4.88}$$

Combining Eqs. (7.4.87) and (7.4.88) leads to

$$(e_{11})^2 + (e_{22})^2 + (e_{33})^2 = (\gamma_1)^2 + (\gamma_2)^2, \tag{7.4.89}$$

$$\gamma_1 = \sqrt{\frac{3}{2}}e_{11}, \quad \gamma_2 = \sqrt{2}\left(e_{22} + \frac{1}{2}e_{11}\right). \tag{7.4.90}$$

Similarly, the deviatoric stress $\hat{\sigma}_{ij}$ with $\hat{\sigma}_{11} + \hat{\sigma}_{22} + \hat{\sigma}_{33} = 0$ takes the form

$$\sigma_1 = \sqrt{\frac{3}{2}}\hat{\sigma}_{11}, \quad \sigma_2 = \sqrt{2}\left(\hat{\sigma}_{22} + \frac{1}{2}\hat{\sigma}_{33}\right),$$
$$\sigma_3 = \sqrt{2}\hat{\sigma}_{12}, \quad \sigma_4 = \sqrt{2}\hat{\sigma}_{23}, \quad \sigma_5 = \sqrt{2}\hat{\sigma}_{31}. \tag{7.4.91}$$

Ilyushin introduces the integral constitutive equation

$$\sigma(s) = \int_0^s K\left[s,\ s',\ k_i(s),\ k_i(s')\right] d\gamma(s')\ (i = 1, 2, 3, 4), \tag{7.4.92}$$

where s' is the integration parameter and k_i represents the geometry of strain trajectory.

Valanis (1980), instead of using the total strain, resorts to the plastic strain with irreversible thermodynamics principle by introducing an intrinsic time η such that

$$d\eta^2 = dQ_{ij}dQ_{ij}, \tag{7.4.93}$$

with

$$Q_{ij} = \gamma_{ij} - G_{ijkm}\sigma_{km}, \tag{7.4.94}$$

where G_{ijkm} represents the fourth-order tensor of material constants.

The constitutive equation consists of deviatoric and volumetric parts:

$$\sigma^*_{ij} = 2\int_{\xi_0}^{\xi} \mu(\eta - \eta')\frac{d\gamma_{ij}}{d\eta'}d\eta', \tag{7.4.95a}$$

$$\sigma_{kk} = 3\int_{\eta_0}^{\eta} K(\eta - \eta')\frac{d\gamma_{kk}}{d\eta'}d\eta'. \tag{7.4.95b}$$

The deviatoric constitutive equation for one dimension takes the form

$$\sigma^* = 2\int_0^z \mu(z - z')\frac{d\gamma}{dz'}dz' = 2\mu_0\int_{z_0}^z \xi(z - z')\frac{d\gamma}{dz'}dz' \tag{7.4.96}$$

or

$$\sigma^* = 2\mu_0\int_0^z \rho(z - z')\frac{dQ}{dz'}dz', \tag{7.4.97}$$

where

$$Q = \gamma - k_1\frac{\hat{\sigma}}{2\mu_0}, \quad \mu(z) = \mu_0\xi(z), \quad \xi(0) = 1. \tag{7.4.98}$$

Using the Laplace transformation we can show that

$$\sigma^* = 2\mu_0\rho_0\frac{dQ(z)}{dz} + 2\mu_0\int_{z_0}^z \rho_1(z - z')\frac{dQ(z')}{dz'}dz'. \tag{7.4.99}$$

This is the constitutive equation of the endochronic theory.

The deviatoric part of the total strain approach of Ilyushin and the plastic strain endochronic method of Valanis constitute the strain trajectory model. In reality, however, the plasticity theories just described are likely to be influenced greatly in thermoelastic and/or viscoelastic environments. This subject is discussed in the next section.

7.5 Thermoviscoelastoplasticity

In retrospect, the thermodynamics of solids introduced in Chap. 4 with the concept of entropy associated with the second law of thermodynamics was seen as a powerful tool leading to the thermomechanically coupled equations of motion and heat conduction for an elastic solid, known as thermoelasticity. In the previous sections of this chapter, we dealt with viscoelasticity and plasticity as separate subjects (Chung, 1973; Chung, 1975). Is it possible to combine them all? This is the subject of thermoviscoelastoplasticity discussed in this section.

The task is theoretically monumental. However, one approach is to combine all different material behavior incrementally in small time segments. We achieve this by introducing the free-energy function consisting of energies contributed from various material properties such as viscoelasticity and plasticity within small discrete-time segments. The resulting governing equations can then be solved numerically as marching in time until steady-state conditions are reached (Chung and Eidson, 1971, 1972, 1973a, 1973b, 1974).

It is proposed that the continuous histories of material behavior be characterized by the free energy Φ, the stress σ_{ij}, the heat flux q, the entropy η, and the plastic strain rate as the functions of elastic strain $\gamma_{ij}^{(e)}$, plastic strains $\gamma_{ij}^{(p)}$, absolute temperature θ, and a set of internal (hidden) variables $J_{ij}^{(r)} (r = 1, 2, \ldots, n)$. The constitutive equations can be written in the following forms:

$$\Phi(\Delta t) = \hat{\Phi}\left[\gamma_{ij}^{(e)}(\Delta t), \gamma_{ij}^{(p)}(\Delta t), \theta(\Delta t), J_{ij}^{(s)}(\Delta t)\right], \tag{7.5.1a}$$

$$\sigma_{ij}(\Delta t) = \hat{\sigma}_{ij}\left[\gamma_{ij}^{(e)}(\Delta t), \gamma_{ij}^{(p)}(\Delta t), \theta(\Delta t), J_{ij}^{(s)}(\Delta t)\right], \tag{7.5.1b}$$

$$\eta(\Delta t) = \hat{\eta}\left[\gamma_{ij}^{(e)}(\Delta t), \gamma_{ij}^{(p)}(\Delta t), \theta(\Delta t), J_{ij}^{(s)}(\Delta t)\right], \tag{7.5.1c}$$

$$q_i(\Delta t) = \hat{q}\left[\gamma_{ij}^{(e)}(\Delta t), \gamma_{ij}^{(p)}(\Delta t), \theta(\Delta t), J_{ij}^{(s)}(\Delta t)\right], \tag{7.5.1d}$$

$$\dot{\gamma}_{ij}^{(p)}(\Delta t) = \hat{\dot{\gamma}}_{ij}^{(p)}\left[\gamma_{ij}^{(e)}(\Delta t), \gamma_{ij}^{(p)}(\Delta t), \theta(\Delta t), J_{ij}^{(s)}(\Delta t)\right], \tag{7.5.1e}$$

$$\dot{J}_{ij}^{(s)}(\Delta t) = \hat{\dot{J}}^{(s)}\left[\gamma_{ij}^{(e)}(\Delta t), \gamma_{ij}^{(p)}(\Delta t), \theta(\Delta t), J_{ij}^{(s)}(\Delta t)\right]. \tag{7.5.1f}$$

Note that smoothness of all the preceding variables is valid only within the small time interval Δt. The internal variables $J_{ij}^{(s)}$ are assumed to represent the time-dependent viscous behavior alone, not viscoplastic. It now follows from the Taylor series expansion, retaining only the quadratic terms of (7.5.1a), that

$$\rho\Phi(\Delta t) = \frac{1}{2}E_{ijkm}\gamma_{ij}^{(e)}(\Delta t)\gamma_{km}^{(e)}(\Delta t) + \frac{1}{2}E_{ijkm}^{*}\gamma_{ij}^{(p)}(\Delta t)\gamma_{km}^{(p)}(\Delta t)$$
$$- B_{ij}T(\Delta t)\gamma_{ij}^{(e)}(\Delta t) - B_{ij}^{*}T(\Delta t)\gamma_{ij}^{(p)}(\Delta t) - \frac{ct^2(\Delta t)}{2T_0}$$
$$- \sum_{s=1}^{n} B_{ij}^{(s)}T(\Delta t)J_{ij}^{(s)}(\Delta t) + \frac{1}{2}\sum_{s=1}^{n}\xi_{ijkm}^{(s)}J_{ij}^{(s)}(\Delta t)J_{km}^{(s)}(\Delta t)$$
$$+ \sum_{s=1}^{n}\xi_{ijkm}^{(s)}J_{ij}^{(s)}(\Delta t)\gamma_{km}(\Delta t),$$

where E_{ijkm} and B_{ij} are tensors of elastic and thermoelastic moduli; E_{ijkm}^{*} and B_{ij}^{*} are tensors of plastic and thermoplastic moduli, their explicit forms to be derived later; c is the heat capacity; T and T_0 are the temperature change and reference temperature; $\xi_{ijkm}^{(s)}$ is the array of material stiffness constants representing any viscoelastic model related to the internal variables. The specific form of $J_{ij}^{(s)}$ is of the form

$$J_{ij}^{(s)}(\Delta t) = \int_0^t \exp\left[\frac{-(t-\tau)}{T_{(s)}}\right]\dot{\gamma}_{ij}(\tau)\, d\tau, \tag{7.5.2}$$

where τ is the time variable and $T_{(s)}$ is the relaxation time.

Recall that the conservation of energy (4.4.12), the Clausius–Duhem inequality (4.4.14), and internal dissipation (4.4.17) may be used to obtain

$$\rho\dot{\Phi} = \sigma_{ij}(\Delta t)[\dot{\gamma}_{ij}^{(e)}(\Delta t) + \dot{\gamma}_{ij}^{(p)}(\Delta t)] - D(\Delta t) - \rho\eta(\Delta t)\dot{T}(\Delta t),$$

$$D = \sigma^{ij}\dot{\gamma}_{ij}^{(e)} + \sigma_{ij}\dot{\gamma}_{ij}^{(p)} - \rho\left[\frac{\partial\Phi}{\partial\gamma_{ij}^{(e)}}\dot{\gamma}_{ij}^{(e)} + \frac{\partial\Phi}{\partial\gamma_{ij}^{(p)}}\dot{\gamma}_{ij}^{(p)}\right.$$
$$\left. + \frac{\partial\Phi}{\partial T}\dot{T} + \frac{\partial\Phi}{\partial J_{ij}^{(e)}}\dot{J}_{ij}^{(e)}\right] - \rho\eta\dot{T} \geq 0. \tag{7.5.3}$$

It follows from Eqs. (4.4.24), (4.4.25), and (4.4.17) combined by (7.5.1) that

$$\sigma_{ij} = E_{ijkm}\gamma_{km}^{(e)} - B^{ij}T + \sum_{s=1}^{n}\xi_{ijkm}^{(s)}J_{km}^{(s)}, \tag{7.5.4}$$

$$\rho\eta = B_{ij}\gamma_{km}^{(e)} - B_{ij}^{*}\gamma_{ij}^{(p)}\frac{cT}{T_0} + \sum_{s=1}^{n}B_{ij}^{(s)}J_{ij}^{(s)}, \tag{7.5.5}$$

$$D = E_{ijkm}^{*}\gamma_{ij}^{(p)}\dot{\gamma}_{km}^{(p)} + B_{ij}^{*}T\dot{\gamma}_{ij} + \sigma_{ij}\dot{\gamma}_{ij}^{(p)} - \sum_{s=1}^{n}\xi_{(s)}^{ijkm}\left(J_{ij}^{(s)}J_{km}^{(s)}\right.$$
$$\left. + \dot{J}_{ij}^{(s)}\gamma_{km} + \dot{\gamma}_{ij}^{(p)}J_{km}^{(s)}\right) + \sum_{s=1}^{n}B_{ij}^{(s)}T\dot{J}_{ij}^{(s)}. \tag{7.5.6}$$

At this point it is interesting to compare Eqs. (7.5.4) and (7.5.5) with Eqs. (4.4.34) and (4.4.35) for isotropic linear elastic solids. Effects of viscoelasticity and plasticity are clearly evident for the stress tensor and entropy. It is also seen that the internal dissipation ($D \neq 0$) depends on the energy associated with plastic strains and internal variables that represent the thermoviscoelastoplastic coupling.

The thermomechanically coupled equations of motion, by use of Eqs. (4.4.29a) and (4.4.29b) with $D = 0$ discussed in Subsection 4.4.5 can be modified to obtain the governing equations of motion and heat conduction for thermoviscoelastoplasticity. Numerical solutions of such equations are shown in Chung and Eidson (1971, 1972, 1973a, 1973b).

PROBLEMS

7.1 Show the Rivlin–Eriksen tensor and material time derivatives in the forms given by Eqs. (7.2.30) and (7.2.21), respectively.

7.2 What is meant by a simple fluid?

7.3 Describe the convected coordinate approach as opposed to the vector space approach for constitutive equations of non-Newtonian fluids.

7.4 Derive the one-dimensional constitutive equation for a Kelvin material. Describe how the material constant may be determined.

7.5 Repeat Problem 7.6 for a Maxwell material.

7.6 Verify the differential equations, creep compliance, and relaxation modulus for the models shown in Table 7.3.1.

7.7 Consider the problem presented in Example 7.3.1. Verify the results for all three cases by filling in gaps and elaborate on your comments on the viscoelastic behavior for different models and loading conditions.

7.8 Draw yield surfaces in the principal stress space, the von Mises circular cylinder, the Tresca hexagonal prism, the deviatoric plane, and the hydrostatic line.

7.9 Verify the stress-space representation given in Fig. 7.4.9 and provide your comments for each theory.

7.10 Derive the expressions given in Eq. (7.4.47).

7.11 Show that the tensor of plastic moduli assumes the form

$$E^*_{ijkl} = \frac{E_{ijmn} Z_{mn} Z_{pq} E_{pqkl}}{E_p + Z_{rs} Z_{rs} E_{rstu}}.$$

8 Electromagnetic Continuum

8.1 Introduction

In the previous chapters we dealt with solids, liquids, and gases. In this chapter, another substance, called plasma, is introduced. The plasma state can be distinguished from the other states of matter in that a significant number of its molecules are in an electrically charged or ionized state. Very low degrees of ionization are sufficient for a gas to exhibit electromagnetic properties; at about 0.1% ionization a gas achieves an electrical conductivity of half its positive maximum and at 1% ionization the conductivity is nearly that of the fully ionized gas. Applications of plasma are diverse: energy conversion, electrical communications, and material sciences, to name a few (Grad, 1967; Holt and Haskell, 1968; Balescu, 1988; Jackson, 1998; Boyd and Sanderson, 2003).

Our objective in this chapter, however, is limited to exploring governing equations for ionized gases from the viewpoint of continuum mechanics. The governing equations for electrodynamics, magnetohydrodynamics, and electromagnetic waves with moderately low frequencies will be examined. Subsequently, in Section 9.4, the subject of electromagnetism will be employed in spacetime geometries for relativistic continuum.

8.2 The Electrodynamics

8.2.1 The Field Variables in Electrodynamics

In electrodynamics the main variables consist of plasma flow velocity **V**, magnetic field **B**, electric field **E**, current density **J**, and accelerating force **F**. We shall identify and define these variables in the magnetohydrodynamic (MHD) generator and plasma accelerator in this section.

The properties of plasma in motion and the interaction between plasma and a magnetic field are illustrated in the MHD generator in Fig. 8.2.1. Here the MHD generator converts the kinetic energy of a flowing plasma into electrical power. The inverse effect is used in the design of the plasma accelerator such as an engine for space vehicles (Fig. 8.2.2). The plasma flow velocity **V** is shown along the x_1 axis with an applied magnetic field $\mathbf{B}_{\text{appl}}$ in the x_2 direction. If electrodes are placed in the walls of the channel with an external circuit to permit a current

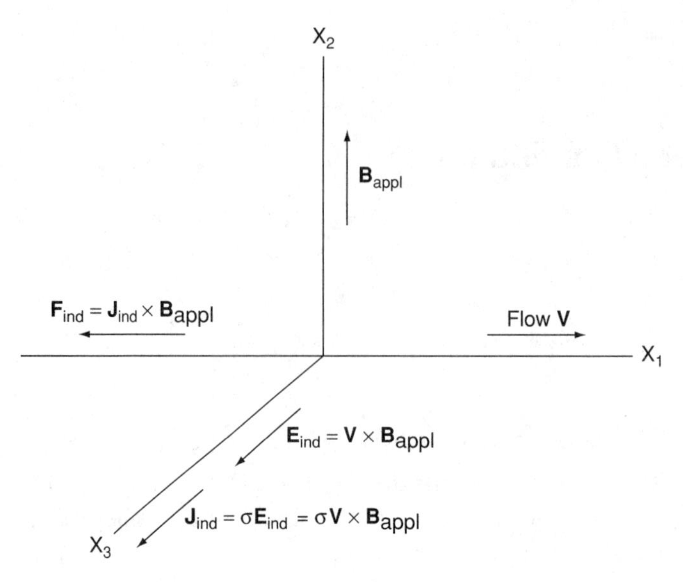

Figure 8.2.1. The principle of the MHD generator.

to flow through the plasma because of this electric field, then an induced current density $\mathbf{J}_{\text{ind}}$ flows across the stream. This, in turn, produces a force $\mathbf{F}_{\text{ind}}$ in the x_1 direction that acts to decelerate the plasma. Thus, because a current passes through an external load connected across the channel electrodes, the plasma loses kinetic energy.

The plasma accelerator requires both applied electric $\mathbf{E}_{\text{appl}}$ and magnetic $\mathbf{B}_{\text{appl}}$ fields, as shown in Fig. 8.2.2. An induced current density $\mathbf{J}_{\text{ind}}$ is present as in the case of the MHD generator, but this is offset by an opposite current density $\mathbf{J}_{\text{cond}}$ driven by the applied electric field. The accelerating force $\mathbf{F}_{\text{accl}}$ that acts on the

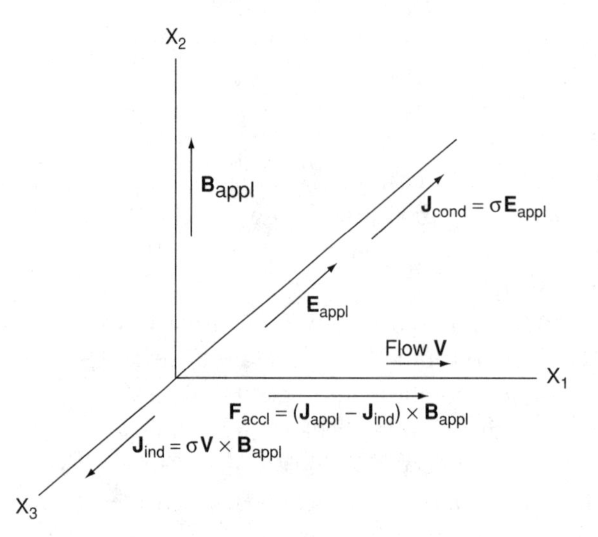

Figure 8.2.2. The principle of the plasma accelerator.

Table 8.2.1. *Units for electromagnetic variables*

Symbol	Name	Units
E_i	Electric-field intensity	$\frac{\text{V}}{\text{m}} = \frac{\text{kg m}}{\text{C s}^2} = \frac{\text{N}}{\text{C}}$
B_i	Magnetic induction	$\frac{\text{Wb}}{\text{m}^2} = \frac{\text{kg}}{\text{C s}} = \frac{\text{N s}}{\text{C m}}$
J_i	Electric current density	$\frac{\text{A}}{\text{m}} = \frac{\text{C}}{\text{m s}}$
P_i	Polarization	$\frac{\text{C}}{\text{m}^2}$
D_i	Electric displacement	$\frac{\text{C}}{\text{m}^2}$
H_i	Magnetic-field intensity	$\frac{\text{A}}{\text{m}^2} = \frac{\text{C}}{\text{m}^2\text{ s}}$

plasma will be positive if the applied electric field is strong enough, as shown in Fig. 8.2.2:

$$\mathbf{F}_{\text{accl}} = (\mathbf{J}_{\text{cond}} - \mathbf{J}_{\text{ind}}) \times \mathbf{B}_{\text{appl}}. \tag{8.2.1}$$

The electric-field intensity **E** and the magnetic induction **B** are defined in terms of the forces that act on a charged particle. It is found experimentally that these forces are of two types: one that is independent of the velocity of the particle, and another that is a function of the particle velocity. The combination of these two forces, which can be written in terms of the electric- and magnetic-field vectors, is defined as the Lorentz force. Let **R** be this combination of forces in units of force per unit mass m. Then the Lorentz force is given by

$$m\mathbf{R} = q\,(\mathbf{E} + \mathbf{V} \times \mathbf{B}), \tag{8.2.2}$$

where q is the electric charge of the particle, **V** is the particle velocity, and m is the particle mass. (See the units of various quantities in Table 8.2.1.)

The charge density η is defined as

$$\eta = \sum N^{(S)} q^{(S)}, \tag{8.2.3}$$

where N is the number density of the sth species. For a plasma containing only one type of singly ionized positive ion, the charge of each ion is $(+e)$ whereas the charge of each electron is $(-e)$ so that

$$\eta = e\,(N_+ - N_-). \tag{8.2.4}$$

Let the current density be J_i and n_i be the direction normal to the surface. Then we have

$$\int_S J_i n_i dA = -\frac{d}{dt}\int_V \eta dV. \tag{8.2.5}$$

Using the Green–Gauss theorem we obtain

$$\frac{\partial \eta}{\partial t} = -J_{i,i}. \tag{8.2.6}$$

The current density J_i can result from the application of an electric field E_i. However, in a plasma a current density can also flow as the result of mass motion or density and thermal gradients. Let us consider a current density that results from the application of an electric field. We assume that a linear relationship exists between the current density and the electric-field intensity,

$$J_i = C_{ij} E_j, \tag{8.2.7}$$

where C_{ij} is called the conductivity tensor. If the plasma is isotropic, then

$$C_{ij} = C\delta_{ij}, \tag{8.2.8}$$

where C is a scalar conductivity. Thus

$$J_i = C E_i \tag{8.2.9}$$

is known as Ohm's law.

Dielectrics are characterized by bound charges as contrasted with the free charges of conductors. When an electric field is applied to a dielectric, the centers of positive and negative charges of the atoms become displaced by some distance, say r_i. Such a displacement creates a dipole moment equal to er_i. The dipole moment per unit volume is given by Ner_i where N is the number density. The amount of charge that will cross an area dA because of this displacement of charge will be $Ner_i n_i dA$.

The net charge crossing a closed surface S enclosing a volume V will be

$$Q_N = \int_S N q r_i n_i dA. \tag{8.2.10}$$

Because the original net charge in V was zero, there will be a net charge left in V equal to $-Q_N$, called the polarization charge,

$$Q_P = -\int_S N q r_i n_i dA = -\int_S P_i n_i dA, \tag{8.2.11}$$

where

$$P_i = N q r_i \tag{8.2.12}$$

is called the polarization and is equal to the dipole moment per unit volume.

We assume that the polarization vector P_i is linearly related to the electric field E_i:

$$P_i = \epsilon_0 \alpha_{ij} E_j, \tag{8.2.13}$$

where α_{ij} is called the electric susceptibility tensor. It is taken to be dimensionless so that ϵ_0, called the permittivity of free space, is a constant that must have the dimensions

$$\epsilon_0 : \frac{\mathrm{C}^2\ \mathrm{s}^2}{\mathrm{m}^3\ \mathrm{kg}} \quad \text{or} \quad \frac{\mathrm{F}}{\mathrm{m}}.$$

We define the electric displacement D_i by

$$D_i = \epsilon_0 E_i + P_i \tag{8.2.14}$$

or

$$D_i = \epsilon_0 E_i + \epsilon_0 \alpha_{ij} E_j = \epsilon_0(\delta_{ij} + \alpha_{ij}) E_j = \epsilon_0 \kappa_{ij} E_j, \tag{8.2.15}$$

where

$$\kappa_{ij} = \delta_{ij} + \alpha_{ij} \tag{8.2.16}$$

is the dielectric tensor.

If the dielectric is isotropic then κ_{ij} reduces to a scalar dielectric constant κ. The quantity $\epsilon = \epsilon_0 \kappa$ is called the permittivity of the medium.

The dimensions of the time derivative of D_i are those of a current density. The total current density is written as C_i so that

$$C_i = J_i + \frac{\partial D_i}{\partial t}. \tag{8.2.17}$$

The magnetization vector M_i is defined as the magnetic moment per unit volume:

$$M_i = N m_i. \tag{8.2.18}$$

We define the magnetic-field intensity H_i by

$$B_i = \mu_0(H_i + M_i), \tag{8.2.19}$$

where μ_0 is a constant called the permeability of free space:

$$\mu_0 : \frac{\mathrm{kg\ m}}{\mathrm{C}^2} \quad \text{or} \quad \frac{\mathrm{H}}{\mathrm{m}}.$$

We assume a linear relationship between the magnetization vector and the magnetic-field intensity,

$$M_i = \chi_{ij} H_j, \tag{8.2.20}$$

where χ_{ij} is called the magnetic susceptibility tensor. Thus,

$$B_i = \mu_0(\delta_{ij} + \chi_{ij}) H_j \tag{8.2.21}$$

or

$$B_i = \mu_{ij} H_j, \tag{8.2.22}$$

where

$$\mu_{ij} = \mu_0(\delta_{ij} + \chi_{ij}) \tag{8.2.23}$$

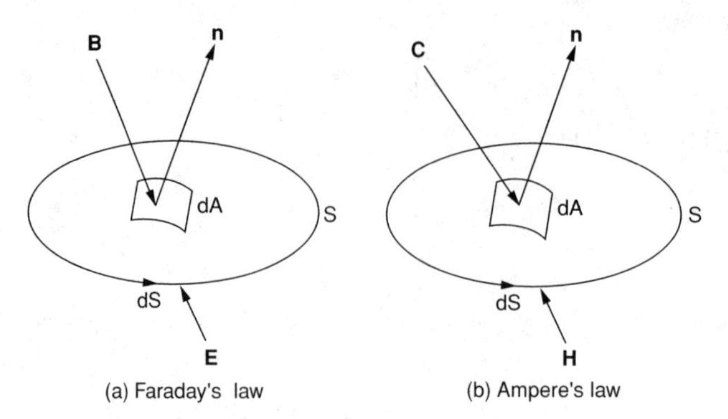

Figure 8.2.3. Principle of (a) Faraday's law and (b) Ampere's law.

is the permeability tensor. If the magnetic field is isotropic, $\mu = \mu_0 \chi$ is called the scalar permeability.

8.2.2 Maxwell's Equations

The Maxwell's equations consist of Faraday's law describing the time rate of change of magnetic flux **B** and Ampere's law defining the electric-current flux **C** in the macroscopic media acting in the direction normal to the infinitesimal area dA, as shown in Fig. 8.2.3.

Specifically, Faraday's law states that every change in magnetic flux that traverses a given surface produces in its boundaries an electromotive force numerically equal to the change in flux but opposite in sign. Using MKSA units, we obtain

$$\frac{d}{dt} \int_A \mathbf{B} \cdot \mathbf{n} \, dA = - \int_S \mathbf{E} \cdot d\mathbf{S}. \tag{8.2.24}$$

With the Stokes theorem, it follows that

$$\frac{d}{dt} \int_A \mathbf{B} \cdot \mathbf{n} \, dA = - \int_A \nabla \times \mathbf{E} \cdot \mathbf{n} \, dA, \tag{8.2.25a}$$

$$\frac{d}{dt} \int_A B_i n_i \, dA = - \int_A \epsilon_{ijk} E_{k,j} n_i \, dA, \tag{8.2.25b}$$

or

$$\int_A (\dot{B}_i + \epsilon_{ijk} E_{k,j}) n_i dA = 0. \tag{8.2.26}$$

For all arbitrary surfaces we must have

$$\frac{\partial \mathbf{B}}{\partial t} + \nabla \times \mathbf{E} = 0, \tag{8.2.27}$$

which is Faraday's law, the first of Maxwell's equations.

From Ampere's law, we obtain for the macroscopic media

$$\int_A \mathbf{C} \cdot \mathbf{n} dA = \int_S \mathbf{H} \cdot d\mathbf{S} = \int_A \nabla \times \mathbf{H} \cdot \mathbf{n} dA, \tag{8.2.28}$$

with

$$\mathbf{C} = \mathbf{J} + \frac{\partial \mathbf{D}}{\partial t}.$$

Thus

$$\nabla \times \mathbf{H} - \frac{\partial \mathbf{D}}{\partial t} = \mathbf{J}, \tag{8.2.29}$$

which is Ampere's law, the second of Maxwell's equations.

Ampere's law in a vacuum ($\mathbf{D} = \epsilon_0 \mathbf{E}$, $B = \mu_0 \mathbf{H}$) takes the form

$$\nabla \times \mathbf{B} - \frac{\partial \mathbf{E}}{\partial t} = \mathbf{J}. \tag{8.2.30}$$

Note that the Maxwell's equations just derived are subject to the following constraints:

$$\nabla \cdot \mathbf{B} = 0, \tag{8.2.31}$$

$$\nabla \cdot \mathbf{E} = \rho. \tag{8.2.32}$$

We can prove these relationships by taking divergences of (8.2.27) and (8.2.30), respectively.

Using the Gaussian units, we can write Eqs. (8.2.27) and (8.2.30) as

$$\frac{1}{c}\frac{\partial \mathbf{B}}{\partial t} + \nabla \times \mathbf{E} = 0, \tag{8.2.33}$$

$$\nabla \times \mathbf{B} - \frac{1}{c}\frac{\partial \mathbf{E}}{\partial t} = \frac{4\pi}{c}\mathbf{J}, \tag{8.2.34}$$

where c is the speed of light.

Similarly, constraint conditions (8.2.31) and (8.2.32) are given by

$$\nabla \cdot \mathbf{B} = 0, \tag{8.2.35}$$

$$\nabla \cdot \mathbf{E} = 4\pi\rho. \tag{8.2.36}$$

Combining (8.2.33) and (8.2.34) and taking derivatives, we obtain the second-order Maxwell's equations in the form

$$\nabla^2 \mathbf{E} - \nabla(\nabla \cdot \mathbf{E}) = \frac{1}{c^2}\frac{\partial^2 \mathbf{E}}{\partial t^2} + \frac{4\pi}{c^2}\frac{\partial J}{\partial t}, \tag{8.2.37}$$

where the MKSA units have been converted to the Gaussian units, as shown in Table 8.2.2.

Table 8.2.2. *Conversion between MKSA and Gaussian units (c = speed of light)*

MKSA	Gaussian
ρ	$4\pi\rho$
J	$\frac{4\pi}{c}J$
B	$\frac{\mathbf{B}}{c}$
D	$\frac{\mathbf{D}}{c}$

8.2.3 Electromagnetic Stress and Energy

It is seen from Table 8.2.1 that we can obtain terms with the dimensions of energy per unit volume per unit time by multiplying (8.2.27) by **H** and (8.2.29) by **E**, leading to

$$\nabla\times\mathbf{E}\cdot\mathbf{H} = -\frac{\partial\mathbf{B}}{\partial t}\cdot\mathbf{H}, \tag{8.2.38}$$

$$\nabla\times\mathbf{H}\cdot\mathbf{E} = \mathbf{J}\cdot\mathbf{E} + \frac{\partial\mathbf{D}}{\partial t}\cdot\mathbf{E}. \tag{8.2.39}$$

Subtracting (8.2.39) from (8.2.38), we have

$$\nabla\cdot(\mathbf{E}\times\mathbf{H}) + \mathbf{J}\cdot\mathbf{E} = -\dot{\mathbf{D}}\cdot\mathbf{E} - \dot{\mathbf{B}}\cdot\mathbf{H}. \tag{8.2.40}$$

Integrating (8.2.36) over the volume and applying the Green–Gauss theorem, we obtain

$$\int_A \mathbf{G}\cdot\mathbf{n}\,dA + \int_V \mathbf{J}\cdot\mathbf{E}\,dV = -\int_V (\dot{\mathbf{D}}\cdot\mathbf{E} + \dot{\mathbf{B}}\cdot\mathbf{H})\,dV, \tag{8.2.41}$$

where **G** is the Poynting vector

$$\mathbf{G} = \mathbf{E}\times\mathbf{H} \quad (\mathrm{W/m^2}), \tag{8.2.42}$$

which is normal to both electric and magnetic intensity vectors. It represents the energy per unit time, crossing a unit area.

Taking the cross product of (8.2.27) with **D** and (8.2.29) with **B** and adding, we obtain

$$T_{rs,s} - \eta E_r = \epsilon_{ris} J_i B_s + \frac{\partial}{\partial t}(\epsilon_{ris} D_i B_s) \tag{8.2.43}$$

with

$$T_{rs} = T_{rs}^{(E)} + T_{rs}^{(M)}, \tag{8.2.44}$$

where $T_{rs}^{(E)}$ and $T_{rs}^{(M)}$ are the electric and magnetic stress tensors, respectively, given by the following equations.

Electric stress tensor

$$T_{rs}^{(E)} = \begin{bmatrix} E_1D_1 - \frac{1}{2}E_jD_j & E_1D_2 & E_1D_3 \\ E_2D_1 & E_2D_2 - \frac{1}{2}E_jD_j & E_2D_3 \\ E_3D_1 & E_3D_2 & E_3D_3 - \frac{1}{2}E_jD_j \end{bmatrix}. \tag{8.2.45}$$

Magnetic stress tensor

$$T_{rs}^{(M)} = \begin{bmatrix} B_1H_1 - \frac{1}{2}B_jH_j & B_1H_2 & B_1H_3 \\ B_2H_1 & B_2H_2 - \frac{1}{2}B_jH_j & B_2H_3 \\ B_3H_1 & B_3H_2 & B_3H_3 - \frac{1}{2}B_jH_j \end{bmatrix}. \tag{8.2.46}$$

The integral form of (8.2.43) may be written as

$$\int_V \eta E_r dV + \int_V \epsilon_{irs} J_i B_s dV = \int_A T_{rs} n_s dA - \mu_0 \epsilon_0 \int_V \frac{\partial G_r}{\partial t} dV. \tag{8.2.47}$$

If the fields do not vary with time, we have

$$\int_V \epsilon_{ris} J_i B_s dV = \int_A T_{rs} n_s dA, \tag{8.2.48}$$

or, using the Green–Gauss theorem, we obtain

$$\epsilon_{ris} J_i B_s = T_{rs,s} \tag{8.2.49}$$

or

$$\mathbf{J} \times \mathbf{B} = T_{rs,s}\mathbf{i}_r, \tag{8.2.50}$$

where it is seen that the forces acting on the charges and currents within the volume V can be expressed as the integral of the electromagnetic stress tensor over the surface enclosing the charges and currents.

8.3 Magnetohydrodynamics

8.3.1 General

In magnetohydrodynamics we study the behavior of the combined system of electromagnetic fields and a conducting liquid or gas. In this system, free electrons move under the action of applied fields. In a solid conductor, however, the electrons are bound, but move considerable distances on the atomic scale before making collisions. Applied fields cause stresses to develop in the lattice structure. On the other hand, for a fluid, the fields act on both electrons and ionized atoms, leading to bulk motion of the medium. Thus, in this case, a coupled system of matter and fields must be considered.

For low frequencies we may neglect the displacement current in Ampere's law. This approximation is applicable to magnetohydrodynamics. For high frequencies, however, the charge separation and displacement must be included as in plasma physics.

8.3.2 Magnetohydrodynamic Equations

With the magnetic force $\mathbf{J} \times \mathbf{B}$ as identified in Eqs. (8.1.1) and (8.2.50), let us now consider the momentum equation in the form,

$$\rho \frac{\partial \mathbf{V}}{\partial t} + \rho(\mathbf{V} \cdot \nabla)\mathbf{V} = -\nabla p + \mathbf{J} \times \mathbf{B} + \tau_{ij,i}\mathbf{i}_j + \rho \mathbf{F}. \tag{8.3.1}$$

In general, the electric displacement $\mathbf{D}$ remains constant in time for MHD flows. Thus, neglecting $\partial \mathbf{D}/\partial \mathbf{t}$, we may write the Maxwell equations in the following forms:

Faraday

$$\nabla \times \mathbf{E} = -\frac{\partial \mathbf{B}}{\partial t}. \tag{8.3.2}$$

Ampere

$$\nabla \times \mathbf{B} = \mu_0 \mathbf{J}. \tag{8.3.3}$$

Ohm's law

$$\mathbf{J} = \sigma(\mathbf{E} + \mathbf{V} \times \mathbf{B}). \tag{8.3.4}$$

It follows from (8.3.1) and (8.3.3) that

$$\begin{aligned}
\rho \frac{\partial V_j}{\partial t} + \rho V_{j,i} V_i &= -p_{,j} + \frac{1}{\mu_0} \epsilon_{ijk} \epsilon_{imn} B_{n,m} B_k + \tau_{ij,i} + \rho F_j \\
&= -p_{,j} + \frac{1}{\mu_0}(\delta_{km}\delta_{jn} - \delta_{kn}\delta_{jm}) B_{n,m} B_k + \tau_{ij,i} + \rho F_j \\
&= -p_{,j} + \frac{1}{\mu_0}(B_{j,k} B_k - B_{k,j} B_k) + \tau_{ij,i} + \rho F_j \\
&= -\left(p + \frac{B^2}{2\mu_0}\right)_{,j} + \frac{1}{\mu_0} B_{j,k} B_k + \tau_{ij,i} + \rho F_j
\end{aligned} \tag{8.3.5}$$

or

$$\rho \frac{\partial \mathbf{V}}{\partial t} + \rho(\mathbf{V} \cdot \nabla)\mathbf{V} = -\nabla\left(p + \frac{B^2}{2\mu_0}\right) + \frac{1}{\mu_0}(\mathbf{B} \cdot \nabla)\mathbf{B} + \tau_{ij,i}\mathbf{i}_j + \rho \mathbf{F}, \tag{8.3.6}$$

with

$$B_{k,j} B_k = \frac{1}{2}(B^2)_{,j}, \quad B^2 = B_k B_k. \tag{8.3.7}$$

Combining (8.3.3) and (8.3.4), we obtain

$$\mathbf{E} = \frac{1}{\mu_0 \sigma} \nabla \times \mathbf{B} - \mathbf{V} \times \mathbf{B}. \tag{8.3.8}$$

Using the relationships given in Eqs. (1.2.27) and (1.2.26), it follows from (8.3.2) and (8.3.8) that

$$\begin{aligned} -\frac{\partial \mathbf{B}}{\partial t} &= \frac{1}{\mu_0 \sigma} \nabla \times (\nabla \times \mathbf{B}) - \nabla \times (\mathbf{V} \times \mathbf{B}) \\ &= \frac{1}{\mu_0 \sigma} (\nabla(\nabla \cdot \mathbf{B}) - \nabla^2 \mathbf{B}) - (\mathbf{B} \cdot \nabla)\mathbf{V} - \mathbf{V}(\nabla \cdot \mathbf{B}) \\ &\quad + \mathbf{B}(\nabla \cdot \mathbf{V}) + (\mathbf{V} \cdot \nabla)\mathbf{B} \end{aligned} \tag{8.3.9}$$

or

$$\frac{\partial B_j}{\partial t} - V_{j,i} B_i + V_{i,i} B_j + B_{j,i} V_i - \frac{1}{\mu_0 \sigma} B_{j,ii} = 0, \tag{8.3.10}$$

where $\nabla \cdot \mathbf{B} = 0$ has been enforced. Thus the Maxwell equations [Eqs. (8.3.2) and (8.3.3)] are now represented by a single equation, Eq. (8.3.10).

Equation (8.3.10) may be written in a conservation form,

$$\frac{\partial B_j}{\partial t} + \frac{\partial}{\partial x_i} (V_i B_j - V_j B_i - \frac{1}{\mu_0 \sigma} B_{j,i}) = 0. \tag{8.3.11}$$

It is seen that, on differentiation of the spatial derivatives, we recover the non-conservation form, Eq. (8.3.10), with $B_{i,i} = 0$.

The conservation form of the continuity, momentum, energy, and Maxwell equations can be written, following the procedure similar to the Navier–Stokes system of equations (Subsection 5.3.2), as

$$\frac{\partial \mathbf{U}}{\partial t} + \frac{\partial \mathbf{F}_i}{\partial x_i} + \frac{\partial \mathbf{G}_i}{\partial x_i} = \mathbf{S}, \tag{8.3.12}$$

where the conservation flow variables $\mathbf{U}$, convection flux variable $\mathbf{F}_i$, diffusion flux variables $\mathbf{G}_i$, and source terms $\mathbf{S}$ are given by

$$\mathbf{U} = \begin{bmatrix} \rho \\ \rho V_j \\ \rho \hat{E} \\ B_j \end{bmatrix}, \tag{8.3.13a}$$

$$\mathbf{F}_i = \begin{bmatrix} \rho V_i \\ \rho V_i V_j + p^* \delta_{ij} \\ (\rho \hat{E} + p^*) V_i \\ V_i B_j - V_j B_i \end{bmatrix}, \tag{8.3.13b}$$

$$\mathbf{G}_i = \begin{bmatrix} 0 \\ -\tau^*_{ij} \\ -\tau^*_{ij} V_j + q_i \\ \dfrac{1}{\mu_0 \sigma} B_{j,i} \end{bmatrix}, \tag{8.3.13c}$$

$$\mathbf{S} = \begin{bmatrix} 0 \\ \rho F_j \\ \rho F_j V_j \\ 0 \end{bmatrix}, \tag{8.3.13d}$$

where $\hat{E}$ is the internal energy density, τ^*_{ij} is the total stress tensor resulting from the various contributions including the viscous stress tensor τ_{ij}, magnetic product tensor $\tau^{(\mathrm{MP})}_{ij}$, electric stress tensor $\tau^{(E)}_{ij}$, and magnetic stress tensor $\tau^{(M)}_{ij}$:

$$\tau^*_{ij} = \tau_{ij} + \tau^{(\mathrm{MP})}_{ij} + \tau^{(E)}_{ij} + \tau^{(M)}_{ij}, \tag{8.3.14}$$

$$\tau_{ij} = 2\hat{\mu}\left(d_{ij} - \frac{1}{3} d_{kk}\delta_{ij}\right) \quad (\hat{\mu} = \text{viscosity constant}),$$

$$d_{ij} = \frac{1}{2}(V_{i,j} + V_{j,i}),$$

$$\tau^{(\mathrm{MP})}_{ij} = \frac{1}{\mu_0} B_i B_j,$$

$$\tau^{(E)}_{ij} = \text{Eq. (8.2.45)},$$

$$\tau^{(M)}_{ij} = \text{Eq. (8.2.46)},$$

$$\hat{E} = \frac{3}{2} N k_B T = \frac{3}{2} p \ (k_B = \text{Boltzmann's constant}).$$

Similarly, the total pressure p^* consists of contributions from the standard hydrodynamic pressure p and the magnetic pressure $p^{(M)}$:

$$p^* = p + p^{(M)}$$

$$p^{(M)} = \frac{1}{2\mu_0} B_k B_k. \tag{8.3.15}$$

All other notations are the same as in Subsection 5.3.2.

Note that we may perform differentiations as implied in (8.3.12) and confirm that we arrive at the nonconservation forms of continuity, momentum, energy, and the Maxwell's equations. It is well known that the use of the conservation form is essential for accurate solutions of the governing equations in computational fluid dynamics, particularly when MHD shock waves and turbulence are involved. In particular, the conservation form of magnetohydrodynamics as given by Eq. (8.3.12) is computationally efficient to determine the various aspects of magnetohydrodynamics such as magnetic reconnection and instabilities associated with plasma resistivity.

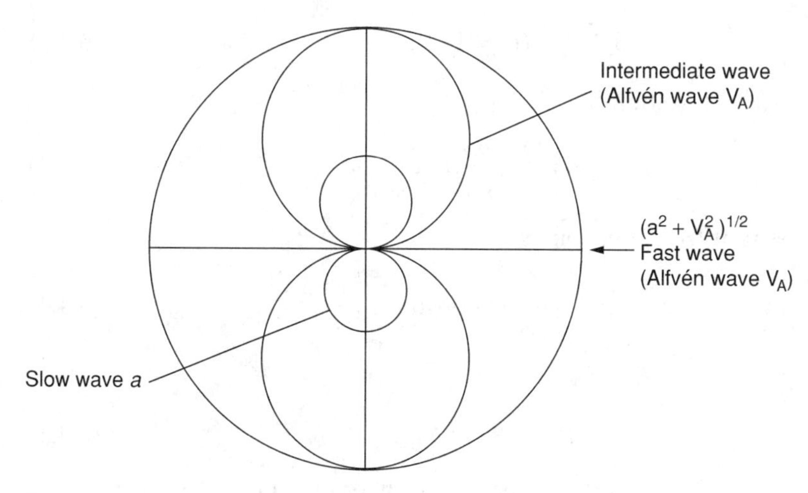

Figure 8.4.1. Phase polar diagram.

Another important aspect of the conservation form given by Eq. (8.3.12) is the explicit Neumann boundary conditions that arise from integration, resulting in

$$\int_{\Gamma} (\mathbf{F}_i + \mathbf{G}_i)\, n_i d\Gamma.$$

This is useful in the numerical simulation of Eq. (8.3.1) as fully described in Section 5.3 and in Chung (2002).

8.4 Electromagnetic Waves

There are three types of electromagnetic waves: slow waves, intermediate waves, and fast waves. Slow waves are defined by acoustic waves, whereas intermediate and fast waves are characterized by Alfvén waves. These waves are shown in Fig. 8.4.1.

The oblique Alfvén waves refer to intermediate waves that can propagate field-aligned currents and carry a charge density having a significant polarization. Magnetoacoustic waves (fast waves, compressional Alfvén waves) carry no current along the magnetic-field direction. Also charge density has no electrostatic polarization.

8.4.1 Alfvén Velocity

Let us assume that the plasma is of infinite extent and that the only spatial variations are in the x_3 direction. The perturbed magnetic field may be given as

$$B_i = B_i^0 + \bar{B}_i. \tag{8.4.1}$$

It follows from (8.4.1) and (8.3.11) with the diffusion term neglected that

$$\dot{\bar{B}}_\alpha = B_3 V_{\alpha,3} \quad (\alpha = 1, 2), \quad \bar{B}_3 = 0. \tag{8.4.2}$$

Similarly, from (8.3.7) we obtain

$$\rho \dot{V}_\alpha = \frac{B_0}{\mu_0} \bar{B}_{\alpha,3}. \tag{8.4.3}$$

The x_3 component of (8.3.7) becomes

$$p + \frac{\bar{B}^2}{2\mu_0} = \text{constant}, \tag{8.4.4}$$

with

$$\bar{B}^2 = \bar{B}_1^2 + \bar{B}_2^2. \tag{8.4.5}$$

The time derivatives of (8.4.2) together with (8.4.5) lead to

$$\ddot{\bar{B}}_\alpha = \frac{B_0^2}{\mu_0 \rho} \bar{B}_{\alpha,33}. \tag{8.4.6}$$

The time derivatives of (8.3.3) with (8.4.4) give

$$\ddot{V}_\alpha = \frac{B_0^2}{\mu_0 \rho} V_{\alpha,33}. \tag{8.4.7}$$

We note that (8.4.6) and (8.4.7) satisfy

$$\bar{B}_\alpha = \bar{B}_\alpha^0 \cdot e^{ik(x_3 - v_A t)}, \tag{8.4.8}$$

$$V_\alpha = V_\alpha^0 + e^{ik(x_3 - v_A t)}, \tag{8.4.9}$$

where v_A is the Alfvén velocity defined as

$$v_A = \frac{B_0}{(\mu_0 \rho)^{1/2}}. \tag{8.4.10}$$

8.4.2 Electromagnetic Waves in a Compressible Conducting Fluid

Let us consider the wave motion of a conducting fluid in the presence of a magnetic field. The fluid is compressible and nonviscous and no gravitation is present. The governing equations are

$$\frac{\partial \rho}{\partial t} + \nabla \cdot (\rho \mathbf{V}) = 0,$$

$$\rho \frac{\partial \mathbf{V}}{\partial t} + \rho (\mathbf{V} \cdot \nabla) \mathbf{V} = -\nabla P - \frac{1}{4\pi} \mathbf{B} \times (\nabla \times \mathbf{B}),$$

$$\frac{\partial \mathbf{B}}{\partial t} = \nabla \times (\mathbf{V} \times \mathbf{B}). \tag{8.4.11}$$

Assuming small-amplitude departures from the equilibrium states, we obtain

$$\rho = \rho_0 + \rho_1,$$
$$p = p_0 + p_1,$$
$$\mathbf{V} = \mathbf{V}_1,$$
$$\mathbf{B} = \mathbf{B}_0 + \mathbf{B}_1. \tag{8.4.12}$$

The linearized equations of (8.4.11) become

$$\frac{\partial \rho_1}{\partial t} + \rho_0 \nabla \cdot \mathbf{V}_1 = 0, \tag{8.4.13a}$$

$$\rho_0 \frac{\partial \mathbf{V}_1}{\partial t} + a^2 \nabla \rho_1 + \frac{\mathbf{B}_0}{4\pi} \times (\nabla \times \mathbf{B}_1) = 0, \tag{8.4.13b}$$

$$\frac{\partial \mathbf{B}_1}{\partial t} - \nabla \times (\mathbf{V}_1 \times \mathbf{B}_0) = 0, \tag{8.4.13c}$$

with $a = (\partial p/\partial \rho)_0$.

By combining the three equations given by (8.4.12) we obtain a single equation for $\mathbf{V}_1$:

$$\frac{\partial^2 \mathbf{V}_1}{\partial t} - a^2 \nabla(\nabla \cdot \mathbf{V}_1) + \mathbf{v}_A \times \nabla \times [\nabla \times (\mathbf{V}_1 \times \mathbf{v}_A)] = 0, \tag{8.4.14}$$

with the vectorial Alfvén velocity defined as $\mathbf{v}_A = \frac{\mathbf{B}_0}{(4\pi\rho_0)^{12}}$. Because $\mathbf{V}_1(\mathbf{x}, t)$ is a plane wave with wave vector $\mathbf{k}$ and frequency ω we can write

$$\mathbf{V}_1(\mathbf{x}, t) = \mathbf{V}_1 e^{i(\mathbf{k}\cdot\mathbf{x} - \omega t)}. \tag{8.4.15}$$

Substituting (8.4.15) into (3.4.14) leads to

$$\begin{aligned} &-\omega^2 \mathbf{V}_1 + \left(a^2 + v_A^2\right)(\mathbf{k} \cdot \mathbf{V}_1)\mathbf{k} + \mathbf{v}_A \cdot \mathbf{K}[(\mathbf{v}_A \cdot \mathbf{k})\mathbf{V}_1 - (\mathbf{v}_A \cdot \mathbf{V}_1)\mathbf{k} \\ &\quad - (\mathbf{k} \cdot \mathbf{V}_1)\mathbf{v}_A] = 0 \end{aligned} \tag{8.4.16}$$

or

$$-\omega^2 \mathbf{V}_1 + \left(a^2 + v_A^2\right)(\mathbf{k} \cdot \mathbf{V}_1)\mathbf{k} = 0, \tag{8.4.17}$$

where $\mathbf{k}$ is assumed to be perpendicular to $\mathbf{v}_A$. Then the solution for $\mathbf{V}_1$ represents a longitudinal magnetosonic wave with a phase velocity

$$u = \left(a^2 + v_A^2\right)^{1/2}. \tag{8.4.18}$$

Rewriting (8.4.17) in the form,

$$\left(k^2 v_A^2 - \omega^2\right)\mathbf{V}_1 + \left(\frac{a^2}{V_a^2} - 1\right) k^2 (\mathbf{v}_A \cdot \mathbf{V}_1)\mathbf{v}_A = 0, \tag{8.4.19}$$

we note that there are two types of wave motion: (1) An ordinary longitudinal wave ($\mathbf{V}_1$ parallel to $\mathbf{k}$ and $\mathbf{v}_A$) with phase velocity equal to the sound velocity, and (2) a transverse wave ($\mathbf{V}_1 \cdot \mathbf{v}_A = 0$) with a phase velocity equal to the Alfvén velocity. This is purely MHD, depending on the magnetic field and the density.

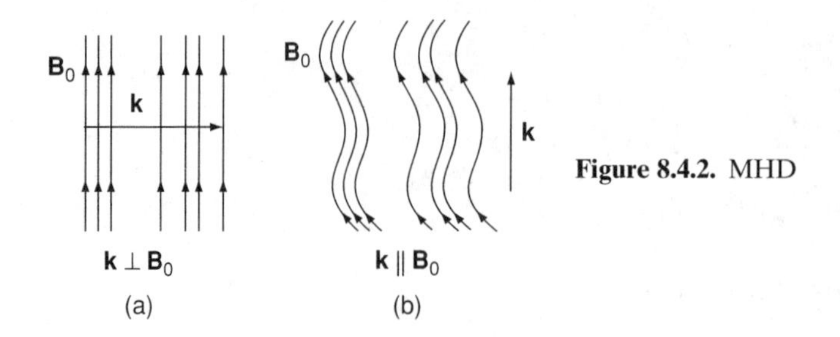

Figure 8.4.2. MHD

Now referring to Fig. 8.4.2 for propagation along the magnetic field, we see that the disturbance V_2 propagates as a transverse Alfvén wave with a phase velocity $\mathbf{v}_A$. On the other hand, the disturbance $\mathbf{V}_3$ propagates as a longitudinal acoustic wave with a phase velocity a. For propagation across the magnetic field, only the disturbance $\mathbf{V}_2$ propagates as a longitudinal magnetosonic wave with a phase velocity.

The magnetic fields of different waves can be found from (8.4.13c):

$$B_1 = \begin{cases} \dfrac{\mathbf{k}}{\omega} V_1 B_0, & \mathbf{k} \perp \mathbf{B}_0 \\ 0, & \mathbf{k} || \mathbf{B}_0 \quad \text{(longitudinal)} \left(a^2 + v_A^2\right)^{1/2}. \\ -\dfrac{\mathbf{k}}{\omega} B_0 \mathbf{V}_1, & \mathbf{k} \, || B_0 \quad \text{(transverse)} \end{cases} \tag{8.4.20}$$

As shown in Fig. 8.4.2(a), compressions and rarefactions in the lines of force arise without changing their directions if the magnetosonic wave moves perpendicular to $\mathbf{B}_0$. On the other hand, the Alfvén wave parallel to $\mathbf{B}_0$ causes the lines of force to oscillate back and forth laterally, as shown in Fig. 8.4.2(b). In both cases the lines of force are "frozen in" and move with the fluid.

In summary, the basic theory of electromagnetism has been introduced in this chapter, including the electromagnetic-field variables, Maxwell's equations, electromagnetic stress and energy, magnetohydrodynamics, and electromagnetic waves. This theory will then be applied to special relativity in Section 9.4.1, general relativity in Section 9.4.2, and gravitohydromagnetics in Section 9.4.3.

PROBLEMS

8.1 Derive the Maxwell equations by using detailed steps as illustrated by sketches.

8.2 Show the procedure of obtaining the electric and magnetic stress tensors.

8.3 Demonstrate the complete processes of obtaining the MHD equations.

8.4 Show how electromagnetic wave equations can be derived for compressible conducting fluids.

9 Differential Geometry Continuum

Differential geometry has numerous applications to engineering and physics (Graustein, 1935; Kreyszig, 1959; Oprea, 1997). Among them are shell geometries in engineering structures and spacetime geometries in physics of relativistic gravitation. The curvilinear coordinates introduced in Section 6.1 are prerequisite to understanding the differential geometry as discussed in this chapter.

Common to applications of the structures of shells and spacetime geometries of relativistic gravitation is the concept of curvatures consisting of first, second, and third fundamental tensors. These tensors arise from the tangent vectors defined on the surface of a domain. Derivatives of these tangent vectors lead to various forms of curvatures such as Riemann, Gauss, and Codazzi. They are the basic foundations for the shell theory and Einstein's relativity theory.

Thus the purpose of this chapter is to demonstrate that the differential geometry continuum indeed constitutes a part of continuum mechanics in which the concept of curvature characterizes a solid material on earth and the spacetime geometries of vacuum and interacting particles in the universe.

Mathematical preliminaries for curvatures presented in Sections 9.1 and 9.2 are followed by applications to shell geometries in Section 9.3 and applications to spacetime geometries in Section 9.4.

9.1 Fundamental Tensors

9.1.1 Curvilinear Coordinates and Base Vectors

Consider a reference surface characterized by a curvilinear system ($\xi^1, \xi^2, \xi^3 = 0$) with an origin located at P by a position vector $\mathbf{r}_0$, as shown in Fig. 9.1.1. The curvilinear coordinate lines ξ^i ($i = 1, 2, 3$) with the superscripted i are adopted here, contrary to the subscripted ξ_i as used in Section 6.1. It is intended that the tradition in literature be followed unless otherwise dictated by technical reasons. Let ξ^3 be the distance along the normal to the reference surface ($\xi^3 = 0$) and $\hat{\mathbf{a}}_3 = \mathbf{n}$ be the unit normal vector. An arbitrary point Q on the ξ^3 coordinate is defined by a position vector $\mathbf{r} = x_i\mathbf{i}_i$, where the x_i's are the Cartesian coordinates ($i = 1, 2, 3$):

$$\mathbf{r} = x_i\mathbf{i}_i = \mathbf{r}_0 + \xi^3\hat{\mathbf{a}}_3 = \mathbf{r}_0 + \xi^3\mathbf{n}. \tag{9.1.1}$$

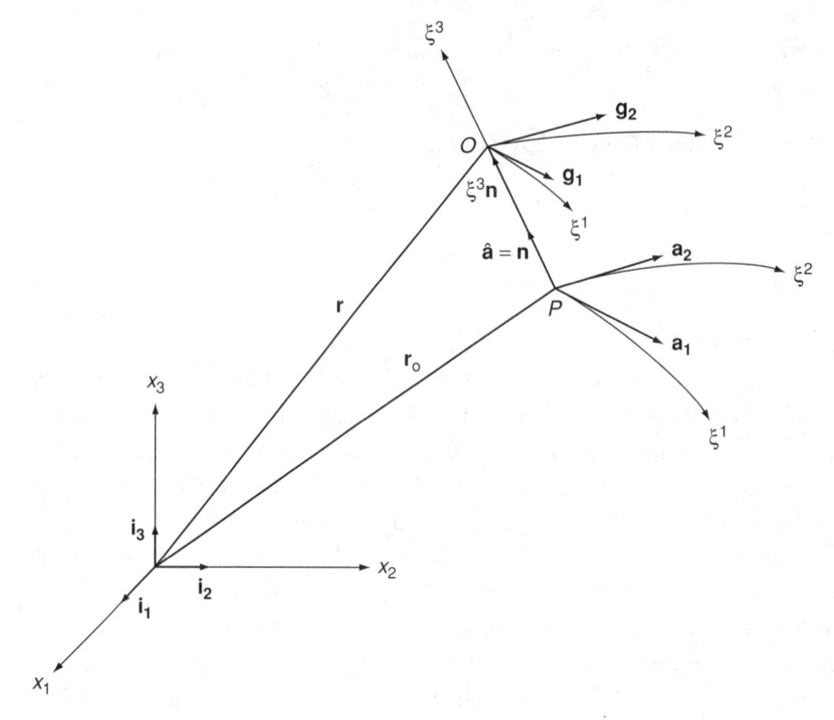

Figure 9.1.1. Coordinate system for a shell.

The tangent vectors or base vectors (used interchangeably) along the curvilinear coordinates ξ^α ($\alpha = 1, 2$) on the reference surface are represented by the partial derivatives of $\mathbf{r}_0$ with respect to ξ^α:

$$\frac{\partial \mathbf{r}_0}{\partial \xi^\alpha} = \mathbf{r}_{0,\alpha} = \mathbf{a}_\alpha. \tag{9.1.2}$$

Here, $\mathbf{a}_\alpha$ is the covariant component of the tangent vector. Likewise, the tangent vectors along ξ^α on the arbitrary surface at $\mathbf{r}$ are

$$\frac{\partial \mathbf{r}}{\partial \xi^\alpha} = \mathbf{r}_{,\alpha} = \mathbf{r}_{0,\alpha} + \xi^3 \mathbf{n}_{,\alpha} = \mathbf{g}_\alpha$$

or

$$\mathbf{g}_\alpha = \mathbf{a}_\alpha + \xi^3 \mathbf{n}_{,\alpha} \tag{9.1.3a}$$

and

$$\mathbf{g}_3 = \mathbf{a}_3 = \hat{\mathbf{a}}_3 = \mathbf{n}. \tag{9.1.3b}$$

The reciprocal base vector or the contravariant component of the tangent vector $\mathbf{a}^i$ has the properties (Fig. 9.1.2)

$$\mathbf{a}^i \cdot \mathbf{a}_j = \delta^i_j, \tag{9.1.4}$$

$$\mathbf{a}^\alpha = a^{\alpha\beta} \mathbf{a}_\beta, \quad \mathbf{a}_\alpha = a_{\alpha\beta} \mathbf{a}^\beta, \tag{9.1.5}$$

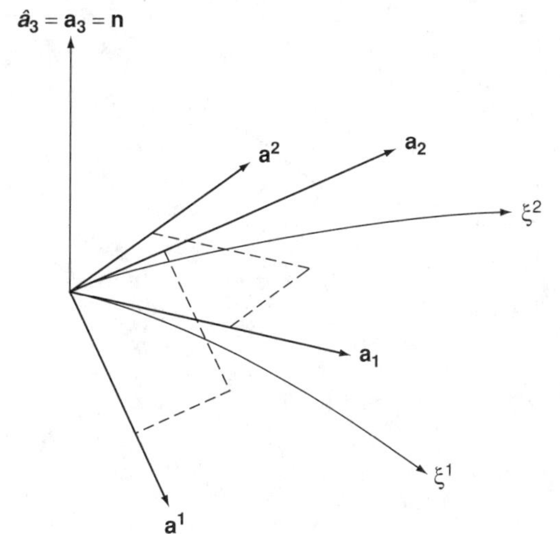

Figure 9.1.2. Covariant and contravariant components of a metric tensor.

in which $a_{\alpha\beta} = \mathbf{a}_\alpha \cdot \mathbf{a}_\beta$ and $a^{\alpha\beta} = \mathbf{a}^\alpha \cdot \mathbf{a}^\beta$ are the covariant and contravariant metric tensor, respectively. Note that $\mathbf{a}^\alpha$ is normal to the ξ^α surface. It also follows that

$$a^{\alpha\beta} = g^{\alpha\beta}(\xi^1, \xi^2, 0), \quad a_{\alpha\beta} = g_{\alpha\beta}(\xi^1, \xi^2, 0), \tag{9.1.6a}$$

$$a^{\alpha\beta} a_{\beta\gamma} = \delta^\alpha_\gamma, \tag{9.1.6b}$$

$$|a_{\alpha\beta}| = a = g(\xi^1, \xi^2, 0), \tag{9.1.6c}$$

$$|a^{\alpha\beta}| = \frac{1}{a}, \tag{9.1.6d}$$

$$a^{11} = \frac{a_{22}}{a}, \quad a^{22} = \frac{a_{11}}{a}, \quad a^{12} = -\frac{a_{12}}{a}.$$

An elemental volume bound by the coordinate surfaces is given by

$$\begin{aligned} d\Omega &= \mathbf{g}_1 d\xi^1 \cdot (\mathbf{g}_2 d\xi^2 \times \mathbf{g}_3 d\xi^3) = \mathbf{g}_1 \cdot \sqrt{g}\epsilon_{23k}\mathbf{g}^k d\xi^1 d\xi^2 d\xi^3 \\ &= \mathbf{g}_1 \cdot \mathbf{g}^1 \sqrt{g}\epsilon_{231} d\xi^1 d\xi^2 d\xi^3 = \sqrt{g}\, d\xi^1 d\xi^2 d\xi^3. \end{aligned} \tag{9.1.7}$$

Recall that the proof of this relation for three dimensions was given in Eq. (2.2.17). Let us now consider a tetrahedron bound by the coordinate surface and the inclined plane with the normal $\mathbf{n}^* = n_i^* \mathbf{g}^i$. If the area of a coordinate surface is ds_i and the area of the inclined face is ds, we have, from Fig. 9.1.3,

$$ds_i \mathbf{e}^i = ds\, \mathbf{n}^*, \tag{9.1.8}$$

where $\mathbf{e}^i$ is the unit normal to ds_i:

$$\mathbf{e}^i = \frac{\mathbf{g}^i}{\sqrt{g^{(ii)}}}.$$

The area ds_3 of the ξ^3 face is

$$ds_3 = \mathbf{e}^3 \cdot (\mathbf{g}_1 \times \mathbf{g}_2 d\xi^1 d\xi^2) = \sqrt{g^{33}}\sqrt{g}\, d\xi^1 d\xi^2 \tag{9.1.9}$$

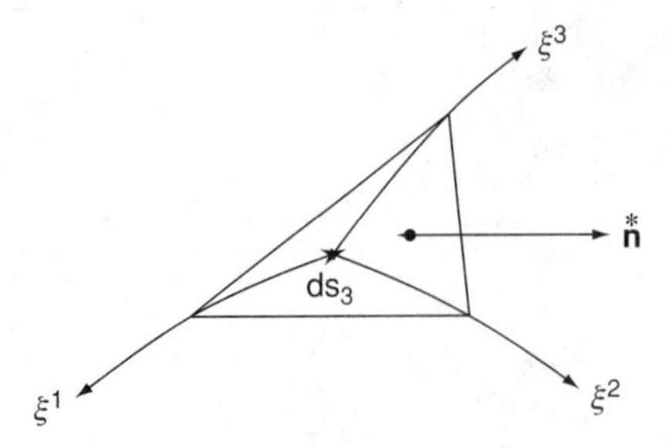

Figure 9.1.3. Inclined surface.

or

$$\sqrt{g}\, d\xi^1 d\xi^2 = \sqrt{g_{33}} ds_3 = n_3^* ds. \tag{9.1.10a}$$

Similarly,

$$\sqrt{g} d\xi^2 d\xi^3 = \sqrt{g_{11}} ds_1 = n_1^* ds, \quad \sqrt{g} d\xi^3 d\xi^1 = \sqrt{g_{22}} ds_2 = n_2^* ds.$$

Note that if $\mathbf{e}^3$ is perpendicular to the surface formed by the coordinates ξ^1 and ξ^2 (i.e., if $\mathbf{e}^3 = \mathbf{g}^3 = \mathbf{a}^3 = \hat{\mathbf{a}}_3 = \mathbf{n}$), then we have

$$\mathbf{n} \cdot (\mathbf{a}_1 \times \mathbf{a}_2) = a, \tag{9.1.10b}$$

$$ds = \sqrt{a} d\xi^1 d\xi^2, \tag{9.1.11}$$

$$\bar{\epsilon}_{\alpha\beta}(\xi^1, \xi^2) = \bar{\epsilon}_{\alpha\beta 3}(\xi^1, \xi^2, 0) = \sqrt{a}\epsilon_{\alpha\beta 3}, \tag{9.1.12a}$$

$$\bar{\epsilon}^{\alpha\beta}(\xi^1, \xi^2) = \bar{\epsilon}^{\alpha\beta 3}(\xi^1, \xi^2, 0) = \frac{1}{\sqrt{a}}\epsilon^{\alpha\beta 3}, \tag{9.1.12b}$$

$$\mathbf{a}_\alpha \times \mathbf{a}_\beta = \bar{\epsilon}_{\alpha\beta}\mathbf{n}, \quad \mathbf{a}^\alpha \times \mathbf{a}^\beta = \bar{\epsilon}^{\alpha\beta}\mathbf{n}, \tag{9.1.12c}$$

$$\mathbf{n} \times \mathbf{a}_\beta = \bar{\epsilon}_{\alpha\beta}\mathbf{a}^\beta, \quad \mathbf{n} \times \mathbf{a}^\alpha = \bar{\epsilon}^{\alpha\beta}\mathbf{a}_\beta. \tag{9.1.12d}$$

To describe the changes of geometry and curvature associated with deformations, we examine the derivatives of the base vectors in the next subsection.

9.1.2 Derivatives of Base Vectors

The curvatures of a surface are defined through scalar products of the base vectors and the derivatives of the base vectors through the Christoffel symbols of the first kind ($\Gamma_{\alpha\beta\gamma}$) and the second kind ($\Gamma^\gamma_{\alpha\beta}$):

$$\Gamma_{\alpha\beta\gamma}(\xi^1, \xi^2, 0) = \mathbf{a}_\gamma \cdot \mathbf{a}_{\alpha,\beta} = \Gamma_{\beta\alpha\gamma}, \tag{9.1.13a}$$

$$\Gamma^\gamma_{\alpha\beta}(\xi^1, \xi^2, 0) = \mathbf{a}^\gamma \cdot \mathbf{a}_{\alpha,\beta} = -\mathbf{a}_\alpha \cdot \mathbf{a}^\gamma_{,\beta} = \Gamma^\gamma_{\beta\alpha}. \tag{9.1.13b}$$

It follows from the discussion in Subsection 6.1.1 that

$$\Gamma_{\alpha\beta\gamma} = \frac{1}{2}(a_{\alpha\gamma,\beta} + a_{\beta\gamma,\alpha} - a_{\alpha\beta,\gamma}), \tag{9.1.14}$$

$$\Gamma_{\alpha\beta\gamma} = a_{\gamma\eta}\Gamma^\eta_{\alpha\beta}, \quad \Gamma^\eta_{\alpha\beta} = a^{\eta\gamma}\Gamma_{\alpha\beta\gamma}. \tag{9.1.15}$$

A scalar product of the normal vector $\mathbf{n}$ and the derivatives of the tangent base vectors is known as a *curvature tensor*:

$$b_{\alpha\beta} = \mathbf{n} \cdot \mathbf{a}_{\alpha,\beta} = -\mathbf{a}_{\alpha} \cdot \mathbf{n}_{,\beta} = \Gamma_{\alpha\beta 3}(\xi^1, \xi^2, 0) = b_{\beta\alpha}, \tag{9.1.16a}$$

$$b^{\alpha}_{\beta} = \mathbf{n} \cdot \mathbf{a}^{\alpha}_{,\beta} = -\mathbf{a}^{\alpha} \cdot \mathbf{n}_{,\beta} = -\Gamma^{\alpha}_{3\beta}(\xi^1, \xi^2, 0). \tag{9.1.16b}$$

Note that $\mathbf{n} \cdot \mathbf{n}_{,\alpha} = 0$, $\Gamma_{3\alpha 3} = \Gamma^3_{3\alpha} = 0$. Combining Eqs. (9.1.13) and (9.1.16) we obtain

$$\mathbf{a}_{\alpha,\beta} = \Gamma^{\gamma}_{\alpha\beta}\mathbf{a}_{\gamma} + b_{\alpha\beta}\mathbf{n} = \Gamma_{\alpha\beta\gamma}\mathbf{a}^{\gamma} + b_{\alpha\beta}\mathbf{n}, \tag{9.1.17a}$$

$$\mathbf{a}^{\alpha}_{,\beta} = -\Gamma^{\alpha}_{\beta\gamma}\mathbf{a}^{\gamma} + b^{\alpha}_{\beta}\mathbf{n}, \tag{9.1.17b}$$

$$\mathbf{n}_{,\beta} = -b_{\alpha\beta}\mathbf{a}^{\alpha} = -b^{\alpha}_{\beta}\mathbf{a}_{\alpha}. \tag{9.1.17c}$$

In view of Eqs. (9.1.13) and (9.1.17), it follows that

$$\mathbf{g}_{\alpha} = \mathbf{a}_{\alpha} - \xi^3 b^{\beta}_{\alpha}\mathbf{a}_{\beta} = \mathbf{a}_{\alpha} - \xi^3 b_{\alpha\beta}\mathbf{a}^{\beta}, \tag{9.1.18}$$

$$g_{\alpha\beta} = a_{\alpha\beta} - 2\xi^3 b_{\alpha\beta} + (\xi^3)^2 b_{\alpha\gamma}b^{\gamma}_{\beta}, \tag{9.1.19a}$$

$$g_{\alpha 3} = 0, \quad g_{33} = 1. \tag{9.1.19b}$$

The changes in the position vector and the normal vector are given by

$$d\mathbf{r}_0 = \mathbf{r}_{0,\alpha}d\xi^{\alpha} = \mathbf{a}_{\alpha}d\xi^{\alpha}, \tag{9.1.20a}$$

$$d\mathbf{n} = \mathbf{n}_{,\alpha}d\xi^{\alpha} = -b_{\alpha\beta}\mathbf{a}^{\beta}d\xi^{\alpha}. \tag{9.1.20b}$$

The scalar products of Eqs. (9.1.20) are

$$d\mathbf{r}_0 \cdot d\mathbf{r}_0 = ds_0^2 = \mathbf{r}_{0,\alpha}d\xi^{\alpha} \cdot \mathbf{r}_{0,\beta}d\xi^{\beta} = a_{\alpha\beta}d\xi^{\alpha}d\xi^{\beta}, \tag{9.1.21a}$$

$$d\mathbf{r}_0 \cdot d\mathbf{n} = \mathbf{a}_{\alpha}d\xi^{\alpha} \cdot \mathbf{n}_{,\beta}d\xi^{\beta} = -b_{\alpha\beta}d\xi^{\alpha}d\xi^{\beta}, \tag{9.1.21b}$$

$$\begin{aligned} d\mathbf{n} \cdot d\mathbf{n} &= \mathbf{n}_{,\alpha}d\xi^{\alpha} \cdot \mathbf{n}_{,\beta}d\xi^{\beta} = \left(-b^{\gamma}_{\alpha}\mathbf{a}_{\gamma}\right) \cdot \left(-b^{\mu}_{\beta}\mathbf{a}_{\mu}\right)d\xi^{\alpha}d\xi^{\beta} \\ &= b^{\gamma}_{\alpha}b^{\mu}_{\beta}a_{\gamma\mu}d\xi^{\alpha}d\xi^{\beta} = b^{\gamma}_{\alpha}b_{\beta\gamma}d\xi^{\alpha}d\xi^{\beta} = c_{\alpha\beta}d\xi^{\alpha}d\xi^{\beta}. \end{aligned} \tag{9.1.21c}$$

Here, $a_{\alpha\beta}$, $b_{\alpha\beta}$, and $c_{\alpha\beta}$ are called the first, second, and third fundamental tensors, respectively.

9.2 Curvature Tensors

9.2.1 Riemann–Christoffel Tensor

From the results obtained in Eqs. (9.1.20), we write

$$\begin{aligned} A_{i|jk} &= (A_{i|j})_{,k} - \Gamma^{r}_{ik}A_{r|j} - \Gamma^{r}_{jk}A_{i|r} \\ &= \left(A_{i,j} - \Gamma^{r}_{ij}A_r\right)_{,k} - \Gamma^{r}_{ik}\left(A_{r,j} - \Gamma^{s}_{rj}A_s\right) - \Gamma^{r}_{jk}\left(A_{i,r} - \Gamma^{s}_{ir}A_s\right) \\ &= A_{i,jk} - \left(\Gamma^{r}_{ij}\right)_{,k}A_r - \Gamma^{r}_{ij}A_{r,k} - \Gamma^{r}_{ik}A_{r,j} + \Gamma^{r}_{ik}\Gamma^{s}_{rj}A_s - \Gamma^{r}_{jk}A_{i,r} + \Gamma^{r}_{jk}\Gamma^{s}_{ir}A_s. \end{aligned} \tag{9.2.1a}$$

Similarly,

$$A_{i|kj} = A_{i,kj} - (\Gamma^r_{ik})_{,j} A_r - \Gamma^r_{ik} A_{r,j} - \Gamma^r_{ij} A_{r,k} + \Gamma^r_{ij}\Gamma^s_{rk} A_s - \Gamma^r_{kj} A_{i,r} + \Gamma^r_{kj}\Gamma^t_{ir} A_t. \tag{9.2.1b}$$

Subtracting Eq. (9.2.1b) from Eq. (9.2.1a) yields

$$\begin{aligned} A_{i|jk} - A_{i|kj} &= \Gamma^r_{ik}\Gamma^s_{rj} A_s - (\Gamma^r_{ij})_{,k} A_r - \Gamma^r_{ij}\Gamma^s_{rk} A_s + (\Gamma^r_{ik})_{,j} A_r \\ &= \left[(\Gamma^r_{ik})_{,j} - (\Gamma^r_{ij})_{,k} + \Gamma^s_{ik}\Gamma^r_{sj} - \Gamma^s_{ij}\Gamma^r_{sk}\right] A_r, \end{aligned} \tag{9.2.2}$$

in which A_i is an arbitrary covariant tensor of rank 1, the difference of the two tensors $A_{i|jk} - A_{i|kj}$ is a covariant tensor of rank 3, and the expression in the brackets of Eq. (9.2.2) is a mixed tensor of rank 4:

$$(\Gamma^r_{ik})_{,j} - (\Gamma^r_{ij})_{,k} + \Gamma^s_{ik}\Gamma^r_{sj} - \Gamma^s_{ij}\Gamma^r_{sk} = R^r_{ijk}. \tag{9.2.3}$$

The order of differentiation on the left-hand side of Eq. (9.2.3) is immaterial, in general, and it follows that

$$R^r_{ijk} = 0 \tag{9.2.4}$$

because A_r is arbitrary. The function R^r_{ijk} is called the Riemann–Christoffel tensor of the second kind. The associated tensor

$$R_{ijkl} = g_{ir} R^r_{jkl}$$

is the Riemann–Christoffel tensor of the first kind, which may be written in the form

$$R_{ijkl} = \frac{1}{2}(g_{il,jk} + g_{jk,il} - g_{ik,jl} - g_{jl,ik}) + g^{mn}(\Gamma_{jkm}\Gamma_{iln} - \Gamma_{jlm}\Gamma_{ikm}). \tag{9.2.5}$$

It can be shown that

$$R_{ijkl} = -R_{jikl} = -R_{ijlk} = R_{klij}, \tag{9.2.6}$$

which implies that R_{ijkl} is skew-symmetric in (ij) and (kl). We also note that there are six different components of R_{ijkl}, namely,

$$R_{3131}, \quad R_{3232}, \quad R_{1212}, \quad R_{3132}, \quad R_{3212}, \quad \text{and} \quad R_{3112}.$$

The Riemann–Christoffel tensors for the reference surface with $\xi^3 = 0$ are of the form

$$R^\lambda_{\alpha\beta\gamma} = \Gamma^\lambda_{\alpha\gamma,\beta} - \Gamma^\lambda_{\alpha\beta,\gamma} + \Gamma^\mu_{\alpha\gamma}\Gamma^\lambda_{\mu\beta} - \Gamma^\mu_{\alpha\beta}\Gamma^\lambda_{\mu\gamma} + \Gamma^3_{\alpha\gamma}\Gamma^\lambda_{3\beta} - \Gamma^3_{\alpha\beta}\Gamma^\lambda_{3\gamma} \tag{9.2.7}$$

or

$$
\begin{aligned}
R^{\lambda}_{\alpha\beta\gamma} &= \bar{R}^{\lambda}_{\alpha\beta\gamma} + \Gamma^{3}_{\alpha\gamma}\Gamma^{\lambda}_{3\beta} - \Gamma^{3}_{\alpha\beta}\Gamma^{\lambda}_{3\gamma} = 0,\\
R^{3}_{\alpha\beta\gamma} &= \Gamma^{3}_{\alpha\gamma,\beta} - \Gamma^{3}_{\alpha\beta,\gamma} + \Gamma^{m}_{\alpha\gamma}\Gamma^{3}_{m\beta} - \Gamma^{m}_{\alpha\beta}\Gamma^{3}_{m\gamma} = 0,\\
\bar{R}^{\lambda}_{\alpha\beta\gamma} &= \Gamma^{3}_{\alpha\beta}\Gamma^{\lambda}_{3\gamma} - \Gamma^{3}_{\alpha\gamma}\Gamma^{\lambda}_{3\beta} = b_{\alpha\beta}\left(-b^{\lambda}_{\gamma}\right) - b_{\alpha\gamma}\left(-b^{\lambda}_{\beta}\right),\\
\bar{R}_{\lambda\alpha\beta\gamma} &= a_{\lambda\mu}R^{\mu}_{\alpha\beta\gamma} = b_{\alpha\gamma}b_{\lambda\beta} - b_{\alpha\beta}b_{\lambda\gamma}.
\end{aligned}
$$

From the symmetry of $\Gamma^{\gamma}_{\alpha\beta}$ and $b_{\alpha\beta}$ we obtain

$$
\begin{aligned}
\bar{R}_{\alpha\gamma\beta\gamma} &= \bar{R}_{\alpha\beta\gamma\gamma} \quad (\alpha, \gamma \text{ not summed}),\\
\bar{R}_{1212} &= \bar{R}_{2121} = -\bar{R}_{2112} = -\bar{R}_{1221}.
\end{aligned}
$$

Hence, every nonzero component of $R_{\alpha\beta\gamma\delta}$ is equal to $\bar{R}_{1212}$ or to $-\bar{R}_{1212}$; thus we obtain

$$
\bar{R}_{1212} = |b_{\alpha\beta}| = b_{11}b_{22} - b^{2}_{12}. \tag{9.2.8}
$$

9.2.2 Gaussian and Codazzi Curvatures

We introduce an invariant K, called the *Gaussian curvature*:

$$
K = \frac{\bar{R}_{1212}}{a} = \frac{|b_{\alpha\beta}|}{|a_{\alpha\beta}|} = |b^{\alpha}_{\beta}| = b^{1}_{1}b^{2}_{2} - b^{12}_{21}. \tag{9.2.9}
$$

We may also write $\bar{R}_{\alpha\beta\gamma\delta} = K\bar{\epsilon}_{\alpha\beta}\bar{\epsilon}_{\gamma\delta}$ because $\bar{\epsilon}_{\alpha\beta} = \sqrt{a}\epsilon_{\alpha\beta}$ and $\bar{\epsilon}^{\alpha\beta} = \epsilon^{\alpha\beta}/\sqrt{a}$. Thus we have

$$
K = \frac{1}{4}\bar{R}_{\alpha\beta\gamma\delta}\bar{\epsilon}^{\alpha\beta}\bar{\epsilon}^{\gamma\delta} = \frac{1}{2}\bar{\epsilon}_{\alpha\gamma}\bar{\epsilon}^{\beta\delta}b^{\alpha}_{\beta}b^{\gamma}_{\delta}. \tag{9.2.10}
$$

Another important invariant, H, called the *mean curvature* of the surface, is of the form

$$
H = \frac{1}{2}a^{\alpha\beta}b_{\alpha\beta} = \frac{1}{2}b^{\alpha}_{\alpha} = \frac{1}{2}\left(b^{1}_{1} + b^{2}_{2}\right). \tag{9.2.11}
$$

Because $R^{3}_{\alpha\beta\gamma}$ also vanishes, we obtain, from Eq. (9.2.11)

$$
\begin{aligned}
R^{3}_{\alpha\beta\gamma} &= \Gamma^{3}_{\alpha\gamma,\beta} - \Gamma^{3}_{\alpha\beta,\gamma} + \Gamma^{m}_{\alpha\gamma}\Gamma^{3}_{m\beta} - \Gamma^{m}_{\alpha\beta}\Gamma^{3}_{m\gamma}\\
&= b_{\alpha\gamma,\beta} - b_{\alpha\beta,\gamma} + \Gamma^{m}_{\alpha\gamma}b_{m\beta} - \Gamma^{m}_{\alpha\beta}b_{m\gamma}.
\end{aligned}
$$

From the identity $b_{\alpha\gamma\,|\beta} = b_{\alpha\gamma,\beta} - b_{\alpha\lambda}\Gamma^{\lambda}_{\gamma\beta} - b_{\lambda\gamma}\Gamma^{\lambda}_{\alpha\beta}$ we have

$$
b_{\alpha\gamma\,|\beta} = b_{\alpha\beta\,|\gamma}, \tag{9.2.12}
$$

which represents either of two equations, namely

$$
b_{11\,|2} = b_{12\,|1} \quad \text{or} \quad b_{21\,|2} = b_{22\,|1}. \tag{9.2.13}
$$

These equations [Eq. (9.2.13)] are called the *Codazzi equations of the surface*, representing the compatibility of deformations of curved geometries, similar to the compatibility of torsional strains of three-dimensional solids discussed in Subsections 6.2.3 and 6.2.4.

9.3 Applications to Flexural Continuum (Shell Geometries)

The purpose of this section is to describe what the differential geometry can offer in the development of kinematics of a shell structure. Thus the traditional full shell theory, including kinetics, is not included. An attempt is made here to demonstrate that the geometric character of shell deformations is similar to the spacetime geometry of Einstein's theory of relativity.

By using the surface geometries that were introduced in Sections 9.1 and 9.2, we will now relate the various strain components of the reference surface to the displacement components. The coordinate through the thickness of a shell is taken normal to the reference surface; its stretching through the thickness is assumed to be zero. This leads to a reduction of the three-dimensional theory to a theory for the two dimensions of the surface, i.e., a formulation in two independent variables, ξ^1 and ξ^2. The most general form of the strain-displacement equations for shells is presented in the next subsection. For complete developments of kinetics of shells, the reader is invited to see Chung, 1988.

9.3.1 Kinematics of Shells

To discuss the kinematics of a shell, it is necessary to consider a deformed surface as well as an undeformed surface; both are shown in Fig. 9.3.1. We define the position vector of an arbitrary point through the thickness as $\mathbf{r} = x_i \mathbf{i}_i$, where $x_i = x_i(\xi^1, \xi^2, \xi^3 = z)$, $\mathbf{i}_i$ are the unit vectors corresponding to the Cartesian coordinates x_i, and ξ^i are the curvilinear coordinates. The position vector $\mathbf{r}$ is also written as

$$\mathbf{r} = \mathbf{r}_0 + \xi^3 \mathbf{n} = \mathbf{r}_0 + z\mathbf{n}, \tag{9.3.1}$$

where $\mathbf{r}_0$ is the position vector of a point on the undeformed middle surface and $\mathbf{n} = \mathbf{n}(\xi^1, \xi^2)$ is a unit normal vector to the undeformed middle surface. The square of the length of the line element is given by

$$\begin{aligned}(ds_0)^2 &= d\mathbf{r} \cdot d\mathbf{r} = g_{ij} d\xi^i d\xi^j \\ &= g_{\alpha\beta} d\xi^\alpha d\xi^\beta + g_{\alpha 3} d\xi^\alpha d\xi^3 + g_{33} d\xi^3 d\xi^3,\end{aligned} \tag{9.3.2}$$

where $d\mathbf{r} = d\mathbf{r}_0 + d(z\mathbf{n}) = d\mathbf{r}_0 + dz\,\mathbf{n} + z\,d\mathbf{n}$, with $\alpha, \beta = 1, 2$, and $i, j = 1, 2, 3$. Thus,

$$\begin{aligned}(ds_0)^2 = {} & d\mathbf{r}_0 \cdot d\mathbf{r}_0 + 2dz\,\mathbf{n} \cdot d\mathbf{r}_0 + 2z\,d\mathbf{n} \cdot d\mathbf{r}_0 + 2z\,dz\,\mathbf{n} \cdot d\mathbf{n} \\ & + z^2 d\mathbf{n} \cdot d\mathbf{n} + (dz)^2.\end{aligned} \tag{9.3.3}$$

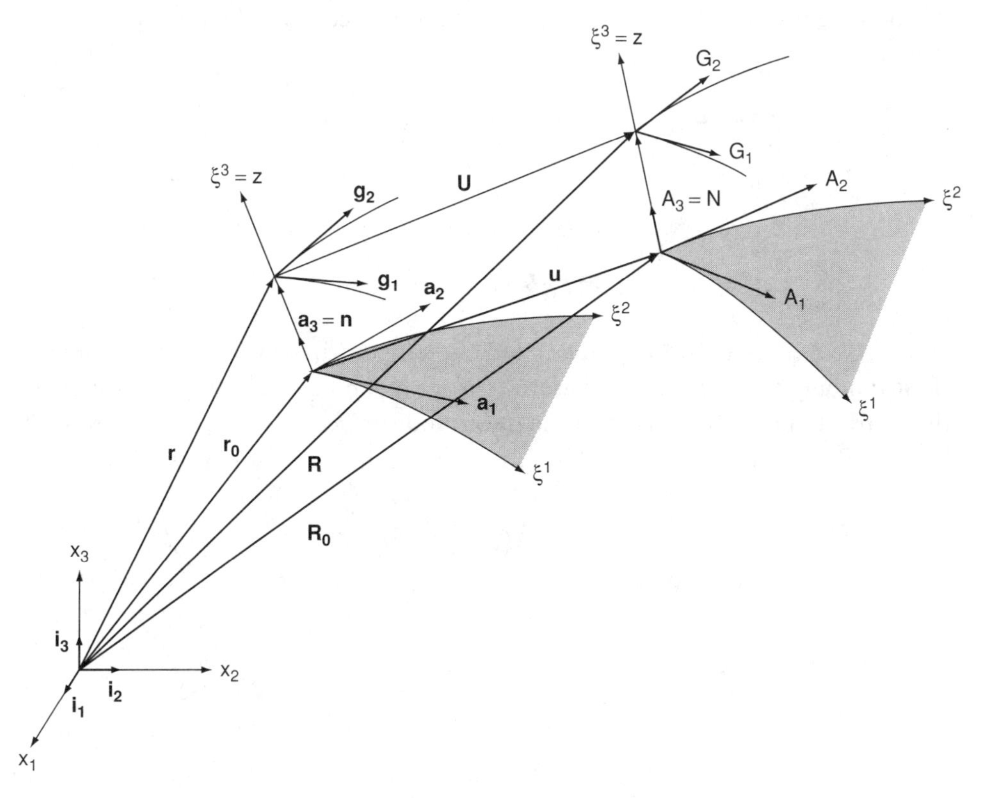

Figure 9.3.1. Undeformed and deformed surfaces.

Introducing the notation discussed in Section 9.2, where $a_{\alpha\beta} = \mathbf{r}_{0,\alpha} \cdot \mathbf{r}_{0,\beta}$ is the first fundamental tensor, $b_{\alpha\beta} = -\mathbf{n}_{,\alpha} \cdot \mathbf{a}_\beta$ is the second fundamental tensor, and $c_{\alpha\beta} = b^\lambda_\alpha b_{\beta\lambda}$ is the third fundamental tensor, and noting that $\mathbf{n} \cdot d\mathbf{r}_0 = \mathbf{n} \cdot d\mathbf{n} = 0$, we may rewrite Eq. (9.3.3) to read

$$(ds_0)^2 = \left(a_{\alpha\beta} - 2zb_{\alpha\beta} + z^2 c_{\alpha\beta}\right) d\xi^\alpha d\xi^\beta + (dz)^2. \tag{9.3.4}$$

In view of Eqs. (9.3.2) and (9.3.4), it follows that

$$g_{\alpha\beta} = a_{\alpha\beta} - 2zb_{\alpha\beta} + z^2 c_{\alpha\beta}, \quad g_{\alpha 3} = g_{3\alpha} = 0, \quad g_{33} = 1. \tag{9.3.5}$$

Now, after deformation, consider the new position vector

$$\mathbf{R} = \mathbf{R}_0 + \xi^3 \mathbf{N} = \mathbf{R}_0 + z\mathbf{N}. \tag{9.3.6}$$

The squared length of the line element on the deformed surface is given by

$$\begin{aligned}(ds)^2 = d\mathbf{R} \cdot d\mathbf{R} = G_{ij} d\xi^i d\xi^j &= G_{\alpha\beta}\, d\xi^\alpha d\xi^\beta + G_{\alpha 3} d\xi^\alpha d\xi^3 + G_{33} d\xi^3 d\xi^3 \\ &= d\mathbf{R}_0 \cdot d\mathbf{R}_0 + 2d\mathbf{R}_0 \cdot \mathbf{N}\, dz + 2d\mathbf{R}_0 \cdot z\, d\mathbf{N} + 2z\, dz\, \mathbf{N} \cdot d\mathbf{N} \\ &\quad + (dz)^2 \mathbf{N} \cdot \mathbf{N} + z^2 d\mathbf{N} \cdot d\mathbf{N}.\end{aligned} \tag{9.3.7}$$

It should be noted here that $\mathbf{N} \cdot d\mathbf{N} \neq 0$ and $d\mathbf{R}_0 \cdot \mathbf{N} \neq 0$, because the vector $\mathbf{N}$

can no longer remain normal to the deformed surface and is no longer unity. We define

$$\begin{aligned}
\mathbf{A}_\alpha &= \mathbf{R}_{0,\alpha} \quad \text{(tangent vector on deformed surface)},\\
A_{\alpha\beta} &= \mathbf{A}_\alpha \cdot \mathbf{A}_\beta,\\
B_{\alpha\beta} &= -\mathbf{N}_{,\alpha} \cdot \mathbf{A}_\beta,\\
C_{\alpha\beta} &= \mathbf{N}_{,\alpha} \cdot \mathbf{N}_{,\beta} = B_\alpha^\lambda B_{\beta\lambda}.
\end{aligned}$$

Here, A_α, $A_{\alpha\beta}$, $B_{\alpha\beta}$, and $C_{\alpha\beta}$ are called the tangent base vector, metric tensor (first fundamental tensor), curvature tensor (second fundamental tensor), and the third fundamental tensor on the deformed surface, respectively. We rewrite Eq. (9.3.7) as follows:

$$\begin{aligned}
(ds)^2 &= (A_{\alpha\beta} - 2zB_{\alpha\beta} + z^2C_{\alpha\beta})d\xi^\alpha d\xi^\beta + 2(\mathbf{A}_\alpha + z\mathbf{N}_{,\alpha}) \cdot N\, d\xi^\alpha dz\\
&\quad + (dz)^2\mathbf{N} \cdot \mathbf{N}.
\end{aligned} \tag{9.3.8}$$

Once again, we may define

$$\begin{aligned}
G_{\alpha\beta} &= A_{\alpha\beta} - 2zB_{\alpha\beta} + z^2C_{\alpha\beta},\\
G_{\alpha 3} &= 2(\mathbf{A}_\alpha + z\mathbf{N}_{,\alpha}) \cdot \mathbf{N},\\
G_{33} &= \mathbf{N} \cdot \mathbf{N}.
\end{aligned}$$

If the unit vector normal to the undeformed surface is assumed to remain normal to the deformed surface, the well-known Kirchhoff hypothesis (Kirchhoff, 1850) may be adopted and we set $\mathbf{N} \cdot \mathbf{N} = 1$, $G_{\alpha 3} = 0$, and $G_{33} = 1$. For generality, however, let us proceed with no simplifying assumptions.

The difference between the squared lengths of the line elements on the deformed and undeformed surfaces is

$$\begin{aligned}
ds^2 - ds_0^2 &= 2\gamma_{ij}d\xi^i d\xi^j = 2\gamma_{\alpha\beta}d\xi^\alpha d\xi^\beta + 2\gamma_{\alpha 3}d\xi^\alpha d\xi^3 + 2\gamma_{33}(dz)^2\\
&= [(A_{\alpha\beta} - a_{\alpha\beta}) - 2z(B_{\alpha\beta} - b_{\alpha\beta}) + z^2(C_{\alpha\beta} - c_{\alpha\beta})]d\xi^\alpha d\xi^\beta\\
&\quad + 2(\mathbf{N} \cdot \mathbf{A}_\alpha + z\mathbf{N} \cdot \mathbf{N}_{,\alpha})d\xi^\alpha dz + (\mathbf{N} \cdot \mathbf{N})(dz)^2 - (dz)^2,
\end{aligned} \tag{9.3.9}$$

where $\gamma_{ij} = (\gamma_{\alpha\beta}, \gamma_{\alpha 3}, \gamma_{33})$ are the components of the strain tensor, defined as follows:

$$\gamma_{\alpha\beta} = \frac{1}{2}[(A_{\alpha\beta} - a_{\alpha\beta}) - 2z(B_{\alpha\beta} - b_{\alpha\beta}) + z^2(C_{\alpha\beta} - c_{\alpha\beta})], \tag{9.3.10a}$$

$$\gamma_{\alpha 3} = \mathbf{N} \cdot \mathbf{A}_\alpha + z\mathbf{N} \cdot \mathbf{N}_{,\alpha}, \tag{9.3.10b}$$

$$\gamma_{33} = \frac{1}{2}(\mathbf{N} \cdot \mathbf{N} - 1). \tag{9.3.10c}$$

We also define

$$e_{\alpha\beta} = \frac{1}{2}(A_{\alpha\beta} - a_{\alpha\beta}), \tag{9.3.11a}$$

$$\chi_{\alpha\beta} = -(B_{\alpha\beta} - b_{\alpha\beta}), \tag{9.3.11b}$$

$$\Lambda_{\alpha\beta} = \frac{1}{2}(C_{\alpha\beta} - c_{\alpha\beta}), \tag{9.3.11c}$$

where $e_{\alpha\beta}$, $\chi_{\alpha\beta}$, and $\Lambda_{\alpha\beta}$ are called the components of the membrane, bending, and higher-order bending strain tensors, respectively. Rewriting Eq. (9.3.10a) we have

$$\gamma_{\alpha\beta} = e_{\alpha\beta} + z\chi_{\alpha\beta} + z^2\Lambda_{\alpha\beta}. \tag{9.3.12}$$

It should be noted that $\gamma_{\alpha\beta}(\gamma_{11}, \gamma_{22}, \gamma_{12})$ represents the normal strains γ_{11}, γ_{22} and shear strain γ_{12} on the reference surface, whereas $\gamma_{\alpha 3}$ $(\gamma_{13}, \gamma_{23})$ represents the *transverse shear strains* acting on the planes through the thickness of the shell. The normal strain through the thickness γ_{33} is zero if we make an assumption that stretching through the thickness is negligible and that $\mathbf{n} \cdot \mathbf{n} = \mathbf{N} \cdot \mathbf{N} = 1$.

9.3.2 Membrane Strain Tensor

The membrane strain tensor is given by

$$e_{\alpha\beta} = \frac{1}{2}(A_{\alpha\beta} - a_{\alpha\beta}). \tag{9.3.13}$$

The tangent base vector $\mathbf{A}$ on the deformed surface reads

$$\mathbf{A}_\alpha = \mathbf{R}_{0,\alpha} = (\mathbf{r}_0 + \mathbf{u})_{,\alpha} = \mathbf{r}_{0,\alpha} + \mathbf{u}_{,\alpha} = \mathbf{a}_\alpha + \mathbf{u}_{,\alpha},$$

where $\mathbf{u}$ is the displacement vector, as shown in Fig. 9.3.1:

$$\mathbf{u} = \mathbf{u}^\beta \mathbf{a}_\beta + u^3\mathbf{n}. \tag{9.3.14}$$

Thus the metric tensor on the deformed surface is

$$A_{\alpha\beta} = a_{\alpha\beta} + \mathbf{a}_\alpha \cdot \mathbf{u}_{,\beta} + \mathbf{a}_\beta \cdot \mathbf{u}_{,\alpha} + \mathbf{u}_{,\alpha} \cdot \mathbf{u}_{,\beta}. \tag{9.3.15}$$

Substituting Eq. (9.3.15) into Eq. (9.3.13) yields

$$e_{\alpha\beta} = \frac{1}{2}(\mathbf{a}_\alpha \cdot \mathbf{u}_{,\beta} + \mathbf{a}_\beta \cdot \mathbf{u}_{,\alpha} + \mathbf{u}_{,\alpha} \cdot \mathbf{u}_{,\beta}). \tag{9.3.16}$$

From Eq. (9.3.14), the derivative of displacements can be written as

$$\mathbf{u}_{,\alpha} = u^\beta_{,\alpha}\mathbf{a}_\beta + u^\beta\mathbf{a}_{\beta,\alpha} + u^3_{,\alpha}\mathbf{n} + u^3\mathbf{n}_{,\alpha}.$$

It follows from Eq. (9.3.17a) that the derivative of a tangent vector takes the form

$$\mathbf{a}_{\beta,\alpha} = \Gamma^\mu_{\beta\alpha}\mathbf{a}_\mu + \Gamma^3_{\beta\alpha}\mathbf{n} = \Gamma^\mu_{\beta\alpha}\mathbf{a}_\mu + b_{\beta\alpha}\mathbf{n}, \tag{9.3.17}$$

with

$$\Gamma^{\mu}_{\beta\alpha} = \mathbf{a}_{\beta,\alpha} \cdot \mathbf{a}^{\mu}, \qquad \Gamma^{3}_{\beta\alpha} = \mathbf{a}_{\beta,\alpha} \cdot \mathbf{n} = -\mathbf{a}_{\beta} \cdot \mathbf{n}_{,\alpha} = b_{\beta\alpha}.$$

Hence,

$$\begin{aligned} u_{,\alpha} &= \left(u^{\mu}_{,\alpha} + \Gamma^{\mu}_{\beta\alpha} u^{\beta}\right) \mathbf{a}_{\mu} + u^{\beta} b_{\alpha\beta}\, \mathbf{n} + u^{3}_{,\alpha}\, \mathbf{n} - u^{3} b^{\mu}_{\alpha} \mathbf{a}_{\mu} \\ &= \left(u^{\mu}_{,\alpha} + \Gamma^{\mu}_{\beta\alpha} u^{\beta} - u^{3} b^{\mu}_{\alpha}\right)\mathbf{a}_{\mu} + \left(u^{\beta} b_{\alpha\beta} + u^{3}_{,\alpha}\right)\mathbf{n} \\ &= \left(u^{\mu}_{|\alpha} - u^{3} b^{\mu}_{\alpha}\right)\mathbf{a}_{\mu} + \left(u^{3}_{,\alpha} + u^{\beta} b_{\alpha\beta}\right)\mathbf{n}, \end{aligned} \tag{9.3.18}$$

where the covariant differentiation $u^{\mu}_{|\alpha}$ is given by

$$u^{\mu}_{|\alpha} = u^{\mu}_{,\alpha} + \Gamma^{\mu}_{\beta\alpha} u^{\beta}. \tag{9.3.19}$$

Substituting Eqs. (9.3.17) and (9.3.18) into Eq. (9.3.16) yields

$$\begin{aligned} e_{\alpha\beta} &= \frac{1}{2}\Big\{\mathbf{a}_{\alpha} \cdot \left[\left(u^{\mu}_{|\beta} - u^{3} b^{\mu}_{\beta}\right)\mathbf{a}_{\mu} + \left(u^{3}_{,\beta} + u^{\mu} b_{\beta\mu}\right)\mathbf{n}\right] \\ &\quad + \mathbf{a}_{\alpha} \cdot \left[\left(u^{\mu}_{|\alpha} - u^{3} b^{\mu}_{\alpha}\right)\mathbf{a}_{\mu} + \left(u^{3}_{,\alpha} + u^{\mu} b_{\alpha\mu}\right)\mathbf{n}\right] \\ &\quad + \left[\left(u^{\mu}_{|\alpha} - u^{3} b^{\mu}_{\alpha}\right)\mathbf{a}_{\mu} + \left(u^{3}_{,\alpha} + u^{\mu} b_{\alpha\mu}\right)\mathbf{n}\right] \cdot \left[\left(u^{\mu}_{|\beta} + u^{3} b^{\mu}_{\beta}\right)\mathbf{a}_{\mu} + \left(u^{3}_{,\beta} + u^{\mu} b_{\beta\mu}\right)\mathbf{n}\right]\Big\} \\ &= \frac{1}{2}\left[u^{\mu}_{|\beta} a_{\alpha\mu} + u^{\mu}_{|\alpha} a_{\beta\mu} - u^{3} b^{\mu}_{\beta} a_{\alpha\mu} - u^{3} b^{\mu}_{\alpha} a_{\beta\mu} + \left(u^{\mu}_{|\alpha} - u^{3} b^{\mu}_{\alpha}\right)\left(u^{\mu}_{|\beta} + u^{3} b^{\mu}_{\beta}\right) a_{\mu\mu}\right] \\ &\quad + \left(u^{3}_{,\alpha} + u^{\lambda} b_{\alpha\lambda}\right)\left(u^{3}_{,\beta} + u^{\gamma} b_{\beta\gamma}\right) \\ &= \frac{1}{2}\left[u_{\alpha|\beta} + u_{\beta|\alpha} - 2u^{3} b_{\alpha\beta} + \left(u_{\mu|\alpha} + u^{3} b_{\mu\alpha}\right)\left(u^{\mu}_{|\beta} - u^{3} b^{\mu}_{\beta}\right)\right. \\ &\quad \left. + \left(u^{3}_{,\alpha} + u^{\lambda} b_{\alpha\lambda}\right)\left(u^{3}_{,\beta} + u^{\gamma} b_{\beta\gamma}\right)\right]. \end{aligned} \tag{9.3.20}$$

This is the most general form of the membrane strain tensor; it retains all possible nonlinear terms.

9.3.3 Bending Strain Tensor

To examine the bending strain tensor, we first set the vector $\mathbf{m}$ equal to the difference between the vectors $\mathbf{N}$ and $\mathbf{n}$:

$$\mathbf{m} = \mathbf{N} - \mathbf{n}, \qquad \mathbf{N} = \mathbf{m} + \mathbf{n}.$$

As a result, the bending strain tensor is written as

$$\begin{aligned} \chi_{\alpha\beta} &= -(B_{\alpha\beta} - b_{\alpha\beta}) = \mathbf{A}_{\alpha} \cdot \mathbf{N}_{,\beta} - \mathbf{a}_{\alpha} \cdot \mathbf{n}_{,\beta} \\ &= (\mathbf{a}_{\alpha} + \mathbf{u}_{,\alpha}) \cdot (\mathbf{m} + \mathbf{n})_{,\beta} - \mathbf{a}_{\alpha} \cdot \mathbf{n}_{,\beta} \\ &= (\mathbf{a}_{\alpha} + \mathbf{u}_{,\alpha}) \cdot \mathbf{m}_{,\beta} + \mathbf{u}_{,\alpha} \cdot \mathbf{n}_{,\beta} \\ &= (\mathbf{a}_{\alpha} + \mathbf{u}_{,\alpha}) \cdot \mathbf{m}_{,\beta} - \mathbf{u}_{,\alpha} \cdot b^{\lambda}_{\beta} \mathbf{a}_{\lambda}. \end{aligned} \tag{9.3.21a}$$

Denoting $\mathbf{m} = m^\alpha \mathbf{a}_\alpha + m\mathbf{n}$ and the derivative of $\mathbf{m}$ as

$$\mathbf{m}_{,\beta} = \left(m^\mu_{|\beta} - mb^\mu_\beta\right)\mathbf{a}_\mu + \left(m_{,\beta} + m^\mu b_{\beta\mu}\right)\mathbf{n},$$

we express the most general form of the bending strain tensor in the form

$$\begin{aligned}
\chi_{\alpha\beta} &= \left[\mathbf{a}_\alpha + \left(u^\mu_{|\alpha} - u^3 b^\mu_\alpha\right)\mathbf{a}_\mu + \left(u^3_{,\alpha} + u^\mu b_{\alpha\mu}\right)\mathbf{n}\right] \\
&\quad \cdot\left[\left(m^\mu_{|\beta} - mb^\mu_\alpha\right)\mathbf{a}_\mu + \left(m_{,\beta} + m^\mu b_{\beta\mu}\right)\mathbf{n}\right] \\
&\quad - \left[\left(u^\mu_{|\alpha} - u^3 b^\mu_\alpha\right)\mathbf{a}_\mu + \left(u^3_{,\alpha} + u^\mu b_{\alpha\mu}\right)\mathbf{n}\right]\cdot b^\mu_\alpha \mathbf{a}_\mu \\
&= m_{\alpha|\beta} - mb_{\alpha\beta} + \left(u_{\mu|\alpha} - u^3 b_{\mu\alpha}\right)\left(m^\mu_{|\beta} - mb^\mu_\beta\right) \\
&\quad + \left(u^3_{,\alpha} + u^\lambda b_{\alpha\lambda}\right)\left(m_{,\beta} + m^\gamma b_{\beta\gamma}\right) - b^\mu_\beta\left(u_{\mu|\alpha} - u^3 b_{\mu\alpha}\right). \qquad (9.3.21b)
\end{aligned}$$

To obtain the bending strain tensor in a more explicit form, we proceed as follows. First, we note that

$$\begin{aligned}
&\mathbf{a}_\alpha \times \mathbf{a}_\beta = \bar{\epsilon}_{\alpha\beta}\mathbf{n}, \\
&\bar{\epsilon}^{\alpha\beta}\mathbf{a}_\alpha \times \mathbf{a}_\beta = \bar{\epsilon}^{\alpha\beta}\bar{\epsilon}_{\alpha\beta}\mathbf{n} = 2\mathbf{n}, \\
&\mathbf{n} = \frac{1}{2}\bar{\epsilon}^{\alpha\beta}\mathbf{a}_\alpha \times \mathbf{a}_\beta = \frac{1}{2\sqrt{a}}\epsilon_{\alpha\beta}\mathbf{a}_\alpha \times \mathbf{a}_\beta, \qquad (9.3.22) \\
&\mathbf{N} = \frac{1}{2}\bar{\epsilon}^{*\alpha\beta}\mathbf{A}_\alpha \times \mathbf{A}_\beta = \frac{1}{2\sqrt{A}}\epsilon_{\alpha\beta}\mathbf{A}_\alpha \times \mathbf{A}_\beta.
\end{aligned}$$

The normal difference vector $\mathbf{m}$ may be written in the form

$$\begin{aligned}
\mathbf{m} = \mathbf{N} - \mathbf{n} &= \frac{1}{2}\left(\bar{\epsilon}^{*\alpha\beta}\mathbf{A}_\alpha \times \mathbf{A}_\beta - \bar{\epsilon}^{\alpha\beta}\mathbf{a}_\alpha \times \mathbf{a}_\beta\right) \\
&= \frac{1}{2}\left[\bar{\epsilon}^{*\alpha\beta}(\mathbf{a}_\alpha + \mathbf{u}_{,\alpha}) \times (\mathbf{a}_\beta + \mathbf{u}_{,\beta}) - \bar{\epsilon}^{\alpha\beta}\mathbf{a}_\alpha \times \mathbf{a}_\beta\right] \\
&= \frac{1}{2}\left[\bar{\epsilon}^{*\alpha\beta}(\mathbf{a}_\alpha + \mathbf{u}_{,\beta} + \mathbf{u}_{,\alpha} \times \mathbf{a}_\beta + \mathbf{u}_{,\alpha} \times \mathbf{u}_{,\beta}) + (\bar{\epsilon}^{*\alpha\beta} - \bar{\epsilon}^{\alpha\beta})\mathbf{a}_\alpha \times \mathbf{a}_\beta\right]. \qquad (9.3.23)
\end{aligned}$$

If large displacements are considered with small strains, it is possible to assume that $\sqrt{A} \approx \sqrt{a}$ and $\bar{\epsilon}^{*\alpha\beta} \approx \bar{\epsilon}^{\alpha\beta}$. This leads to

$$\begin{aligned}
\mathbf{m} &= \frac{1}{2}\bar{\epsilon}^{\alpha\beta}(\mathbf{A}_\alpha \times \mathbf{A}_\beta - \mathbf{a}_\alpha \times \mathbf{a}_\beta) \\
&= \frac{1}{2}\bar{\epsilon}^{\alpha\beta}\left[(\mathbf{a}_\alpha + \mathbf{u}_{,\alpha}) \times (\mathbf{a}_\beta + \mathbf{u}_{,\beta}) - \mathbf{a}_\alpha \times \mathbf{a}_\beta\right] \\
&\quad - \left[u^3_{,\alpha} + u^\beta b_{\alpha\beta} + \epsilon^{\lambda\beta}\epsilon_{\alpha\gamma}\left(u^\gamma_{|\lambda} - b^\gamma_\lambda u^3\right)\left(u^3_{,\beta} + b_{\beta\eta}u^\eta\right)\right]\mathbf{a}^\alpha \\
&\quad + \left[u^\alpha_{|\alpha} - b^\alpha_\alpha u^3 + \frac{1}{2}\bar{\epsilon}^{\alpha\beta}\bar{\epsilon}_{\gamma\lambda}\left(u^\gamma_{|\alpha} - b^\gamma_\alpha u^3\right)\left(u^\lambda_{|\beta} - b^\lambda_\beta u^3\right)\right]\mathbf{n}. \qquad (9.3.24)
\end{aligned}$$

From (9.3.24), we note that the components of the normal difference vector in the directions $\mathbf{a}^\alpha$ and $\mathbf{n}$ are, respectively,

$$m_\alpha = -u^3_{,\alpha} - u^\beta b_{\alpha\beta} - \bar{\epsilon}^{\lambda\beta}\bar{\epsilon}_{\alpha\gamma}\left(u^\gamma_{|\lambda} - b^\gamma_\lambda u^3\right)\left(u^3_{,\beta} + b_{\beta\eta}u^\eta\right), \quad (9.3.25a)$$

$$m = u^\alpha_{|\alpha} - u^3 b^\alpha_\alpha + \frac{1}{2}\bar{\epsilon}^{\alpha\beta}\bar{\epsilon}_{\gamma\lambda}\left(u^\gamma_{|\alpha} - b^\alpha_\alpha u^3\right)\left(u^\lambda_{|\beta} - b^\lambda_\beta u^3\right). \quad (9.3.25b)$$

The last terms of Eqs. (9.3.25a) and (9.3.25b) are of higher order and are negligible in comparison with other terms, and therefore they may be dropped:

$$m_\alpha = -\left(u^3_{,\alpha} - u^\beta b_{\alpha\beta}\right), \quad (9.3.25c)$$

$$m = u^\alpha_{|\alpha} - u^3 b^\alpha_\alpha. \quad (9.3.25d)$$

The bending strain tensor assumes an explicit form if we substitute Eqs. (9.3.25a) and (9.3.25c) into Eqs. (9.3.21):

$$\begin{aligned}\chi_{\alpha\beta} = {} & -\left(u^3_{,\alpha} + u^\lambda b_{\alpha\lambda}\right)_{|\beta} - \left(u^\lambda_{|\lambda} - b^\lambda_\lambda u^3\right) b_{\alpha\beta} \\ & + \left(u_{\mu|\alpha} - u^3 b_{\mu\alpha}\right) \left[\left(-u^3_{,\lambda} - u^\mu b^\lambda_\lambda\right)_{|\beta} - \left(u^\lambda_{|\lambda} - b^\lambda_\lambda u^3\right) b^\mu_\beta\right] \\ & + \left(u^3_{,\alpha} + u^\mu b_{\alpha\mu}\right) \left[\left(u^\gamma_{|\gamma} - b^\gamma_\gamma u^3\right)_{,\beta} + \left(-u^3_{,\lambda} a^{\lambda\gamma} - u^\gamma b^\lambda_\lambda\right) b_{\beta\gamma}\right] \\ & - b^\mu_\beta\left(u_{\mu|\alpha} - u^3 b_{\mu\alpha}\right). \end{aligned} \quad (9.3.26)$$

At this point, it is quite evident that the full expansion of Eq. (9.3.26) is extremely complex, even with the nonlinear terms in Eqs. (9.3.25a) and (9.3.25b) eliminated. The literature in the field for the past century indicates that researchers have proposed various forms of bending strain–displacement equations that drop some of the less-significant terms in Eq. (9.3.26). The decision to retain or remove various terms was not fully justified. Prejudices and inconsistencies prevailed.

The approximations $\bar{\epsilon}^{*\alpha\beta} - \bar{\epsilon}^{\alpha\beta} = 0$ and $\epsilon_{\alpha\beta}(1/\sqrt{A} - 1/\sqrt{a}) = 0$ already made in deriving Eqs. (9.3.25a) and (9.3.25b) as well as further simplifications in Eqs. (9.3.25c) and (9.3.25d) would lead to an argument that retention of any term in Eq. (9.3.26) with an order of accuracy equal to or lower than that which we have in Eqs. (9.3.25a)–(9.3.25d) is inconsistent. Unfortunately, there has been a considerable amount of controversy in the past, as mentioned in Love (1934), Koiter (1959), Reissner (1941), Flügge (1960), Naghdi (1963), Novozhilov (1961), Sanders (1959), Timoshenko and Woinowsky-Krieger (1959), Vlasov (1964), Kraus (1967), and others. Some numerical applications are found in Chung and Jenkins (1973) and Chung and Rush (1973).

With the kinematics of shells now available, the next step is to implement kinetics of shells into the equations of motion, Eq. (7.1.18) [see Chung (1988), *Continuum Mechanics*, for details]. This process is described for the simplified geometry such as that in plates next.

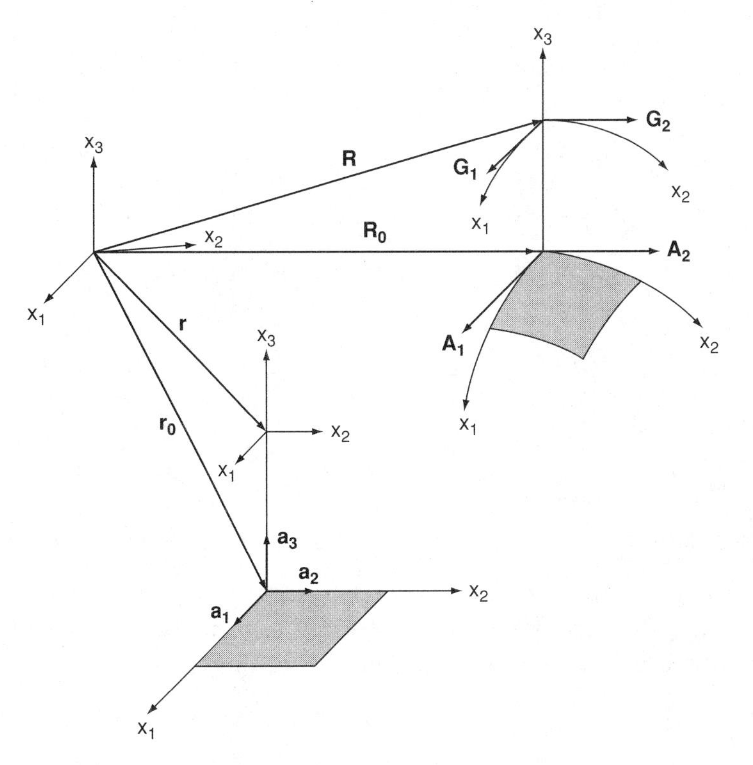

Figure 9.3.2. Coordinate system for a plate.

9.3.4 Plates Derived From Shell Theory

Although the governing equations for a plate may be derived independently, it is obvious that a shell with no curvatures results in a plate and that the mathematics for plates can be deduced from a shell theory. A membrane shell was shown to be a shell without bending. It may now appear proper to define a plate as a flat shell without curvatures. Indeed, this definition can be accepted if the plate is not loaded in plane and if bending of the plate does not produce excessive curvatures. Thus the governing equations for a plate are much the same as those for a shell except for the absence of initial curvatures (Fig. 9.3.2).

Because $a_{\alpha\beta} = \delta_{\alpha\beta}$ and $b_{\alpha\beta} = c_{\alpha\beta} = 0$ for a flat surface with rectangular Cartesian coordinates, the strain displacement equations for a plate are, from Eq. (9.3.12),

$$\gamma_{\alpha\beta} = e_{\alpha\beta} + \xi^3 \chi_{\alpha\beta} + (\xi^3)^2 \Lambda_{\alpha\beta}, \tag{9.3.27a}$$

$$\gamma_{\alpha 3} = \frac{1}{2}\mathbf{A}_\alpha \cdot \mathbf{A}_3, \tag{9.3.27b}$$

$$\gamma_{33} = \frac{1}{2}(\mathbf{A}_3 \cdot \mathbf{A}_3 - 1), \tag{9.3.27c}$$

in which

$$e_{\alpha\beta} = \frac{1}{2}(A_{\alpha\beta} - \delta_{\alpha\beta}), \tag{9.3.28a}$$

$$\chi_{\alpha\beta} = -B_{\alpha\beta}, \tag{9.3.28b}$$

$$\Lambda_{\alpha\beta} = \frac{1}{2}(\mathbf{A}_{3,\alpha} \cdot \mathbf{A}_{3,\beta}). \tag{9.3.28c}$$

Under the Kirchhoff hypothesis, and neglecting higher-order bending strains, we obtain

$$\gamma_{\alpha\beta} = e_{\alpha\beta} + \xi^3 \chi_{\alpha\beta}, \tag{9.3.29}$$

in which

$$e_{\alpha\beta} = \frac{1}{2}(u_{\alpha,\beta} + u_{\beta,\alpha} + u_{\lambda,\alpha}u_{\lambda,\beta}), \tag{9.3.30}$$

$$\chi_{\alpha\beta} = -u_{3,\alpha\beta} - \left(u^3_{,\lambda}u_{\lambda,\alpha}\right)_{,\beta} + u_{3,\alpha}u_{\lambda,\lambda\beta}. \tag{9.3.31}$$

For plate bending we neglect the curvatures on the deformed surface and strains are assumed to be small. Furthermore, in the absence of in-plane loadings and for a negligible membrane action resulting from bending, we simplify as follows:

$$e_{\alpha\beta} = 0, \tag{9.3.32}$$

$$\chi_{\alpha\beta} = -u^3_{,\alpha\beta}, \tag{9.3.33}$$

$$m^{\alpha\beta}_{,\alpha\beta} - P = 0, \tag{9.3.34}$$

where $m^{\alpha\beta}$ is the moment and P is the static applied transverse load.

In physical components ($x_1 = x, x_2 = y, f^3 = P, u^3 = w$), Eqs. (9.3.33) and (9.3.34) assume the form

$$\chi_{11} = -\frac{\partial^2 w}{\partial x^2}, \quad \chi_{22} = -\frac{\partial^2 w}{\partial y^2}, \quad \chi_{12} = -\frac{\partial^2 w}{\partial x \partial y}, \tag{9.3.35}$$

$$\frac{\partial^2 m^{11}}{\partial x^2} + 2\frac{\partial^2 m^{12}}{\partial x \partial y} + \frac{\partial^2 m^{22}}{\partial y^2} + P = 0. \tag{9.3.36}$$

The constitutive equations for a plate are given by

$$m^{\alpha\beta} = D[v\delta^{\alpha\beta}\delta^{\lambda\gamma} + (1-v)\delta^{\alpha\lambda}\delta^{\beta\gamma}]\chi_{\lambda\gamma}.$$

Thus we obtain

$$m^{11} = D(\chi_{11} + v\chi_{22}), \quad m^{22} = D(\chi_{22} + v\chi_{11}), \quad m^{12} = D(1-v)\chi_{12}.$$

Substituting these and Eqs. (9.3.35) into Eq. (9.3.36), we obtain

$$\frac{\partial^4 w}{\partial x^4} + 2\frac{\partial^4 w}{\partial x^2 \partial y^2} + \frac{\partial^4 w}{\partial y^4} = \frac{P}{D}, \tag{9.3.37}$$

where $D = Eh^3/12(1+v^2)$ is called the *flexural modulus*. Equation (9.3.37) represents the biharmonic equation for a plate flexural action.

Recall that the left-hand side of the plate equation, Eq. (9.3.37), was studied in Eq. (1.3.10), Example 4.3.3, and Eq. (5.6.10). Boundary conditions required for the solution of a biharmonic equation discussed earlier can be applied for the solution of Eq. (9.3.37).

In summary, the basic and the most rigorous mathematical foundations for the theory of shells and plates have been presented. Fundamental tensors and Riemann's curvature tensors led to accurate details of kinematics. Assessments of physical behavior with both kinematics and kinetics (equations of motion) and the solution procedures for the governing equations, however, are beyond the scope of the present text. Specialized books on plates and shells should be consulted for details.

Remarks

As we end this section, the reader may inquire as to why flexural continuum for shell geometries and the relativistic continuum for spacetime geometries are related through the subject of differential geometry. The most significant common ground is the first fundamental tensor or membrane strain tensor ($a_{\alpha\beta}$, $e_{\alpha\beta}$) and the second fundamental tensor or bending strain tensor ($b_{\alpha\beta}$, $\chi_{\alpha\beta}$) in shell structures that will be shown to be similar to the intrinsic three-geometry metric tensor (g_{ij}) and the extrinsic curvature tensor (K_{ij}) in the theory of general relativity. This signifies the analogy between the flexural continuum (shell geometries) and relativistic continuum (curved spacetime geometries). The mathematics dealing with these two subjects is identical, governed by the concept of differential geometry. Solutions of the resulting differential equations for a shell will lead to the strains and stresses of the deformed shell whereas those for general relativity exhibit various astrophysical phenomena in the universe. Applications to relativistic continuum are presented next.

9.4 Applications to Relativistic Continuum (Spacetime Geometries)

Applications of differential geometry to shell structures were discussed in the previous section. Another important application of differential geometry is the spacetime geometry of relativity. It is well known that Einstein's theory of gravitation is geometry itself. Thus the differential geometry by means of Riemann curvatures and various curvature tensors describing the kinematics of shell structures can be shown to lay down the foundations of spacetime geometries of Einstein's theory of relativity as well. It is the continuous media through which the concept of mass, velocity, acceleration, stress, momentum, and energy can be defined whether the continuum is a solid, fluid, electromagnetic field, or gravitational field of spacetime.

In this section, however, the intention is not to describe the detailed physics of gravitation or theory of relativity. Rather, our focus is on the mathematical concepts of spacetime geometries from the viewpoints of differential geometry and continuum mechanics.

Governing equations for relativity are presented, followed by kinematics and dynamics of general relativity. The emphasis is on providing the conservation forms of relativity equations in line with those Newtonian fluids discussed in Chap. 5, thus establishing the common ground on which the relativity continuum is founded as are all other previous subjects covered in this book.

Although the differential geometry is applied only to general relativity, it is necessary that we begin with special relativity so that a proper sequence can be maintained to general relativity.

For convenience we use the following notations interchangeably: partial derivatives, $\partial_\beta A_\alpha \equiv A_{\alpha,\beta}$; covariant derivatives, $D_\beta A_\alpha \equiv A_{\alpha|\beta}$, as traditionally practiced in the literature.

9.4.1 Special Relativity

Newtonian mechanics prevailed until the theory of relativity became available at the dawn of the twentieth century. In Newtonian mechanics one presupposes the existence of an absolute space of three dimensions in which one can choose rigid reference frames; furthermore, one assumes that time is a universal independent variable. All frames moving uniformly with respect to this fundamental system are inertial frames. Thus the dynamical behavior of all mechanical systems governed by Newton's law is identical in all inertial frames; that is, in Newtonian mechanics all inertial frames are dynamically equivalent, satisfying the Galilean transformation, $x_i' = x_i - v_i t$, $t' = t$, where the system x_i' is moving uniformly with velocity v_i relative to a system x_i. Because x_i' does not accelerate with respect to x_i, an isolated body moving with constant velocity in x_i will appear to move with different constant velocity in x_i, requiring that the mechanical force be the same in both systems, $\mathbf{f} = m\ddot{\mathbf{x}} = m\ddot{\mathbf{x}}'$. Thus Newton's laws of motion are invariant under a Galilean transformation. Therefore the dynamical behaviors of all mechanical systems governed by Newton's laws are identical in all inertial frames (Moller, 1972; Mihalas and Mihalas, 1984).

Maxwell's theory of electromagnetism led to the conclusion that light travels with a unique speed c in a vacuum, or a luminiferous ether, which is at rest relative to Newton's absolute space. A consequence of applying the Galilean transformation to Maxwell's equations is that the velocity of light measured by an observer should depend on that observer's motion relative to ether, contrary to sensitive experimental measurements. Thus it is concluded that the Newtonian concept of space and time is faulty.

Subsequently Einstein's theory of special relativity resolved these difficulties by two fundamental principles: (1) All inertial frames are completely equivalent for performing all physical experiments, requiring the covariance of all physical laws; and (2) speed of light is the same in all inertial frames, independent of the motion of the source in spacetime, contrary to the Galilean transformation in which space and time are independent.

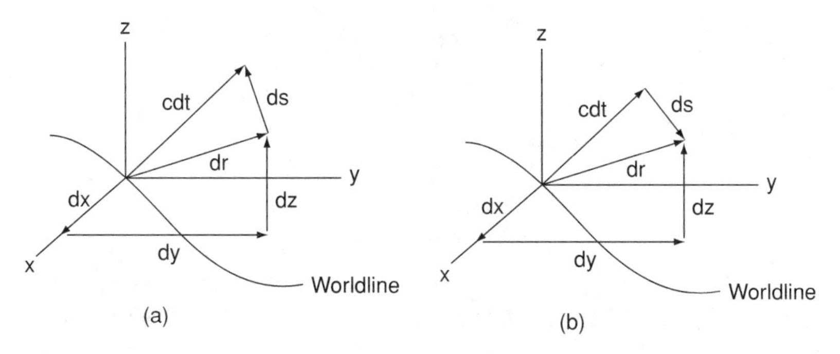

Figure 9.4.1. Worldline displacements. Two options, either (a) or (b), are used for convenience.

It should be noted that the word "covariance" is used here to imply "invariance" with "multiple" physical laws. This should be distinguished from "covariant component of a vector" or "covariant derivatives of vectors and tensors" first introduced in Chap. 6. It is an unfortunate tradition that the same word represents different physical notions.

In special relativity, the spacetime geometry is characterized by the trajectory of a particle in four-dimensional spacetime, called the worldline (Fig. 9.4.1). The spacetime interval is given, in terms of the speed of light c in a vacuum, by

$$ds^2 = c^2dt^2 - dx^2 - dy^2 - dz^2 \tag{9.4.1a}$$

or

$$ds^2 = -c^2dt^2 + dx^2 + dy^2 + dz^2, \tag{9.4.1b}$$

depending on the convenience for applications. A particle traveling from the past to a point and departing this point to the future follows the worldline within the light cone, as shown in Fig. 9.4.2. The squares of line segments or displacements can be defined as $ds^2 > 0$, $ds^2 = 0$, and $ds^2 < 0$ representing, respectively, timelike, lightlike, and spacelike metrics.

Consider the case of $ds^2 = 0$ or

$$c^2t^2 - x^2 - y^2 - z^2 = 0 \tag{9.4.2a}$$

or, disregarding the z direction,

$$c^2t^2 - x^2 - y^2 = 0. \tag{9.4.2b}$$

This represents the light cone as shown in Fig. 9.4.2 (y axis perpendicular to the plane), with the upper ($t > 0$) and lower ($t < 0$) parts indicating the future and past light cones, respectively. Here, particles leaving the origin have worldlines inside the future light cone, whereas particles arriving at the origin have worldlines inside the past light cone.

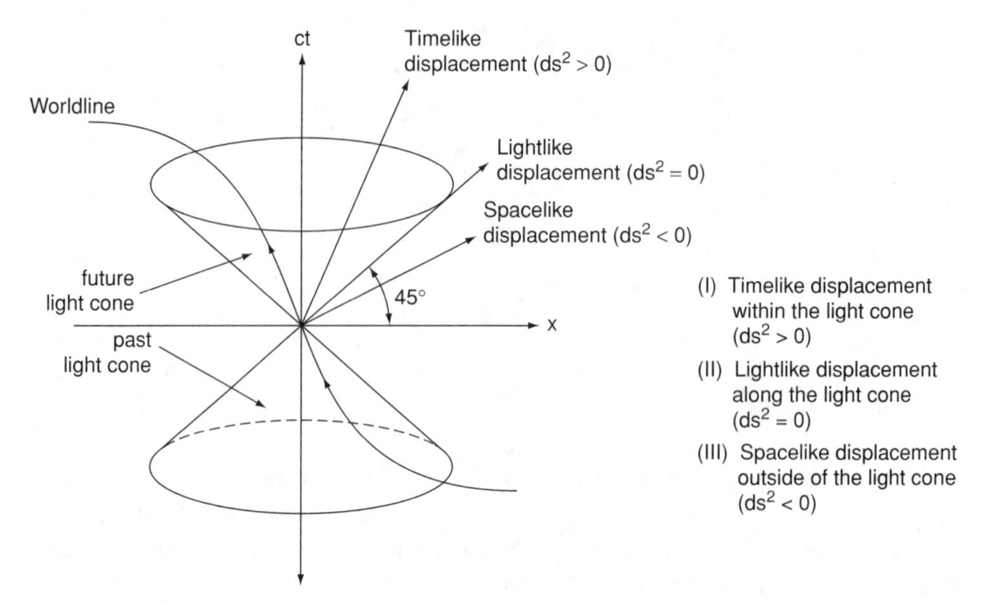

Figure 9.4.2. The light cone showing timelike, lightlike, and spacelike displacements. The worldlines are contained within the future light cone or past light cone with particles leaving the origin or particles arriving at the origin, respectively.

9.4.1.1 Lorentz Transformation

The spacetime interval given by (9.4.1) must be independent of the coordinates. Let the origin of the $\bar{x}$, $\bar{y}$, $\bar{z}$ coordinates move with velocity v along the x axis of the x, y, z coordinates so that Eq. (9.4.2a) is written as

$$c^2\bar{t}^2 - \bar{x}^2 - \bar{y}^2 - \bar{z}^2 = 0. \tag{9.4.3}$$

Here we assume that

$$\bar{x} = \gamma\,(x - vt), \tag{9.4.4}$$

with

$$\gamma = (1 - \eta^2)^{-1/2}, \tag{9.4.5}$$

$$\eta = \frac{v}{c}. \tag{9.4.6}$$

Similarly, we may assume that

$$\bar{t} = at + bx. \tag{9.4.7}$$

Substituting (9.4.4) and (9.4.7) into (9.4.3) and using (9.4.2a), we obtain

$$\gamma^2 - b^2c^2 = 1, \tag{9.4.8a}$$

$$a^2c^2 - \gamma^2 v^2 = c^2, \tag{9.4.8b}$$

$$abc^2 + \gamma^2 v = 0. \tag{9.4.8c}$$

Solving these equations simultaneously for a, b, and γ, we obtain

$$\gamma = a = \left(1 - \frac{v^2}{c^2}\right)^{-1/2}, \tag{9.4.9}$$

$$b = -\gamma \frac{v}{c^2}. \tag{9.4.10}$$

The preceding process is known as the Lorentz transformation:

$$\bar{t} = \gamma \left(t - \eta \frac{x}{c}\right), \tag{9.4.11a}$$

$$\bar{x} = \gamma(x - vt), \tag{9.4.11b}$$

$$\bar{y} = y, \tag{9.4.11c}$$

$$\bar{z} = z. \tag{9.4.11d}$$

Writing (9.1.11) in tensor notation, we express

$$\bar{x}^\alpha = L^\alpha_\beta x^\beta \quad (\alpha, \beta = 0, 1, 2, 3), \tag{9.4.12}$$

where

$$L^\alpha_\beta = \begin{bmatrix} \gamma & -\gamma\eta & 0 & 0 \\ -\gamma\eta & \gamma & 0 & 0 \\ 0 & 0 & 1 & 0 \\ 0 & 0 & 0 & 1 \end{bmatrix}. \tag{9.4.13}$$

Note that in these equations the natural unit is used, $c = 1 = 3 \times 10^8\ \mathrm{m\,s^{-1}}$, $1\ \mathrm{s} = 3 \times 10^8\ \mathrm{m}$, $1\ \mathrm{m} = (3 \times 10^8)^{-1}\ \mathrm{s}$, leading to

$$\gamma = (1 - v^2)^{-1/2}, \tag{9.4.14a}$$

$$\eta = v. \tag{9.4.14b}$$

Because Eq. (9.4.12) is not meant to be for curvilinear coordinates, according to the rules established in Section 1.2 for the Cartesian coordinates, Eq. (9.4.12) should have been written as

$$\bar{x}_\alpha = L_{\alpha\beta} x_\beta. \tag{9.4.15}$$

Recall that in Subsection 6.1.1 covariant and contravariant components arise only when curvilinear coordinates are used. Repeated indices appear diagonally only when curvilinear coordinates are used. Unfortunately, tensor algebra notation in relativity literature has been inconsistent in that all tensor notation is written as if all coordinates are treated as curvilinear. With this caution we proceed with the traditional notation as practiced in relativity, adopting (9.4.12) instead of (9.4.15), with all repeated indices appearing diagonally.

From (9.4.12), we write

$$\frac{\partial \bar{x}^\alpha}{\partial x^\beta} = L^\alpha_\beta, \tag{9.4.16}$$

with

$$x^\alpha = \left(J^\alpha_\beta\right)^{-1} \bar{x}^\beta, \tag{9.4.17}$$

where

$$\left(L^\alpha_\beta\right)^{-1} = \begin{bmatrix} \gamma & \gamma\eta & 0 & 0 \\ \gamma\eta & \gamma & 0 & 0 \\ 0 & 0 & 1 & 0 \\ 0 & 0 & 0 & 1 \end{bmatrix}. \tag{9.4.18}$$

Substituting (9.4.17) into (9.4.12) yields

$$x^\alpha = L^\alpha_\gamma L^\gamma_\beta x^\beta = \delta^\alpha_\beta x^\beta = x^\alpha. \tag{9.4.19}$$

We write (9.4.2a) in the form

$$ds^2 = \eta_{\alpha\beta} dx^\alpha dx^\beta \quad (\alpha, \beta = 0, 1, 2, 3),$$

with

$$\eta_{\alpha\beta} = \begin{bmatrix} 1 & 0 & 0 & 0 \\ 0 & -1 & 0 & 0 \\ 0 & 0 & -1 & 0 \\ 0 & 0 & 0 & -1 \end{bmatrix}. \tag{9.4.20}$$

This is the timelike metric, known as the Minkowski metric tensor. Note that the spacelike metric arises if the diagonal signs are reversed.

The matrix L^α_β for a general Lorentz transformation can be shown to satisfy

$$L^\alpha_\mu L^\beta_\nu \eta_{\alpha\beta} = \eta_{\mu\nu}. \tag{9.4.21}$$

Let us now define the proper time interval $d\tau$ associated with the worldline of a particle set equal to ds along the worldline of the particle:

$$d\tau^2 = \eta_{\mu\nu} dx^\mu dx^\nu = dt^2 - dx^2 - dy^2 - dz^2 = (1 - v^2)dt^2, \tag{9.4.22}$$

where $v = (dx^2 + dy^2 + dz^2)^{1/2}/dt$ is the particle velocity. Note that $d\tau = dt$ in the particle's rest frame. The proper time $d\tau$ represents the time interval measured by a clock that moves along with the same velocity as the particle.

Given that the proper time is a world scalar, we define the four-velocity as

$$V^\alpha = \frac{dx^\alpha}{d\tau} = (\gamma, \gamma v_x, \gamma v_y, \gamma v_z), \tag{9.4.23}$$

which is the correct four-dimensional generalization of v. In components,

$$V^{(0)} = c\frac{dt}{d\tau} = \gamma c,$$

$$V^i = \frac{dt}{d\tau}\frac{dx^i}{dt} = \gamma v^i \quad (i = 1, 2, 3),$$

with

$$V^\alpha = \gamma(c, v),$$
$$V_\alpha V^\alpha = -c^2. \tag{9.4.24}$$

In ordinary three-space the acceleration is $\mathbf{a} = d\mathbf{v}/dt$. This expression is not Lorentz covariant, but if we define

$$A^\alpha = \frac{dV^\alpha}{d\tau},$$

then A^α is a contravarient four-vector whose space components reduce to the ordinary three-acceleration $\mathbf{a}$ in the limit $\mathbf{v}/c \ll 1$. Differentiating (9.4.24) with respect to proper time we have

$$\frac{d(V_\alpha V^\alpha)}{d\tau} = 2V_\alpha \frac{dV^\alpha}{d\tau} = 0;$$

hence

$$V_\alpha A^\alpha = 0, \tag{9.4.25}$$

indicating that the four-acceleration is orthogonal to the four-velocity in spacetime.

In summary, it has been shown that the relativity theory, when the Lorentz transformation is used, not only overcomes the deficiency of Newtonian theory, but also the Newtonian theory is indeed recovered in the limit $v/c \ll 1$.

9.4.1.2 The Energy-Momentum Tensor

The energy-momentum four-vector, defined as the product of the particle's mass m and the four-velocity V^μ, is given by

$$\begin{aligned} p^\mu &= mV^\mu \\ &= (m\gamma,\ m\gamma v_x,\ m\gamma v_y,\ m\gamma v_z) \\ &= (E, p_x, p_y, p_z), \end{aligned} \tag{9.4.26}$$

where E is the energy with p_x, p_y, p_z denoting three components of the momentum.

The energy-momentum tensor is defined as

$$T^{\mu\nu} = \rho_0 V^\mu V^\nu, \tag{9.4.27}$$

with

$$\rho_0 = n_0 m \quad \text{(proper mass density)}, \tag{9.4.28}$$

$$T^{00} = \frac{nm}{\sqrt{1-v^2}} \quad \text{(energy density)}, \tag{9.4.29}$$

$$T^{k0} = T^{0k} = \frac{nmv^k}{\sqrt{1-v^2}} \quad \text{(k momentum density)}, \tag{9.4.30}$$

$$T^{kl} = \frac{nmv^k v^l}{\sqrt{1-v^2}} \quad \text{(k, l momentum flux density)}. \tag{9.4.31}$$

For the system not subjected to external forces, the conservation law of charge leads to

$$\partial_{\mu} T^{\mu\nu} = 0. \tag{9.4.32}$$

This relationship is similar to the derivation of the conservation of momentum by use of the Green–Gauss theorem discussed in Subsection 3.3.1:

$$\int_{\Gamma} T^{\mu\nu} n_{\mu} d\Gamma = \int_{\Omega} T^{\mu\nu}{}_{,\mu} d\Omega = 0, \tag{9.4.33}$$

where n_{μ} is the vector normal to the surface $d\Gamma$ of the domain $d\Omega$. Note, however, that, because the volume is four dimensional, its surface is actually a three-dimensional hypersurface. The vector $n_{\mu} d\Gamma$ has a magnitude equal to the corresponding three-dimensional volume element and a direction perpendicular to the hypersurface.

9.4.1.3 Ideal Fluids in Special Relativity

Conservation-of-Mass and Stress-Energy Tensor. In special relativity we consider two different time frames: An inertial (lab) frame t and proper (comoving) time τ. Thus in relativistic flows the Eulerian substantial derivative discussed in Subsection 2.2.2 is revised to read

$$\begin{aligned}\frac{D}{D\tau} &= \frac{dt}{d\tau}\frac{\partial}{\partial t} + \frac{\partial x^i}{d\tau}\frac{\partial}{\partial x^i} \\ &= \frac{V^0}{c}\frac{\partial}{\partial t} + V^i \frac{\partial}{\partial x^i} \quad (i = 1, 2, 3)\end{aligned}$$

or

$$\frac{D}{D\tau} = V^{\alpha}\frac{\partial}{\partial x^{\alpha}} \quad (\alpha = 0, 1, 2, 3), \tag{9.4.34}$$

which is manifestly Lorentz covariant. In the limit $v/c \ll 1$, $D/D\tau$ reduces to D/Dt. For curvilinear coordinates Eq. (9.4.34) becomes

$$\frac{Df}{D\tau} = V^{\alpha} f_{|\alpha}, \tag{9.4.35}$$

where f is any differentiable function (see Section 6.3 for covariant derivatives in curvilinear coordinates). The derivative $D/D\tau$ in (9.4.35) is called the intrinsic derivative with respect to proper time in a four-dimensional spacetime.

The density of the proper mass in the laboratory frame is

$$\rho = nm_0 = \gamma\rho_0, \tag{9.4.36}$$

whereas the density of relative mass is

$$\rho' = nm = \gamma n_0 \gamma m_0 = \gamma^2 \rho_0, \tag{9.4.37}$$

which is useful in writing the momentum density as $\rho' \mathbf{v}$, where $\mathbf{v}$ is the ordinary three-velocity of the fluid.

Using the Newtonian continuity equation

$$\rho_{,t} + (\rho v^i)_{,i} = 0 \tag{9.4.38}$$

and substituting (9.4.36) into (9.4.38), we obtain

$$(\gamma\rho_0)_{,t} + \left(\gamma\rho_0 v^i\right)_{,i} = 0. \tag{9.4.39}$$

Because $\gamma \approx c \approx V^0 = 1$ for $V \ll c$, we may write (9.4.39) as

$$\left(\rho_0 V^\alpha\right)_{,\alpha} = 0, \tag{9.4.40}$$

which is manifestly covariant under Lorentz transformation.

For curvilinear coordinates, continuity equation (9.4.40) becomes

$$\left(\rho_0 V^\alpha\right)_{|\alpha} = 0. \tag{9.4.41}$$

The material stress-energy tensor is involved in the conservation of momentum and energy. In Newtonian fluid dynamics the three equations of momentum are formulated as

$$(\rho v^i)_{,t} + \Pi^{ij}_{,j} = f^i, \tag{9.4.42}$$

where Π^{ij} is the momentum flux density tensor,

$$\Pi^{ij} = \rho v^i v^j + p\delta^{ij}, \tag{9.4.43}$$

and f^i is the force density required for the nonideal flow. Our objective is to construct a similar formulation in spacetime and obtain an appropriate expression for the material stress-energy tensor. To this end, we define

$$\rho^* = \rho_0 \left(1 + \frac{\epsilon}{c^2}\right), \tag{9.4.44}$$

where ϵ is the specific internal energy. The stress-energy tensor is written as

$$M^{\alpha\beta} = \rho^{**} V^\alpha V^\beta + p\eta^{\alpha\beta}, \tag{9.4.45}$$

with

$$\rho^{**} = \rho^* + \frac{p}{c^2} = \rho_0 \left(1 + \frac{h}{c^2}\right), \tag{9.4.46}$$

where h is the specific enthalpy.

The stress-energy tensor $M^{\alpha\beta}$ has the following properties:

(1) it is symmetric;
(2) it is covariant, and hence must contain only world scalars, four-vectors, and four-tensors;
(3) it gives the correct fluid energy density and hydrostatic pressure in the comoving frame;
(4) it yields the correct nonrelativistic equations in the laboratory frame when $v/c \ll 1$.

The first three requirements are obvious. To check the fourth criterion, we note that $\rho = \gamma\rho_0$, as in (9.4.36), and also that $\epsilon/c^2 \ll 1$ and $p/\rho_0 c^2 \ll 1$ for a nonrelativistic fluid. Let us now expand (9.4.45) in powers of v/c:

$$\begin{aligned} M^{ij} &= \gamma^2 \rho_0 v^i v^j [1 + (\rho_0 \epsilon + p)/\rho_0 c^2] + p\delta^{ij} \\ &= \rho v^i v^j + p\delta^{ij} + O(v^2/c^2) \\ &= \Pi^{ij} + O(v^2/c^2), \end{aligned} \tag{9.4.47}$$

$$\begin{aligned} M^{0i} &= M^{i0} = \gamma^2 \rho_0 c v^i [1 + (\rho_0 \epsilon + p)/\rho_0 c^2] \\ &= c\rho v^i + O(v^2/c^2), \end{aligned} \tag{9.4.48}$$

$$\begin{aligned} M^{00} &= \gamma^2 \rho_0 (c^2 + \epsilon + p v^2/\rho_0 c^2) \\ &= \gamma\rho(c^2 + \epsilon) \\ &= \rho c^2 + \frac{1}{2}\rho v^2 + \rho\epsilon + O(v^2/c^2). \end{aligned} \tag{9.4.49}$$

These expressions indicate that the relativistic stress-energy tensor (9.4.45) reduces to the nonrelativistic components as $v/c \ll 1$.

The four-force density is given by a Lorentz covariant generalization of the right-hand side of (9.4.42):

$$F^\mu = \Phi^\mu/\delta V_0 = (\gamma/\delta V_0)(\boldsymbol{\Phi} \cdot \mathbf{v}/c, \boldsymbol{\Phi})^\mu, \tag{9.4.50}$$

where Φ^μ and $\boldsymbol{\Phi}$ are, respectively, the four-force and the Newtonian force vector acting in the proper volume δV_0. The ordinary force density is defined as

$$\mathbf{f} = \boldsymbol{\Phi}/\delta V.$$

Thus we rewrite (9.4.50) as

$$\begin{aligned} F^\mu &= (\gamma\delta V/\delta V_0)(\mathbf{f} \cdot v/c,\ \mathbf{f})^\mu \\ &= (\mathbf{f} \cdot \mathbf{v}/c, \mathbf{f})^\mu. \end{aligned} \tag{9.4.51}$$

It is seen that the four-force density has space components equal to the rate of increase of the momentum and a time component equal to $1/c$ times the rate of increase of the energy of the material.

Conservation of Energy and Momentum. The relativistic fluid dynamical equations analogous to Newtonian equation (9.4.42) may be written as

$$M^{\alpha\beta}_{|\beta} = F^\alpha \tag{9.4.52}$$

or

$$(\rho^{**} V^\alpha V^\beta + p\eta^{\alpha\beta})_{|\beta} = F^\alpha. \tag{9.4.53}$$

This leads to the energy equation:

$$V_\alpha V^\alpha (\rho^{**} V^\beta)_{|\beta} + \rho^{**} V^\beta (V_\alpha V^\alpha_{|\beta}) + \eta^{\alpha\beta} V_\alpha p_{,\beta} = V_\alpha F^\alpha, \tag{9.4.54}$$

with

$$V_\alpha V^\alpha = -c^2,$$
$$V_\alpha V^\alpha_{|\beta} = 0.$$

Hence (9.4.54) reduces to

$$c^2(\rho^{**}V^\alpha)_{|\alpha} - V^\alpha p_{,\alpha} = -V_\alpha F^\alpha \tag{9.4.55}$$

or

$$\rho_0\left[\frac{D\epsilon}{D\tau} + p\frac{D}{D\tau}\left(\frac{1}{\rho_0}\right)\right] = -V_\alpha F^\alpha. \tag{9.4.56}$$

Similarly, Euler's equation of motion takes the form

$$\gamma\rho^{**}\frac{D\mathbf{v}}{D\tau} = \mathbf{f} - \nabla p - c^{-2}\mathbf{v}(p_{,t} + \mathbf{f}\cdot\mathbf{v})$$

or, for the comoving frame ($v = 0$), we have

$$\rho^{**}\frac{D\mathbf{v}}{D\tau} = \rho^{**}\left(\frac{D\mathbf{v}}{D\tau}\right)_0 = \mathbf{f}_0 - \nabla p. \tag{9.4.57}$$

This is the relativistic inviscid equation of motion in its simplest form.

9.4.1.4 Nonideal Fluids in Special Relativity

Velocity-Gradient Tensor. In this section we consider a relativistic viscous and heat conducting fluid. First we define the projection tensor:

$$C^\alpha_\beta = \eta^\alpha_\beta + c^{-2}V^\alpha V_\beta \tag{9.4.58}$$

or

$$C_{\alpha\beta} = \eta_{\alpha\beta} + c^{-2}V_\alpha V_\beta. \tag{9.4.59}$$

By use of (9.4.58) and (9.4.59) the following relations can be shown:

$$C^\alpha_\beta C^\beta_\gamma = C^\alpha_\gamma, \tag{9.4.60}$$

$$C^{\alpha\beta} = C^{\alpha\gamma}C^\beta_\gamma, \tag{9.4.61}$$

$$C^{\alpha\beta}C_{\alpha\beta} = 3. \tag{9.4.62}$$

It follows from (9.4.24) that

$$V_\alpha C^\alpha_\beta = V^\beta C^\alpha_\beta = 0. \tag{9.4.63}$$

This indicates that W^α_β leads to a projection orthogonal to V^α in spacetime. Let us decompose any vector A^α into a scalar a that gives its projection onto the proper time axis and a vector a^α that is the projection of A^α into proper space.

Let us define

$$a = -c^{-2} V_\alpha A^\alpha, \tag{9.4.64}$$

$$a^\alpha = C^\alpha_\beta A^\beta; \tag{9.4.65}$$

then the acceleration takes the form

$$A^\alpha = aV^\alpha + a^\alpha. \tag{9.4.66}$$

Similarly, we may decompose any tensor W^α_β into its proper components by defining the scalar

$$W = c^{-4} V_\alpha V_\beta W^{\alpha\beta}, \tag{9.4.67}$$

$$W^\alpha = -c^{-2} C^\alpha_\beta W^{\beta\gamma} V_\gamma, \tag{9.4.68}$$

$$W^{\alpha\beta} = C^\alpha_\gamma C^\beta_\delta W^{\gamma\delta}. \tag{9.4.69}$$

Thus

$$W^{\alpha\beta} = WV^\alpha V^\beta + W^\alpha V^\beta + W^\beta V^\alpha + W^{\alpha\beta}. \tag{9.4.70}$$

This is known as the Eckert decomposition theorem.

Consider the fluid four-acceleration A_α, defined as

$$A_\alpha = \frac{\delta V_\alpha}{\delta\tau} = V_{\alpha|\beta} V^\beta, \tag{9.4.71}$$

where δ refers to the intrinsic derivative and

$$V_{\alpha|\beta} = V_{\alpha|\gamma} C^\gamma_\beta - c^{-2} A_\alpha V_\beta, \tag{9.4.72}$$

with

$$V^\beta V_{\alpha|\beta} = \left(V^\beta C^\gamma_\beta\right) V_{\alpha|\gamma} - c^{-2}(V^\beta V_\beta) A_\alpha = A_\alpha, \tag{9.4.73}$$

$$C^\gamma_\beta V_{\alpha|\gamma} = \left(C^\gamma_\delta C^\delta_\gamma\right) V_{\alpha|\beta} - c^{-2}\left(C^\gamma_\beta V_\gamma\right) A_\alpha = C^\delta_\beta V_{\alpha|\delta}. \tag{9.4.74}$$

Let us decompose the right-hand side of (9.4.73) and define the antisymmetric rotation tensor as

$$\Omega_{\alpha\beta} = \frac{1}{2}\left(V_{\alpha|\gamma} C^\gamma_\beta - V_{\beta|\gamma} C^\gamma_\alpha\right) \tag{9.4.75}$$

and the symmetric shear tensor (rate-of-deformation tensor) as

$$E_{\alpha\beta} = \frac{1}{2}\left(V_{\alpha|\gamma} C^\gamma_\beta + V_{\beta|\gamma} C^\gamma_\alpha\right) \tag{9.4.76}$$

in the comoving frame. Note that these tensors reduce to the covariant generalizations of the Newtonian expressions for these quantities:

$$E_{\alpha\beta} + \Omega_{\alpha\beta} = V_{\alpha|\gamma} C^\gamma_\beta, \tag{9.4.77}$$

with

$$V_{\alpha|\beta} = E_{\alpha\beta} + \Omega_{\alpha\beta} - c^{-2} A_\alpha V_\beta. \tag{9.4.78}$$

The deviatoric rate-of-deformation tensor is of the form

$$D_{\alpha\beta} = E_{\alpha\beta} - \frac{1}{3}\theta\, C_{\alpha\beta}, \tag{9.4.79}$$

with

$$\theta = V^{\alpha}_{|\alpha}. \tag{9.4.80}$$

Here $D_{\alpha\beta}$ is the covariant generalization of the Newtonian rate-of-deformation tensor. Using these relations, we may write

$$V_{\alpha|\beta} = D_{\alpha\beta} + \Omega_{\alpha\beta} + \frac{1}{3}\theta\, C_{\alpha\beta} - c^{-2} A_\alpha V_\beta. \tag{9.4.81}$$

Furthermore, we can also show that

$$V^\alpha D_{\alpha\beta} = V_\alpha D^{\alpha\beta} = 0, \tag{9.4.82}$$

$$V^\alpha \Omega_{\alpha\beta} = V_\alpha \Omega^{\alpha\beta} = 0, \tag{9.4.83}$$

$$\begin{aligned} C_{\alpha\beta} D^{\alpha\beta} &= C^{\alpha\beta} D_{\alpha\beta} = \frac{1}{2}(V_{\alpha,\gamma} C^{\alpha\gamma} - V_{\beta|\gamma} C^{\beta\gamma}) - \theta \\ &= V_{\alpha|\gamma}(\eta^{\alpha\gamma} + c^{-2} V^\alpha V^\gamma) - \theta = V^{\gamma}_{|\gamma} - \theta = 0. \end{aligned} \tag{9.4.84}$$

These results will be subsequently used in derivations of the equations of energy and momentum.

The Stress-Energy Tensor. Using suitable covariant generalizations of the usual Newtonian expressions, we can write the viscous terms in the stress-energy tensor as

$$M^{(v)}_{\alpha\beta} = 2\mu D_{\alpha\beta} + \lambda\theta\, C_{\alpha,\beta}. \tag{9.4.85}$$

where μ and λ are shear and bulk viscosity, respectively.

The stress-energy tensor for heat flow is of the form

$$M^{(q)}_{\alpha\beta} = c^{-2}(V_\alpha Q_\beta + Q_\alpha V_\beta), \tag{9.4.86}$$

where Q_α is the four-vector generalization of q. When all contributions are added, the stress-energy tensor becomes

$$M_{\alpha\beta} = \rho^{**} V_\alpha V_\beta + p C_{\alpha\beta} - 2\mu D_{\alpha\beta} - \lambda\theta C_{\alpha\beta} + c^{-2}(Q_\alpha V_\beta + V_\alpha Q_\beta), \tag{9.4.87}$$

with

$$V_\alpha Q^\alpha = 0. \tag{9.4.88}$$

The Energy Equation. We can obtain the general equations of motion by differentiating each term of (9.4.87) as follows:

$$(\rho^* V^\alpha V^\beta)_{|\beta} = \left(\frac{D\rho^*}{D\tau} + \rho^*\theta\right) V^\alpha + \rho^* A^\alpha, \tag{9.4.89}$$

$$(pC^{\alpha\beta})_{|\beta} = p_{|\beta} C^{\alpha\beta} + pC^{\alpha\beta}_{|\beta}, \tag{9.4.90}$$

$$\lambda(\theta\, C^{\alpha\beta})_{|\beta} = \lambda\theta_{|\beta} C^{\alpha\beta} + \lambda\theta\, C^{\alpha\beta}_{|\beta}, \tag{9.4.91}$$

$$(Q^\alpha V^\beta)_{|\beta} = \theta\, Q^\alpha + \frac{DQ^\alpha}{D\tau}, \tag{9.4.92}$$

$$\begin{aligned}(V^\alpha Q^\beta)_{|\beta} &= V^\alpha Q^\beta_{|\beta} + Q^\beta \eta^{\alpha\gamma} V_{\gamma|\beta} \\ &= V^\alpha Q^\beta_{|\beta} + Q_\beta\left(D^{\alpha\beta} + \Omega^{\alpha\beta} + \frac{1}{3}\theta\, C^{\alpha\beta}\right).\end{aligned} \tag{9.4.93}$$

Using the relation

$$C^{\alpha\beta}_{|\beta} = (\eta^{\alpha\beta} + c^{-2} V^\alpha V^\beta)_{|\beta} = c^{-2}(A^\alpha + \theta\, V^\alpha), \tag{9.4.94}$$

we arrive at

$$\begin{aligned}M^{\alpha\beta}_{|\beta} = &\left\{\frac{D\rho^*}{D\tau} + \theta\left[\rho^* + c^{-2}(p - \lambda\theta)\right]\right\} V^\alpha \\ &+ \left[\rho^* + c^{-2}(p - \lambda\theta)\right] A^\alpha + C^{\alpha\beta}(p - \lambda\theta)_{,\beta} - 2\mu\, D^{\alpha\beta}_{|\beta} \\ &+ c^{-2}\left[\frac{DQ^\alpha}{Dt} + \frac{4}{3}\theta Q^\alpha + V^\alpha Q^\beta_{|\beta} + Q_\beta(D^{\alpha\beta} + \Omega^{\alpha\beta})\right] = F^\alpha.\end{aligned} \tag{9.4.95}$$

Multiplying (9.4.95) by V_α, we obtain

$$c^2 \frac{D\rho^*}{D\tau} + (\rho^* c^2 + p)\theta = -2\mu V_\alpha D^{\alpha\beta}_{|\beta} + \lambda\theta^2 - \left(Q^\beta_{|\beta} - c^{-2} V_\alpha \frac{DQ^\alpha}{D\tau}\right), \tag{9.4.96}$$

where we have used $V_\alpha F^\alpha = 0$.

We may write the left-hand side of (9.4.96) in the form

$$V^\alpha \rho^*_{,\alpha} + \rho^{**} V^\alpha_{|\alpha} = V^\alpha \rho^{**}_{,\alpha} + \rho^{**} V^\alpha_{|\alpha} - c^{-2} V^\alpha p_{,\alpha}. \tag{9.4.97}$$

Similarly, for the right-hand side of (9.4.96), using (9.4.98), we have

$$V_\alpha \frac{DQ^\alpha}{D\tau} = \frac{DV_\alpha Q^\alpha}{D\tau} - Q^\alpha A_\alpha = -Q^\alpha A_\alpha. \tag{9.4.98}$$

By means of (9.4.82) we obtain

$$V_\alpha D^{\alpha\beta}_{|\beta} = (V_\alpha D^{\alpha\beta})_{|\beta} - D^{\alpha\beta} V_{\alpha|\beta} = -D^{\alpha\beta} V_{\alpha|\beta}. \tag{9.4.99}$$

It follows from (9.4.81), (9.4.84), and (9.4.83) that

$$D^{\alpha\beta} V_{\alpha|\beta} = D^{\alpha\beta} D_{\alpha\beta}. \tag{9.4.100}$$

Thus the energy equation finally takes the form

$$(\rho^{**}c^2V^\alpha)_{|\alpha} - V^\alpha P_{,\alpha} = 2\mu D^{\alpha\beta}D_{\alpha\beta} + \lambda\theta^2 - \left(Q^\alpha_{|\alpha} + c^{-2}Q^\alpha A_\alpha\right). \qquad (9.4.101)$$

By use of the procedure for the ideal flow and the second law of thermodynamics, it follows from Eq. (9.4.101) that

$$\rho_0 T\frac{D\eta}{Dt} = \rho_0\left[\frac{D\epsilon}{D\tau} + p\frac{D}{D\tau}\left(\frac{1}{\rho_0}\right)\right] = 2\mu D^{\alpha\beta}D_{\alpha\beta} + \lambda\theta^2 - \left(Q^\alpha_{|\alpha} + \frac{Q^\alpha}{c^2}A_\alpha\right), \qquad (9.4.102)$$

where η is the specific entropy. It is interesting to note the similarity of Eq. (9.4.102) to Eq. (5.3.55) for the nonrelativistic Newtonian equation.

The Momentum Equation. We can extract the equations of motion for a relativistic nonideal fluid from (9.4.95) by taking the space components and calculating $C_{\alpha\gamma}M^{\gamma\beta}_{|\beta}$. Note that

$$C_{\alpha\gamma}A^\gamma = A_\alpha, \qquad (9.4.103)$$

$$C_{\alpha\gamma}F^\gamma = F_\alpha. \qquad (9.4.104)$$

It follows from (9.4.95) that

$$[\rho^* + c^{-2}(p-\lambda\theta)]\frac{DV_\alpha}{D\tau} = F_\alpha - C^\beta_\alpha(p-\lambda\theta)_{,\beta} + 2\mu C_{\alpha\gamma}D^{\gamma\beta}_{|\beta} - c^{-2}\left[C_{\alpha\gamma}\frac{DQ^\gamma}{D\tau} + \frac{4}{3}\theta Q_\alpha + C_{\alpha\gamma}Q_\beta(D^{\gamma\beta} + \Omega^{\gamma\beta})\right]. \qquad (9.4.105)$$

It is readily seen that (9.4.105) reduces to the nonrealistic limit by taking $c \to \infty$, $\frac{D}{D\tau} \to \frac{D}{Dt}$, $C_{ij} \to \delta_{ij}$, resulting in the standard Navier–Stokes equation.

9.4.1.5 Relativistic Electrodynamics

Invariance of Electric Charge and Covariance of Electrodynamics. Lorentz and Poincaré showed the invariance of the equations of electrodynamics even before Einstein's special theory of relativity. The invariance of form or covariance of the Maxwell and Lorentz force equations requires that the various quantities such as ρ, $\mathbf{J}$, $\mathbf{E}$, and $\mathbf{B}$, as discussed in Chap. 8, transform under Lorentz transformations. To show this, let us consider the Lorentz force

$$\frac{d\mathbf{P}}{dt} = q\left(\mathbf{E} + \frac{\mathbf{v}}{c}\times\mathbf{B}\right), \qquad (9.4.106)$$

where $\mathbf{P}$ transforms as the space part of the four-vector of energy and momentum,

$$P^\alpha = (P_0, \mathbf{P}) = m(U_0, \mathbf{U}),$$

where $P_0 = E/C$ and U^α is the four-velocity.

Equation (9.4.106) can be written with the proper time τ:

$$\frac{d\mathbf{P}}{d\tau} = \frac{q}{c}(U_0 E + \mathbf{U} \times \mathbf{B}). \tag{9.4.107}$$

This is the space part of a four-vector. The time component equation is given by

$$\frac{dp_0}{d\tau} = \frac{q}{c}\mathbf{U} \cdot \mathbf{E}. \tag{9.4.108}$$

The experimental invariance of electric charge and the requirement of Lorentz covariance of Lorentz force equations (9.4.107) and (9.4.108) determine the Lorentz transformation properties of the electromagnetic field. For example, the requirement from (9.4.108) that $\mathbf{U} \cdot \mathbf{E}$ be the time component of a four-vector establishes that the components of $\mathbf{E}$ are the time–space parts of a second-order tensor $F^{\alpha\beta}$, or $\mathbf{E} \cdot \mathbf{U} = F^{\alpha\beta} U_\beta$. To discuss the Lorentz transformation properties of the Maxwell equations [(8.2.33)–(8.2.36)], we write the field strength tensor $F^{\alpha\beta}$ as

$$F^{\alpha\beta} = \begin{bmatrix} 0 & -E_x & -E_y & -E_x \\ E_x & 0 & -B_z & B_y \\ E_y & B_z & 0 & -B_x \\ E_z & -B_y & B_x & 0 \end{bmatrix}, \tag{9.4.109}$$

which is the antisymmetric tensor.

Let us further define that

$$J^\upsilon = (c\rho, J_x, J_y, J_z). \tag{9.4.110}$$

Then, it is is seen that the Maxwell equations presented in Eqs. (8.2.33)–(8.2.36) are reproduced by

$$\partial_\mu F^{\mu\nu} = \frac{4\pi}{c} J^\upsilon, \tag{9.4.111}$$

$$\partial^\alpha F^{\mu\nu} + \partial^\mu F^{\nu\alpha} + \partial^\upsilon F^{\alpha\mu} = 0, \tag{9.4.112}$$

with Eq. (9.4.111) equivalent to Eqs. (8.2.36) and (8.2.34),

$$\nabla \cdot \mathbf{E} = 4\pi\rho, \tag{9.4.113}$$

$$\nabla \times \mathbf{B} - \frac{1}{c}\frac{\partial \mathbf{E}}{\partial t} = \frac{4\pi}{c}\mathbf{J}, \tag{9.4.114}$$

and Eq. (9.4.112) equivalent to Eqs. (8.2.35) and (8.2.33),

$$\nabla \cdot \mathbf{B} = 0, \tag{9.4.115}$$

$$\nabla \times \mathbf{E} + \frac{1}{c}\frac{\partial \mathbf{B}}{\partial t} = 0. \tag{9.4.116}$$

From these results we conclude that the covariance of electromagnetism has been demonstrated.

To complete the covariance properties we substitute Lorentz force equation (9.4.107) and rate of change of energy equation (9.4.108):

$$\frac{dp^{\alpha}}{d\tau} = m\frac{dU^{\alpha}}{d\tau} = \frac{q}{c}F^{\alpha\beta}U_{\beta}. \tag{9.4.117}$$

Note that this is manifestly in covariant form.

The Lagrangian Equations for Fields. The Lagrangian formalism for a system of fields is a generalization of the formalism for a system of particles. The Lagrangian formalism rests on the assumption that the dynamical equations of motion can be derived from Hamilton's variational principle. Let us define action I as the integral

$$I = \int_{t_1}^{t_2} L\left[q_i(t),\ \dot{q}_i(t)\right] dt \quad (i = 1, 2, \ldots, N), \tag{9.4.118}$$

where the Lagrangian $L(q_i,\ \dot{q}_i)$ is some function with the generalized coordinates q_i and the velocity $\dot{q}_i = dq_i/dt$. The equations of motion are then obtained from the requirement that the action remain stationary for infinitesimal variations of the functions $q_i(t)$. These variations are arbitrary except for the constraint that they vanish at times t_1 and t_2:

$$\begin{aligned}\delta I &= \int \sum_{i=1}^{N} \left(\frac{\partial L}{\partial q_i}\delta q_i + \frac{\partial L}{\partial \dot{q}_i}\delta \dot{q}_i \right) dt \\ &= \int \sum_{i=1}^{N} \left(\frac{\partial L}{\partial q_i}\delta q_i + \frac{\partial L}{\partial \dot{q}_i}\frac{d}{dt}\delta q_i \right) dt.\end{aligned}$$

Integrating by parts and applying the end condition $\delta q_i = 0$ at $t = t_1$ and $t = t_2$, we obtain

$$\delta I = \int \sum_{i=1}^{N} \left(\frac{\partial L}{\partial q_i} + \frac{d}{dt}\frac{\partial L}{\partial \dot{q}_i} \right) \delta q_i \, dt. \tag{9.4.119}$$

For arbitrary choices of δq_i, we obtain the Euler–Lagrange equations:

$$\frac{d}{dt}\frac{\partial L}{\partial \dot{q}_i} - \frac{\partial L}{\partial q_i} = 0. \tag{9.4.120}$$

We find that the Euler–Lagrange equations have the consequence that the Hamiltonian, defined as

$$H = \sum_{i=1}^{N} \dot{q}_i \frac{\partial L}{\partial \dot{q}_i} - L, \tag{9.4.121}$$

is a constant of the motion by noting that

$$\frac{dH}{dt} = \sum_{i=1}^{N}\left(\ddot{q}_i \frac{\partial L}{\partial \dot{q}_i} + \dot{q}_i \frac{d}{dt}\frac{\partial L}{\partial \dot{q}_i}\right) - \sum_{i=1}^{N}\left(\frac{\partial L}{\partial q_i}\dot{q}_i + \frac{\partial L}{\partial \dot{q}_i}\ddot{q}_i\right)$$

$$= \sum_{i=1}^{N}\dot{q}_i\left(\frac{d}{dt}\frac{\partial L}{\partial \dot{q}_i} - \frac{\partial L}{\partial q_i}\right) = 0. \tag{9.4.122}$$

For illustration, let the Lagrangian of a free particle be given by

$$L = -m(1 - \dot{x}^2 + \dot{y}^2 + \dot{z}^2)^{1/2}. \tag{9.4.123}$$

Substituting this into (9.4.121), we show that

$$H = \frac{m}{[1 - (\dot{x}^2 + \dot{y}^2 + \dot{z}^2)]^{1/2}}, \tag{9.4.124}$$

representing the total energy (rest mass plus kinetic energy) of the particle.

9.4.1.6 Relativistic Charged Particles in External Electromagnetic Fields

We intend to extend the Lagrangian and Hamiltonian formalism to relativistic particle motion, leading, for charged particles in external fields, to the equations of motion:

$$\frac{d\mathbf{P}}{dt} = e\left[\mathbf{E} + \frac{\mathbf{v}}{c} \times \mathbf{B}\right], \tag{9.4.125}$$

$$\frac{dE}{dt} = e\mathbf{v}\cdot\mathbf{E}, \tag{9.4.126}$$

or, in the covariant form,

$$\frac{dU^{\alpha}}{d\tau} = \frac{e}{mc}F^{\alpha\beta}U_{\beta}, \tag{9.4.127}$$

where $U^{\alpha} = (\gamma c,\ \gamma v) = P^{\alpha}/m$ is the four-velocity of the particle.

To obtain a relativistic Lagrangian for a particle in external fields we first consider the question of the Lorentz transformation properties of the Lagrangian. First, we introduce the particle's proper time τ into (9.4.118), setting $I = A$:

$$A = \int_{\tau_1}^{\tau_2} \gamma L \, d\tau. \tag{9.4.128}$$

Because proper time is invariant, the condition that I also be invariant requires that γL be Lorentz invariant.

The Lagrangian for a free particle can be a function of the velocity of the particle and its mass, but cannot depend on its position. The only Lorentz invariant function of the velocity is

$$U_{\alpha}U^{\alpha} = c^2. \tag{9.4.129}$$

Thus the Lagrangian for a free particle is

$$L_{\text{free}} = -mc^2\gamma^{-1} = -mc^2\sqrt{1-\left(\frac{v}{c}\right)^2}. \tag{9.4.130}$$

It follows from (9.4.120) and (9.4.130) that the free-particle equation of motion is of the form

$$\frac{d}{dt}(\gamma mv) = 0. \tag{9.4.131}$$

The general requirement that γL be Lorentz invariant allows us to determine the Lagrangian for a relativistic charged particle in an external electromagnetic field, provided we know the Lagrangian (or equations of motion) for nonrelativistic motion in static fields.

Because γ times the Lagrangian must be translationally invariant as well as Lorentz invariant, it cannot involve the coordinates explicitly. Hence the interaction Lagrangian must be

$$L_{\textbf{int}} = -\frac{e}{\gamma c}U_\alpha I^\alpha = -V + \frac{e}{c}\mathbf{v}\cdot A = -e\Phi + \frac{e}{c}\mathbf{v}\cdot\mathbf{A}, \tag{9.4.132}$$

with $V = e\Phi$ being the potential energy of interaction. The combination of (9.4.130) and (9.4.132) yields the complete relativistic Lagrangian for a charged particle:

$$L = -mc^2\sqrt{1-\left(\frac{u}{c}\right)^2} + \frac{e}{c}\mathbf{v}\cdot\mathbf{A} - e\Phi. \tag{9.4.133}$$

We can show that (9.4.133) leads to the Lorentz force equation by using the substantial derivative and the standard definition of fields in terms of the potentials.

The canonical momentum $\mathbf{P}$ conjugate to the position coordinates x is obtained by the definition

$$P_i = \frac{\partial L}{\partial v_i} = \gamma m v_i + \frac{e}{c}A_i. \tag{9.4.134}$$

Thus the conjugate momentum is

$$\bar{\mathbf{P}} = \mathbf{P} + \frac{e}{c}\mathbf{A}, \tag{9.4.135}$$

where $P = \gamma mv$ is the ordinary kinetic momentum. The Hamiltonian H is a function of the coordinate $\mathbf{x}$ and its conjugate momentum $\bar{\mathbf{P}}$ and is a constant of the motion if the Lagrangian is not an explicit function of time. The Hamiltonian is given by

$$H = \bar{\mathbf{P}}\cdot\mathbf{v} - L. \tag{9.4.136}$$

Solving $\mathbf{v}$ from (9.4.134) and (9.4.135), we obtain

$$\mathbf{v} = \frac{c\bar{\mathbf{P}} - e\mathbf{A}}{\sqrt{\left(\bar{\mathbf{P}} - \frac{e\mathbf{A}}{c}\right)^2 + m^2c^2}}. \tag{9.4.137}$$

It follows from (9.4.137) and (9.4.133) that

$$H = [(c\bar{\mathbf{P}} - \mathbf{e}A)^2 + m^2c^4]^{1/2} + e\Phi \tag{9.4.138}$$

or

$$(H - e\Phi)^2 - (c\bar{\mathbf{P}} - e\mathbf{A})^2 = (mc^2)^2, \tag{9.4.139}$$

which represents the four-vector scalar product.

To show a manifestly covariant description of the relativistic Lagrangian, the customary variables $\mathbf{x}$ and $\mathbf{v}$ are replaced with the four-vectors x^α and u^α. The free-particle Lagrangian (9.4.130) can be written as

$$L_{\text{free}} = -\frac{mc}{\gamma}\sqrt{u_\alpha u^\alpha}. \tag{9.4.140}$$

Then action integral (9.4.128) becomes

$$I = -mc\int_{\tau_1}^{\tau_2}\sqrt{u^\alpha u_\alpha}d\tau, \tag{9.4.141}$$

with

$$\sqrt{u^\alpha u_\alpha}d\tau = \sqrt{\frac{dx_\alpha}{d\tau}\frac{dx_\alpha}{d\tau}}d\tau = \sqrt{\eta^{\alpha\beta}dx_\alpha dx_\beta}.$$

Thus

$$I = -mc\int_{s_1}^{s_2}\sqrt{\eta^{\alpha\beta}\frac{dx_\alpha}{ds}\frac{dx_\beta}{ds}}ds, \tag{9.4.142}$$

with

$$\sqrt{\eta^{\alpha\beta}\frac{dx_\alpha}{ds}\frac{dx_\beta}{ds}}\,ds = c\,d\tau. \tag{9.4.143}$$

From (9.4.132) the Euler–Lagrange equations arise:

$$mc\frac{df}{ds} = 0,$$

with

$$f = \frac{dx_\alpha}{ds}\Bigg/\sqrt{\eta^{\alpha\beta}\frac{dx_\alpha}{ds}\frac{dx_\beta}{ds}}$$

or

$$m\frac{d^2x^\alpha}{d\tau^2} = 0$$

for free-particle motion.

For a charged particle in an external field the covariant form of the action integral takes the form

$$A = -\int_{s_1}^{s_2}\left[mc\sqrt{\eta^{\alpha\beta}\frac{dx_\alpha}{ds}\frac{dx_\beta}{ds}} + \frac{c}{e}\frac{dx_\alpha}{ds}A^\alpha(x)\right]ds. \tag{9.4.144}$$

The Euler–Lagrange equations from Hamilton's principle lead to

$$\frac{d}{ds}\left[\frac{\partial \tilde{L}}{\partial\left(\dfrac{dx_\alpha}{ds}\right)}\right] - \partial^\alpha \tilde{L} = 0, \tag{9.4.145}$$

with

$$\tilde{L} = -\left[mc\sqrt{\eta^{\alpha\beta}\frac{dx_\alpha}{ds}\frac{dx_\beta}{ds}} + \frac{e}{c}\frac{dx_\alpha}{ds}A^\alpha(x)\right]. \tag{9.4.146}$$

It follows from (9.4.145) and (9.4.143) that

$$m\frac{d^2x^\alpha}{d\tau^2} + \frac{e}{c}\frac{dA^\alpha(x)}{d\tau} - \frac{e}{c}\frac{dx_\beta}{d\tau}\partial^\alpha A^\beta(x) = 0$$

or

$$m\frac{d^2x^\alpha}{d\tau^2} = \frac{e}{c}(\partial^\alpha A^\beta - \partial^\beta A^\alpha)\frac{dx_\beta}{d\tau}, \tag{9.4.147}$$

which is covariant equation of motion (9.4.127).

The transition to the conjugate momenta and a Hamiltonian can be performed with the conjugate momentum four-vector,

$$P^\alpha = -\frac{\partial \tilde{L}}{\partial\left(\dfrac{dx_\alpha}{ds}\right)} = mU^\alpha + \frac{e}{c}A^\alpha, \tag{9.4.148}$$

and a Hamiltonian,

$$\tilde{H} = \frac{1}{2}(P_\alpha U^\alpha + \tilde{L}). \tag{9.4.149}$$

Using (9.4.148), we obtain

$$\tilde{H} = \frac{1}{2m}\left(P_\alpha - \frac{eA_\alpha}{c}\right)\left(P^\alpha - \frac{eA^\alpha}{c}\right) - \frac{1}{2}mc^2 = 0. \tag{9.4.150}$$

This leads to Hamilton's equations,

$$\frac{dx^\alpha}{d\tau} = \frac{\partial \tilde{H}}{\partial P_\alpha} = \frac{1}{m}\left(P^\alpha - \frac{e}{c}A^\alpha\right), \tag{9.4.151}$$

$$\frac{dP^\alpha}{d\tau} = -\frac{\partial \tilde{H}}{\partial x_\alpha} = \frac{e}{mc}\left(P_\beta - \frac{eA_\beta}{c}\right)\partial_\alpha A^\beta. \tag{9.4.152}$$

Note that these equations are equivalent to (9.4.147). The relativistic effects are evident, as seen from (9.4.127) and (9.4.147).

Lagrangian Description of the Electromagnetic Field. Recall that we considered the Lagrangian formulation of the equations of motion of a charged particle in an external electromagnetic field. Let us now examine a Lagrangian description of the electromagnetic field interacting with specified external sources of charge and current. Here we replace the finite number of coordinates $q_i(t)$ and $\dot{q}_i(t)$ $(i = 1, 2, \ldots, n)$ with an infinite number of degrees of freedom. The generalized coordinates q_i are replaced with a continuous field $\phi_k(x)$ $(k = 1, 2, \ldots, n)$ and a continuous index x^α. The generalized velocity $\dot{q}_i$ is replaced with the four-vector gradient $\partial^\beta \phi_k$. These correspondences are shown in the table.

Discrete Point Particles	Infinite Number of Degrees of Freedom
i	x^α, k
q_i	$\phi_k(x)$
$\dot{q}_i$	$\partial^\alpha \phi_k(x)$
$\mathbf{L} = \sum_i \mathbf{L}_i(q_i, \dot{q}_i)$	$\int \mathbf{L}(\phi_k, \partial^\alpha \phi_k)\, d^3x$
$\frac{d}{dt}\left(\frac{\partial \mathbf{L}}{\partial \dot{q}_i}\right) = \frac{\partial \mathbf{L}}{\partial q_i}$	$\partial^\beta \frac{\partial \mathbf{L}}{\partial(\partial^\beta \phi_k)} = \frac{\partial \mathbf{L}}{\partial \phi_k}$

L is the Lagrange density, with the action integral defined as

$$A = \int\int \mathbf{L}\, d^3x\, dt = \int \mathbf{L}\, d^4x, \tag{9.4.153}$$

with

$$\mathbf{L} = -\frac{1}{16\pi} F_{\alpha\beta} F^{\alpha\beta} - \frac{1}{c} J_\alpha A^\alpha. \tag{9.4.154}$$

Here the coefficient and sign of the interaction terms are chosen to agree with (9.4.132); the sign and scale of the free Lagrangian are set by the definitions of the field strength and the Maxwell equations:

$$\mathbf{L} = -\frac{1}{16\pi} \eta_{\lambda\mu} \eta_{\nu\sigma} (\partial^\mu A^\sigma - \partial^\sigma A^\mu)(\partial^\lambda A^\nu - \partial^\nu A^\lambda) - \frac{1}{c} J_\alpha A^\alpha, \tag{9.4.155}$$

$$\frac{\partial L}{\partial(\partial^\beta A^\alpha)} = -\frac{1}{16\pi} \eta_{\lambda\mu} \eta_{\nu\sigma} \left(\delta^\mu_\beta \delta^\sigma_\alpha F^{\lambda\nu} - \delta^\sigma_\beta \delta^\mu_\alpha F^{\lambda\nu} + \delta^\lambda_\beta \delta^\nu_\alpha F^{\mu\sigma} - \delta^\nu_\beta \delta^\lambda_\alpha F^{\mu\sigma}\right). \tag{9.4.156}$$

The symmetry $\eta_{\alpha\beta}$ and the antisymmetry of $F^{\alpha\beta}$ make all four terms equal, leading to

$$\frac{\partial \mathbf{L}}{\partial(\partial^\beta A^\alpha)} = -\frac{1}{4\pi} F_{\beta\alpha} = \frac{1}{4\pi} F_{\alpha\beta}, \tag{9.4.157}$$

$$\frac{\partial \mathbf{L}}{\partial I^\alpha} = -\frac{1}{c} J_\alpha. \tag{9.4.158}$$

The equation of motion of the electromagnetic fields becomes

$$\frac{1}{4\pi} \partial^\beta F_{\beta\alpha} = \frac{1}{c} J_\alpha. \tag{9.4.159}$$

We recognize this to be a covariant form of inhomogeneous Maxwell equations (9.4.112).

The electrodynamics for special relativity discussed in this section is extended to general relativity in Subsection 9.4.3.3, with effects of gravitation taken into account.

9.4.2 General Relativity

In general relativity we consider gravity that is neglected in special relativity. The role of gravity becomes significant in most astrophysical applications. General relativity addresses curved spacetime geometry, thus being naturally placed under differential geometry. The crucial mathematical foundation for general relativity is the Riemann geometry describing various concepts of curvatures. In the following subsection basic derivations of the Riemann curvature tensor are described, followed by the nonlinear theory of gravitation and kinematics and dynamics of general relativity. Literature on general relativity is preponderant, beginning with Einstein (1915a, 1915b, and 1945) and followed by Misner, Thorne, and Wheeler (1973); Landau and Lifshitz (1975); Adler, Bazin, and Schiffer (1975), Wald (1984), and Ohanian and Ruffini (1994), among others.

9.4.2.1 Tangent Vectors

The notion of curved spacetime geometry forms the building block of Einstein's general relativity theory through the concept of Riemann curvatures. The curvilinear coordinates introduced in Chap. 6 are used, but because of the traditional notation established in general relativity, some adjustments are made in this chapter. For example, in Chap. 6, the tangent vectors $\mathbf{g}_i$ on any point in the curved surface with reference to the unit vectors $\mathbf{i}$ on the Cartesian coordinates, Eq. (6.1.1), as defined in Fig. 6.1.2, are given by

$$\mathbf{g}_i = \frac{\partial z_j}{\partial \xi_i}\mathbf{i}_j \quad (i, j = 1, 2, 3), \tag{9.4.160}$$

where z_j are the old coordinates and ξ_i are the new coordinates. Following the traditional practice in general relativity, This is revised as

$$\mathbf{g}_\alpha = \frac{\partial x^\beta}{\partial \bar{x}^\alpha}\mathbf{i}_\beta \quad (\alpha, \beta = 0,\ 1, 2, 3), \tag{9.4.161}$$

where $\mathbf{i}_\beta$ are unit vectors and $\mathbf{g}_\alpha$ represent the four-dimensional covariant tangent vectors along the curvilinear coordinates (Fig. 6.1.1). The contravariant tangent vectors $\mathbf{g}^\alpha$ are shown in Fig. 6.1.2. These vectors are perpendicular to covariant tangent vectors such that

$$\mathbf{g}^\alpha = \frac{\partial \bar{x}^\alpha}{\partial x^\beta}\mathbf{i}_\beta. \tag{9.4.162}$$

The rate of change of the covariant tangent vector with respect to the new curvilinear coordinates is known as the parallel transport:

$$\mathbf{g}_{\alpha,\beta} = \frac{\partial \mathbf{g}_\alpha}{\partial x^\beta} = \frac{\partial}{\partial \bar{x}^\beta}\frac{\partial x^\mu}{\partial \bar{x}^\alpha}\mathbf{i}_\mu = \frac{\partial^2 x^\mu}{\partial \bar{x}^\beta \partial \bar{x}^\alpha}\frac{\partial \bar{x}^\nu}{\partial x^\mu}\mathbf{g}_\nu \tag{9.4.163}$$

or

$$\mathbf{g}_{\alpha,\beta} = \Gamma^{\nu}_{\beta\alpha}\mathbf{g}_{\nu}, \tag{9.4.164}$$

where $\Gamma^{\nu}_{\beta\alpha}$ is the Christoffel symbol of the second kind:

$$\Gamma^{\nu}_{\beta\alpha} = \frac{\partial^2 x^{\mu}}{\partial \bar{x}^{\beta}\partial \bar{x}^{\alpha}}\frac{\partial \bar{x}^{\nu}}{\partial x^{\mu}}. \tag{9.4.165}$$

This is the product of the second derivative of the old coordinate with respect to the new coordinate and the first derivative of the new coordinate with respect to the old coordinate.

Let us now consider the rate of change of the contravariant tangent vector. First we note that

$$\mathbf{g}^{\alpha} \cdot \mathbf{g}_{\beta} = \delta^{\alpha}_{\beta}, \tag{9.4.166}$$

$$\frac{\partial}{\partial \bar{x}^{\alpha}}(\mathbf{g}^{\beta} \cdot \mathbf{g}_{\mu}) = \mathbf{g}_{\mu} \cdot \mathbf{g}^{\beta}_{,\alpha} + \mathbf{g}^{\beta} \cdot \mathbf{g}_{\mu,\alpha} = 0, \tag{9.4.167}$$

or

$$\mathbf{g}_{\mu} \cdot \mathbf{g}^{\beta}_{,\alpha} = -\mathbf{g}^{\beta} \cdot \mathbf{g}_{\mu,\alpha} = -\mathbf{g}^{\beta} \cdot g_{\nu}\Gamma^{\nu}_{\mu\alpha} = -\Gamma^{\beta}_{\mu\alpha}. \tag{9.4.168}$$

Thus the derivative of contravariant tangent vectors is of the form

$$\frac{\partial \mathbf{g}^{\beta}}{\partial x^{\alpha}} = \mathbf{g}^{\beta}_{,\alpha} = -\Gamma^{\beta}_{\alpha\mu}\mathbf{g}^{\mu}. \tag{9.4.169}$$

This equation may be rewritten as

$$d\mathbf{g}^{\beta} = -\Gamma^{\beta}_{\alpha\mu}\mathbf{g}^{\mu}dx^{\alpha}. \tag{9.4.170}$$

From (9.4.162), dx^{α} may be considered as the component corresponding to $\mathbf{g}^{\alpha}$ so that (9.4.169) can be written (exchanging α and β) as

$$d(dx^{\alpha}) = -\Gamma^{\alpha}_{\beta\mu}dx^{\mu}dx^{\beta}. \tag{9.4.171}$$

We now proceed with a step-by-step construction of a geodesic curve by parallel displacement of a small segment of the curve and divide by a parameter ds^2. This process leads to the differential equation for the geodesic:

$$\frac{d^2x^{\alpha}}{ds^2} + \Gamma^{\alpha}_{\beta\gamma}\frac{dx^{\gamma}}{ds}\frac{dx^{\beta}}{ds} = 0. \tag{9.4.172}$$

We shall reexamine this geodesic equation in the next subsection.

9.4.2.2 Metric Tensors

Tangent vectors discussed in the previous subsection form the basis for metric tensors. First, consider the dot product of covariant and contravariant tangent vectors (see Section 6.1):

$$\mathbf{g}_\alpha \cdot \mathbf{g}^\beta = \delta^\beta_\alpha = \frac{\partial x_\mu}{\partial \bar{x}_\alpha}\mathbf{i}_\mu \cdot \frac{\partial \bar{x}_\beta}{\partial x_\nu}\mathbf{i}_\nu = \frac{\partial x_\mu}{\partial \bar{x}_\alpha}\frac{\partial \bar{x}_\beta}{\partial x_\nu}\delta_{\mu\nu} = \frac{\partial x_\mu}{\partial \bar{x}_\alpha}\frac{\partial \bar{x}_\beta}{\partial x_\mu}.$$

The fact that the dot product $\mathbf{g}_\alpha \cdot \mathbf{g}^\beta$ is a Kronecker delta arises from our choice of contravariant tangent vectors directed perpendicular to the planes constructed by any two covariant tangent vectors, as shown in Fig. 6.1.2.

On the other hand, if we take a dot product of covariant tangent vectors, then we have

$$\mathbf{g}_\alpha \cdot \mathbf{g}_\beta = \frac{\partial x_\mu}{\partial \bar{x}_\alpha}\frac{\partial x_\mu}{\partial \bar{x}_\beta} = g_{\alpha\beta}. \tag{9.4.173}$$

This is known as the covariant metric tensor and is symmetric.

Similarly, we obtain the contravariant metric tensor $g^{\alpha\beta}$ (this is also symmetric):

$$\mathbf{g}^\alpha \cdot \mathbf{g}^\beta = \frac{\partial \bar{x}_\alpha}{\partial x^\mu}\frac{\partial \bar{x}_\beta}{\partial x^\mu} = g^{\alpha\beta}. \tag{9.4.174}$$

The following relationships can easily be found:

$$g^\alpha = g^{\alpha\beta}g_\beta, \qquad g_\alpha = g_{\alpha\beta}g^\beta g; \tag{9.4.175a}$$

$$g^{\alpha\beta} = (g_{\alpha\beta})^{-1}, \quad g_{\alpha\gamma}g^{\gamma\beta} = \delta^\beta_\alpha. \tag{9.4.175b}$$

In terms of the covariant metric tensor, the Christoffel symbol of the first kind is given by

$$\Gamma_{\alpha\beta\mu} = g_{\mu\nu}\Gamma^\nu_{\alpha\beta}. \tag{9.4.176}$$

The definitions as given in Section 6.1 for spherical coordinates in terms of standard notations for relativity are

$$\begin{aligned}
\bar{x}^0 &= ct = t \quad (c = 1),\\
\bar{x}^1 &= r,\\
\bar{x}^2 &= \theta,\\
\bar{x}^3 &= \phi,\\
x^0 &= ct = t \quad (c = 1),\\
x^1 &= r\sin\theta\cos\phi,\\
x^2 &= r\sin\theta\sin\phi,\\
x^3 &= r\cos\theta.
\end{aligned} \tag{9.4.177}$$

The position vector $\mathbf{R}$ is of the form

$$\mathbf{R} = \bar{x}^\alpha \mathbf{i}_\beta = ti_0 + r\sin\theta\cos\phi\mathbf{i}_1 + r\sin\theta\sin\phi\mathbf{i}_2 + r\cos\theta i_3. \tag{9.4.178}$$

Thus the tangent vectors are

$$\mathbf{g}_0 = \frac{\partial x^\mu}{\partial \bar{x}^0}\mathbf{i}_\mu = 1,$$

$$\mathbf{g}_1 = \frac{\partial x^\mu}{\partial \bar{x}^1}\mathbf{i}_\mu = \sin\theta\cos\phi\mathbf{i}_1 + \sin\theta\sin\phi\mathbf{i}_2 + \cos\theta\mathbf{i}_3,$$

$$\mathbf{g}_2 = \frac{\partial x^\mu}{\partial \bar{x}^2}\mathbf{i}_\mu = r\cos\theta\cos\phi\mathbf{i}_1 + r\cos\theta\sin\phi\mathbf{i}_2 - r\sin\theta\mathbf{i}_3,$$

$$\mathbf{g}_3 = \frac{\partial x^\mu}{\partial \bar{x}^3}\mathbf{i}_\mu = -r\sin\theta\sin\phi\mathbf{i}_1 + r\sin\theta\cos\phi\mathbf{i}_2. \tag{9.4.179}$$

From these the metric tensor is calculated:

$$g_{\alpha\beta} = \mathbf{g}_\alpha \cdot \mathbf{g}_\beta = \begin{bmatrix} 1 & 0 & 0 & 0 \\ 0 & -1 & 0 & 0 \\ 0 & 0 & -r^2 & 0 \\ 0 & 0 & 0 & -r^2\sin^2\theta \end{bmatrix}. \tag{9.4.180}$$

This is based on our sign convention that the time component of the metric tensor is positive whereas the diagonal space components are negative.

The expression for an interval corresponding to (9.4.180) becomes

$$\begin{aligned} ds^2 &= g_{\alpha\beta}dx^\alpha dx^\beta \\ &= dt^2 - dr^2 - r^2\,d\theta^2 - r^2\sin^2\theta\,d\phi^2. \end{aligned} \tag{9.4.181}$$

The tensor relationships given by (9.4.160)–(9.4.176) lead to important results connecting the metric tensors and Christoffel symbols (see similar exercises in Section 6.1):

$$\Gamma_{\alpha\beta\mu} = \frac{1}{2}(g_{\alpha\mu,\beta} + g_{\beta\mu,\alpha} - g_{\alpha\beta,\mu}), \tag{9.4.182}$$

$$\Gamma^\mu_{\alpha\beta} = \frac{1}{2}g^{\mu\nu}(g_{\nu\alpha,\beta} + g_{\nu\beta,\alpha} - g_{\alpha\beta,\nu}), \tag{9.4.183}$$

$$\Gamma^\alpha_{\beta\mu} = \frac{1}{2}g^{\alpha\nu}g_{\alpha\nu,\beta}, \quad \Gamma^\alpha_{\alpha\beta} = (\ln\sqrt{g})_{,\beta}, \quad g = |g_{\alpha\beta}|, \tag{9.4.184}$$

$$g^{\alpha\beta} = \frac{1}{g}\frac{\partial g}{\partial g_{\alpha\beta}}, \tag{9.4.185}$$

$$g^{\alpha\beta}_{,\alpha} = 0, \tag{9.4.186}$$

where the commas indicate important partial derivatives with respect to the curvilinear coordinates (new coordinates).

To illustrate calculations of Christoffel symbols, let us assume that

$$ds^2 = e^{a(r)}dt^2 - e^{b(r)}dr^2 - r^2 d\theta^2 - r^2 \sin^2\theta d\phi^2, \tag{9.4.187}$$

$$g_{\alpha\beta} = \begin{bmatrix} e^a & 0 & 0 & 0 \\ 0 & -e^b & 0 & 0 \\ 0 & 0 & -r^2 & 0 \\ 0 & 0 & 0 & -r^2\sin^2\theta \end{bmatrix}, \tag{9.4.188}$$

where a and b are the functions of r.

Using the definition given in (9.4.183) we obtain

$$\begin{aligned}
&\Gamma^0_{01} = \Gamma^0_{10} = \frac{1}{2}a', && \Gamma^1_{00} = \frac{1}{2}a'e^{a-b}, && \Gamma^1_{11} = \frac{1}{2}b', \\
&\Gamma^1_{22} = -re^{-b}, && \Gamma^1_{33} = -r\sin^2\theta e^{-b}, && \Gamma^2_{12} = \Gamma^2_{21} = \frac{1}{r}, \\
&\Gamma^2_{33} = -\sin\theta\cos\theta, && \Gamma^3_{13} = \Gamma^3_{31} = \frac{1}{r}, && \Gamma^3_{23} = \Gamma^3_{32} = \cot\theta,
\end{aligned}$$

where

$$a' = \frac{\partial a}{\partial r}, \quad b' = \frac{\partial b}{\partial r}. \tag{9.4.189}$$

These relations will be useful in the nonlinear gravitation, presented in the next subsection.

It is interesting to note that we can rederive the geodesic equation shown in (9.4.172) in terms of the proper time $d\tau$ by using the metric geometry and invoking the Euler–Lagrange variational process:

$$\delta \int_{P_1}^{P_2} ds = 0 \tag{9.4.190}$$

or

$$\delta \int_{P_1}^{P_2} \left(g_{\alpha\beta} \frac{dx^\alpha}{d\tau}\frac{dx^\beta}{d\tau} d\tau \right)^{1/2} d\tau = 0. \tag{9.4.191}$$

Performing the differentiation, we obtain

$$\frac{d}{d\tau}\left(g_{\alpha\beta}\frac{dx^\beta}{d\tau} \right) - \frac{1}{2} g_{\mu\upsilon,\alpha} \frac{dx^\mu}{d\tau}\frac{dx^\upsilon}{d\tau} = 0 \tag{9.4.192}$$

or

$$g_{\alpha\beta}\frac{d^2x^\beta}{d\tau^2} + g_{\alpha\beta,\mu}\frac{dx^\mu}{d\tau}\frac{dx^\beta}{d\tau} - \frac{1}{2} g_{\mu\nu,\alpha}\frac{dx^\mu}{d\tau}\frac{dx^\nu}{d\tau} = 0. \tag{9.4.193}$$

This leads to

$$\frac{d^2x^\alpha}{d\tau^2} + \frac{1}{2} g^{\alpha\beta}(2g_{\beta\mu,\nu} - g_{\mu\nu,\beta})\frac{dx^\nu}{d\tau}\frac{dx^\mu}{d\tau} = 0, \tag{9.4.194}$$

where we have used the symmetry of $dx^\nu dx^\mu$ in multiplication with $g_{\beta\mu,\nu}$.

Finally, we obtain

$$\frac{d^2 x^\alpha}{d\tau^2} + \Gamma^\alpha_{\mu\nu} \frac{dx^\mu}{d\tau} \frac{dx^\nu}{d\tau} = 0. \tag{9.4.195}$$

It may now be concluded that the geodesics of the curved spacetime are the worldlines of freely falling particles through the parallel transport process. Consider a local free-falling reference frame such as the rest frame of a freely falling clock. In this frame, coordinates are rectangular and thus the Christoffel symbols are zero, $\Gamma^\alpha_{\mu\nu} = 0$, leading to $g_{\alpha\beta} = \eta_{\alpha\beta}$. Consider a particle near the origin in free fall, having zero acceleration, and both the particle and the reference frame accelerating at the same rate. This is known as the *principle of equivalence*. The equation of motion results from (9.4.195) as

$$\frac{d^2 x^\alpha}{d\tau^2} = \frac{d^2 x^\alpha}{dt^2} = 0. \tag{9.4.196}$$

Thus the worldlines of freely falling particles coincide with the geodesics.

9.4.2.3 The Riemann Curvature Tensors

The Riemann curvature tensors are instrumental in Einstein's general relativity (Eisenhart, 1949). The basic idea is the covariant derivatives of a vector field **V**. The covariant derivative is designated by

$$\nabla = \mathbf{g}^\beta \frac{\partial}{\partial \bar{x}^\beta}. \tag{9.4.197}$$

This implies that the derivative with respect to the new coordinate $\bar{x}^\beta$ is guided by the contravariant tangent vector $\mathbf{g}^\beta$.

As we discussed in Subsection 6.1.2 for the three-dimensional case, let us now consider the covariant derivative of **V** for the spacetime geometries,

$$\begin{aligned}
\nabla \cdot \mathbf{V} &= \mathbf{g}^\beta \frac{\partial}{\partial \bar{x}^\beta} \cdot (V^\alpha \mathbf{g}_\alpha) \\
&= V^\alpha_{,\beta} \mathbf{g}^\beta \cdot \mathbf{g}_\alpha + \mathbf{g}_{\alpha,\beta} \cdot \mathbf{g}^\beta V^\alpha \\
&= V^\alpha_{,\beta} \delta^\beta_\alpha + \Gamma^\mu_{\alpha\beta} \mathbf{g}_\mu \cdot \mathbf{g}^\beta V^\alpha \\
&= V^\alpha_{,\alpha} + \Gamma^\alpha_{\alpha\beta} V^\beta = V^\alpha_{|\alpha}.
\end{aligned} \tag{9.4.198}$$

Physically this represents the diffusion representing the translational motion.

To obtain the covariant derivative of a contravariant vector, we must begin with the curl of a vector:

$$\begin{aligned}
\nabla \times \mathbf{V} &= \mathbf{g}^\beta \frac{\partial}{\partial \bar{x}^\beta} \times (V^\alpha \mathbf{g}_\alpha) \\
&= \mathbf{g}^\beta V_{\alpha,\beta} \times \mathbf{g}^\alpha + \mathbf{g}^\beta V_\alpha \times \left(-\Gamma^\alpha_{\beta\mu} \mathbf{g}^\mu\right) \\
&= \frac{1}{\sqrt{g}} \epsilon^{\beta\alpha\nu} \mathbf{g}_\nu \left(V_{\alpha,\beta} - \Gamma^\mu_{\alpha\beta} V_\mu\right) \\
&= \frac{1}{\sqrt{g}} \epsilon^{\beta\alpha\nu} \mathbf{g}_\nu V_{\alpha|\beta}.
\end{aligned} \tag{9.4.199}$$

Thus the covariant derivative of a vector becomes

$$V_{\alpha|\beta} = V_{\alpha,\beta} - \Gamma^{\mu}_{\alpha\beta} V_{\mu}. \tag{9.4.200}$$

The derivation given in (9.4.199) leads to the physically correct vortex motion. Note that $\epsilon^{\beta\alpha\nu}\mathbf{g}_\nu/\sqrt{g}$ dictates rotational motions and $V_{\alpha|\beta}$ represents the rate of change of velocity characteristic of the curvature corresponding to the rotational motion. As mentioned in Subsection 6.1.2, the occurrences of the Christoffel symbols in Eqs. (9.4.198) and (9.4.200) represent the parallel transports, commonplace in general relativity.

Using the similar observation given by (9.4.198) and (9.4.200), we consider the second-order tensors constructed in various different ways:

$$\mathbf{V}\cdot\mathbf{W} = V^{\alpha}\mathbf{g}_{\alpha}\cdot V^{\beta}\mathbf{g}_{\beta} = A^{\alpha\beta}\mathbf{g}_{\alpha}\cdot\mathbf{g}_{\beta} = A^{\alpha\beta}g_{\alpha\beta}, \tag{9.4.201}$$

$$\mathbf{V}\cdot\mathbf{W} = V_{\alpha}\mathbf{g}^{\alpha}\cdot V_{\beta}\mathbf{g}^{\beta} = A_{\alpha\beta}\mathbf{g}^{\alpha}\cdot\mathbf{g}^{\beta} = A_{\alpha\beta}g^{\alpha\beta}, \tag{9.4.202}$$

$$\mathbf{V}\cdot\mathbf{W} = V_{\alpha}\mathbf{g}^{\alpha}\cdot V^{\beta}\mathbf{g}_{\beta} = A^{\beta}_{\alpha}\mathbf{g}^{\alpha}\cdot\mathbf{g}_{\beta} = A^{\beta}_{\alpha}\delta^{\alpha}_{\beta}. \tag{9.4.203}$$

Physically the diffusion process for V^{α} and the rotational process for V_{α} result in, respectively, the contravariant second-order tensor $A^{\alpha\beta}$ and the covariant second-order tensor $A_{\alpha\beta}$. The mixed second-order tensor A^{β}_{α} arises when diffusion and rotation are mixed.

Covariant derivatives of the second-order tensor can be carried out similarly as those for the first-order tensor:

$$\frac{\partial}{\partial\bar{x}^{\mu}}(A^{\alpha\beta}\mathbf{g}_{\alpha}\cdot\mathbf{g}_{\beta}) = A^{\alpha\beta}_{|\mu}g_{\alpha\beta}, \tag{9.4.204}$$

$$A^{\alpha\beta}_{|\mu} = A^{\alpha\beta}_{,\mu} + \Gamma^{\alpha}_{\nu\mu}A^{\nu\beta} + \Gamma^{\beta}_{\nu\mu}A^{\alpha\nu}, \tag{9.4.205}$$

$$\frac{\partial}{\partial\bar{x}^{\mu}}(A_{\alpha\beta}\mathbf{g}^{\alpha}\cdot\mathbf{g}^{\beta}) = A_{\alpha\beta|\mu}g^{\alpha\beta}, \tag{9.4.206}$$

$$A_{\alpha\beta|\mu} = A_{\alpha\beta,\mu} - \Gamma^{\nu}_{\alpha\mu}A_{\nu\beta} - \Gamma^{\nu}_{\beta\mu}A_{\alpha\nu}. \tag{9.4.207}$$

Similarly,

$$\frac{\partial}{\partial\bar{x}^{\mu}}(A^{\beta}_{\alpha}\mathbf{g}_{\beta}\cdot\mathbf{g}^{\alpha}) = A^{\alpha}_{\beta|\mu}\delta^{\beta}_{\alpha}, \tag{9.4.208}$$

$$A^{\alpha}_{\beta|\mu} = A^{\alpha}_{\beta,\mu} - \Gamma^{\nu}_{\alpha\mu}A^{\alpha}_{\nu} + \Gamma^{\alpha}_{\nu\mu}A^{\nu}_{\beta}. \tag{9.4.209}$$

Proceeding in this manner, we can obtain any-order covariant derivatives of any-order covariant and contravariant tensors:

$$\begin{aligned} A^{\beta_1\ldots\beta_m}_{\alpha_1\ldots\alpha_n}\Big|_{\mu} &= A^{\beta_1\ldots\beta_m}_{\alpha_1\ldots\alpha_n,\mu} - \Gamma^{\nu}_{\alpha_1\mu}A^{\beta_1\ldots\beta_m}_{\nu\alpha_2\ldots\alpha_n} - \Gamma^{\nu}_{\alpha^2\mu}A^{\beta_1\ldots\beta_m}_{\alpha_1\nu\alpha_3\ldots\alpha_n} \\ &\quad - \cdots - \Gamma^{\nu}_{\alpha_n\mu}A^{\beta_1\ldots\beta_m}_{\alpha_1\ldots\alpha_n} + \Gamma^{\beta_1}_{\nu\mu}A^{\nu\beta_2\ldots\beta_m}_{\nu\alpha_1\ldots\alpha_n} \\ &\quad + \Gamma^{\beta_2}_{\nu\mu}A^{\beta_1\nu\ldots\beta_m}_{\alpha_1\ldots\alpha_n} + \cdots + \Gamma^{\beta_m}_{\nu\mu}A^{\beta_1\ldots\nu}_{\alpha_1\ldots\alpha_n}. \end{aligned} \tag{9.4.210}$$

Physically, this implies the mixed translational higher-order diffusional motions (terms with positive signs) and higher-order rotational motions (terms with negative signs) associated with any-order covariant and contravariant tensors.

Riemann Tensor of the Second Kind. The Riemann's curvature tensors arise when we consider the covariant second-order tensor representing the rotational motions given by (9.4.207). To this end, let us first rewrite (9.4.200) in the form

$$A_{\alpha|\beta} = A_{\alpha,\beta} - \Gamma^{\mu}_{\alpha\beta} A_{\mu} = B_{\alpha\beta}. \tag{9.4.211}$$

The second-order covariant derivative of A_α is written as

$$\begin{aligned} A_{\alpha|\beta\mu} &= B_{\alpha\beta|\mu} = B_{\alpha\beta,\mu} - \Gamma^{\nu}_{\alpha\mu} B_{\nu\beta} - \Gamma^{\nu}_{\beta\mu} B_{\alpha\beta} \\ &= (A_{\alpha|\beta})_{,\mu} - \Gamma^{\nu}_{\alpha\mu} A_{\nu|\beta} - \Gamma^{\nu}_{\beta\mu} A_{\alpha|\nu} \\ &= \left(A_{\alpha,\beta} - \Gamma^{\nu}_{\alpha\beta} A_{\nu}\right)_{,\mu} - \Gamma^{\nu}_{\alpha\mu}\left(A_{\nu,\beta} - \Gamma^{\gamma}_{\nu\beta} A_{\gamma}\right) - \Gamma^{\nu}_{\beta\mu}\left(A_{\alpha,\nu} - \Gamma^{\gamma}_{\alpha\nu} A_{\gamma}\right) \\ &= A_{\alpha,\beta\mu} - \left(\Gamma^{\nu}_{\alpha\beta}\right)_{,\mu} A_{\nu} - \Gamma^{\nu}_{\alpha\beta} A_{\nu,\mu} - \Gamma^{\nu}_{\alpha\mu} A_{\nu,\beta} \\ &\quad + \Gamma^{\nu}_{\alpha\mu}\Gamma^{\gamma}_{\nu\beta} A_{\gamma} - \Gamma^{\nu}_{\beta\mu} A_{\alpha,\nu} + \Gamma^{\nu}_{\beta\mu}\Gamma^{\gamma}_{\alpha\nu} A_{\gamma}. \end{aligned} \tag{9.4.212}$$

Similarly, by interchanging the derivative indices β and μ, we have

$$\begin{aligned} A_{\alpha|\mu\beta} &= A_{\alpha,\mu\beta} - \left(\Gamma^{\nu}_{\alpha\mu}\right)_{,\beta} A_{\nu} - \Gamma^{\nu}_{\alpha\mu} A_{\nu,\beta} - \Gamma^{\nu}_{\alpha\beta} A_{\nu,\mu} \\ &\quad + \Gamma^{\nu}_{\alpha\beta}\Gamma^{\gamma}_{\nu\mu} A_{\gamma} - \Gamma^{\nu}_{\mu\beta} A_{\alpha,\mu} + \Gamma^{\nu}_{\mu\beta}\Gamma^{\gamma}_{\alpha\nu} A_{\gamma}. \end{aligned} \tag{9.4.213}$$

Subtracting (9.4.213) from (9.4.212) yields

$$\begin{aligned} A_{\alpha,|\mu\beta} - A_{\alpha|\mu\beta} &= \left[\left(\Gamma^{\nu}_{\alpha\mu}\right)_{,\beta} - \left(\Gamma^{\nu}_{\alpha\beta}\right)_{,\mu} + \Gamma^{\gamma}_{\alpha\mu}\Gamma^{\nu}_{\gamma\beta} - \Gamma^{\gamma}_{\alpha\beta}\Gamma^{\nu}_{\gamma\mu}\right] A_{\nu} \\ &= R^{\nu}_{\alpha\beta\mu} A_{\nu}, \end{aligned} \tag{9.4.214}$$

where $R^{\nu}_{\alpha\beta\mu}$ is known as the Riemann tensor of the second kind,

$$R^{\nu}_{\alpha\beta\mu} = \Gamma^{\nu}_{\gamma\mu,\beta} - \Gamma^{\nu}_{\alpha\beta,\mu} + \Gamma^{\gamma}_{\alpha\mu}\Gamma^{\nu}_{\gamma\beta} - \Gamma^{\gamma}_{\alpha\beta}\Gamma^{\nu}_{\gamma\nu}. \tag{9.4.215}$$

Note that the order of differentiation is irrelevant. Equation (9.4.214) vanishes for all arbitrary values of A_ν in a flat surface:

$$R^{\nu}_{\alpha\beta\mu} = 0. \tag{9.4.216}$$

This has 256 components, but only 20 of them are independent because of symmetry and antisymmetry.

The Riemann tensor may be written in various alternative forms as follows:

Riemann tensor of the first kind

$$R_{\alpha\beta\mu\nu} = g_{\alpha\gamma} R^{\gamma}_{\beta\mu\nu}. \tag{9.4.217}$$

Ricci tensor

$$R_{\alpha\beta} = R^{\nu}_{\alpha\beta\nu} = \Gamma^{\nu}_{\alpha\nu,\beta} - \Gamma^{\nu}_{\alpha\beta,\nu} + \Gamma^{\nu}_{\mu\beta}\Gamma^{\mu}_{\alpha\nu} - \Gamma^{\nu}_{\mu\nu}\Gamma^{\mu}_{\alpha\beta} = 0. \tag{9.4.218}$$

Scalar curvature tensor

$$R = g^{\alpha\beta} R_{\alpha\beta} - R^{\alpha}_{\alpha} = R^{\alpha\beta}_{\alpha\beta}. \tag{9.4.219}$$

Bianchi identities

$$\left(R^{\mu}_{\nu} - \frac{1}{2}\delta^{\mu}_{\nu} R \right)_{|\mu} = 0. \tag{9.4.220}$$

The various forms of Riemann tensor will be used extensively in the nonlinear theory of gravitation discussed in Subsection 9.4.2.5.

9.4.2.4 Geodesic Deviation and Tidal Forces

The Riemann curvature tensor derived in the preceding section is the tidal force field, referring to the acceleration between two particles in free fall.

Consider now a curve in spacetime or a geodesic characterized by the contravariant vector field A^{μ}. It follows from (9.4.198) that the rate of change of A^{μ} with respect to proper time is expressed as

$$\frac{DA^{\mu}}{D\tau} = \frac{dA^{\mu}}{d\tau} + \Gamma^{\mu}_{\alpha\beta} A^{\alpha} \frac{dx^{\beta}}{d\tau}, \tag{9.4.221}$$

$$\begin{aligned}
\frac{D^2 A^{\mu}}{D\tau^2} &= \frac{D}{D\tau}\frac{DA^{\mu}}{D\tau} = \frac{d}{d\tau}\frac{DA^{\mu}}{D\tau} + \Gamma^{\mu}_{\alpha\beta} A^{\alpha} \frac{DA^{\mu}}{D\tau}\frac{dx^{\beta}}{d\tau} \\
&= \frac{d}{d\tau}\left(\frac{dA^{\mu}}{d\tau} + \Gamma^{\mu}_{\alpha\beta} A^{\alpha} \frac{dx^{\beta}}{d\tau} \right) + \Gamma^{\mu}_{\alpha\beta} \left(\frac{dA^{\alpha}}{d\tau} + \Gamma^{\alpha}_{\eta\lambda} A^{\eta} \frac{dx^{\lambda}}{d\tau} \right) \frac{dx^{\lambda}}{d\tau} \\
&= \frac{d^2 A^{\mu}}{d\tau^2} + \Gamma^{\mu}_{\alpha\beta,\nu} \frac{dx^{\nu}}{d\tau} A^{\alpha} \frac{dx^{\beta}}{d\tau} + 2\Gamma^{\mu}_{\alpha\beta} \frac{dA^{\alpha}}{d\tau}\frac{dA^{\beta}}{d\tau} \\
&\quad - \Gamma^{\mu}_{\alpha\beta} A^{\alpha} \Gamma^{\beta}_{\eta\lambda} \frac{dx^{\eta}}{d\tau}\frac{dx^{\lambda}}{d\tau} + \Gamma^{\mu}_{\alpha\beta}\Gamma^{\alpha}_{\eta\lambda} A^{\eta} \frac{dx^{\lambda}}{d\tau}\frac{dx^{\beta}}{d\tau}.
\end{aligned} \tag{9.4.222}$$

Using (9.4.172) with $x^{\mu} = x^{\mu}(\tau) + s^{\mu}(\tau)$, we write

$$\frac{d^2(x^{\mu} + s^{\mu})}{d\tau^2} + \Gamma^{\mu}_{\alpha\beta}(x+s)\left(\frac{dx^{\alpha}}{d\tau} + \frac{ds^{\alpha}}{d\tau} \right)\left(\frac{dx^{\beta}}{d\tau} + \frac{ds^{\beta}}{d\tau} \right) = 0. \tag{9.4.223}$$

Let us approximate

$$\Gamma^{\mu}_{\alpha\beta}(x+s) \cong \Gamma^{\mu}_{\alpha\beta}(x) + \Gamma^{\mu}_{\alpha\beta,\gamma} s^{\gamma} \tag{9.4.224}$$

and subtract (9.4.172) from (9.4.223). This leads to

$$\frac{d^2 s^{\mu}}{d\tau^2} = -\Gamma^{\mu}_{\alpha\beta,\gamma} s^{\gamma} \frac{dx^{\alpha}}{d\tau}\frac{dx^{\beta}}{d\tau} - 2\Gamma^{\mu}_{\alpha\beta} \frac{ds^{\alpha}}{d\tau}\frac{dx^{\beta}}{d\tau}. \tag{9.4.225}$$

Using (9.4.225) in (9.4.222) and calculating the second derivative of s^{μ} along the geodesic, we obtain

$$\frac{D^2 s^{\mu}}{D\tau^2} = R^{\mu}_{\alpha\gamma\beta} s^{\gamma} \frac{dx^{\alpha}}{d\tau}\frac{dx^{\beta}}{d\tau}. \tag{9.4.226}$$

This is known as the equation of geodesic deviation, representing the relativistic generalization of the Newtonian result for the tidal force.

To determine the Newtonian limit, we set

$$\frac{d\bar{x}^{\alpha}}{d\tau} \cong (1,\ 0,\ 0,\ 0), \tag{9.4.227}$$

$$\frac{d^2\bar{s}^{\mu}}{d\bar{t}^2} \cong -\bar{R}^{k}_{0l0}\bar{s}^{l}. \tag{9.4.228}$$

The tidal force is

$$f^k \cong -m\bar{R}^{k}_{0j0}\bar{s}^{j},$$

with

$$\Gamma^{\alpha}_{\beta\mu} = \frac{\kappa}{2}\eta^{\alpha\gamma}(h_{\alpha\beta,\mu} + h_{\mu\gamma,\beta} - h_{\beta\mu,\gamma}),$$

$$R^{\alpha}_{\beta\mu\nu} \approx -\frac{\kappa}{2}\eta^{\alpha\gamma}(h_{\mu\gamma,\beta\nu} - h_{\beta\mu,\gamma\nu} - h_{\nu\gamma,\beta\mu} + h_{\beta\nu,\gamma\mu}),$$

$$R^{k}_{0j0} \approx \frac{\kappa}{2}(h_{jk,00} - h_{0j,k0} - h_{0k,0j} + h_{00,kj}).$$

Now setting

$$\frac{1}{2}\kappa h_{00} = \Phi, \tag{9.4.229}$$

we find

$$R^{k}_{0j0} = \frac{\partial^2\Phi}{\partial x^k \partial x^j}. \tag{9.4.230}$$

This is the tidal force tensor at the Newtonian limit, representing components of the Riemann curvature tensor $R^{\mu}_{\nu\alpha\beta}$ (Riemann, 1868).

9.4.2.5 The Nonlinear Theory of Gravitation

Einstein's Field Equation. The interaction of the geometry and matter are examined in this subsection, leading to Einstein's equation for the gravitational field, known as geometrodynamics.

The proper time $d\tau$ is given by

$$d\tau^2 = g_{\mu\nu}dx^{\mu}dx^{\nu} \tag{9.4.231}$$

or

$$d\tau^2 = (\eta_{\mu\nu} + \kappa h_{\mu\nu})dx^{\mu}dx^{\nu}, \tag{9.4.232}$$

where $\eta_{\mu\nu}$ is the Minkowski tensor, $h_{\mu\nu}$ is the gravitational field tensor, and κ is the gravitational coupling constant.

Einstein's field equation relates the Ricci tensor $R_{\mu\nu}$, the scalar Riemann tensor R, and the energy-momentum tensor $T_{\mu\nu}$ of matter in the form

$$\begin{aligned} R_{\mu\nu} - \frac{1}{2}\eta_{\mu\nu}R = \frac{1}{2}\big(&\partial^{\alpha}\partial_{\alpha}g_{\mu\nu} - \partial^{\alpha}\partial_{\nu}g_{\mu\alpha} - \partial^{\alpha}\partial_{\mu}g_{\nu\alpha} \\ &+ \partial_{\mu}\partial_{\nu}g^{\beta}_{\beta} - g_{\mu\nu}\partial_{\alpha}\partial^{\alpha}g^{\beta}_{\beta} + \eta_{\mu\nu}\partial^{\alpha}\partial^{\beta}g_{\alpha\beta}\big) = -\frac{1}{2}\kappa^2 T_{\mu\nu}. \end{aligned} \tag{9.4.233}$$

Setting

$$\kappa^2 = 16\pi G, \quad G = \text{gravitational constant}, \tag{9.4.234}$$

and raising the index, we obtain

$$R^{\nu}_{\mu} - \frac{1}{2}\delta^{\nu}_{\mu} R = -8\pi G T^{\nu}_{\mu}. \tag{9.4.235}$$

It is seen that the left-hand side satisfies the Bianchi identity

$$\left(R^{\nu}_{\mu} - \frac{1}{2}\delta^{\nu}_{\mu} R\right)_{|\nu} = 0. \tag{9.4.236}$$

Thus we have

$$T^{\nu}_{\mu|\nu} = 0 \tag{9.4.237}$$

or

$$\partial_{\nu} T^{\nu}_{\mu} + \Gamma^{\nu}_{\alpha\nu} T^{\alpha}_{\mu} - \Gamma^{\alpha}_{\mu\nu} T^{\nu}_{\alpha} = 0. \tag{9.4.238}$$

The Einstein equation may be modified to include the cosmological term considering the effect of an extra energy momentum associated with empty (matter-free) space. To this end we utilize the minimal coupling principle in which we start from (9.4.237) to obtain Einstein's equation. Thus we begin by assuming

$$G_{\mu\nu} = -8\pi G T_{\mu\nu}. \tag{9.4.239}$$

This implies that

$$G^{\nu}_{\mu|\nu} = 0, \tag{9.4.240}$$

with

$$G_{\mu\nu} = a R_{\mu\nu} + b g_{\mu\nu} R + \Lambda g_{\mu\nu}, \tag{9.4.241}$$

where a, b, and Λ are constants. Because $g^{\nu}_{\mu|\nu} = 0$, condition (9.4.240) reduces to

$$\left(a R^{\nu}_{\mu} + b\delta^{\nu}_{\mu} R\right)_{|\nu} = 0. \tag{9.4.242}$$

Applying the Bianchi identity (9.4.220) to Eq. (9.4.242), we arrive at

$$b = -\frac{1}{2}a. \tag{9.4.243}$$

This allows the field equation to be written in the form

$$a\left(R_{\mu\nu} - \frac{1}{2}g_{\mu\nu} R\right) + \Lambda g_{\mu\nu} = -8\pi G T_{\mu\nu}. \tag{9.4.244}$$

For (9.4.244) to agree with the linear, nonrelativistic limit, we must have $a = 1$ and $\Lambda = 0$, leading to the Einstein equation.

The quantity $\Lambda g_{\mu\nu}$ in (9.4.244) is known as the cosmological term, and the constant Λ is called the cosmological constant. If Λ is small, but not zero, then Einstein's field equation takes the form

$$R_{\mu\nu} - \frac{1}{2} g_{\mu\nu} R + \Lambda g_{\mu\nu} = -8\pi G T_{\mu\nu} \tag{9.4.245}$$

or

$$R_{\mu\nu} - \frac{1}{2} g_{\mu\nu} R = -8\pi G (T_{\mu\nu} + \hat{T}_{\mu\nu}), \tag{9.4.246}$$

where

$$\hat{T}_{\mu\nu} = \frac{\Lambda}{8\pi G} g_{\mu\nu} \tag{9.4.247}$$

is known as the extra energy-momentum tensor associated not with matter, but with empty space.

The effects of the Λ term will be most noticeable in the large-scale motion of the universe. Hence cosmological data set the best limits on Λ. Our universe is expanding, and the attractive Newtonian gravitational force between the masses in the universe tends to decelerate this expansion if Λ is negative and accelerate if Λ is positive. Consequently a large negative value of Λ is positive, and consequently a large negative value of Λ implies a short age for the universe and a large value implies a long age. From observations on globular clusters of stars, astronomers can place a firm lower limit of 10^{10} years on the age of the Universe.

Einstein's field equation as previously introduced may be solved analytically for vacuum (Schwarzschild, 1916) and for rotating mass (Kerr, 1963; Kerr and Schild, 1965). Subsequently they were then applied to evaluate physical phenomena such as occur in black holes (Bardeen, Carter, and Hawking, 1973; Chandrasekhar, 1983), gravitational waves, and cosmology. Such analytical solutions, however, are severely limited, and we must resort to computer simulations (known as numerical relativity) for more practical and real situations. To this end, it is necessary to modify Einstein's field equations conducive to numerical simulations and introduce appropriate initial and boundary conditions for various real situations (Subsection 9.4.2.6).

Isometries of Spacetime. The standard transformation between the metric tensor $\bar{g}_{\alpha\beta}$ and $g_{\alpha\beta}$ may be performed as

$$g_{\alpha\beta} = \frac{\partial \bar{x}^{\mu}}{\partial x^{\alpha}} \frac{\partial \bar{x}^{\nu}}{\partial x^{\beta}} g_{\mu\nu}. \tag{9.4.248}$$

Let us consider an infinitesimal coordinate transformation,

$$\bar{x}^{\mu} = x^{\mu} + \varepsilon \xi^{\mu} \quad \text{with} \quad \varepsilon \to 0. \tag{9.4.249}$$

Inserting (9.4.249) into (9.4.248) gives

$$g_{\alpha\beta} = \left(\delta^{\mu}_{\alpha} + \varepsilon \frac{\partial \xi^{\mu}}{\partial x^{\alpha}} \right) \left(\delta^{\nu}_{\beta} + \varepsilon \frac{\partial \xi^{\nu}}{\partial x^{\beta}} \right) \bar{g}_{\mu\nu}[\bar{x}(x)]$$

or

$$g_{\alpha\beta} \cong \bar{g}_{\alpha\beta} + \varepsilon\xi^{\mu}\frac{\partial g_{\alpha\beta}}{\partial x^{\mu}} + \varepsilon\frac{\partial \xi^{\mu}}{\partial x^{\alpha}} g_{\mu\beta} + \varepsilon\frac{\partial \xi^{\mu}}{\partial x^{\beta}} g_{\alpha\mu}, \tag{9.4.250}$$

with terms of order ε^2 neglected. Also, the second term of the right-hand side of (9.4.250) arises because of the Taylor series expansion of $\bar{g}_{\alpha\beta}[\bar{x}(x)]$,

$$\bar{g}_{\alpha\beta}[\bar{x}(x)] \cong \bar{g}_{\alpha\beta}(x) + \varepsilon\xi^{\mu}\frac{\partial \bar{g}_{\alpha\beta}}{\partial x^{\mu}} \cong \bar{g}_{\alpha\beta}(x) + \varepsilon\xi^{\mu}\frac{\partial g_{\alpha\beta}}{\partial x^{\mu}}. \tag{9.4.251}$$

In terms of the covariant components, $\xi_\alpha = g_{\nu\alpha}\xi^{\nu}$, and using the condition for the invariance of the metric tensor under the infinitesimal transformation, we obtain

$$\xi^{\mu}\frac{\partial g_{\alpha\beta}}{\partial x^{\mu}} + \left(\frac{\partial \xi_\beta}{\partial x^{\alpha}} - \xi^{\mu}\frac{\partial \xi_{\mu\beta}}{\partial x^{\alpha}}\right) + \left(\frac{\partial \xi_\alpha}{\partial x^{\beta}} - \xi^{\nu}\frac{\partial \xi_{\alpha\nu}}{\partial x^{\beta}}\right) = 0, \tag{9.4.252}$$

which may be written with Christoffel symbols as

$$\frac{\partial \xi_\beta}{\partial x^{\alpha}} + \frac{\partial \xi_\alpha}{\partial x^{\beta}} - 2\xi_\mu\Gamma^{\mu}_{\alpha\beta} = 0. \tag{9.4.253}$$

Using the covariant derivatives, we obtain

$$\xi_{\beta|\alpha} + \xi_{\alpha|\beta} = 0 \tag{9.4.254a}$$

or

$$\nabla_\alpha\xi_\beta + \nabla_\beta\xi_\alpha = 0. \tag{9.4.254b}$$

This is known as Killing's equation. The variables ξ_α obtained as a solution of this differential equation are called Killing vectors, leading to the symmetry transformation of the metric, both the infinitesimal and finite transformations (Rindler, 1969).

If the spacetime geometry is time independent, so $g_{\alpha\beta}$ does not depend on x^0, then $\xi^\alpha = (a, 0, 0, 0)$ is the corresponding Killing vector. The momentum conservation requires that

$$\xi_\alpha P^\alpha = \xi_\alpha m\frac{dx^\alpha}{d\tau} = \text{constant}. \tag{9.4.255}$$

This conservation law also applies to photons moving in the curved spacetime, ensuring that the photon energy is constant.

The infinitesimal symmetry transformation may be expressed also by the Lie derivative. To this end, we rewrite (9.4.250) in the form

$$g_{\alpha\beta} - \bar{g}_{\alpha\beta} = \epsilon\xi^{\mu}\frac{\partial g_{\alpha\beta}}{\partial x^{\mu}} + \epsilon\frac{\partial \xi^{\mu}}{\partial x^{\alpha}} g_{\mu\beta} + \epsilon\frac{\partial \xi^{\nu}}{\partial x^{\beta}} g_{\alpha\nu}. \tag{9.4.256}$$

Using the infinitesimal transformation relation given by Eq. (9.4.256), the Lie derivative $L_\xi g_{\alpha\beta}$ is defined as

$$L_\xi g_{\alpha\beta} = \lim_{\epsilon\to 0}\frac{g_{\sigma\beta} - \bar{g}_{\alpha\beta}}{\epsilon} = \xi^{\mu}\frac{\partial g_{\alpha\beta}}{\partial x^{\mu}} + \frac{\partial \xi^{\mu}}{\partial x^{\alpha}} g_{\mu\beta} + \frac{\partial \xi^{\nu}}{\partial x^{\beta}} g_{\alpha\nu}. \tag{9.4.257}$$

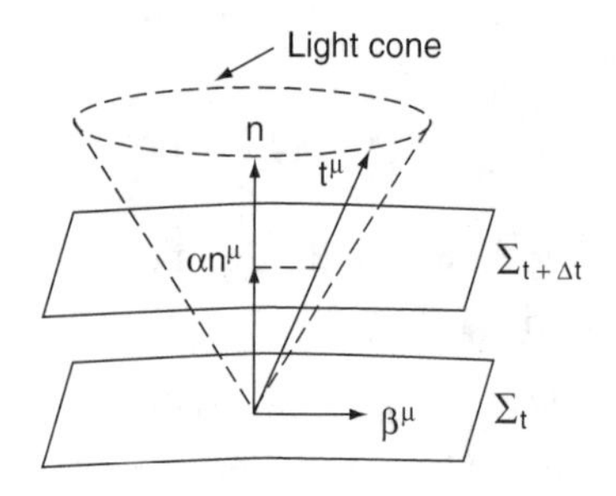

Figure 9.4.3. Foliation of Cauchy surfaces between two slices, Σ_t and $\Sigma_{t+\Delta t}$ parameterized by a global time function t with the unit normal vector n^{μ} and a vector field decomposing into lapse function α and shift vector β^{μ}.

Lie derivatives for tensors of other ranks can also be defined similarly by the change of the tensor under an infinitesimal transformation (Stachel, 1962, 1969). Utilizations of Lie derivatives will be shown in kinematics and dynamics of general relativity in the following subsection.

9.4.2.6 Kinematics and Dynamics of General Relativity

The Einstein's field equations given by Eq. (9.4.245) represent a total of 10 nonlinear simultaneous partial differential equations. Analytical solutions for some simple cases are available (Schwarzchild, 1916; Kerr, 1963). However, solutions to many of the important astrophysical problems are yet to be explored. To this end, we must examine kinematics and dynamics of general relativity for possible numerical simulations.

A convenient scheme in which Einstein's field equations can be recast in the Cauchy formalism is the $3+1$ dimensional decomposition of the Einstein equation (Arnowitt, Deser, and Misner, 1962), known as the ADM decomposition. By Cauchy formalism is meant the reconstruction of differential equations in such a form that appropriate initial and boundary conditions may be properly applied to obtain desired solutions. In this process we study the observer's kinematics (velocities and accelerations) in terms of the geometry of their congruencies of worldlines relative to families of time slicings of spacetime. The gravitational field enters in nonlinear form through the presence of curvature in the equations.

A classical gravitational field is the time history of the geometry of a spacelike hypersurface, consisting of the initial-value problem, prescribing a reference system and integrating the dynamical equations along the trajectories of the reference system. To perform this program we require a space – plus – time $(3+1)$ system. Typically, the equations are three-dimensionally covariant, elliptic, and linear in the kinematical functions known as the lapse function and shift vector. In what follows these kinematical functions and coordinate conditions are described.

Cauchy Formalism for General Relativity. In the three-dimensional Cauchy (ADM or $3+1$) formulation of Einstein's equations we establish n^{μ} as the future-pointing timelike unit normal to the time slice within a light cone geometry, as shown in Fig. 9.4.3. We then foliate the manifold M with a parameterized (with parameter t) set of spacelike, three-dimensional hypersurfaces Σ_t and choose coordinates x^i $(i = 1, 2, 3)$ to label points on each one, $x^\mu = (t, x^i)$.

Let us consider the time derivative in the direction normal and tangential to the manifold,

$$\frac{\partial}{\partial t} = \frac{\partial}{\partial n}\frac{\partial n}{\partial t} + \frac{\partial}{\partial x_i}\frac{\partial x_i}{\partial t} = \alpha\frac{\partial}{\partial n} + \beta^i\frac{\partial}{\partial x_i}, \tag{9.4.258}$$

where $\alpha = (\partial n/\partial t)$ and $\beta^i = (\partial x_i/\partial t)$ are defined as the lapse function and shift vector, respectively. We may now choose n^μ such that

$$n^\mu \equiv -\alpha\nabla^\mu t, \tag{9.4.259}$$

with the lapse function α defining the proper interval measured by observers traveling normal to the hypersurface.

The complementary or dual aspect of time is described by the four-vector along which the data on the slice Σ_t are to be evolved. The use of t^μ (coordinate observers) corresponds in hydrodynamics to the use of mixed Euler–Lagrange trajectories because in general t^μ lies along neither the Eulerian trajectories nor the Lagrangian (matter) trajectories. In this case, besides α, we require the shift vector β^μ to relate the two:

$$t^\mu = \alpha n^\mu + \beta^\mu, \qquad \beta^\mu n_\mu = 0, \tag{9.4.260}$$

with $\beta^\mu = (0, \beta^i)$.

If the Euclidian metric of each Σ_t is given by g_{ij} (the pullback of $g_{\alpha\beta}$ onto Σ_t) the spacetime metric becomes

$$ds^2 = -\alpha^2 dt^2 + g_{ij}(dx^i + \beta^i dt)\,(dx^j + \beta^j dt) \tag{9.4.261a}$$

or

$$ds^2 = g_{\alpha\beta}dx^\alpha dx^\beta. \tag{9.4.261b}$$

Comparing (9.4.261a) and (9.4.261b) leads to

$$\begin{bmatrix} g_{00} & g_{0k} \\ g_{i0} & g_{ik} \end{bmatrix} = \begin{bmatrix} \beta_r\beta^r & \alpha^2 \\ \beta_i & g_{ik} \end{bmatrix}. \tag{9.4.262}$$

Thus the reciprocal four-metric takes the form

$$\begin{bmatrix} g^{00} & g^{0m} \\ g^{k0} & g^{km} \end{bmatrix} = \begin{bmatrix} -\dfrac{1}{\alpha^2} & \dfrac{\beta^m}{\alpha^2} \\ \dfrac{\beta^k}{\alpha^2} & g^{km} - \dfrac{\beta^k\beta^m}{\alpha^2} \end{bmatrix}. \tag{9.4.263}$$

The description of the embedding of Σ_t in spacetime requires the extrinsic curvature tensor K_{ij} in addition to the three-metric tensor g_{ij} induced by $g_{\alpha\beta}$ on Σ_t. To this end, we consider a dot product of a tangent base vector $\mathbf{g}_i$ and a normal vector **n**. Taking a derivative of this product, we obtain

$$\begin{aligned} K_{ij} &= \mathbf{n} \cdot \mathbf{g}_{i,j} = -\mathbf{n}_{,j} \cdot \mathbf{g}_i = -n_{i,j} + n_k \Gamma^k_{ij} \\ &= -n_{i|j} = -\nabla_j n_i = -\frac{1}{2}\frac{\partial}{\partial n} g_{ij} = -\alpha \Gamma^0_{ij} \end{aligned} \tag{9.4.264a}$$

or

$$K_{ij} = -\alpha\left(g^{00}\Gamma_{0ik} + g^{0s}\Gamma_{sij}\right). \tag{9.4.264b}$$

Using the relations given by (9.4.262) and (9.4.263), we may write the extrinsic curvature tensor as

$$\begin{aligned} K_{ij} &= \frac{1}{\alpha}\left(\Gamma_{0ij} - \beta^s \Gamma_{sij}\right) \\ &= \frac{1}{2\alpha}\left(\frac{\partial \beta_i}{\partial x^j} + \frac{\partial \beta_j}{\partial x^i} - \frac{\partial g_{ij}}{\partial t} - 2\Gamma_{sij}\beta^s\right) \\ &= \frac{1}{2\alpha}\left(\nabla_j \beta_i + \nabla_i \beta_j - \frac{\partial g_{ij}}{\partial t}\right). \end{aligned} \tag{9.4.265}$$

This represents the extrinsic curvature obtained by use of the ADM lapse and shift functions.

Note that the mathematical structure of extrinsic curvature is similar to the curvature tensor $b_{\alpha\beta}$ introduced in Eqs. (9.1.16) that is subsequently used in the bending strain tensor, Eq. (9.3.21), for a shell structure. Thus we may think of deformations of the slice in spacetime as being analogous to deformations of a curved thin shell in ordinary space used in engineering fields. This is one of the most eloquent manifestations of the power of tensor analysis describing the physical phenomena between physics and engineering.

It follows from Codazzi conditions (9.2.12) and Lie derivatives (9.4.257) that the time derivatives of the metric tensor g_{ij} and extrinsic curvature tensor K_{ij} are given by

$$(\partial_t - L_\beta) g_{ij} = -2\alpha K_{ij}, \tag{9.4.266}$$

$$(\partial_t - L_\beta) K_{ij} = -D_i D_j \alpha + \alpha\left(R_{ij} - 2K_{im}K^m_j + K K_{ij}\right), \tag{9.4.267}$$

when $K = K^i_i$ and R_{ij} is the Ricci tensor. The Lie derivatives of g_{ij} and K_{ij} are given by

$$L_\beta g_{ij} = \nabla_i \beta_j + \nabla_j \beta_i, \tag{9.4.268}$$

$$L_\beta K_{ij} = \beta^m \nabla_m K_{ij} + K_{im} \nabla_j \beta^m + K_{mj} \nabla_i \beta^m. \tag{9.4.269}$$

Hence, it is seen that g_{ij} and K_{ij} are the set of initial data that must be specified for a Cauchy evolution of Einstein's equations. Equations (9.4.267) and (9.4.268) constitute the evolution equations that are used to obtain the spacetime to the future of the initial hypersurface. These equations are to be coupled with two constraint equations: the Hamiltonian (scalar) constraint,

$$R + K^2 - K_{ij}K^{ij} = 0, \tag{9.4.270}$$

and the momentum (vector) constraint,

$$D_j(K^{ij} - g^{ij}K) = 0. \tag{9.4.271}$$

The Cauchy or $3 + 1$ formulation is often called ADM (for Arnowitt, Deser, and Misner, 1962). There are various approaches using ADM available in literature (York, 1971, 1972, and 1979; Smarr, 1979).

Coordinate Conditions, Lapse Functions, and Shift Vectors. Lapse functions and shift vectors are useful for numerical simulations of the Einstein equations, but in general this is not possible without prior knowledge of the expected dynamics.

In dynamic spacetimes there are no timelike symmetries to suggest a preferred choice of slices or the lapse function α. For $\alpha \to 1$ and $\beta^\mu \to 0$, t^μ reduces to a constant time translation at spatial infinity, with t^μ approaching n^μ, which is appropriate in the case in which the curved spacetime Eulerian observers are to be identified asymptotically with the standard observers of Minkowski spacetime. In spacetimes with compact time slices, the preceding reasoning does not apply. Here, the time slicing must be constructed wholly from the spacetime dynamics itself without appealing to an external standard reference system such as observers at infinity.

Lapse Functions: Maximal Slicings. As introduced by Lichnerowicz (1944 and 1967) and York (1971), the method of maximal slicings maximizes the three-volume of the slices. This condition translates into slices that effectively deform so that $K = g^{ij}K_{ij} \equiv F(t)$, which in turn implies a nonuniform α. A straightforward evaluation of (9.4.269) with the Hamiltonian constraint to reexpress the Ricci tensor in terms of K_{ij} provides the elliptic equation for α,

$$\nabla^i \nabla_i \alpha = \alpha K_{ij}K^{ij} - K_{,t} \tag{9.4.272}$$

or

$$\frac{1}{\sqrt{g}}\left(\sqrt{g}g^{ij}\alpha_{,j}\right)_{,i} = \alpha K_{ij}K^{ij} - K_{,t}. \tag{9.4.273}$$

Note that from (9.4.266) we obtain

$$\partial_t(\log \sqrt{g}) = -\alpha K + D_i\beta^i, \tag{9.4.274}$$

which describes the evolution of the determinant of g_{ij}. In this case, in which $\beta^i = 0 = K$, we can avoid the singularity as the variation of the local volume remains fixed. This slows down the evolution in regions of strong curvature, whereas the simulation proceeds in the future regions. However, as the evolution proceeds, the slices pile up in regions of high curvature. The sequence of slices that result is considerably bent and large numerical gradients are induced (grid stretching). As the evolution proceeds, these gradients become large and ultimately the evolution crashes. We may prevent this by choosing $K = 0$.

Shift Vectors: Minimal Strain and Minimal Distortion. A shift condition known as minimal strain was introduced by Smarr and York (1978) through a set of elliptic equations obtained by means of a minimization of the hypersurface strain. Minimizing an action defined with g_{ij} and $L_\beta g_{ij}$ with respect to β^i yields the elliptic set of equations,

$$D_i D^i \beta^j + D_i D^j \beta^i - 2D_i(\alpha K^{ij}) = 0. \tag{9.4.275}$$

We obtain a related condition known as minimal distortion by considering a different action defined in terms of a distortion tensor:

$$D^j D_j \beta^i + \frac{1}{3} D^i D_j \beta^j + R^i_j \beta^j - 2D_j \left[\alpha \left(K^{ij} - \frac{1}{3} K \right) \right] = 0. \tag{9.4.276}$$

We have extended the minimal strain prescription by minimizing the action with respect to both α and β^i, obtaining (9.4.260) and the lapse condition,

$$K^{ij}(-2\alpha K_{ij} + 2D_i \beta_j) = 0. \tag{9.4.277}$$

York (1979) has shown that with an appropriate choice of slice and boundary conditions the rate of change along $(\partial_t)^\mu$ can be minimized, particularly when the metric variables vary slowly in the resulting coordinates.

Initial and Boundary Conditions. The Cauchy initial-value problem requires prescribing g_{ij} and K_{ij} on an initial hypersurface. However, not all these variables are independent. Namely, there are four constraints to be satisfied and so only 8 out of 12 in the (g_{ij}, K_{ij}) pair need be specified.

Additionally, when one is dealing with spacetimes containing singularities, either the solution is renormalized, effectively factoring out the divergent part, or a region containing the singularity is excised, which requires introducing an inner boundary where data must be provided as well.

In dealing with black-hole spacetimes and the presence of singularities, one may use a slicing that effectively freezes the evolution near the singularities (such as the maximal slicing condition). The so-called cosmological censorship requires that singularities be hidden inside the event horizons. Thus an inner boundary could be chosen to lie inside the event horizon surrounding the singularity. Because the concept of an event horizon is a global one, it can be found only after the evolution has been carried over. To obtain a local notion, trapped surfaces must be found, referred to as the apparent horizon. This will always lie inside the event horizon. Thus the apparent horizon is used as a marker and the region inside it is excised from the computational domain, defining an inner boundary.

The spacelike slices in $3+1$ implementation extend to spacelike infinity. In general, the spacetime is globally hyperbolic. Thus data on a given initial hypersurface completely determine the unique geometry to the future of it. To handle these infinitely large hypersurfaces, one can compact the spacetime to deal with a finite domain and gain access to infinity so that the concept of asymptotically flatness can be used to provide boundary data.

It is possible to write the evolution equations and constraint conditions in terms of a first-order flux – conservation system so that the propagation of characteristic fields in an inhomogeneous background can be considered (Bona, 1996). Numerous other strategies have been proposed as reviewed in Font (2000), Baumgarte and Shapiro (1998), and Lehner (2001). Hyperbolic conservation formulations for ideal and nonideal relativistic hydrodynamics will be discussed in Subsection 9.4.3.

We conclude this section on Einstein's nonlinear theory of gravitation by pointing out that, because Einstein's field equations are difficult to solve, we introduced the Cauchy formalism, making it possible to produce evolution equations for metric tensor g_{ij} and extrinsic curvature tensor K_{ij} along with momentum and energy (Hamiltonian) constraints. As mentioned earlier (Remarks, Section 9.3), g_{ij} and K_{ij} correspond to the membrane and bending strain tensors for deformations of shell structures in engineering, thus being ideally treated under the differential geometry continuum together with the relativistic continuum.

Remarks

Our goal in general relativity is to solve the Einstein's field equations, (9.4.239) or (9.4.245). However, their analytical solutions are extremely difficult. For this reason, using the ADM approximations, we may reconstruct Einstein's field equations into the evolution equations for metric tensor g_{ij} and extrinsic curvature tensor K_{ij} together with momentum and energy constraint equations. Solutions of these equations by use of numerical methods have been successful. The conservation form of relativistic hydrodynamic equations, however, leads to numerically more expedient strategies. This topic is presented next.

9.4.3 Conservation Hydrodynamic Equations in Numerical Relativity

Einstein's evolution equations presented in Subsection 9.4.2.6 may be solved numerically. However, as the velocity becomes very high, the numerical solutions will be unstable, and it is necessary that the governing equations be written in a conservation form, as discussed in Subsection 5.3.2. It is well known that the conservation form enables certain numerical schemes to resolve physical discontinuities such as shock waves in black-hole accretion disks and gravitational radiation (Hawking, 1971; DeWitt, 1967 and 1979; Deruelle and Piran, 1982).

To find the conservation form we begin first with the nonconservation form. We then construct the conservation form such that, on differentiation, we must recover the nonconservation form. We discuss both ideal fluids and nonideal fluids, followed by MHD general relativity equations.

9.4.3.1 Ideal Fluids

The equations describing the evolution of a relativistic fluid are local conservation laws of the stress energy, $T^{\mu\nu}$, and the matter current density, J^{μ}, given by covariant derivatives as follows:

$$T^{\mu\nu}_{|\mu} = 0 \quad (\mu, \nu = 0, 1, 2, 3), \tag{9.4.278}$$

$$J^{\mu}_{|\mu} = 0, \tag{9.4.279}$$

with

$$J^{\mu} = \rho u^{\mu}, \tag{9.4.280}$$

where ρ is the rest mass density and u^{μ} is the four-velocity of the fluid.

For a perfect fluid (viscosity and thermal conduction are neglected) the energy-momentum tensor can be written as

$$T^{\mu\nu} = \rho h u^{\mu} u^{\nu} + p g^{\mu\nu}, \tag{9.4.281}$$

with $g^{\mu\nu}$ being the four-vector describing the spacetime and h the specific enthalpy defined as

$$h = 1 + \epsilon + \frac{P}{\rho}, \tag{9.4.282}$$

where ϵ is the specific internal energy and P is the isotropic pressure. The preceding system of equations can be closed by the normalization condition for the four-velocity,

$$g_{\mu\nu} u^{\mu} u^{\mu} = -1, \tag{9.4.283}$$

and the equation of state,

$$p = p(\rho, \epsilon). \tag{9.4.284}$$

Dynamics of the gravitational field in general relativity theory is described by Einstein's field equation,

$$G_{\mu\nu} = 8\pi T_{\mu\nu}, \tag{9.4.285}$$

where $G_{\mu\nu}$ is the Einstein tensor associated with the 10 metric components $g_{\mu\nu} = g_{\nu\mu}$ of the spacetime and the stress-energy tensor $T_{\mu\nu}$.

Instead of using the covariant derivatives, it is convenient to formulate the numerical process by means of a coordinate system ($x^{\mu} = t, x^1, x^2, x^3$) and express (9.4.278) and (9.4.279) in terms of coordinate derivatives. To this end, we write the governing conservation equations in the form (Font, 2000; Richardson and Chung, 2002)

$$\frac{\partial U}{\partial t} + \frac{\partial F^i}{\partial x^i} = B, \tag{9.4.286}$$

$$U = \begin{bmatrix} D \\ s_i \\ \tau \end{bmatrix} = \begin{bmatrix} \sqrt{g} W \rho \\ \sqrt{g} \rho h W^2 V_j \\ \sqrt{g}(\rho h W^2 - p - W\rho) \end{bmatrix}, \tag{9.4.287}$$

where g is the determinant of g_{ij}, V_j is the fluid three-velocity, and W is the Lorentz factor,

$$W = (1 - g_{ij} V^i V^j)^{-1/2}, \tag{9.4.288}$$

where we invoke the natural unit

$$G = c = 1,$$

and the spatial components of the four-velocity u^i are related by the three-velocity V_i by

$$u^i = V^i - \frac{\beta^i}{\alpha}. \tag{9.4.289}$$

The convection flux vector F^i and the source vector B are given by

$$F^i = \begin{bmatrix} \alpha\left(V^i - \dfrac{\beta^i}{\alpha}\right) D \\ \alpha\left(V^i - \dfrac{\beta^i}{\alpha}\right) S_j + \sqrt{g} p \delta^i_j \\ \alpha\left(V^i - \dfrac{\beta^i}{\alpha}\right) \tau + \sqrt{g} V^i p \end{bmatrix}, \tag{9.4.290}$$

$$B = \begin{bmatrix} 0 \\ \alpha\sqrt{g} T^{\mu\nu} - g_{\mu\nu} \Gamma^{\sigma}_{\mu j} \\ \alpha\sqrt{g}(T^{\mu 0} \alpha_{,\mu} - \alpha T^{\mu\nu} \Gamma^{0}_{\mu\nu}) \end{bmatrix}, \tag{9.4.291}$$

where it is seen that the lapse function α and the shift vector β^i are associated with the primitive variables (ρ, V_i, ϵ) through the conservation variables (D, S_i, τ). Once the conservation variables are calculated, then the primitive variables can be extracted from the conservation variables.

9.4.3.2 Nonideal Fluids

The effect of viscosity and thermal conduction is undoubtedly an important consideration, particularly in accretion disks with rotating Kerr black holes where angular momentum can be significant (Richardson and Chung, 2002). The basic equations for nonideal fluids in special relativity were presented in Subsection 9.4.1.4. They can be applied to general relativity as well.

The stress-energy tensor (9.4.94) may be rewritten with more convenient notation for general relativity as

$$T^{\alpha\beta} = \rho(1+\epsilon)u^\alpha u^\beta + \left(p - \zeta u^\mu_{|\mu}\right) f^{\alpha\beta} - 2\eta d^{\alpha\beta} + q^\alpha u^\beta + q^\beta u^\alpha, \qquad (9.4.292)$$

with the following terms.

Rate-of-deformation tensor

$$d^{\alpha\beta} = \frac{1}{2}(u^\alpha_{|\mu} f^{\mu\beta} + u^\beta_{|\mu} f^{\mu\alpha}) - \frac{1}{3}\theta\, f^{\alpha\beta}. \qquad (9.4.293)$$

Spatial projection tensor

$$f^{\alpha\beta} = u^\alpha u^\beta + g^{\alpha\beta}. \qquad (9.4.294)$$

Heat flux

$$q^\alpha = -K f^{\alpha\beta}(\tau_{,\beta} + \tau\, a_\beta). \qquad (9.4.295)$$

Four-acceleration

$$a_\alpha = u_{\alpha|\beta} u^\beta. \qquad (9.4.296)$$

Enthalpy

$$h = 1 + \epsilon + \frac{p}{\rho}. \qquad (9.4.297)$$

Temperature

$$\tau = \frac{1}{c_v}\left(h - 1 - \frac{p}{\rho} - \frac{1}{2}V_j V_j\right). \qquad (9.4.298)$$

Four-velocity

$$u^\mu = \alpha W v^\mu \quad \alpha = -(g^{tt})^{-1/2}, \qquad (9.4.299)$$

$\eta =$ shear viscosity,

$\zeta =$ bulk viscosity,

$k =$ thermal conductivity.

The equations for continuity, momentum, and energy are as follows.

Continuity

$$\frac{\partial}{\partial t}(\sqrt{-g}\rho u^0) + \frac{\partial}{\partial x_i}(\sqrt{-g}\rho u^i) = 0. \qquad (9.4.300)$$

Momentum

$$\frac{\partial}{\partial t}(\rho h u^0 u^j + p g^{j0}) + \frac{1}{\sqrt{-g}}\frac{\partial}{\partial x_i}\sqrt{-g}(\rho h u^i u^j + p g^{ij} - \zeta\theta f^{ij})$$
$$+ \frac{1}{\sqrt{-g}}\frac{\partial}{\partial x_i}\sqrt{-g}\left[\frac{2\eta}{3}\theta f^{ij} - \eta(u^i_{|\gamma} f^{\gamma j} + u^j_{|\gamma} f^{\gamma i})\right.$$
$$\left. + W(q^i v^j + q^j v^i)\right] + T^{\mu\alpha}\Gamma^j_{\alpha\mu} = 0. \qquad (9.4.301)$$

Energy

$$\frac{\partial}{\partial t}(\rho h u^0 u^0 + p g^{00}) + \frac{1}{\sqrt{-g}}\frac{\partial}{\partial x_i}\sqrt{-g}(\rho h u^0 u^i + p g^{0j} - \zeta\theta f^{0i})$$
$$+ \frac{1}{\sqrt{-g}}\frac{\partial}{\partial x_i}\sqrt{-g}\left[\frac{2\eta}{3}\theta f^{0i} - \eta\left(u^0_{|\gamma} f^{\gamma i} + u^i_{|\gamma} f^{\gamma 0}\right) q^i W\right] + T^{\mu\alpha}\Gamma^0_{\alpha\mu} = 0. \tag{9.4.302}$$

As indicated earlier, the numerical solutions for the nonconservation form such as those given by (9.4.300), (9.4.301), and (9.4.302), are difficult, particularly for high Lorentz factors. Thus we convert these nonconservation equations into a conservation form as follows:

$$\frac{\partial U}{\partial t} + \frac{\partial F^i}{\partial x_i} + \frac{\partial G^i}{\partial x_i} = B, \tag{9.4.303}$$

where the conservation variables U, convection flux variables F^i, diffusion flux variables G_i, and the source terms B are given by

$$U = \begin{bmatrix} \sqrt{-g}\rho\alpha W \\ \sqrt{-g}(\rho h\alpha^2 W^2 V^1) \\ \sqrt{-g}(\rho h\alpha^2 W^2 V^2) \\ \sqrt{-g}(\rho h\alpha^2 W^2 V^3 + p g^{03}) \\ \sqrt{-g}(\rho h\alpha^2 W^2 + p g^{00}) \end{bmatrix}, \tag{9.4.304}$$

$$F^i = \begin{bmatrix} \sqrt{-g}\rho h\alpha W V^i \\ \sqrt{-g}\left(\rho h\alpha^2 W^2 V^i V^1 + p g^{11}\delta^i_1\right) \\ \sqrt{-g}\left(\rho h\alpha^2 W^2 V^i V^2 + p g^{22}\delta^i_2\right) \\ \sqrt{-g}\left(\rho h\alpha^2 W^2 V^i V^3 + p g^{33}\delta^i_3\right) \\ \sqrt{-g}(\rho h\alpha^2 W^2 V^i + p g^{30} V^i) \end{bmatrix}, \tag{9.4.305}$$

$$G^i = \begin{bmatrix} 0 \\ \sqrt{-g}\left\{\zeta\theta P^{i1} - 2\eta\left[\frac{1}{2}\left(V^i_{|\gamma}P^{\gamma 1} + V^1_{|\gamma}P^{\gamma i}\right) - \frac{1}{3}V^\gamma_{|\gamma}P^{i1}\right] + Q^i V^1 + Q^1 V^i\right\} \\ \sqrt{-g}\left\{-\zeta\theta P^{i2} - 2\eta\left[\frac{1}{2}\left(V^i_{|\gamma}P^{\gamma 2} + V^2_{|\gamma}P^{\gamma i}\right) - \frac{1}{3}V^\gamma_{|\gamma}P^{i2}\right] + Q^i V^2 + Q^2 V^i\right\} \\ \sqrt{-g}\left\{-\zeta\theta P^{i3} - 2\eta\left[\frac{1}{2}\left(V^i_{|\gamma}P^{\gamma 3} + V^3_{|\gamma}P^{\gamma i}\right) - \frac{1}{3}V^\gamma_{|\gamma}P^{i3}\right] + Q^i V^3 + Q^3 V^i\right\} \\ \sqrt{-g}\left\{-\zeta\theta P^{i0} - 2\eta\left[\frac{1}{2}\left(V^0_{|\gamma}P^{\gamma i} + V^i_{|\gamma}P^{\gamma 0}\right) - \frac{1}{3}V^\gamma_{|\gamma}P^{i0}\right] + Q^i W\right\} \end{bmatrix}, \tag{9.4.306}$$

$$B = \begin{bmatrix} 0 \\ \sqrt{-g}T^{\mu\alpha}\Gamma^1_{\alpha\mu} \\ \sqrt{-g}T^{\mu\alpha}\Gamma^2_{\alpha\mu} \\ \sqrt{-g}T^{\mu\alpha}\Gamma^3_{\alpha\mu} \\ \sqrt{g}T^{\mu\alpha}\Gamma^0_{\alpha\mu} \end{bmatrix}. \qquad (9.4.307)$$

It is seen that, on differentiation of (9.4.304), (9.4.305), and (9.4.306) as indicated in (9.4.294), we recover the nonconservation forms, (9.4.300) – (9.4.302). As in the case of an ideal flow (Subsection 9.4.3.1), it is necessary that the primitive variables ρ, V^i, P, h, τ, and W be extracted from the conservation variables. This requires the solution of fourth-degree polynomial equations. Details can be found in Richardson (2000).

9.4.3.3 Magnetohydrodynamic General Relativity

It is well known that physical phenomena in general relativity involve electromagnetic environments. This is particularly the case for black-hole accretion disks, gamma-ray bursts, and jet formation.

The conservation form for nonideal fluids given by (9.4.303) and MHD equations (Section 7.3) may be combined to obtain the conservation form of the relativistic gravitomagnetohydrodynamic equations,

$$\frac{\partial \bar{U}}{\partial t} + \frac{\partial \bar{F}^i}{\partial x_i} + \frac{\partial \bar{G}^i}{\partial x_i} = \bar{B}, \qquad (9.4.308)$$

where

$$\bar{U} = \begin{bmatrix} \sqrt{-g}W\rho \\ \sqrt{-g}\rho h W^2 V_j \\ \sqrt{-g}(\rho h W^2 + P g^{00}) \\ \sqrt{-g}B \end{bmatrix}, \qquad (9.4.309)$$

$$\bar{F}^i = \begin{bmatrix} \sqrt{-g}\left(V^i - \frac{1}{\alpha}\beta^i\right)\rho h \alpha W \\ \sqrt{-g}\left[\left(V^i - \frac{1}{\alpha}\beta^i\right)\rho h \alpha^2 W^2 V_j + \bar{P}\delta^i_1\right] \\ \sqrt{-g}\left[\left(V^i - \frac{1}{\alpha}\beta^i\right)(\rho h \alpha^2 W^2 + P g^{30}) + V^i \bar{P}\right] \\ \sqrt{-g}(V^i B^j - V^j B^i) \end{bmatrix}, \qquad (9.4.310)$$

$$\bar{G}^i = \begin{bmatrix} 0 \\ -\sqrt{-g}\tau^{ij} \\ \sqrt{-g}(-\bar{\tau}^{ij}V_j + q^i) \\ \sqrt{-g}\dfrac{1}{\mu_0\sigma}B^j_{|i} \end{bmatrix}, \tag{9.4.311}$$

$$\bar{B} = \begin{bmatrix} 0 \\ \sqrt{g}\alpha T^{\mu\alpha}g_{\nu\sigma}\Gamma^{\sigma}_{\mu\nu} \\ \sqrt{g}\alpha T^{\mu\alpha}\Gamma^{0}_{\alpha\mu} \\ 0 \end{bmatrix}, \tag{9.4.312}$$

with

$$\bar{p} = p + \frac{1}{2\mu}B^k B_k, \tag{9.4.313}$$

$$\bar{\tau}^{ij} = \tau^{ij} + \tau^{ij}_{(m)}, \tag{9.4.314}$$

$$\tau^{ij}_{(m)} = \frac{1}{\mu_0}B^i B^j, \quad \mu_0 = \text{permeability}, \tag{9.4.315}$$

$$\Gamma^{\alpha}_{\mu\nu} = \frac{1}{2}g^{\alpha\beta}(g_{\nu\beta,\mu} + g_{\mu\beta,\nu} - g_{\mu\nu,\beta}). \tag{9.4.316}$$

The conservation form of general relativistic equations given by Eq. (9.4.303) represents the most rigorous approach conducive to numerical solutions for relativistic gravitohydromagnetics. Investigations into the properties of the plasma-filled magnetospheres of black holes, collapsing stellar cores, colliding black holes, neutron stars, and gravitational radiation can be pursued. Numerical simulations for the equations of relativistic hydrodynamics written in conservation forms, Eqs. (9.4.286), (9.4.303), and (9.4.308), may be carried out by use of finite differences, finite elements, or finite volumes (Chung, 2002).

Summary. In this subsection, Applications to Relativistic Continuum (Spacetime Geometries), we examined Einstein's field equations (9.4.239) or (9.4.245) representing the most elegant law of physics that governs the mechanics of universe (stars, black holes, etc.). Analytical solutions (Schwarzshield, Kerr) for these equations have been obtained for simple cases. We can solve more detailed and realistic cases numerically by converting Einstein's field equations into Cauchy problems (commonly known as 3 + 1 equations) given by two evolution equations (9.4.266) and (9.4.267) and two constraint equations (9.4.270) and (9.4.271). For more convenient and rigorous numerical solutions these equations can be written in conservation forms [(9.4.286) for ideal fluid, (9.4.303) for nonideal fluid, and (9.4.308) for gravitohydromagnetics].

Special efforts are made to demonstrate the interesting analogy of the first fundamental form ($a_{\alpha\beta}, g_{ij}$) and the second fundamental form ($b_{\alpha\beta}, K_{ij}$) between mechanisms of engineering shell structures and Einstein's spacetime geometries. Thus the bending of a shell structure on earth is similar to the curvatures displayed by the spacetimes of universe. This is the reason why both subjects are placed under the same title, "Differential Geometry Continuum." They all constitute the mechanics of continuum.

PROBLEMS

9.1 Prove the expression for the Riemann–Christoffel tensor, Eq. (9.2.3).

9.2 Prove the expressions for strain tensors including the membrane, bending, and higher-order bending strain components, Eq. (9.3.12)

9.3 Derive the governing equations for ideal fluids in special relativity.

9.4 Derive the governing equations for nonideal fluids in special relativity.

9.5 Prove Einstein's field equations, Eqs. (9.4.233) and (9.4.235).

Epilogue

The attempt was made in this book to show the analogy between the physical behavior of motions of solids and fluids on earth and that of astrophysical objects in the universe. We saw the conservation laws arising from the thermodynamics laws for solids in Chap. 4 and for fluids in Chap. 5 in a very similar fashion. The curvilinear continuum of Chap. 6 was extended to nonlinear mechanics for large deformations in Chap. 7. We witnessed the electromagnetic continuum of Chap. 8 combined with the spacetime geometries to arrive at the relativistic gravitohydromagnetics in Chap. 9. Differential geometries with curvilinear continuum introduced in Chap. 6 were applied in Chap. 9 to shell geometries for engineering and to spacetime geometries of Einstein's theory of relativity. Thus we have seen actions of nature on earth and throughout the universe under one roof, as called in this book, *General Continuum Mechanics*. Thus the energy and momentum contained in a solid bar and in fluids under external disturbances are controlled by the same principles: the first and second laws of thermodynamics. Similarly, deformations of a solid shell in engineering on earth and spacetimes of Einstein's theory of relativity in the universe are controlled by the same principles: the first and second fundamental tensors. They all belong to the mechanics of continuum.

References

Ackerman, C. C. and Berman, B. (1966). "Second sound in solid helium," *Phys. Rev. Lett.* **16**, 789–91.

Adkins, J. E. and Rivlin, R. S. (1952). "Large elastic deformations of isotropic materials, IX. The deformation of thin shells," *Philos. Trans. R. Soc. London Ser. A* **244**, 505–31.

Adler, R., Bazin, M., and Schiffer, M. (1975). *Introduction to General Relativity*, 2nd ed. McGraw-Hill, New York.

Alexander, H. (1968). "A constitutive relation for rubber-like materials," *Int. J. Eng. Sci.* **6**, 549–63.

Alfvén, H. (1940). *Ark. Mat. Astron. Och Fys.* **27A** (22).

Alfvén, H. (1951). *Cosmical Electrodynamics*. Oxford University Press, New York.

Alfvén, H. and Fälthammer, C.-G. (1963). *Cosmical Electrodynamics*, 2nd ed. Clarendon, Oxford.

Allis, W. P., Buchsbaum, S. I., and Bers, A. (1963). *Waves in Anisotropic Plasmas*. MIT Press, Cambridge, MA.

Ampère, A. M. (1825). *Mem. Acad. Sci. Inst. France* **6**, 175.

Anderson, J. L. (1967). *Principles of Relativity Physics*. Academic, New York.

Appell, P. E. and Lacour, E. (1897). *Principes de la Theorie des Fonctions Elliptiques et Applications*. Gauthier-Villars, Paris.

Aris, R. (1962). *Vectors, Tensors, and the Basic Equations of Fluid Mechanics*. Prentice-Hall, Englewood Cliffs, NJ.

Arnowitt, R., Deser, S., and Misner, C. W. (1962). *The Dynamics of General Relativity in Gravitation: An Introduction to Current Research*, edited by L. Witten. Wiley, New York.

Ashtekar, A. and Geroch, R. P. (1974). "Quantum theory of gravitation," *Rep. Prog. Phys.* **37**, 1211–56.

Atiyah, M. F. and Ward, R. S. (1977). "Instantons and algebraic geometry," *Commun. Math. Phys.* **55**, 117–24.

Avis, S. J. and Isham, C. J. (1980). "Generalized spin structures on four dimensional spacetimes," *Commun. Math. Phys.* **72**, 103–18.

Balescu, R. (1988). *Transport Processes in Plasmas*. North-Holland, Amsterdam.

Bardeen, J. M., Carter, B., and Hawking, S. W. (1973). "The four laws of black hole mechanics," *Commun. Math. Phys.* **31**, 161–70.

Baumgarte, T. W. and Shapiro, S. L. (1998). "Numerical integration of Einstein's field equations," *Phys. Rev. D* **59**, 024007.

Bejan, A. (1984). *Convection Heat Transfer*. Wiley, New York.

Bergmann, P. G. (1976). *Introduction to the Theory of Relativity*. Dover, New York.

Biot, M. A. (1956). "Thermoelasticity and irreversible thermodynamics," *Appl. Phys.* **27**, 240–53.

Birkhoff, G. D. (1923). *Relativity and Modern Physics*. Harvard University Press, Cambridge, MA.

Bland, D. R. (1960). *The Theory of Linear Viscoelasticity*. Pergamon, Oxford.

Boley, B. A. and Weiner, J. H. (1960). *Theory of Thermal Stresses*. Wiley, New York.

Bona, C. (1996). "Four lectures on numerical relativity," in *Relativity and Scientific Computing*, edited by F. W. Hehl, R. A. Puntigam, and H. Ruder. Springer, New York.

Born, M. and Green, H. S. (1949). *A General Kinetic Theory of Liquids*. Cambridge University Press, London.

Boyd, T. J. M. and Sanderson, J. J. (1969). *Plasma Dynamics*. Thomas Nelson, London.

Boyd, T. J. M. and Sanderson, J. J. (2003). *The Physics of Plasmas*. Cambridge University Press.

Bozorth, R. M. (1951). *Ferromagnetism,* Part II. Van Nostrand, Princeton, NJ, pp. 595–712.

Budiansky, B. (1959). "A reassessment of deformation theories of plasticity," *J. Appl. Mech.* **26**, 259–64.

Byrd, P. F. and Friedman, M. D. (1954). *Handbook of Elliptic Integrals for Engineers and Physicists*. Springer-Verlag, Berlin.

Carter, B. (1968). "Hamilton–Jacobi and Schrödinger separable solutions of Einstein's equations," *Commun. Math. Phys.* **10**, 280–310.

Chandrasekhar, S. (1983). *The Mathematical Theory of Black Holes*. Clarendon, Oxford.

Christensen, R. M. (1971). *Theory of Viscoelasticity*. Academic, New York.

Christensen, R. M. and Naghdi, P. (1967). "Linear nonisothermal viscoelastic solids," *Acta Mech.* **3**(1).

Christoffel, E. B. (1869a). "Über die Transformation der homogenen Differentialausdrücke zweiten Grades betreffendes Theorem," *J. Reine Angew. Math.* **70**, 46–70.

Christoffel, E. B. (1869). "Über die Transformation der homogenen Differentialausdrücke zweiten Grades betreffendes Theorem," *J. Reine Angew. Math.* **70**, 241–5.

Chung, T. J. (1973). "Viscoelastoplastic response of axisymmetric shells under impulsive loading," *AIAA J.* **11**, 478–82.

Chung, T. J. (1975). "Thermoviscoplasticity of composites," *Dev. Mech.* **8**, 71–88.

Chung, T. J. (1988). *Continuum Mechanics*. Prentice-Hall, Englewood Cliffs, NJ.

Chung, T. J. (ed.). (1993). *Numerical Modeling in Combustion*. Taylor & Francis, Washington, D.C.

Chung, T. J. (1996). *Applied Continuum Mechanics*. Cambridge University Press, New York.

Chung, T. J. (2002). *Computational Fluid Dynamics*. Cambridge University Press, New York.

Chung, T. J. and Eidson, R. L. (1971). "Analysis of viscoelastoplastic structural behavior of anisotropic shells by the finite element method," in *Proceedings of the First International Conference on Structural Mechanics in Reactor Technology*. Bundesant für Materialprüfung, Berlin.

Chung, T. J. and Eidson, R. L. (1973a). "The incremental theory of three-dimensional transient thermoelastoplasticity – Formulation and solution," *Dev. Mech.* **7**, 843–56.

Chung, T. J. and Eidson, R. L. (1973b). "Viscoelastoplastic response of axisymmetric shells under impulsive loadings," *AIAA J.* **11**(4).

Chung, T. J. and Eidson, R. L. (1974). "Thermodynamic behavior of viscoelastoplastic fiber reinforced structures," Paper No. 2227, presented at the American Society of Civil Engineers National Structural Engineering Meeting, Cincinnati, OH.

Chung, T. J. and Jenkins, J. F. (1973). "Dynamic stability of shells submerged in fluids," in R. Szilard, ed., *Hydrodynamically Loaded Shells*. University of Hawaii Press, Honolulu, HI, pp. 581–90.

Chung, T. J. and Kim, J. Y. (1984). "Two-dimensional, combined-mode heat transfer by conduction, convection and radiation in emitting, absorbing, and scattering media," *ASME Trans. J. Heat Transfer* **106**, 448–52.

Chung, T. J. and Rush, R. H. (1976). "Dynamically coupled motion of surface-fluid shell systems," *ASME J. Appl. Mech.* **43**, 507–8.

Coleman, B. D. (1964). "Simple liquid crystals," *Arch. Ration. Mech. Anal.* **20**, 41–58.

Coleman, B. D., Markovit, H., and Noll, W. (1966). *Viscometric Flows of Non-Newtonian Fluids*. Springer-Verlag, New York, Chap. 2.

Coleman, B. D. and Noll, W. (1959). "On certain steady flows of general fluid," *Arch. Ration. Mech. Anal.* **3**(4), 289–303.

Coleman, B. D. and Noll, W. (1961). "Foundations of linear viscoelasticity," *Rev. Mod. Phys.* **33**, 239–49.

Coulomb, C. A. de (1785–9). *Mem. Acad. Sci. Inst. France*.

Cox, J. P. and Giuli, T. R. (1968). *Principle of Stellar Structure*. Gordon & Breach, New York.

Dafalias, Z. and Popov, E. P. (1976). "A model of nonlinear hardening materials for complex loading," *Acta Mech.*, **21**, 173.

Davis, H. T. (1960). *Introduction to Non-Linear Differential and Integral Equations*. U.S. Atomic Energy Commission, Washington, D.C.

de Groot, S. R., van Leeuwen, W. A., and van Weert, Ch.G. (1981). *Relativistic Kinetic Theory*. North-Holland, Amsterdam.

Deligianni, D. D., Maris, A., and Missirlis, Y. F. (1966). "Stress relaxation behavior of trabecular bone specimens," *J. Biomech.* **27**, 1469–76.

Deruelle, N. and Piran, T. (eds.). (1982). *Gravitational Radiation*. North-Holland, Amsterdam.

DeWitt, B. S. (1967). "Quantum theory of gravity. II. The manifestly covariant theory," *Phys. Rev.* **162**, 1195–239.

DeWitt, B. S. (1967). "Quantum theory of gravity. III. Applications of the covariant theory," *Phys. Rev.* **162**, 1239–56.

DeWitt, B. S. (1979). "Quantum gravity: The new synthesis," in *General Relativity, an Einstein Centenary Survey*, edited by S. W. Hawking and W. Israel. Cambridge University Press, Cambridge.

Drozdov, A. D. and Kalamkarov, A. L. (1995). "Optimization of the winding for composite pressure vessels," *Int. J. Pressure Vessels Pipes* **62**, 69–81.

Drozdov, A. D. and Kolmanovskii, A. L. (1994). *Stability in Viscoelasticity*. North-Holland, Amsterdam.

Drucker, D. C. (1950). "Stress–strain relations in the plastic range – A survey of theory and experiment," Report to the Office of Naval Research, under contract N7-onr-358, Division of Applied Mathematics, Brown University, Providence, RI.

Drucker, D. C. (1959). "A definition of stable inelastic material," *J. Appl. Mech.* **26**, 101–6.

Duong, C. N. and Knauss, W. G. (1993). "The effect of thermoviscoelastic residual stresses in adhesive bonds: 1. Stress Analysis," SM Rep. 93–37B; "2. Fracture Analysis," SM Rep. 93–38. California Institute of Technology, Pasadena, CA.

Dutt, P. (1988). "Stable boundary conditions and difference schemes for Navier–Stokes equations," *SIAM J. Num. Anal.* **25**, 245–67.

Eckart, C. (1940). "The thermodynamics of irreversible processes. III. Relativistic theory of the simple fluid," *Phys. Rev.* **58**, 919.

Eckert, E. R. G. and Drake, R. M. (1987). *Heat and Mass Transfer*, 2nd ed. McGraw-Hill, New York.

Ehlers, J. (1971). In *General Relativity and Cosmology*, edited by R. K. Sachs. Academic, New York, pp. 1–70.

Einstein, A. (1915a). "Zur Allgemeinen Relativitätstheorie," *Preuss. Akad. Wiss. Berlin: Sitzungsber.*, 778–86.

Einstein, A. (1915b). "Der Feldgleichungen der Gravitation," *Preuss. Akad. Wiss. Berlin: Sitzungsber.*, 844–7.

Einstein, A. (1945). "A generalization of the relativistic theory of gravitation," *Ann. Math.* **46**, 578–84.

Einstein, A., Lorentz, H. A., Minkowski, H., and Weyl, H. (1952). *The Principle of Relativity*. Dover, New York.

Eisenhart, L. P. (1949). *Riemannian Geometry*. Princeton University Press, Princeton, NJ.

Elshabka, A. M. and Chung, T. J. (1999). "Numerical salution of three-dimensional stream function vector components of vorticity transport equations," *Comp. Meth. Appl. Mech. Eng.* **170**, 131–53.

Engler, H. (1991). "A matrix Volterra integrodifferential equation occurring in polymer rheology," *Pac. J. Math.* **149**, 25–60.

Erickson, J. L. (1960). "Tensor fields," in S. Flügge, ed., *Encyclopedia of Physics*. Springer-Verlag, Berlin. Vol. 3-3, Appendix, pp. 794–858.

Eringen, A. C. (1962). *Nonlinear Theory of Continuous Media*. Academic, New York.

Eringen, A. C. (1967). *Mechanics of Continua*. Wiley, New York.

Faraday, M. (1951). *Experimental Researches in Electricity*. London, Vol. 1–3, pp. 1839–55.

Ferziger, J. H. (1977). "Numerical simulation of turbulent flow," *AIAA J.* **15**, 1261–71.

Ferziger, J. H. (1983). "High level simulations of turbulent flows," in *Computational Methods for Turbulent, Transonic, and Viscous Flows*. J. A. Essers (ed.), 93–182.

Ffowcs, Williams, J. E. and Hawkings, D. L. (1969). "Sound generated by turbulence and sources in arbitrary motion," *Phil. Trans. R. Soc. A* 264, 342–42.

Finlayson, B. A. (1970). *The Method of Weighted Residuals and Variational Principles*. Academic, New York.

Font, J. A. (2000). *Numerical Hydrodynamics in General Relativity. Living Reviews in Relativity*. Max-Planck Institute.

Flügge, W. (1960). *Stresses in Shells*. Springer-Verlag, Berlin.

Flügge, W. (1967). *Viscoelasticity*. Blaisdell, Waltham, MA.

Foster, H. O. (1965). "Some problems in the nonlinear theory of highly elastic symmetrically loaded membrane structures," Ph.D. dissertation, University of Delaware.

Franklin, B. (1951). "New experiments and observations on electricity," in *Autobiography*. Heritage Press.

Frederick, D. and Chang, T. S. (1965). *Continuum Mechanics*. Allyn & Bacon, Boston.

Fredrickson, A. G. (1964). *Principles and Applications of Rheology*. Prentice-Hall, Englewood Cliffs, NJ.

Fung, Y. C. (1977). *A First Course in Continuum Mechanics*. Prentice-Hall, Englewood Cliffs, NJ.

Gatewood, B. E. (1957). *Thermal Stresses*. McGraw-Hill, New York.

Gauss, C. F. *Carl Friedrich Gauss Werke*, E. J. Schering, ed. Göttingen, Vol. 5, p. 629.

Goodier, J. N. and Hodge, P. G. Jr. (1958). *Elasticity and Plasticity*. Wiley, New York.

Grad, H. (1967). *Magneto-Fluid and Plasma Dynamics*, Vol. 18, edited by H. Grad. American Mathematical Society, Providence, RI.

Graustein, W. C. (1935). *Differential Geometry*. Macmillan, New York.

Green, A. E. and Adkins, J. E. (1960). *Large Elastic Deformations and Nonlinear Continuum Mechanics*. Oxford University Press, London.

Green, A. E. and Naghdi, P. M. (1965). "A general theory of an elastic-plastic continuum," *Arch. Ration. Mech. Anal.* **18**, 251–81.

Green, A. E. and Rivlin, R. S. (1957). "The mechanics of nonlinear materials with memory. I," *Arch. Ration. Mech. Anal.* **1**, 1–21.

Green, A. E., Rivlin, R. S., and Spencer, A. J. M. (1959). "The mechanics of nonlinear materials with memory. II," *Arch. Ration. Mech. Anal.* **3**, 82–90.

Green, A. E. and Shield, R. T. (1950). "Finite elastic deformation of incompressible isotropic bodies," *Proc. R. Soc. London Ser. A* **202**, 407–10.

Green, A. E. and Zerna, W. (1954). *Theoretical Elasticity*. Oxford University Press, London.

Green, G. (1841). "On the propagation of light in crystallized media," *Trans. Cambridge Philos. Soc.* **1**, 121–40.

Greenberg, P. J. (1975). "The equations of hydrodynamics for a thermally conducting, viscous, compressible fluid in special relativity," *Astrophys. J.* **195**, 761.

Gustafsson, B. and Sundström, A. (1978). "Incompletely parabolic problems in fluid dynamics," *SIAM J. Appl. Math.* **35**, 343–57

Gurtin, M. E. (1963). "A note on the principle of minimum potential energy for linear anisotropic elastic solids," *Q. Appl. Math.* **20**, 379–382.

Gurtin, M. E. (1981). *Principles of Continua*. Wiley, New York.

Gurtin, M. and Sternberg, E. (1962). "On the linear theory of viscoelasticity," *Arch. Ration. Mech. Anal.* **II**, 291–356.

Hanjalic, K. and Launder, B. E. (1976). "Contribution towards a Reynolds-stress closure for low-Reynolds-number turbulence," *J. Fluid Mech.* **74**, 593–610.

Hart-Smith, L. J. (1966). "Elastic parameters for finite deformations of rubber-like materials," *Z. Angew. Math. Phys.* **17**, 608–25.

Hawking, S. W. (1971). "Gravitational radiation from colliding black holes," *Phys. Rev. Lett.* **26**, 1344–6.

Helmholtz, H. (1858). "Über Integrale der hydrodynamischen Gleichungen, welche den Wirbelbewegungen entsprechen," *J. Reine Angew. Math.* **55**, 25–55.

Henry, J. (1832). *Am. J. Sci.* **22**, 408; *Proc. Am. Philos. Soc.* **II**, 193.
Hertz, H. (1884). *Ann. Phys.* **23**, 84; (1887). *Ann. Phys.* **31**, 421; (1888). *Ann. Phys.* **34**, 373, 551, 610; (1889). **36**, 1, 769; (1890). *Ann. Phys.* **40**, 577; (1890). *Ann. Phys.* **41**, 369; (1890). Göttingen Nachr., p. 106.
Hill, R. (1950). *The Mathematical Theory of Plasticity*. Clarendon, Oxford.
Hinze, J. O. (1975). *Turbulence*. McGraw-Hill, New York.
Hirschfelder, C. E., Curtis, C. E., and Bird, R. B. (1954). *Molecular Theory of Gases and Liquids*. Wiley, New York.
Hodge, Jr., P. G. (1970). *Continuum Mechanics*. Wiley, New York.
Hoff, N. J. (1955). "The accuracy of Donnell's equations," *J. Appl Mech.* **22**, 329–34.
Holt, E. H. and Haskell, R. E. (1968). *Foundations of Plasma Dynamics.* Macmillan. New York.
Huilgol, R. R. (1968). "On the construction of motions with constant stretch history, I: Superposable viscometric flows," U.S. Army Mathematics Center, Rep. 954. University of Wisconsin.
Huilgol, R. R. and Phan-Thien, N. (1986). "Recent advances in the continuum mechanics of viscoelastic liquids," *Int. J. Eng. Sci.* **24**, 161–251.
Hunter, S. C. (1983). *Mechanics of Continuous Media*, 2nd ed. Ellis Horwood/Wiley, New York.
Huser, A. and Biringin, S. (1993). "Direct numerical simulation of turbulent flow in a square duct," *J. Fluid Mech.* **257**, 65–95.
Ilyushin, A. A. (1963). "Plasticity," *Izdatel'stvo Akad. Nauk SSSR.*
Isham, C. J., Penrose, R., and Sciama, D. W. (eds.). (1975). *Quantum Gravity*. Clarendon, Oxford.
Jacobi, C. G. J. (1927). *Fundamenta Nova Theoria Functionam Ellipticarum*. Cambridge University Press, Cambridge.
Jackson, J. D. (1975). *Classical Electrodynamics*. Wiley, New York.
Jaunzemas, W. (1967). *Continuum Mechanics*. Macmillan, New York.
Johnson, W. and Mellor, P. B. (1962). *Plasticity for Mechanical Engineers*. Van Nostrand, Princeton, NJ.
Joule, J. P. (1841). *Philos. Mag. London, Edinburgh, Dublin* **20**, 98: (1843). *Ibid.* **22**, 204.
Kerr, R. P. (1963). "Gravitational field of a spinning mass as an example of algebraically special metrics," *Phys. Rev. Lett.* **11**, 237–8.
Kerr, R. P. and Schild, A. (1965). "A new class of vacuum solutions of the Einstein field equations," in *Proceedings of the Galileo Galilei Centenary Meeting on General Relativity, Problems of Energy and Gravitational Waves*, edited by G. Barbera. Comitato Nazionale per le Manifestazione Celebrative, Florence.
Kirchhoff, G. (1850). "Über das Gleichgewicht und die Bewegung einer elastichen Scheibe," *J. Reine Angew. Math.* **40**, 51–88.
Koiter, W. T. (1953). "Stress–strain relations, uniqueness and variational theorems for elastic-plastic materials with a singular yield surface," *Q. Appl Math.* **11**, 350–54.
Koiter, W. T. (1959). "A consistent first approximation in the general theory of thin elastic shells," in *Proceedings of the Symposium on the Theory of Thin Elastic Shells*. North-Holland, Amsterdam, pp. 12–33.
Koiter, W. T. (1966). "On the nonlinear theory of thin elastic shells," *K. Ned. Akad. Wet.* **69**, 19.
Klosner, J. M. and Segal, A. (1969). "Mechanical characterization of a natural rubber," PIPAL, Rep. 69–42. Polytechnic Institute of Brooklyn, Brooklyn, New York.
Kraus, H. (1967). *Thin Elastic Shells*. Wiley, New York.
Kreyszig, E. (1959). *Differential Geometry*. University of Toronto Press, Toronto, Canada.
Lagrange, J. L. (1867). *Oeuvres de Lagrange*. Gaauthier-Villars, Paris. Vols. 1–14; (1892). *Mécanique Analytique*, Gauthier-Villars, Paris. Vols. 11 and 12.
Lai, W. M., Rubin, G., and Krempl, E. (1995). *Introduction to Continuum Mechanics*, 2nd ed. Pergamon, New York.
Lamb, H. (1879). *A Treatise on the Mathematical Theory of the Motion of Fluids*. 6th ed. published in 1932 by Cambridge University Press, Cambridge.
Lamb, H. (1959). *Hydrodynamics*, Reprint of 6th ed. Dover, New York.

Lamé, G. (1841). "Mémoire sur les surfaces isostatiques dans les corps solide homogènes en équilibre d'élasticité," *J. Math. Pures Appl.* **6**, 37–60.

Lanczos, C. (1949). *The Variational Principles of Mechanics*. University of Toronto Press, Toronto, Canada.

Landau, L. D. and Lifshitz, E. M. (1962). *The Classical Theory of Fields*. Pergamon, Oxford.

Landau, L. D. and Lifshitz, E. M. (1975). *The Classical Theory of Fields*. 4th ed. Pergamon, Oxford.

Langhaar, H. L. (1953). "The principles of complementary energy in nonlinear elasticity theory," *J. Franklin Inst.*

Langhaar, H. L. (1962). *Energy Methods in Applied Mechanics*. Wiley, New York.

Lawden, D. F. (1968). *An Introduction to Tensor Calculus and Relativity*. Chapman & Hall, London.

Lee, E. H. (1960). "Viscoelastic stress analysis," in J. N. Goodier and N. J. Hoff, eds., *Structural Mechanics*. Pergamon, Oxford, pp. 456–82.

Lee, E. H. (1966). "Elastic-plastic waves of one-dimensional strain," in *Proceedings of the Fifth U.S. National Congress on Applied Mechanics*. American Society of Mechanical Engineers, New York.

Lehner, L. (2001). "Numerical relativity," *Class. Quantum Grav.* **18**, R25–R86.

Leigh, D. C. (1968). *Nonlinear Continuum Mechanics*. McGraw-Hill, New York.

Leighton, R. B. (1959). *Principles of Modern Physics*. McGraw-Hill, New York.

Lekhnitskii, S. G. (1963). *Theory of Elasticity of an Anisotropic Elastic Body*. Holden-Day, San Francisco.

Lichnerowicz, A. (1944). "L'integration des equations de la gravitation relativiste et le probleme des *n* corps," *J. Math. Pures Appl.* **23**, 37–63.

Lichnerowicz, A. (1967). *Relativistic Hydrodynamics and Magnetohydrodynamics*. Benjamin, New York.

Liepmann, H. W. and Roshko, A. (1957). *Elements of Gas Dynamics*. Wiley, New York.

Lighthill, M. J. (1952). "On sound generated aerodynamically. Part I: General theory," *Proc. R. Soc. London Ser. A* 241, 564–87.

Lighthill, M. J. (1954). "On sound generated aerodynamically. Part II: Turbulence as a source of sound," *Proc. R. Soc. London A* 222, 1–32.

Lightman, A. P., Press, W. H., Price, R. H., and Teukolsky, S. A. (1975). *Problem Book in Relativity and Gravitation*. Princeton University Press, Princeton, NJ.

Lipschitz, R. (1869). "Untersuchungen in Betrefft der ganzen homogenen Funktionen von *n* Variablen," *J. Reine Angew. Math.* **70**, 71–102.

Little, R. W. (1973). *Elasticity*. Prentice-Hall, Englewood Cliffs, NJ.

Lodge, A. S. (1964). *Elastic Liquids*. Academic, New York.

Lord, E. A. (1976). *Tensors, Relativity, and Cosmology*. Tata McGraw-Hill, New Delhi.

Lorentz, H. A. (1877). "Over de theorie der terugkaatsing en breking van het licht," *Z. Math. Phys.* **22**, 1, 205; (1890). *Arch. Neerland* **25**, 363; (1878). *Verh. Akad. Wet. Amsterdam, Deel* **18**.

Love, A. E. H. (1892). *A Treatise on the Mathematical Theory of Elasticity*. 4th ed. published in 1934 by Cambridge University Press, Cambridge.

Love, A. E. H. (1934). *The Mathematical Theory of Elasticity*, Cambridge University Press, London.

Lur'e, A. I. (1964). *Three-Dimensional Problems of the Theory of Elasticity*. Wiley Interscience, New York.

Malvern, L. E. (1969). *Introduction to the Mechanics of a Continuous Medium*. Prentice-Hall, Englewood Cliffs, NJ.

Maxwell, J. C. (1860). "Illustrations of the dynamical theory of gases," *Philos. Mag. London, Edinburgh, Dublin* **19**, 19–32; (1864). *Trans. Cambridge Philos. Soc.* **10**, 27; (1936). *Proc. Cambridge Philos. Soc.* **32**, 695; Scientific Papers.

May, M. M. and White, R. H. (1966). *Phys. Rev.* **141**, 1232.

May, M. M. and White, R. H. (1967). In *Methods in Computational Physics*. Academic, New York, Vol. 7.

McConnell, A. J. (1957). *Applications of Tensor Analysis*, Dover, New York.

McDonald, P. H. (1996). *Continuum Mechanics.* PWS Publishing, Boston.
Mihalas, D. and Mihalas, B. W. (1984). *Foundations of Radiation Hydrodynamics*. Oxford University Press, London.
Mikhlin, S. G. (1964). *Variational Methods in Mathematical Physics* (translated by T. Boddington from the 1957 Russian edition). Macmillan, New York.
Milne-Thompson, L. M. (1960). *Theoretical Hydrodynamics*, 4th ed. Macmillan, New York.
Misner, C. W., Thorne, K. S., and Wheeler, J. A. (1973). *Gravitation*. Freeman, San Francisco.
Mohr, O. (1914). *Abhandlungen aus dem Gebiete der technischen Mechanik*, 2nd ed. Willhelm Ernest und Sohn, Berlin.
Moller, C. (1972). *The Theory of Relativity*. Oxford University Press, London.
Mooney, M. (1940). "A theory of large elastic deformation," *J. Appl. Phys.* **11**, 582–92.
Morland, L. W. and Lee, E. H. (1960). "Stress analysis for linear viscoelastic materials with temperature variation," *Trans. Soc. Rheol.* **4**, 233–63.
Mroz, Z. (1967). "On the description of anisotropic work hardening," *J. Mech. Phys. Solids* **15**, 163.
Murdoch, D. C. (1957). *Linear Algebra for Undergraduates*. Wiley, New York.
Murnaghan, F. D. (1951). *Finite Deformations of an Elastic Solid.* Wiley, New York.
Nadeau, G. (1964). *Introduction to Elasticity*. Holt, Rinehart & Winston, New York.
Naghdi, P. M. (1957). "The effect of transverse shear deformation on the bending of elastic shells of revolution," *J. Appl. Math.* **15**, 41–52.
Naghdi, P. M. (1960). "Stress–strain relations in plasticity and thermoplasticity," in E. H. Lee and P. Symonds, eds., *Plasticity*. Pergamon, Oxford, pp. 121–69.
Naghdi, P. M. (1963). "Foundations of elastic shell theory," in I. N. Sneddon and R. Hill, eds., *Progress in Solid Mechanics*. North-Holland, Amsterdam.
Navier, C. L. M. H. (1821). "Sur les lois du mouvements des fluids, en ayant égard à l'adhésion des molecules," *Ann. Chim.* **19**, 244–260.
Noll, W. (1955). "On the continuity of the solid and fluid states," *J. Ration. Mech. Anal.* **4**, 3–81.
Noll, W. (1965). "A mathematical theory of the mechanical behavior of continuous media," *J. Ration. Mech. Anal.* **4**, 3–81.
Novozhilov, V. V. (1953). *Foundations of the Nonlinear Theory of Elasticity*. Graylock Press, Baltimore, MD.
Novozhilov, V. V. (1959). *The Theory of Thin Shells.* Noordhoff, Groningen, The Netherlands.
Novozhilov, V. V. (1961). *Theory of Elasticity* (translated by J. K. Lusher from the Russian). Pergamon, New York.
Nowacki, W. (1962). *Thermoelasticity.* Addison-Wesley, Reading, MA.
Oden, J. T. (1972). *Finite Elements in Nonlinear Continua*. McGraw-Hill, New York.
Oden, J. T. and Reddy, J. N. (1976). *Variational Methods in Theoretical Mechanics*. Springer-Verlag, Berlin.
Ogden, R. W. (1984). *Nonlinear Elastic Deformations*. Dover, New York.
Ohanian, H. and Ruffini, R. (1994). *Gravitation and Spacetime*. Norton, New York.
Ohm, G. S. (1826). "Die Galvanische Kette mathematisch bearbeitet," *Ann. Phys. Chem.* **VI**, 459; *ibid.* **VII**, 45, 117.
Oldroyd, J. G. (1950). "On the formulation of rheological equations of state," *Proc. R. Soc. A* **200**, 523–41.
Onsager, L. (1931). "Reciprocal relations in irreversible processes," I. *Phys. Rev.* **37**, 405–26; II, *Phys. Rev.* **38**, 2265–79.
Ørsted, H. C. (1820). "Experimenta circa effect urn conflict us electrici in acum magneticam," Copenhagen; (1820). *J. Chem. Phys.* **29**, 575.
Pai, S.-I. (1954). *Introduction to the Theory of Compressible Fluid Flow*. Ronald Press, New York, 2 vols.
Pai, S.-I. (1957). *Viscous Flow Theory*: Vol. I, Laminar Flow; Vol. 2, Turbulent Flow. Van Nostrand, Princeton, NJ.

Parkus, H. (1962). "Thermal stresses," in W. Flügge, ed., *Handbook of Engineering Mechanics.* McGraw-Hill, New York, Chap. 43.

Parkus, H. (1968). *Thermoelasticity*. Blaisdell, Waltham, MA.

Pearson, C. E. (1959). *Theoretical Elasticity*. Harvard University Press, Cambridge, MA.

Penrose, R. (1972). *Techniques of Differential Topology in Relativity*. Society for Industrial and Applied Mathematics, Philadelphia.

Perera, M. G. N. and Walters, K. (1980). "Long-range memory effects in flows involving abrupt changes in geometry, I: Flows associated with L-shaped and T-shaped geometries," *J. Non-Newtonian Fluid Mech.* **2**, 49–81.

Phillips, A. (1956). *Introduction to Plasticity*. Ronald Press, New York.

Pierce, A. (1981). *Acoustics*. McGraw-Hill, New York.

Piola, G. (1825). "Sull' applicazione de' principi della meccanica analitica del Lagrange ai principali problemi," Regia Stamparia, Milan.

Pipkin, A. C. and Rivlin, R. S. (1959). "The formulation of constitutive equations in continuum mechanics," I. *Arch. Ration. Mech. Anal.* **4**, 129–44.

Pipkin, A. C. and Wineman, A. S. (1963). "Material symmetry restrictions on non-polynomial constitutive equations," *Arch. Ration. Mech. Anal.* **12**, 420–6.

Poisson, S. D. (1831). "Mémoire sur les équations générales de l'équilibre et du movement des corps solide élastique et des fluids," *J. Ec. Polytech.* **14**(20), 1–174.

Pomraning, G. C. (1973). *The Equations of Radiation Hydrodynamics*. Pergamon, Oxford.

Prager, W. (1945). "Strain hardening under combined stress," *J. Appl. Phys.* **16**, 837–40.

Prager, W. (1956). "A new method of analyzing stresses and strains in work hardening plastic solids," *J. Appl. Mech.* **23**, 493–6.

Prager, W. (1961). *Introduction to Mechanics of Continua*. Ginn, Lexington, MA.

Prandtl, L. (1924): Spannungverteilung in plastischen korpen, in *Proceedings of the First International. Congress of Applied Mechanics*, Delft, The Netherlands, Vol. 43.

Prigogine, I. (1961). *Introduction to the Thermodynamics of Irreversible Processes*. 2nd ed. Wiley-Interscience, New York.

Rankine, W. J. M. (1851). "Laws of the elasticity of solid bodies," *Cambridge Dublin Math. J.* **6**, 41–80.

Rayleigh, J. W. S. (1945). *The Theory of Sound* 2nd ed., rev. (1st ed. published 1877). Dover, New York.

Reiner, M. (1945). "A mathematical theory of dilatancy," *Am. J. Math.* **67**, 305–62.

Reissner, E. (1941). "A new derivation of the equations for the deformation of elastic shells," *Am. J. Math.* **63**, 177–84.

Reissner, E. (1950). "On a variational theorem in elasticity," *J. Math. Phys.* **29**, 90–5.

Reissner, E. (1950). "On axisymmetrical deformation of thin shells of revolution," *Proc. Symp. Appl. Math.* **3**, 27.

Reissner, E. (1967). "On the foundations of generalized linear shell theory," presented at the IUTAM symposium, Copenhagen.

Reuss, A. (1930). Berrucksichtingung der elastischen formanderungen in der plastizitatstheorie, *Z. Angew. Math. Mech.* **10**, 266.

Ricci, G. and Levi-Civita, T. (1901). "Méthodes de calcul différential absolu et leurs applications," *Math. Ann.* **54**, 125–37.

Richardson, G. A. (2000). "The development and the application of the finite element general relativistic astrophysical flow and shock solver," Ph.D. dissertation, University of Alabama, Huntsville, AL.

Richardson, G. A. and Chung, T. J. (2002). "Computational relativistic astrophysics using the flowfield-dependent variation theory," *Astrophys. J. Suppl. Ser.* **139**, 539–63.

Riemann, G. (1866). "Über die Hypothesen, welche der Geometrie zu Grunde liegen," *Abh. Gesellschafte Wiss. Göttingen* **13**, 272–87.

Riemann, G. (1868). "Über die Hypothesen, welche der Geometrie zu Grunde liegen," *Abh. Gesammelte Wiss. Göttingen* **13**, 133–50.

Rindler, W. (1969). *Essential Relativity*. Van Nostrand Reinhold, New York.
Ritz, W. (1909). "Über eine neue Methods zur losung gewisser Variations–Probleme der mathematishen Physik," *J. Reine Angew. Math.* **135**, 1.
Rivlin, R. S. (1948). "The hydrodynamics of non-Newtonian fluids," *Proc. R. Soc. London Ser. A* **193**, 260–81.
Rivlin, R. S. (1948). "Large elastic deformations of isotropic materials, IV. Further developments of the general theory," *Philos. Trans. R. Soc. London Ser. A* **241**, 379–97.
Rivlin, R. S. (1960). "The formulation of constitutive equations in continuum physics. II.," *Arch. Ration. Mech. Anal.* **4**, 262–72.
Rivlin, R. S. (1964). "Nonlinear viscoelastic solids," *SIAM Rev.* **7**, 323–40.
Rivlin, R. S. (1966). "The fundamental equations in nonlinear continuum mechanics," in S.-I. Pai, ed., *Dynamics of Fluids and Plasmas (Burgers Anniversary Volume)*. Academic, New York.
Rivlin, R. S. and Ericksen, J. L. (1955). "Stress-deformation relations for isotropic materials," *J. Ration. Mech. Anal.* **4**, 323–425.
Rivlin, R. S. and Saunders, D. W. (1951). "Large elastic deformations of isotropic materials, VII. Experiments on the deformations of rubber," *Philos. Trans. R. Soc. Ser. A* **243**, 251–88.
Robertson, J. M. (1965). *Hydrodynamics in Theory and Applications*. Prentice-Hall, Englewood Cliffs, NJ.
Saint-Vénant, A. J. C. Barré de (1864). "Théorie de l'élasticité des solides, ou cinématique de leurs déformations," *L'Institut* **32**, 389–90.
Sanders, J. L. (1959). "An improved first approximation theory for thin shells," NASA-TR-R24.
Schapery, R. A. (1966). "A theory of nonlinear thermoviscoelasticity based on irreversible thermodynamics," in *Proceedings of the Fifth U.S. National Congress on Applied Mechanics*, American Society of Mechanical Engineers, New York.
Schlichting, H. (1979a). *Boundary Layer Theory*. McGraw-Hill, New York.
Schlichting, H. (1979b). *Turbulence*. McGraw-Hill, New York.
Schouten, J. A. (1954). *Ricci-Calculus*. Springer-Verlag, New York.
Schutz, B. F. (1985). *Introduction to Special Relativity*. McGraw-Hill, New York.
Schwarzschild, K. (1916). "Über das Graitationsfeld eines Massenpunktes nach der Einsteinschen Theorie," *Sitzungsber. Dtsch. Akad. Wiss. Berlin, Kl. Math. Phys. Tech.* 189–96.
Scipio, L. A. (1967). *Principles of Continua*. Wiley, New York.
Sedov, L. I. (1971). *A Course in Continuum Mechanics*. Walters/Noordhoff, Groningen, The Netherlands. Vols. 1 and 2; (1972). Vol. 3.
Segel, C. A. (1977). *Mathematics Applied to Continuum Mechanics*. Macmillan, New York.
Shapiro, A. H. (1954). *The Dynamics and Thermodynamics of Compressible Fluid Flow*. Ronald Press, New York, 2 vols.
Smarr, L. L. (ed.). (1979). *Sources of Gravitational Radiation*. Cambridge University Press, New York.
Smarr, L. L. and York, J. W. (1978). "Kinematical conditions in the construction of spacetime," *Phys. Rev. D.* **17**, 2529–51.
Sneddon, I. N. (1957). *Elements of Partial Differential Equations*. McGraw-Hill, New York.
Sneddon, I. N. and Berry, D. S. (1958). "The classical theory of elasticity," in S. Flügge, ed., *Encyclopedia of Physics*. Spinger-Verlag, Berlin. Vol. 6, pp. 1–126.
Sneddon, I. N. and Hill, R. (eds.). (1960). *Progress in Solid Mechanics*. North-Holland, Amsterdam, Vol. 1; *ibid*. (1961), Vol. 2; *ibid*. (1963), Vols. 3 and 4.
Sokolnikoff, I. S. (1956). *Mathematical Theory of Elasticity*, 2nd ed. McGraw-Hill, New York, pp. 177–84.
Sokolnikoff, I. S. (1958). *Tensor Analysis: Theory and Applications*. Wiley, New York.
Southwell, R. V. (1941). *An Introduction to the Theory of Elasticity for Engineers and Physicists*. Oxford University Press, London.

Sparrow, E. M. and Cess, R. D. (1966). *Radiation Heat Transfer*. Brooks/Cole, Monterey, CA.
Stachel, J. (1962). "Lie derivatives and the Cauchy problem in the general theory of relativity," Ph.D. dissertation, Stevens Institute of Technology, Hoboken, NJ.
Stachel, J. (1969). "Covariant formulation of the Cauchy problem in generalized electrodynamics and general relativity," *Acta Phys. Pol.* **35**, 689–709.
Staverman, A. J. and Schwarzl, F. (1952). "Time–temperature dependence of linear viscoelastic behavior," *J. Appl. Phys.* **23**, 838–43.
Stoker, J. J. (1969). *Differential Geometry*. Wiley-Interscience, New York.
Stokes, G. G. (1842). "On the steady motion of incompressible fluids," *Trans. Cambridge Philos. Soc.* **7**, 439–53.
Strickwerda, J. C. (1977). "Initial boundary value problems for incompletely parabolic system," *Commun. Pure Appl. Math.* **30**, 797–822.
Synge, J. L. (1957). *The Relativistic Gas*. North-Holland, Amsterdam.
Synge, J. L. and Schild, A. (1949). *Tensor Calculus*. University of Toronto Press, Toronto.
Taub, A. H. (1978). "Relativistic fluid mechanics," *Annu. Rev. Fluid Mech.* **10**, 301.
Temkin, S. (1981). *Elements of acoustics*. Wiley, New York.
Thomas, L. H. (1930). "The radiation field in a fluid in motion," *Q. J. Math.* **1**, 239.
Thompson, W. (Baron Kelvin). (1845). *Cambridge Dublin Math. J.* **1**, 75; (1847). *Ibid.* **2**, 230; (1851). Papers on Electricity and Magnetism; *Philosophical Magazine of London, Edinburgh and Dublin; Philos. Trans. R. Soc.*, pp. 243, 269.
Timoshenko, S. and Goodier, J. N. (1951). *Theory of Elasticity*, 2nd ed. McGraw-Hill, New York.
Timoshenko, S. and Woinowsky-Krieger, S. (1959). *Theory of Plates and Shells*, 2nd ed. McGraw-Hill, New York, Chap. 3.
Treloar, L. R. G. (1944). "Stress–strain data for vulcanized rubber under various types of deformation," *Trans. Faraday Soc.* **40**, 59.
Treloar, L. R. G. (1958). *The Physics of Rubber Elasticity*, 2nd ed. Oxford University Press, London.
Tresca, H. (1864). Mémire sur l'Ecoulement des corps solids soumis a de fortes pressions comptes rendua academie des sciences. Paris, France, Vol. 59, p 754.
Truesdell, C. (1952). "The mechanical foundations of elasticity and fluid dynamics," *J. Ration. Mech. Anal.* **1**, 125–300.
Truesdell, C. (1965). *The Elements of Continuum Mechanics*. Springer-Verlag, New York.
Truesdell, C. and Noll, W. (1965). "The non-linear theories of mechanics," in S. Flügge, ed., *Encyclopedia of Physics*. Springer-Verlag, Berlin. Vol. 3–3.
Truesdell, C. and Toupin, R. A. (1960). "The classical field theories," in S. Flügge, ed., *Encyclopedia of Physics*. Springer-Verlag, Berlin. Vol. 3–1, pp. 226–793.
Valanis, K. C. (1971). "A theory of viscoplasticity without a yield surface," *Arch. Mech.* **23**, 517.
Valanis, K. C. (1980). "Fundamental consequences of a new intrinsic time measure – Plasticity as a limit of the endochronic theory," *Arch. Mech.* **32**, 171.
Valanis, K. C. (1990). "Back stress and Jaumann rates in finite elasticity," *Int. J. Plasticity* **6**, 353.
Vlasov, V. Z. (1964). *General Theory of Shells and Their Applications in Engineering*. NASA Tech. Transl. IT F-99. Originally publihsed in 1949 by Gosudarstvennoye Izdatel'stvo Tekhniko-Teoreticheskoy Literatury, Moscow.
Volta, Alessandro, Count (1793). *Philos. Trans.* **83**, 10, 27; *Philos. Mag.* **4**, 59, 163, 306.
Volterra, V. (1909). "Sulle equazione integro-differenziale della elasticità," *Rend. Lincei Ser. 5a* **18**, 295–391.
Volterra, V. (1959). *Theory of Functionals and of Integral and Integro-Differential Equations*. Dover, New York.
von Mises, R. (1930). "Über die bisherigen Ansätze in der lassischen Mechanik der Kontinua," in *Proceedings of the Third International Congress for Applied Mechanics*. Vol. 2, pp. 1–9.
Wald, R. M. (1984). *General Relativity*. University of Chicago Press, Chicago.
Wang, C. T. (1953). *Applied Elasticity*. McGraw-Hill, New York.
Walters, K. (1975). *Rheometry*. Chapman & Hall, London.

Walters, K. and Schowalter, W. R. (eds.). (1972). *Progress in Heat and Mass Transfer*. Pergamon, Oxford, Vol. 5.

Washizu, K. (1968). *Variational Methods in Elasticity and Plasticity*. Pergamon, Oxford.

Weinberg, S. (1971). "Entropy generation and the survival of protogalaxies in an expanding universe," *Astrophys. J.* **168**, 175.

Weinberg, S. (1972). *Gravitation and Cosmology: Principles and Applications of the General Theory of Relativity*. Wiley, New York.

Wempner, G. A. (1973). *Mechanics of Solids*. McGraw-Hill, New York.

Westergaard, H. M. (1952). *Theory of Elasticity and Plasticity*. Harvard University Press, Cambridge, MA.

Whittaker, E. T. and Watson, G. N. (1960). *A Course of Modern Analysis*. Cambridge University Press, Cambridge.

Williams, M. L. F., Landel, F. R., and Ferry, J. D. (1955). "The temperature dependence of relation mechanisms in amorphous polymers and other glass forming liquids," *J. Am. Chem. Soc.* **77**, 3701–7.

Wills, A. P. (1938). *Vector Analysis With an Introduction to Tensor Analysis*. Prentice-Hall, Englewood Cliffs, NJ.

Wilson, J. R. (1979). In *Sources of Gravitational Radiation*. edited by L. Smarr. Cambridge University Press, Cambridge.

Yoon, W. S. and Chung, T. J. (1994). "Entropy wave instability analysis," *J. Acoust. Soc. Am.* **9692**, Pt 1, 1096–103.

York, J. W. (1971). "Gravitational degrees of freedom and the initial-value problem," *Phys. Rev. Lett.* **28**, 1656–8.

York, J. W. (1972). "Role of conformal three-geometry in the dynamics of gravitation," *Phys. Rev. Lett.* **26**, 1082–5.

York, J. W. (1979). "Kinematics and dynamics of general relativity," in *Sources of Gravitational Radiation*, edited by L. Smarr, Cambridge University Press, Camrbidge.

Ziegler, H. (1959). "A modification of Prager's hardening rule," *Q. Appl. Math.* **17**, 55–64.

Index

Abstract object, 5
Acoustics, 110, 127, 164, 165
  Absorption, 127
  Combustion wave, 165
  Monochromatic wave, 167
  Oscillations, 127
  Plane wave, 167
  Pressure mode, 171
  Rigid reflector, 170
  Sound (acoustic) wave, 164, 165
  Sound emission, 127
  Standing wave, 170
  Traveling wave, 170
  Vorticity mode, 172
  Wave (acoustic) equation, 166
  Wave number, 167
Adiabatic process, 110
Antisymmetry, 114

Barotropic fluid, 110
Bernoulli equation, 110, 138, 139
Biharmonic differential equation, 20, 89
Boundary conditions, 17, 21, 86, 100, 126
  Dirichlet (essential), 20, 21, 82, 86, 87, 89, 90, 126
  Neumann (natural), 20, 21, 82, 86, 87, 88, 90, 126
Boundary-layer flow, 110, 150, 151
  Buffer layer, 155
  Core, 155
  Laminar, 151, 153, 155, 163
  Liquid metal, 154
  Thermal, 154
  Turbulent, 154, 155
  Velocity (hydrodynamic), 154
  Viscous sublayer, 155
  Wall shear, 151

Calorically perfect gas, 110
Cauchy problem, 20
Cauchy's laws of motion, 58
  First law of motion, 58, 82, 93, 105
  Second law of motion, 59
Cayley–Hamilton equation, 11, 44, 61
Characteristic length, 4, 152, 153
Circulation, 140
  Kelvin, 140
Clapeyron formula, 71
Clausius–Duhem inequality, 93, 95
Compatibility equation, 214, 215, 219, 220
Compressibility, 109, 164, 165
Conservation laws, 28
  Angular, 59, 165, 180
  Continuity (mass), 28, 30, 32, 52, 58, 130, 176
  Energy, 93, 130, 165, 180
  Momentum, 57, 59, 93
Constitutive equations, 70, 72, 82, 97
  Linear elastic, 70
  Theory of, 94
    Determinism, 95, 110
    Equipresence, 95, 97, 110
    Form invariant, 95
    Local action (neighborhood), 95, 110
    Material histories, 96
    Material objectivity (material frame indifference), 95, 110, 113, 115, 116, 146
    Material symmetry, 95, 111
    Memory effects, 96
    Physical admissibility, 95
    Unimodular transformation, 96
Convective terms, 132
Coordinate
  Convective, 27, 32, 40, 53
  Eurerian, 26, 30, 32, 117
  Lagrangian, 26, 28, 32, 55, 57, 91, 117
  Reference (Cartesian), 5, 11
  Spatial, 30, 41
  Transformation, 10, 41, 60

Curvilinear continuum, 191
  Cauchy's motion, 210
    First, 210
    Second, 211
  Christoffel symbols, 197
    First kind, 197, 200
    Second kind, 197, 200, 204
  Contravariant tangent vectors, 192
  Covariant derivative, 198
  Cramer's rule, 198
  Curvilinear coordinates, 192
  Cylindrical coordinates, 201, 204
  Green–Gauss theorem, 209
  Laplace equation, 199
  Mixed tensor, 199
  Metric tensor, 201
    Covariant, 201
    Contravariant, 182
  Navier equation, 212
  Parallel transport, 198, 200
  Permutation symbol, 194
  Riemann–Christoffel tensor, 215, 216
  Spherical coordinates, 203, 204
  Strain tensor, 206
  Tangent (base) vector, 191

Damping constant, 98
Deformation gradient, 27, 29, 37
Del operator, 15
Density, 4, 29
  Function, 30
  Internal energy density, 91, 118
Differential geometry continuum, 339
  Base (tangent) vector, 339, 340
  Christoffel symbols, 342
  Codazzi curvature, 345, 346
  Covariant and contravariant metric tensor, 340
  Curvature tensors, 343
  Curvilinear coordinates, 339, 340
  Fundamental tensor, 339
  Fundamental tensors, 343
    First, 343
    Second, 343
    Third, 343
  Gausssian curvature, 345, 346
  Mean curvature, 345
  Riemann–Christoffel tensor, 343, 344
Direction cosine, 10, 43, 44, 49

Einstein equations, 388
Eigenvalues, 44, 61
Engenvectors, 44
Elastic potential, 71, 80
Elasticity, 25
  Linear, 70
  Nonlinear, 25
Electromagnetic continuum
  Accelerating force, 323
  Alfvén waves, velocity, 335, 336, 337
  Ampere's law, 328, 332
  Boltzmann's constant, 334
  Charge density, 325
  Conservation flux variables, 333
  Current density, 323, 324, 325
  Dielectric, 326
  Diffusion flux variables, 333
  Dipole moment, 326
  Displacement current, 331
  Electric current flux, 328
  Electric displacement, 327, 332
  Electrodynamics, 323
  Electric field, 323
  Electric-field density, 326
  Electric stress tensor, 329, 334
  Electric susceptibility tensor, 326
  Electromagnetic stress and energy, 330
  Electromagnetic waves, 323, 335
  Faraday's law, 328, 332
  Gaussian units, 329
  Gravitohydromagnetics, 338
  Green–Gauss theorem, 326, 330, 331
  Internal energy density, 334
  Lorentz force, 325
  Magnetic field, 323
  Magnetic-field intensity, 327
  Magnetic force, 332
  Magnetic pressure, 334
  Magnetic product tensor, 334
  Magnetic stress tensor, 331, 334
  Magnetic vector, 327
  Magneto acoustic waves, 365
  Magnetohydrodynamics, 323, 331, 332
  Maxwell's equations, 328
  MHD generator, 323, 324
  MKSA units, 329
  Navier–Stokes system of equations, 333
  Neumann boundary conditions, 96
  Number density, 324
  Ohm's law, 332
  Permittivity, 326, 327, 328
  Plasma accelerator, 323, 324
  Plasma flow velocity, 323
  Plasma, 324
  Poyarization charge, 326

Polarization vector, 326
Poynting vector, 330
Source terms, 333
Stokes theorem, 328
Susceptibility tensor, 327
Total stress tensor, 334
Viscosity constant, 334
Viscous stress tensor, 334
Energy, 83, 85, 86, 91
Dissipation, 93, 146
External work, 85, 88
strain, 71, 88
Gibbs, 127
Heat, 91, 92, 118
Helmholtz free, 93, 94, 96, 97, 127
Internal, 92, 93, 94, 118
Kinetic, 92, 117
Mechanical, 92, 118
Potential, 83, 85, 88, 114
Total (stagnation), 119, 120
Variational principle, 83, 85, 86, 87, 88, 89, 90
Virtual work, 83, 88
Enthalpy, 118, 127, 160
Gradient, 160, 161
Entropy, 83, 91, 93, 163
Gradient, 161
Production, 93, 94, 97, 127, 159, 162
Entropy mode, 174
Entropy wave, 175
Entropy-controlled instability, 175, 176
Equation of state, 110, 120, 121
Euler equation, 123, 139
Eulerian acceleration, 31

Failure criteria, 61
Ffowcs-Williams–Hawkings equation, 174
Fiber composites, 101
Fick's law, 178
First, 117, 178
Second, 179
Finite (nonlinear) elasticity
Cauchy stress tensor, 238
Hydrostatic pressure, 239
Internal energy density, 238
Kirchhoff stress tensor, 238, 239, 240
Lagrange multiplier, 239
Mooney material, 240
Principal strain invariant, 238
Flexural continuum (shell geometries), 346
Flexural modulus, 1972
Fundamental tensors, 347, 348
Kinematics of shells, 346
Kirchhoff hypothesis, 348
Membrane, bending, higher-order bending strain tensors, 349, 350, 351, 355
Metric tensor, 349
Plates, 353, 354
Position vector, 346
Transverse shear, 349
Flow, 130
Compressible, 130
Ideal, 132
Incompressible, 34
Inviscid, 132, 163
Viscous, 130
Fluid mechanics, historical, 111
Cauchy, 111
Coleman–Noll, 111
Eriksen, 111
Green–Rivlin, 111
Lamb, 111
Navier, 111
Newton, 111
Noll, 111
Poisson, 111, 114
Reiner, 111
Reiner–Rivlin, 112, 146
Rivlin, 111
Saint-Venant, 111
Stokes, 111
Truesdell–Noll, 112
Force, 118
Body, 118
Surface, 111
Fourier heat conduction law, 94, 98
Full potential equation, 158
Crocco's equation, 160

Gas constant, 110
Specific, 110
Universal, 177
General dissipation inequality, 94
Gravitational acceleration, 138, 157
Green–Gauss theorem, 17, 18, 32, 57, 58, 84, 92, 93
Green's function, 171

Heat transfer, 151
Conductive, 109, 156
Convective, 110, 155, 156
Force convection, 156, 157
Natural (free) convection, 156, 157
Radiative, 110, 156
Vaporization (evaporation), 157

Heat, 91
Capacity, 97
Conduction, 96
Flux, 91, 94, 109, 118
Supply, 91, 118
High-speed aerodynamics, 110, 158
Hooke's law, 70
Anisotropic, 72
Generalized, 72, 98
Hypervelocity impact, 36

Ideal (perfect) gas, 120, 121, 147
Ideal flow, 132, 138, 140
Gas, 159, 177
Incompressibility, 109, 115, 116, 132, 139, 140, 141, 157, 163, 164, 165
Index, 6
Free, 6
Repeated (dummy), 6
Inner product, 20, 21, 83
Internal dissipation, 94, 97
Invariants, 5, 44
Energy, 78
Principle strains, 44, 115, 116
Principle stress, 61, 62, 146
Irreversible process, 94, 97, 186
Irrotational flow, 159

Jacobian, 29, 37

Kinematics, 25
Kinetics, 55
Knudsen number, 4
Kronecker delta, 6, 7, 27, 37

Lagrange multiplier, 43, 44, 61, 63
Lamé constant, 78, 79, 114
Laplace equation, 90

Material label, 26
Material nonlinearity (see plasticity), 61
Material, 73
Isotropic, 76, 78, 99
Monotropic, 73, 74
Orthotropic, 74, 103, 105
Stable, 80
Transversely isotropic, 75, 77, 101, 103, 105
Maxwell equations, 128, 328
Mean free path, 3
Memory fuction, 146
Modulus, 70
Bulk, 80
Dilatational, 78, 80
Elasticity, 80, 97
Shear, 78, 80
Thermoelastic, 97, 104
Young's, 70, 78
Mohr circle, 49, 50, 62, 63

Navier equation, 82, 105, 144
Navier–Stokes system of equations, 119, 120, 145
Conservation form of Navier–Stokes system of equations (CNS), 120, 121, 123, 155, 163
Control-volume–control-surface equations (CVS), 119, 124
First law of thermodynamics equations (FLT), 119, 120
Nonconservation form, 119, 124, 155
Newtonian fluids, 108, 116
Non-Newtonian fluids, 240
Cauchy tensor, 249
Cayley–Hamilton theorem, 252
Constant-stretch history, 250
Convective coordinate approach, 248, 259
Convective subjectivity, 259
Coordinate invariance, 240
Determinism, 240
Direct tensor, 242
Eigenvalues, 242
Fading memory, 240, 256
Left Cauchy–Green matrix, 243, 254
Local action, 241
Material objectivity, 240, 256, 258
Metric tensor, 249
Nonexistence of a natural state, 241
Polar decomposition theorem, 242, 243, 246, 256, 258
Polymeric liquids, 258
Rheological material, 240, 241
Right Cauchy–Green matrix, 243, 254
Right stretch and left stretch, 242, 244
Rivlin–Ericksen tensor, 247, 250, 255, 261
Simple fluids, 255
Spin tensor, 245
Stress relaxation, 258
Stretch and rotation, 241
Viscoinelastic equation, 265
Viscometric flow, 253
Nonlinear continuum, 231
Cauchy's first law of motion, 237
Cauchy stress tensor, 234, 235, 236
Contravariant stress tensor, 233
Contravariant stress vector, 233
Contravariant tangent vector, 232
Kirchhoff stress tensor, 234

Lagrangian coordinate system, 231
Piola–Kirchhoff stress tensor, 235, 236
Nontrivial solution, 44
Number, 153
Lewis, 184
Mach, 158, 159, 161
Prandtl, 154
Reynolds, 153

Particle path, 34
Permutation symbol, 7
Second order, 132
Third order, 7, 8
Pfaffian form, 35, 135
Plasticity, 296
Bauschinger effect, 298
Dafalias and Popov model, 317
Dislocation, 296
Drucker's stability and normality rule, 308
Internal variable, 318
Isotropic hardening, 297, 312
Kinematic hardening, 298, 314
Lagrange multiplier, 311
Maximum distortion-energy theory, 301
Maximum shear theory, 298, 299
Maximum stress theory, 298
Maximum strain theory, 299
Maximum yield distortion theory, 302
Mohr circle representation, 300
Mroz's hardening rule, 316
Multiphase models, 316
Octahedral shear stress, 303
Plastic potential function, 307, 310, 313, 323
Plastically stable material, 310
Prandtl–Reuss flow rule, 311
Pure shear condition, 304
Rankine theory, 298
Stability, 308
Strain-hardening material, 310
Strain trajectory models, 318
Stress-space representation, 304
Tresca theory, 299, 304
Von Mises theory, 301, 303
Yield stress, 299
Yield surface, 298
Poisson equation, 90, 140, 141
Poisson's ratio, 78
Polar moment of inertia, 219
Potential, 138
Vector, 138
Velocity, 161, 167
Pressure, 55
Hydrodynamic, 55, 115, 117
Mean, 80, 115
Mean hydrostatic, 80
Thermodynamic, 112, 115
Principal strains, 42, 44
Axes, 42
Direction, 42
Planes, 42
Proper orthogonal matrix, 14

Rate-of-deformation tensor, 40, 51, 53, 115, 116
Rayleigh–Ritz method, 89
Reacting flows, 110, 176
Conservation of mass for mixtures and species, 179
Dalton's law, 182
Diffusion velocity, 177, 180
Dufour effect, 181
Mass-average velocity, 177
Mass concentration, 175, 176
Mass diffusion flux, 178
Mass diffusivity, 178
Mass fraction, 176
Mass rate of production, 179
Molar concentration, 176
Molar diffusion flux, 178
Mole fraction, 176
Partial pressure, 182
Reaction rate, 180
Shvab–Zel'dovich, 97, 182, 184, 185, 186
Soret effect, 181
Specific reaction–rate constant, 178
Stochiometric coefficient, 178
Reference frame, 5
Relationships between Lagrangian and Eulerian coordinates, 34
Relativistic continuum for general relativity, 377, 396
Accretion disk, 400
ADM equations, 390
Apparent horizon, 394
Bending strain tensor, 392
Bianchi identity, 385, 387
Black holes, 388, 397, 400
Cauchy formalism, 390, 392, 394, 395
Christoffel symbol, 378, 379, 380, 381, 382, 383, 389
Codazzi condition, 392
Collapsing steller, 401
Congruencies of worldline, 390
Cosmological censorship, 394
Cosmological constant, 388
Covariant derivative, 382, 389
Differential geometry, 377

Relativistic continuum for general relativity (*cont.*)
- Distortion tensor, 394
- Einstein's general relativity, 382, 383, 386, 387, 388, 390, 395, 396
- Empty (matter-free) space, 387
- Energy-momentum tensor, 386, 388
- Euler–Lagrange variational process, 381
- Event horizon, 394
- Extrinsic curvature, 391, 395
- Free-falling reference frames, 382
- Gamma-ray burst, 400
- Geodesic equation, 378, 382, 385, 386
- Gravitation, 377, 386
- Gravitational constant, 387
- Gravitational coupling constant, 386
- Gravitational waves, 388
- Gravitomagnetohydrodynamics, 400, 401
- Grid stretching, 394
- Hamiltonian, 393, 395
- Hyperbolic conservation formulations, 395
- Hypersurface strain, 394, 395
- Isometries of spacetime, 388
- Jet formation, 400
- Kerr solution, 388
- Killing equation, 389
- Killing vector, 389
- Kinematics and dynamics of general relativity, 377, 390
- Lapse function, 390, 391, 393, 397
- Lie derivative, 389, 392
- Lorentz factor, 397
- Maximal slicing, 393
- Metric tensor, 378, 379
- Minimal distortion, 394
- Minimal strain, 394
- Minkowski tensor, 386, 393
- Natural unit, 397
- Neutron star, 401
- Newtonian gravitational force, 388
- Newtonian limit, 386
- Nonrelativistic limit, 387
- Numerical relativity, 388, 395
- Parallel transport, 377, 382
- Permeability, 400
- Photon energy, 389
- Primitive variables, 397
- Principle of equivalence, 382
- Proper time, 385, 386
- Riemann curvature tensor, 377, 382, 384, 386
- Ricci tensor, 384, 386, 392
- Riemann geometry, 377
- Rindler transformation, 389
- Scalar curvature tensor, 385
- Schwarzschild solution, 388
- Spacetime geometry, 382
- Spacetime metric, 391
- Spherical coordinates, 379
- Slicing of spactime, 390, 393
- Shift vector, 390, 391, 393, 394, 397
- Stress-energy tensor (energy momentum tensor), *see* energy momentum tensor
- Tangent vectors, 377, 378
- Tidal force, 385
  - proper time, 385, 386
- 3+1 equation, 390

Relativistic continuum for special relativity, 356
- Comoving frame, 363, 365
- Conjugate momentum, 373, 375
- Conservation law of charge, 362
- Contravalent four-vector, 361
- Covariance, 357
- Covariance of electromagnetism, 370
- Covariant component of a vector, 357
- Deviatoric rate of deformation tensor, 367
- Eckert decomposition, 366
- Energy density, 362
- Energy-momentum tensor, 361, 363, 367, 396
- Euler equations of motion, 365
- Euler-Lagrange equation, 371, 374
- Four-acceleration, 366
- Four-velocity, 369
- Galilean transformation, 356
- Green–Gauss theorem, 362
- Hamilton's variational principle, 371, 373, 375
- Hypersurface, 362
- Ideal fluids in special relativity, 362
- Inertial frames, 356
- Intrinsic derivative, 361, 366
- Lagrangian, 371, 376
- Light cone, 357
  - Past, 357
  - Future, 357
- Lorentz covariance, 361, 364, 370, 372, 373
- Lorentz transformation, 358, 360, 363, 372
- Luminiferous ether, 356
- Maxwell equations, 370, 376
- Minkowski metric tensor, 360
- Momentum density, 362, 363
- Natural unit, 359
- Nonideal fluids in special relativity, 365
- Proper mass, 362
- Proper mass density, 362

Projection tensor, 365
Proper time, 360, 372
Proper volume, 364
Relativistic charged particle, 372
Relativistic electrodynamics, 369
Relativistic Lagrangian, 374
Rest frame, 360
Specific enthalpy, 363
Specific internal energy, 363
Speed of light, 356, 357
Stress-energy tensor (energy momentum tensor), *see* energy momentum tensor
Worldline, 357, 360
World scalar, 360
Reversible process, 94, 96, 99, 160
Rigid-body motion, 114
Rotational flow, 110, 113, 132, 139

Saint-Venant principle, 216, 220
Second sound, 100
Shock wave, 162
Airfoil, 161
Compression, 163
Contact surface, 161
Expansion, 164
Incident, 163
Nonisentropic, 162
Pressure coefficient, 163
Pressure discontinuity, 162, 164, 165
Reattachment, 164
Separation, 164
Slipstream, 161
Small-perturbation, 161, 163
Vortex sheet, 161
Solenoidal flow, 138
Solid angle, 120
Sonic vorticity, 161
Space conservation law, 34
Specific heat, 194
Constant pressure, 118
Constant volume, 97, 118, 185
Specific volume, 110
Spectral radiation intensity, 119, 120
Speed of sound, 159, 165
Hypersonic, 158, 163
Subsonic, 158, 162
Supersonic, 158, 162, 163
Transonic, 158, 162
State variables, 90
Stoke's hypothesis, 115
Strain-displacement relationship, 37, 38
Strain, 25
Deviatoric, 25, 51, 53, 80, 115, 116
Dilatational, 25, 51
Intermediate principal, 45
Invariant, 41, 45
Large, 39, 52
Major principal, 45
Maximum shear, 49
Minor principal, 45
Normal, 38
Plane, 80, 81, 82, 98
Principal, 42, 45, 61
Principal direction, 45–47
Small, 38, 39, 52
Small shear, 39
Tensor shear, 38, 40
Total (engineering) shear, 37, 38, 40
Volumetric, 51
Stream function, 20, 22
Line, 35, 132, 134, 136, 137, 160
Three-dimensional, 138
Stress, 55
Deviatoric, 66, 80, 117
Excess, 117, 145
Maximum shear, 64, 66
Minimum shear, 64
Octahedral shear, 67, 68
Plane, 80, 83, 98
Principal, 61, 62, 66
Principal plane, 66
Residual, 71
Reynolds, 145, 148, 149
Shear, 63
Total, 115
Substantial (material) derivative, 31, 40, 117, 118
Surface integral, 17

Temperature, 90
Absolute, 91
Reference, 91
Tensor, 5
Direct, 58
Direct stress, 58
Elastic modulus, 71, 72, 73
Excess stress, 114
First-order, 11
Fourth-order, 11, 60, 78
Isotropic, 14, 114
Metric, 4, 29, 37, 53
Rate-of-deformation, 40, 41, 78, 112, 115, 116
Rotational, 84
Second-order, 11
Shear stress, 114
Spin, 84
Strain, 37, 52, 71, 72, 78
Stress, 57, 72, 97, 112

Tensor (*cont.*)
  Third-order, 11
  Zeroth-order, 11
Thermal conductivity tensor, 98
  Isotropic, 98
  Thermoelastic constant, 98
Thermal diffusivity, 146, 155, 181, 183
Thermal expansion, 94
  Coefficient, 94, 128, 129, 157
  Fluids, 127
Thermal, 109
  Conduction, 109
  Convection, 110
  Radiation, 110
Thermodynamics, 70
  First law, 70, 91, 117, 127
  Fluids, 70
  Heat conduction, 70, 99
  Irreversible, 127
  Maxwell relations, 128
  Reversible, 127
  Second law, 70, 93, 96, 126, 127, 160
  Solids, 70, 90, 91, 117, 127
  Thermomechanically coupled equations of motion, 70, 83, 96, 99
Thermoelasticity, 97, 98
Thermoviscoelastoplasticity, 320
  Clausius–Duhem inequality, 321
  Internal dissipation, 321
  Internal variables, 320
Thermoviscous dissipation, 132
Torsion, 216
Torsional deformation, 220
Torsional moment, 218
Transformation matrix, 10
Transverse displacement, 20
Turbulence, 110, 142, 143
  Closure, 149
  Correlation, 144, 146
  Direct numerical simulation, 151
  Eddy heat conductivity, 149
  Eddy-transport coefficient, 146
  Ensemble average, 144
  Fluctuation, 146
  Homogeneous, 147
  Isotropic, 147
  Kinetic energy, 147, 148, 149
  Kinetic theory, 149
  Large-eddy simulation, 150
  Models, 150
  Phenomenological, 149
  Prandtl mixing length, 149
  Stress, 147, 155
  Subgrid, 151
  Time-average, 142, 143
  Time scale, 143
  Transport, 148
Types of differential equations, 126
  Elliptic, 126
  Hyperbolic, 126
  Parabolic, 126

Vector, 5
  Position, 26, 30
  Rotational, 140
  Tangent, 27, 28, 53
  Vorticity, 157
Velocity potential function, 132
Velocity, 30
  Eurerian, 30, 35
  Lagrangian, 30, 35, 36
Viscoelasticity, 266
  Asymptotic elastic modulus, 268
  Creep and recovery, 269, 274
  Correspondence principle, 288
  Dirac delta function, 269
  Irreversible thermodynamics, 290
  Kelvin solids, 267, 274, 278
  Laplace transform, 272
  Maxwell model, 269, 276
  Reduced time, 292
  Relaxation time, 267, 275
  Spring–dashpot models, 267
  Superposition integral, 280
  Shift function, 291
  Thermorheologically simple material, 296
  Unit step function, 269
Viscosity, 109
  Bulk, 117
  Dilatational, 114, 115
  Dynamic, 117, 149
  Eddy, 145, 146, 149
  Kinematic, 117, 153
  Viscosity constant, 109
Volume of fluid, 34
Von Mises, 68
Vortex, 140
  Line, 140
  Sheet, 140
  Streets, 140
Vorticity transport, 20, 110, 140, 141, 142

Warping function, 220
Wave length, 119, 120

Young's modulus, 70, 78